Novel Technologies in Food Science

ISEKI-Food Series

Series editor: Kristberg Kristbergsson *University of Iceland*
Reykjavík, Iceland

Volume 1	FOOD SAFETY: A Practical and Case Study Approach Edited by Anna McElhatton and Richard J. Marshall
Volume 2	ODORS IN THE FOOD INDUSTRY Edited by Xavier Nicolay
Volume 3	UTILIZATION OF BY-PRODUCTS AND TREATMENT OF WASTE IN THE FOOD INDUSTRY Edited by Vasso Oreopoulou and Winfried Russ
Volume 4	PREDICTIVE MODELING AND RISK ASSESSMENT Edited by Rui Costa and Kristberg Kristbergsson
Volume 5	EXPERIMENTS IN UNIT OPERATIONS AND PROCESSING OF FOODS Edited by Maria Margarida Cortez Vieira and Peter Ho
Volume 6	CASE STUDIES IN FOOD SAFETY AND ENVIRONMENTAL HEALTH Edited by Maria Margarida Cortez Vieira and Peter Ho
Volume 7	NOVEL TECHNOLOGIES IN FOOD SCIENCE: Their Impacts on Products, Consumer Trends, and the Environment Edited by Anna McElhatton and Paulo José do Amaral Sobral
Volume 8	FOOD PROCESSING Edited by Kristberg Kristbergsson and Semih Ötles
Volume 9	APPLIED STATISTICS FOR FOOD AND BIOTECHNOLOGY Edited by Gerhard Schleining, Peter Ho and Severio Mannino
Volume 10	TRADITIONAL FOODS: General and Consumer Aspects Edited by Kristberg Kristbergsson and Jorge Oliveira
Volume 11	MODERNIZATION OF TRADITIONAL FOOD PROCESSES AND PRODUCTS Edited by Anna McElhatton, Paulo José do Amaral Sobral, Mustapha Missbah El Idrissi and Ferruh Erdogdu
Volume 12	FUNCTIONAL PROPERTIES OF TRADITIONAL FOODS Edited by Kristberg Kristbergsson and Semih Ötles
Volume 13	PROCESS ENERGY IN FOOD PRODUCTION Edited by Winfried Russ, Barbara Sturm and Kristberg Kristbergsson

For further volumes:
http://www.springer.com/series/7288

Anna McElhatton · Paulo José do Amaral Sobral
Editors

Novel Technologies in Food Science

Their Impact on Products, Consumer Trends and the Environment

Editors
Anna McElhatton
University of Malta
Msida MSD2090, Malta
anna.mcelhatton@um.edu.mt

Paulo José do Amaral Sobral
Department of Food Engineering
Faculty of Animal Science
and Food Engineering
University of São Paulo
Pirassununga, SP, Brazil
pjsobral@usp.br

Series Editor
Kristberg Kristbergsson
University of Iceland
Reykjavík, Iceland
kk@hi.is

ISBN 978-1-4419-7879-0   e-ISBN 978-1-4419-7880-6
DOI 10.1007/978-1-4419-7880-6
Springer New York Dordrecht Heidelberg London

Library of Congress Control Number: 2011938280

© Springer Science+Business Media, LLC 2012
All rights reserved. This work may not be translated or copied in whole or in part without the written permission of the publisher (Springer Science+Business Media, LLC, 233 Spring Street, New York, NY 10013, USA), except for brief excerpts in connection with reviews or scholarly analysis. Use in connection with any form of information storage and retrieval, electronic adaptation, computer software, or by similar or dissimilar methodology now known or hereafter developed is forbidden.
The use in this publication of trade names, trademarks, service marks, and similar terms, even if they are not identified as such, is not to be taken as an expression of opinion as to whether or not they are subject to proprietary rights.

Printed on acid-free paper

Springer is part of Springer Science+Business Media (www.springer.com)

Series Preface

The ISEKI-Food book series was originally planned to consist of six volumes of texts suitable for food science students and professionals interested in food safety and environmental issues related to sustainability of the food chain and the well-being of the consumer. As the work progressed, it soon became apparent that the interest in and need for texts of this type exceeded the topics covered by the first six volumes published by Springer in 2006–2009. The series originated in work conducted by the European thematic network ISEKI-Food, an acronym for "integrating safety and environmental knowledge into food studies." Participants in the ISEKI-Food network come from most countries in Europe, and most of the institutes and universities involved with food science education at the university level are represented. Some international companies and nonteaching institutions have also participated in the program. The network was expanded in 2008 with the ISEKI_Mundus program, with 37 partners from 23 countries outside Europe joining the consortium, and it continues to grow, with approximately 200 partner institutions from 50 countries from all over the world in 2011. Both networks are coordinated by Prof. Cristina Silva at the Catholic University of Portugal, College of Biotechnology (Escola), in Porto. The program has a Web site at https://www.iseki-food.eu/drupal/. The main objectives of ISEKI-Food have been to improve the harmonization of studies in food science and engineering in Europe and to develop and adapt food science curricula, emphasizing the inclusion of safety and environmental topics. The program has been further expanded into the ISEKI Food Association (IFA) (https://www.iseki-food.net/), an independent organization devoted to the objectives of the ISEKI consortium to further the safety and sustainability of the food chain through education.

The ISEKI-Food book series is now continued with several new volumes to be published soon. The seventh volume in the series will be *Novel Technologies in Food Science: Their Impact on Products, Consumer Trends and The Environment*, edited by Anna McElhatton and Paulo Sobral. The book is intended for food scientists and engineers and readers interested in the new and emerging food processing technologies that are intended to provide foods that are safe, but maintain most of their original freshness. All 16 chapters are written from a safety and environmental standpoint with respect to the emerging technologies. Volume eight in the series will be *Food Processing*, a textbook edited by Kristberg Kristbergsson and Semih Ötles, intended for senior undergraduate students and junior graduate students, providing a comprehensive introduction to food processing. The book should also be useful to professionals and scientists interested in food processing from both the equipment and the process approach as well as in the physicochemical aspect of food processing. The book will contain five sections, starting with chapters on the basic principles and physicochemical properties of foods, followed by sections with chapters on conversion operations, preservation operations, and food processing operations, with separate chapters on most common food commodities. The final section will be devoted to postprocessing operations. The ninth volume, *Applied Statistics for Food Technology and Biotechnology*, edited by Gerhard Schleining, Peter Ho, and Saverio Mannino, is intended for graduate students and industry personnel who need a guide for setting up experiments so that the results will be statistically valid. The book will provide numerous samples and case studies on how to use statistics in food and biotechnology research and testing. It will contain chapters on data collection, data analysis and presentation, handling of multivariate data, statistical process control, and experimental design.

Followed by these professional-oriented texts, there will be a trilogy on traditional foods that will be written for food science professionals as well as for the interested general public. The trilogy will be volumes 10, 11, and 12 in the ISEKI-Food book series and will be in line with the recent internationalization of the ISEKI consortium and will offer more than 140 chapters dedicated to different traditional foods from more than 40 countries all over the world. The trilogy will start with a text offering general descriptions of different traditional foods and topics related to consumers and sensory aspects. Volume 10 will be entitled *Traditional Foods; General and Consumer Aspects*, edited by Kristberg Kristbergsson and Jorge Oliveira. The second book in the trilogy will be *Modernization of Traditional Food Processes and Products*, edited by Anna McElhatton, Paulo Sobral, Mustapha Missbah El Idrissi, and Ferruh Erdogdu. The chapters will be devoted to recent changes and modernizations that have been made in the processing of specific traditional foods, focusing on the processing and engineering aspects of the processes. Finally, *Functional Properties of Traditional Foods* will be devoted to functional and biochemical aspects of traditional foods and the beneficial effects of bioactive components that may be found in some traditional foods.

The ISEKI-Food book series draws on expertise form universities and research institutions all over the world, and we sincerely hope that it may offer interesting topics to students, researchers, professionals, and the general public.

Kristberg Kristbergsson

Series Preface to Volumes 1–6

The single most important task of food scientists and the food industry as a whole is to ensure the safety of foods supplied to consumers. Recent trends in global food production, distribution, and preparation call for increased emphasis on hygienic practices at all levels and for increased research in food safety in order to ensure a safer global food supply. The ISEKI-Food book series is a collection of books where various aspects of food safety and environmental issues are introduced and reviewed by scientists specializing in the field. In all of the books special emphasis was placed on including case studies applicable to each specific topic. The books are intended for graduate students and senior undergraduate students as well as professionals and researchers interested in food safety and environmental issues applicable to food safety.

The idea and planning of the books originates from two working groups in the European thematic network ISEKI-Food, an acronym for "integrating safety and environmental knowledge into food studies." Participants in the ISEKI-Food network come from 29 countries in Europe, and most of the institutes and universities involved with food science education at the university level are represented. Some international companies and nonteaching institutions have also participated in the program. The ISEKI-Food network is coordinated by Prof. Cristina Silva at the Catholic University of Portugal, College of Biotechnology (Escola), in Porto. The program has a Web site at http://www.esb.ucp.pt/iseki/. The main objectives of ISEKI-Food have been to improve the harmonization of studies in food science and engineering in Europe and to develop and adapt food science curricula, emphasizing the inclusion of safety and environmental topics. The ISEKI-Food network started on October 1, 2002, and has recently been approved for funding by the EU for renewal as ISEKI-Food 2 for another 3 years. ISEKI has its roots in an EU-funded network formed in 1998 called Food Net, where the emphasis was on casting a light on the different food science programs available at the various universities and technical institutions throughout Europe. The work of the ISEKI-Food network was organized into five different working groups with specific tasks all aiming to fulfill the main objectives of the network.

The first four volumes in the ISEKI-Food book series come from WG2 coordinated by Gerhard Schleining at Boku University in Austria and Kristberg Kristbergsson.

The main task of WG2 was to develop and collect materials and methods for teaching safety and environmental topics in the food science and engineering curricula. The first volume is devoted to food safety in general, with a practical and a case-study approach. The book is composed of 14 chapters which were organized into three sections on preservation and protection, benefits and risk of microorganisms, and process safety. All of these issues have received high public interest in recent years and will continue to be in the focus of consumers and regulatory personnel for years to come. The second volume in the series is devoted to the control of air pollution and treatment of odors in the food industry. The book is divided into eight chapters devoted to defining the problem, recent advances in analysis, and methods for prevention and treatment of odors. The topic should be of special interest to industry personnel and researchers owing to recent and upcoming regulations by the EU on air pollution from food processes. Other countries will likely follow suit with more stricter regulations on the level of odors permitted to enter the environment from food processing operations. The third volume in the series is devoted to utilization and treatment of waste in the food industry. Emphasis is placed on sustainability of food sources and how waste can be turned into by-products rather than pollution or landfills. The book is composed of 15 chapters, starting with an introduction to problems related to the treatment of waste, and an introduction to the ISO 14001 standard used for improving and maintaining environmental management systems. The book then continues to describe the treatment and utilization of both liquid and solid waste, with case studies from many different food processes. The last book from WG2 is on predictive modeling and risk assessment in food products and processes. Mathematical modeling of heat and mass transfer as well as reaction kinetics is introduced. This is followed by a discussion of the stoichiometry of migration in food packaging, as well as the fate of antibiotics and environmental pollutants in the food chain using mathematical modeling and case study samples for clarification.

Volumes five and six come from work in WG5 coordinated by Margarida Vieira at the University of Algarve in Portugal and Roland Verhé at Ghent University in Belgium. The main objective of the group was to collect and develop materials for teaching food-safety-related topics at the laboratory and pilot plant level using practical experimentation. Volume five is a practical guide to experiments in unit operations and processing of foods. It is composed of 20 concise chapters each describing different food processing experiments, outlining theory, equipment, and procedures, and containing applicable calculations and questions for students or trainees, followed by references. The book is intended to be a practical guide for the teaching of food processing and engineering principles. The final volume in the ISEKI-Food book series is a collection of case studies in food safety and environmental health. It is intended to be a reference for introducing case studies into traditional lecture-based safety courses as well as being a basis for problem-based learning. The book consists of 13 chapters containing case studies that may be used, individually or in a series, to discuss a range of food safety issues. For convenience the book was divided into three main sections, with the first devoted to case studies in a more general framework with a number of specific issues in safety and health ranging from acrylamide and nitrates to botulism and listeriosis. The second section is devoted to some well-known outbreaks related to food intake in different countries. The final section of the book considers food safety from the perspective of the researcher. Cases are based around experimental data and examine the importance of experimental planning, design, and analysis.

The ISEKI-Food book series draws on expertise from close to 100 universities and research institutions all over Europe. It is the hope of the authors, editors, coordinators, and participants in the ISEKI-Food network that the books will be useful to students and colleagues to further their understanding of food safety and environmental issues.

Kristberg Kristbergsson

Series Acknowledgements

ISEKI_Food 2 is a thematic network on food studies, funded by the European Union (EU) as project no. 226032-CP-I-2005-1-PT-ERASMUS-TN. It is a part of the EU program in the field of higher education called ERASMUS, which is the higher-education action of the SOCRATES II program of the EU. ISEKI_Mundus is project no. 136263 - EM - 1–2007 - 1 - PT - ERA MUNDUS - EM4EATN and was established to foster the internationalization and enhance the quality of European higher education in food studies, promote good communication and understanding between European countries and the rest of the world, and extend the work of the ERASMUS thematic network ISEKI_Food throughout the world.

Socrates   **Erasmus Mundus**

Preface

Novel Technologies in Food Science: Their Impact on Products, Consumer Trends and the Environment is the seventh volume of the ISEKI-Food book series. This book describes how novel food technologies are connected to the production of foods and how the products and processes impinge on our daily lives. Topics include waste and waste management, hazard analysis and critical control points, safety considerations in the use of nutraceuticals and functional foods, and the use and development of green(er) technologies to produce novel products. Consumers' selection and behavior towards products placed on the market have also been considered in this book to emphasize the fact that products must be both safe and shown to be of good quality.

The issues addressed in this book were selected to showcase situations that are encountered in modern food production. The book seeks to inform readers about both the issues and the means used to address them. The major themes running through all the chapters are the search for quality processes that are sustainable and the search for processes and products that the consumers would feel most comfortable using. Furthermore, if products are to succeed in a highly volatile and competitive market, they must most certainly appeal to the end users.

Like the other books in the series, this volume has chapters written by scientists specializing in the field. This book is intended for graduate students and senior undergraduate students as well as professionals and researchers interested in both food and environmental issues applicable to sustainable food production.

Anna McElhatton

Contents

Part I Environmental Aspects

1. **Waste and Its Rational Management** 3
   Anna McElhatton and Anton Pizzuto

2. **Implementation of Hazard Analysis and Critical Control Points System in the Food Industry: Impact on Safety and the Environment** 21
   Sueli Cusato, Paula Tavolaro, and Carlos Augusto Fernandes de Oliveira

3. **Food By-products for Biofuels** 39
   Cecilia Hodúr, Zsuzsanna László, and Giovana Tommaso

4. **Integrated Management Methods for the Treatment and/or Valorization of Olive Mill Wastes** 65
   Katerina Stamatelatou, Paraskevi S. Blika, Ioanna Ntaikou, and Gerasimos Lyberatos

Part II Safety and Quality Considerations

5. **Safety Considerations of Nutraceuticals and Functional Foods** 121
   Semih Otles and Ozlem Cagindi

6. **Consumer Behavior: Determinants and Trends in Novel Food Choice** 137
   Mona Elena Popa and Alexandra Popa

Part III Novel Process Technologies with a Green/Environmental Slant

7   Recent Advances in the Microencapsulation of Oils
    High in Polyunsaturated Fatty Acids 159
    S. Drusch, M. Regier, and M. Bruhn

8   Biocontrol of Foodborne Bacteria 183
    Lynn McIntyre, J. Andrew Hudson, Craig Billington,
    and Helen Withers

9   Plant Extracts as Natural Antifungals: Alternative
    Strategies for Mold Control in Foods 205
    Virginia Fernández Pinto, Andrea Patriarca, and Graciela Pose

10 Reduction of Mycotoxin Contamination by Segregation
    with Sieves Prior to Maize Milling 219
    Ana M. Pacin and Silvia L. Resnik

11 Rational Use of Novel Technologies: A Comparative
    Analysis of the Performance of Several New Food
    Preservation Technologies for Microbial Inactivation 235
    Stella M. Alzamora, Jorge Welti-Chanes, Sandra N. Guerrero,
    and Paula L. Gómez

12 Emerging Technologies to Improve the Safety
    and Quality of Fruits and Vegetables 261
    Elisabete M.C. Alexandre, Teresa R.S. Brandão,
    and Cristina L.M. Silva

13 Novel Technologies for the Preservation of Chilled
    Aquatic Food Products 299
    Carmen A. Campos, María F. Gliemmo, Santiago P. Aubourg,
    and Jorge Barros Velázquez

14 Use of Natural Preservatives in Seafood 325
    Carmen A. Campos, Marcela P. Castro, Santiago P. Aubourg,
    and Jorge Barros Velázquez

15 Edible Films: Use of Lycopene as Optical Properties Enhancer 361
    Rosemary A. de Carvalho, Carmen Silvia Fávaro-Trindade,
    and Paulo J.A. Sobral

16 Clean Strategies for the Management of Residues
    in Dairy Industries 381
    Giovana Tommaso, Rogers Ribeiro,
    Carlos Augusto Fernandes de Oliveira, Katerina Stamatelatou,
    Georgia Antonopoulou, Gerasimos Lyberatos, Cecilia Hodúr,
    and József Csanádi

Index 413

Contributors

Elisabete M.C. Alexandre Escola Superior de Biotecnologia, Universidade Católica Portuguesa, Rua Dr. António Bernardino de Almeida, 4200–072, Porto, Portugal

Santiago P. Aubourg Instituto de Investigaciones Marinas de Vigo (CSIC), C/Eduardo Cabello, 6. Vigo. Pontevedra. E-36208, Spain

Stella M. Alzamora Universidad de Buenos Aires, Buenos Aires, Argentina
CONICET , Buenos Aires, Argentina

Georgia Antonopoulou School of Chemical Engineering, National Technical University of Athens, Athens 15780, Greece
Institute of Chemical Engineering and High Temperature Chemical Processes, Patras, Greece

Craig Billington Institute of Environmental Science and Research Ltd, Christchurch 8540, New Zealand

Paraskevi S. Blika Department of Chemical Engineering, University of Patras, Vas. Sofi as 12, Xanthi 67100, Greece

Teresa R.S. Brandão Escola Superior de Biotecnologia, Universidade Católica Portuguesa, Rua Dr. António Bernardino de Almeida, 4200–072, Porto, Portugal

M. Bruhn University of Kiel, Kiel, Germany

Ozlem Cagindi Department of Food Engineering, Celal Bayar University, Manisa, Turkey

Carmen A. Campos Departamento de Industrias, Facultad de Ciencias Exactas y Naturales, Universidad de Buenos Aires, Intendente Guiraldes 2160, C.A.B.A., 1428, Argentina

Rosemary A. de Carvalho Department of Food Engineering, Faculty of Animal Science and Food Engineering, University of São Paulo, Pirassununga, SP, Brazil

Marcela P. Castro Universidad Nacional del Chaco Austral, Comandante, Fernández 755, P.R. Sáenz Peña, Chaco, Argentina

József Csanádi Faculty of Engineering, University of Szeged, Mars sq. 7, 6724 Szeged, Hungary

Sueli Cusato Department of Food Engineering, School of Animal Science and Food Engineering, University of São Paulo, Pirassununga, SP, Brazil

S. Drusch Beuth University of Applied Sciences, Berlin, Germany

Carmen Silvia Fávaro-Trindade Department of Food Engineering, Faculty of Animal Science and Food Engineering, University of São Paulo, Pirassununga, SP, Brazil

María F. Gliemmo Departamento de Industrias, Facultad de Ciencias Exactas y Naturales, Universidad de Buenos Aires, Intendente Guiraldes 2160, C.A.B.A, 1428, Argentina

Paula L. Gómez Universidad de Buenos Aires, Buenos Aires, Argentina

CONICET, Buenos Aires, Argentina

Sandra N. Guerrero Universidad de Buenos Aires, Buenos Aires, Argentina

CONICET, Buenos Aires, Argentina

Cecilia Hodúr Faculty of Engineering, University of Szeged, Mars sq. 7, 6724 Szeged, Hungary

J. Andrew Hudson Institute of Environmental Science and Research Ltd, PO Box 29 181, Christchurch 8540, New Zealand

Zsuzsanna László Faculty of Engineering, University of Szeged, Mars sq. 7, 6724 Szeged, Hungary

Gerasimos Lyberatos School of Chemical Engineering, National Technical University of Athens, Athens 15780, Greece

Institute of Chemical Engineering and High Temperature Chemical Processes, Patras, Greece

Anna McElhatton University of Malta, Msida MSD2090, Malta

Lynn McIntyre Harper Adams University College, Edgmond, Newport, Shropshire, TF10 8NB, UK

Ioanna Ntaikou School of Chemical Engineering, National Technical University of Athens, Athens 15780, Greece

Institute of Chemical Engineering and High Temperature Chemical Processes, Patras, Greece

Contributors

Carlos Augusto Fernandes de Oliveira Department of Food Engineering, School of Animal Science and Food Engineering, University of São Paulo, Pirassununga, Brazil

Semih Otles Department of Food Engineering, Ege University, Bornova, Turkey

Ana M. Pacin CIC, Fundación de Investigaciones Científicas Teresa Benedicta de la Cruz, Buenos Aires, Argentina

Andrea Patriarca Universidad de Buenos Aires, Buenos Aires, Argentina

Virginia Fernández Pinto Universidad de Buenos Aires, Buenos Aires, Argentina

Anton Pizzuto University of Malta, Msida MSD2090, Malta

Alexandra Popa University of Agronomic Sciences and Veterinary Medicine Bucharest, Bucharest, Romania

Mona Elena Popa University of Agronomic Sciences and Veterinary Medicine Bucharest, Bucharest, Romania

Graciela Pose Universidad Nacional de Quilmes, Buenos Aires, Argentina

M. Regier University of Applied Sciences, Trier, Germany

Silvia L. Resnik CIC, Universidad de Buenos Aires, Buenos Aires, Argentina

Rogers Ribeiro School of Animal Science and Food Engineering, University of São Paulo, Pirassununga, Brazil

Cristina L.M. Silva Escola Superior de Biotecnologia, Universidade Católica Portuguesa, Rua Dr. António Bernardino de Almeida, 4200–072, Porto, Portugal

Paulo J. A. Sobral Department of Food Engineering, Faculty of Animal Science and Food Engineering, University of São Paulo, Pirassununga, SP, Brazil
pjsobral@usp.br.

Katerina Stamatelatou School of Chemical Engineering, National Technical University of Athens, Athens 15780, Greece

Institute of Chemical Engineering and High Temperature Chemical Processes, Patras, Greece

Paula Tavolaro Department of Food Engineering, School of Animal Science and Food Engineering, University of São Paulo, Pirassununga, SP, Brazil

Giovana Tommaso Department of Food Engineering, School of Animal Science and Food Engineering, University of São Paulo, Pirassununga, Brazil

Jorge Barros Velázquez Department of Analytical Chemistry, Nutrition and Food Science, LHICA, School of Veterinary Sciences, University of Santiago de Compostela, E-27002, Lugo, Spain

Laboratory of Biotechnology, College of Pharmacy, University of Santiago de Compostela, E-15782, Santiago, Spain

Jorge Welti-Chanes Instituto Tecnológico y de Estudios Superiores de Monterrey, Monterrey, NL, Mexico

Helen Withers AgResearch MIRINZ, Ruakura Research Centre, Private Bag 3123, Hamilton, New Zealand

Part I
Environmental Aspects

Chapter 1
Waste and Its Rational Management

Anna McElhatton and Anton Pizzuto

1.1 Introduction

The management of waste as been an issue that has affected societies, where people have lived in organized and in sufficient numbers to cause stress to local resources. In the past, in most countries, and presently in poorer countries, domestic and industrial waste could be dealt with by removal to nonengineered dumps, where it could be buried, eaten by animals, and burned. Progress taking the form of increased financial and technical complexity has led to increased pressure on the environment. Progress has also been the catalyst for significant increase in awareness of environmental issues at both national and international levels which has led to the development of waste management policy and practice in industrialized countries, mostly in the second half of the twentieth century. These policies were intended to safeguard both public and occupational health and to ensure that environmental resources are used rationally. Indeed, waste management policies have evolved to take on board the social, economic, and environmental dimensions of sustainable development (Strange 2002).

Waste comes in many different forms, from many different sources, and is discharged in many different ways. The common end result is this: material that is no longer of use is set aside or otherwise ends up in the natural environment and affects the general quality of life. Some sources of waste can be measured, but the impact on the environment is less easily quantified. Two significant sources of waste generation are *municipal solid waste* and *hazardous waste* from industrial sources (Calvert-Henderson 2006).

Waste is directly linked to human development, both technologically and socially. The composition of different wastes has varied over time and with location,

---

A. McElhatton (✉) • A. Pizzuto
University of Malta, Msida MSD2090, Malta
e-mails: anna.mcelhatton@um.edu.mt; anton.pizzuto@gmail.com

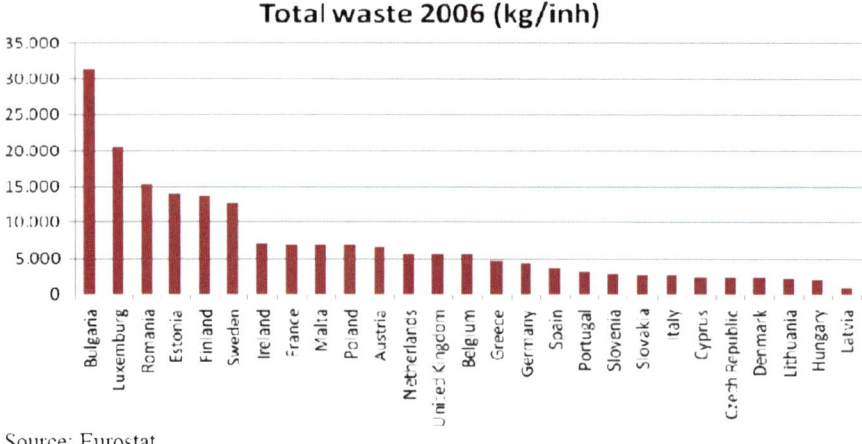

Source: Eurostat

Fig. 1.1 Total waste generation per inhabitant in 2006 (kilograms per inhabitant)

with industrial development and innovation being directly associated with the quantity and type of waste materials generated. Examples of this include plastics and nuclear technology.

In the EU, every year, some two billion metric tons of waste – including hazardous waste – is produced in the member states, and this figure is rising steadily (European Institutions 2010). Furthermore, 2006 data from EUROSTAT show that there is a great variation in the generation of waste per inhabitant across the EU (Fig. 1.1).

The US environmental protection agency (EPA) reports waste generated per capita in a somewhat different manner. The EPA reports categories of waste and is not obliged to publish overall generated waste statistics; indeed most data are found published in a sectorial fashion. In 2008 alone, 249.6 million tons of municipal solid waste was generated, of which 82.9 million tons (33.2%) was recovered (EPA 2008).

Hazardous wastes are somewhat more complicated to define and quantify (EPA 2008b). The EPA has a classification scheme that categorizes waste according to source and type of material or product and has a specific process of identification based on principles drafted and implemented as a consequence of the Resource Conservation and Recovery Act, an amendment to the Solid Waste Disposal Act, that was enacted in 1976 to address the general waste issue that was, as in other developed societies, growing in the USA (EPA 2009).

The developed economies such as the US and European economies are based on a high level of resource consumption. This includes raw materials (such as metals, construction minerals, and wood), energy, and land. The main driving forces of Europe's resource consumption are economic growth, technological developments, and changing consumption and production patterns. About one third of resources used are turned into waste and emissions. Around 4 tons of waste per capita is generated every year in the EEA member countries. Every European citizen, on average, throws away 520 kg of household waste per year, and this figure is expected to increase.

In the EU-15, use of materials has changed little over the past two decades, and remains at about 14.8–15.8 tons per capita per year. However, this number differs considerably from country to country, from some 11.8 tons per capita in Italy to 37.4 tons per capita in Finland. Construction materials have the largest share in this, followed by fossil fuels and biomass. Efficiency of resource use is several times higher in the EU-15 than in the new EU member states or the countries of southeastern Europe. Projections to 2020 indicate that resource use in the EU will continue to increase. Resource use is also increasing in other regions of the world. This is partly due to increased consumption of goods and services in Europe, often based on resources extracted from these other regions (EEA 2010).

The alternative scenario is either prevent the production of such waste or reintroduce it into the product cycle through recycling of component parts using ecologically friendly and economically viable processes (European Institutions 2010).

1.1.1 What Is Waste?

Waste is sometimes a subjective concept, because items that some people discard may have value to others. It is widely recognized that waste materials are a valuable resource, whereas there is debate as to how this value is best exploited and managed.

Waste includes all items that people no longer have any use for, which they either intend to get rid of or have already discarded. Additionally, wastes are such items which people are obliged to discard because legislation has categorized them as hazardous. Many items can be considered waste, from household rubbish, sewage sludge, wastes from manufacturing activities, packaging items, discarded cars, old televisions, garden waste, and old paint to containers, etc. All our daily activities can effectively be a source of different wastes arising from different sources.

Over 1.8 billion tons of waste is generated each year in Europe, which may be expressed as 3.5 tons per person and which is mainly made up of waste coming from households, commercial activities (e.g., shops, restaurants, hospitals), industry (e.g., pharmaceutical companies, clothes manufacturers), agriculture (e.g., slurry), construction and demolition projects, mining and quarrying activities, and the generation of energy. With such vast quantities of waste being produced, it is of vital importance that it is managed in such a way that it does not cause any harm to either human health or the environment (Eionet 2010).

Waste and its management are governed by a series of definitions intended to describe the process. Waste may therefore be defined as any substance or object which the holder discards or intends or is required to discard. Waste management is considered as the collection, transport, recovery, treatment, and disposal of waste.

The term also encompasses the supervision of such operations and the aftercare of disposal sites; these processes also include actions taken as a dealer or broker. Prevention may be considered as those measures taken before a substance, material, or product has become waste. Recovery is defined as any operation of which the principal outcome is that the material retrieved from the waste serves a useful purpose. Recycling is any recovery operation by which waste materials are reprocessed into products, materials, or substances whether for the original or other purposes. (European Parliament 2008).

1.1.2 Types of Waste

There are numerous definitions available for what constitutes waste, and also many classifications that attempt to segregate and categorize waste materials, the most common of which is based on classification according to the source of the waste materials. Thus, one may say that there are agricultural, industrial, civic amenity, household, commercial, and sewage wastes (Read et al. 1998).

Waste with a chemical composition or other property that makes it capable of causing illness, death, or harm to humans and other life forms when mismanaged or released into the environment is termed "hazardous waste." Uncontrolled dumping of wastes, including hazardous industrial wastes, was commonplace in history. To combat this habit, the European Commission issued a directive (91/689/EEC) on the controlled management of such waste which contained a definition for hazardous waste based on a listing of materials called the List of Wastes, formerly known as the European Waste Catalogue, drawn up under that directive (Eionet 2009). The *Global Waste Management Market Report 2007* states that rapid increase in the volume and types of solid and hazardous waste as a result of continuous economic growth, urbanization, and industrialization is becoming a huge problem for national and local governments to ensure effective and sustainable management of waste.

Waste is produced by all activities of industry and commerce, with important waste streams including construction/demolition, mining, quarrying, manufacturing, and municipal waste (Fig. 1.2).

Municipal solid waste is the most widespread waste stream and is produced by millions of people. It requires major financial and logistical resources to collect it, recycle it, and arrange final disposal. Industrial waste generally has a greater tonnage than municipal solid waste, but its management is the responsibility of relatively small and specific sectors of society. Environmentally acceptable waste management practices are essential if damaging consequences are to be avoided, such as those due to toxic/hazardous waste, greenhouse-gas emissions, water pollution, air pollution, and noise/visual impact (of recycling/waste disposal facilities). Incinerators provide an effective means of reducing the bulk of municipal waste, but it is important that they do not emit harmful gases, compounds, and particles. (UNEP 2009).

1 Waste and Its Rational Management

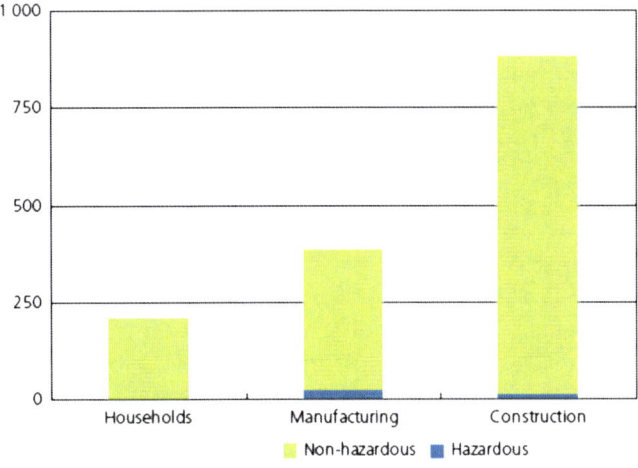

Fig. 1.2 Generation of waste by source, EU-27, 2004 (million tonnes)

1.1.2.1 Liquid

Liquid waste includes any liquid material that has been generated and or modified as a consequence of human activity. Wastewater consists of liquid waste discharged by domestic residences, commercial properties, industry, and/or agriculture and can include a wide range of potential contaminants and concentrations from a variety of sources. This term is commonly taken to mean the municipal wastewater that is generated through the mixing of wastewaters from different sources. On the other hand, sewage is specifically a type of wastewater that is contaminated with feces or urine, but the term is often used to mean any wastewater. Sewage may also be considered as any wastewater which includes domestic, municipal, or industrial liquid waste products that require removal, through the use of a physical infrastructure consisting of pipes, pumps, screens, channels, etc., to treatment facilities. European legislation such as Council Directive 91/271/EEC of 21 May 1991 deals specifically with the requirements for urban wastewater treatment. Over the years this has led to improved general water quality as a consequence of lesser quantities of components that promote eutrophication of waters and its adverse effect on the environment (EEA 2005).

1.1.2.2 Solid Waste

Solid waste may be considered as all discarded material that is not water-based. This material may originate from various sources, but generally municipal solid waste is the major contributor.

1.1.2.3 Industrial Waste

Industrial waste includes all the by-product material that is generated through manufacturing and processing activities. Whether or not the materials are considered hazardous depends on the waste material composition and the legislation governing this form of waste.

1.1.2.4 Municipal Waste

Increased levels of municipal solid waste may not have the catastrophic potential of either global warming or stratospheric ozone depletion, but they have long posed threats to environmental quality and human health that are reasonably well understood and typically of great local and immediate concern. Huge quantities of municipal solid waste are being generated around the world. Although much of it is collected and disposed of through controlled incineration or burial in sanitary landfills, a good deal of the rest continues to be burned in the open or dumped haphazardly, especially in developing countries. Such practices are putting increasing pressure on land, air, and water quality, and posing threats to human health that will be exacerbated by projected increases in total waste generation (Beede and Bloom 1995).

1.1.2.5 Household Waste

"Household waste" is at best a very vague term. The reason for this is that most households utilize a wide variety of substances as a direct consequence of the varied activities that occur in homes. Many hazardous and nonhazardous materials from the kitchen, laundry, garage, and garden shed can be identified as potentially hazardous to human health and safety and/or the environment. Household hazardous materials are products that are likely to be poisonous (toxic), corrosive, flammable, explosive, and reactive. Paints, pesticides, waste oils, cleaners, solvents, pool chemicals, drain cleaners, degreasers and other car care products, batteries, and polishes are all examples of household materials that could be hazardous if stored, used, or disposed of improperly. The used containers and leftover contents of such consumer products are known as household hazardous wastes. Such waste, if carelessly managed, can result in risks to human health and the environment. In the same way that the products should be used carefully, it is extremely important for everyone to manage these hazardous wastes in a responsible manner to minimize the risks.

1.1.2.6 Sources of Waste and Its Classification

Municipal waste is part of the total waste stream; it consists of waste collected by or on behalf of municipal authorities and disposed of through the waste management system. Waste generated by households is an important part of municipal waste, but depending on the national waste management system, it can also include part of the waste generated by commerce, offices, and public institutions. Wastes which are considered particularly dangerous for humans and the environment have been classified as hazardous waste. Specific rules are established for the collection, handling, and recycling of this type of waste (Schäfer 2008).

1.2 How Is Waste Managed

Waste management has evolved from primitive origins through the development of open dumps in ancient Rome to the sophisticated collection and disposal systems that are in use today. Waste management is the collection, transport, processing, recycling or disposal, and monitoring of waste materials. The term usually relates to materials produced by human activity, and waste management is generally undertaken to reduce their effect on health, the environment, or aesthetics. Waste management is also performed to recover resources. Waste management can involve solid, liquid, gaseous, or radioactive substances, with different methods and fields of expertise for each.

Waste management practices differ for developed and developing nations, for urban and rural areas, and for residential and industrial producers. Management of nonhazardous residential and institutional waste in metropolitan areas is usually the responsibility of local government authorities, whereas management of nonhazardous commercial and industrial waste is usually the responsibility of the generator.

There are a number of different options available for the treatment and management of waste, including prevention, minimization, reuse, recycling, energy recovery, and disposal. Under EU policy, land filling is seen as the last resort and should only be used when all the other options have been exhausted, i.e., only material that cannot be prevented, reused, recycled, or otherwise treated should be landfilled. (Eionet 2010).

1.3 An Overview of Means of Disposal

1.3.1 Recycling

Recycling is the reprocessing of materials into new products. Recycling prevents useful material resources from being wasted, reduces the consumption of raw materials, and reduces energy usage and, therefore, greenhouse-gas emissions. Recycling reduces the volume of garbage that is sent for disposal (Misra et al. 2006). Waste materials that can be recycled include newspapers, cardboard, plastic, aluminum, steel, glass, and polystyrene, to name just a few items. To have a successful recycling program it is necessary to collect the recyclable component of municipal solid waste, separate the materials by type (before or after collection), process the materials into a reusable form, and purchase and use the goods made from reprocessed materials. Obviously, separation of the recyclable fraction of the waste disposed of has to be conducted under stringent cost and quality standards to make the process attractive (Nemerow et al. 2009).

1.3.2 Landfills

Since time immemorial human settlements and communities have discarded their waste in dumps and landfills, usually some distance away from the living quarters. This method of "waste management" is both easy and inexpensive, especially where land is available. However, limitations in available areas for landfill purposes have led waste managers to look more closely and modify land filling practices. Despite the efforts of modern waste management practices such as the introduction of low-waste technology, the increase in recycling capabilities, and the implementation of incineration, some of the waste produced has to be disposed of in landfills. Initially, landfills consisted of a hole or geographical depression where all kinds of mixed waste were dumped. However, with the realization that there were inherent dangers to health resulting from leachates contaminating water tables, obnoxious fumes emanating from organic decomposition and spontaneous burning, and the proliferation of vermin, it soon became imperative to modify land filling practices.

Landfills may be placed into three main categories: the inert landfills, the sanitary landfills, and the mono landfills. The inert landfills are used primarily for the deposit of construction and demolition waste and other carefully screened inert materials.

Sanitary or engineered landfills make use of a sophisticated lining system with long-term reliability. For this reason, as a minimum, four basic conditions should be met by any site design and operation before it can be regarded as a sanitary landfill:

1. *Full or partial hydrogeological isolation*: If a site cannot be located on land which naturally contains leachate security, additional lining materials should be brought to the site to reduce leakage from the base of the site (leachate) and help reduce contamination of groundwater and surrounding soil. If a liner – soil or synthetic – is provided without a system of leachate collection, all leachate will eventually reach the surrounding environment. Leachate collection and treatment must be stressed as a basic requirement.
2. *Formal engineering preparations*: Designs should be developed from local geological and hydrogeological investigations. A waste disposal plan and a final restoration plan should also be developed.
3. *Permanent control*: Trained staff should be based at the landfill to supervise site preparation and construction, the depositing of waste, and the regular operation and maintenance.

1 Waste and Its Rational Management

4. *Planned waste emplacement and covering*: Waste should be spread in layers and compacted. A small working area which is covered daily helps make the waste less accessible to pests and vermin.

Mono landfills are those facilities usually reserved for waste from industry; they include waste of one separated waste stream only. This restriction makes such a structuring easier to design, monitor, and maintain. If the need arises to move or "mine" the contents, knowing the characteristics of that one type of waste makes the operation that much simpler and potentially safer.

In Germany and Austria and other European countries, landfill bans on biodegradable and recyclable wastes have been implemented to encourage more waste separation, leading to more composting and recycling. One of the most successful stories of landfill bans is that in Germany, which had 50,000 uncontrolled landfills in 1972, but in 2007 had only about 120 pretreated waste landfills thanks to legislation passed in 1996 (Weimer 2007).

In England and Wales, there are no fewer than 461 official landfills. Of these, 24 are for hazardous waste 265 for nonhazardous waste, and 172 for inert waste; there are also 254 licensed landfills in Scotland (Anon 2010).

The environmental impact of landfills is such that in April 1999 The EU issued Council Directive 1999/31/EC on the landfill of waste.

The landfill impact on the environment can be summarized as follows:

1. Total destruction of flora and fauna in the area.
2. Risk of underground water pollution caused by leaching.
3. Risk of underground water pollution caused by runoff after rainfall.
4. Production biogas, which could be explosive.
5. Production of toxic gases as a result of spontaneous/combustion.
6. Odors, particulate matter, and scattering of light waste caused by wind.
7. Proliferation of rodents and insects.

The directive's overall aim is "to impede or diminish as far as possible negative effect that landfills have on the general environment with special emphasis on contamination and pollution of surface water, groundwater, soil and air. The directive also encompasses effects on the global environment including the greenhouse effect and any other risk to human health that arise from the land filling of waste during the whole life cycle of the landfill" (DEFRA 2009).

1.3.3 Thermal Technologies

When considering the appropriate technology to employ to treat a specific type of waste, the first step is to determine whether the waste is hazardous or if it is simply a solid waste. The technologies employed to manage hazardous and solid wastes fall into four general categories: thermal treatment, biological treatment, physical/chemical treatment, and methods for containment/disposal. The effectiveness of the application of each of these technology groups to a specific type of waste differs depending on the type of waste, the concentration and mixture of individual components of the waste, the physical phase (solid or liquid) of the material, the medium (if any) in which the waste is contained, the desired level of treatment, and the final method of disposal of any remaining residue. Another consideration in selecting a treatment technology is where the wastes are to be treated. Wastes may be treated in place (in situ), within the confines of the site, or at a facility off-site (ex situ). The final major consideration is whether the waste is being treated as a virtually homogeneous stream emanating directly from a manufacturing process or is being removed from an earlier disposal site for additional treatment.

Other than the application of the traditional containment methods (land filling and capping), use of some type of thermal process has, until recently, been the most common form of treatment for hazardous wastes. Thermal treatments have lost some popularity recently because of the threat of emissions from incomplete combustion. Except for vitrification, thermal technologies are ex situ processes, requiring the wastes to be transported to the processing unit.

1.3.3.1 Incineration

When most people think of thermal treatment, they think of incineration, which may be defined as the burning of substances by controlled flame in an enclosed area. Incineration detoxifies hazardous wastes by destroying organic compounds, reduces the volume of the wastes, and converts liquid wastes to solids by vaporizing any fluids present in the wastes. Incinerators have been extremely capable of destroying organic compounds in waste. Removal efficiencies as high as 99.9999% have routinely been achieved (this is often referred to as the "six-9's" treatment level).

Incinerators can be designed to handle wastes in any physical state and have proven effective in treating solids, liquids, sludges, slurries, and gases. The effectiveness of an incinerator depends on three factors:

1. Temperature of the combustion chamber.
2. Residence time of the material in the chamber.
3. Amount of mixing of the material with air while the material is in the chamber.

Normal combustion temperatures range between 900 and 1,500°C, and in some instances are much higher. Many incinerators for hard-to-burn compounds employ two combustion chambers. The first chamber converts the compounds to gas and initiates the combustion process. In the second chamber, combustion of the gases is completed. The inert portion of the wastes remains as ash after incineration. For liquids, the amount of ash remaining is generally insignificant. For solid wastes, the volume of ash is as much as 30% of the original volume. If the ash contains metals or radioactive material, it must be further treated prior to disposal. The most frequently employed method of treating the ash remaining from the incineration process is solidification/stabilization.

Several types of incinerators are available for treating wastes. The most common is the rotary kiln. The kiln of a rotary kiln incinerator is a cylindrical shell mounted on its side on a slight incline. As the kiln rotates, the wastes pass through and are combusted. Rotary kiln incinerators are capable of accepting wastes in all phases. Liquid injection incinerators introduce the material under high pressure through a nozzle which atomizes the wastes. This allows air to mix with the waste and the combustion process to take place.

Fluidized bed incinerators burn finely ground solids or liquids in a bed of inert material suspended above the floor of the combustion chamber. Any ash remaining after combustion is removed when the bed is changed.

A recent development has been the use of infrared incineration technology. This system uses electrically powered silicon carbide rods to raise the wastes to combustion temperatures in the primary combustion chamber. Infrared incinerators can accept solids and sludges. Liquids must be mixed with sand or soil prior to introduction into the combustion chamber.

Although incineration remains one of the most effective methods of treating organic wastes, its application has always been the subject of debate and sometimes resistance. This is mainly due to the public's fear of hazardous emissions escaping from the incinerator's chimneys and being deposited in the surrounding communities.

1.3.3.2 Thermal Desorption

Thermal desorption is the process of heating a waste in a controlled environment, thereby volatizing any organic components. Thermal desorption works especially well for volatile organic compounds but can also be employed for semivolatile organic compounds. Removal efficiencies ranging between 65% and 99% have been achieved depending on the type of waste.

The waste is screened to eliminate coarse pieces prior to it entering the thermal desorption unit. Any excess moisture is also removed. The wastes are then passed to a furnace operating in a temperature range of 300–600°C. The gaseous organic compounds volatilized by this process are then either collected on a medium such as activated charcoal or passed through an incinerator connected in-line with the thermal desorption unit.

Because of its low energy requirements and lower running costs, this technology is gaining popularity over conventional incineration. Public acceptance of thermal desorption has been better since combustion does not occur and the final treatment of the compound may take place off-site, thereby lessening the fear of release of dangerous compounds.

1.3.3.3 Pyrolysis

Pyrolysis is a chemical change brought about by the action of heat. This differs from incineration, which is the combustive destruction of a material in direct flame in the presence of oxygen. Pyrolysis can be thought of as destructive distillation in the absence of oxygen or other oxidant. It converts wastes containing organic material to combustible gas, charcoal, organic liquids, and ash/metal residues. The organic liquid fraction produced during pyrolytic action has the potential to form the basis of synthetic crude oil. An advantage of this technique is that most of the gases resulting from pyrolysis do not pose a hazard to workers.

Pyrolysis units run under optimal conditions at a temperature ranging between 500°C and 800°C have achieved 99.99% destruction removal efficiencies and volume reductions in excess of 50%.

1.3.3.4 Plasma Torch

Plasma torch technology applies the principles of pyrolysis but at high temperatures (5,000–15,000°C). The wastes are fed into the thermal plasma and become disassociated into their basic atomic components. The atoms recombine in the reaction chamber to form carbon monoxide, nitrogen, hydrogen, and small quantities of methane and ethane. Acid gases that are formed in the process can be removed by scrubbers. Any solids produced are either incorporated into the molten bath at the bottom of the chamber or removed from exhaust gases by scrubbers or filters. Plasma torch technology is currently applicable only to fine particulate wastes, liquids, and pumpable wastes.

1.3.3.5 Vitrification

The vitrification process can be applied to both organic and inorganic waste streams. Unlike other thermal treatments, vitrification is versatile and can be applied both in situ and ex situ.

The principles applied in vitrification are essentially those underlying the production of glass. High-temperature electrodes are used to melt the wastes. Organic matter is transformed by pyrolysis, collected, and destroyed in secondary processes. The inorganic component is immobilized in the resulting glass matrix.

Ex situ applications closely resemble typical glass production plants. The wastes are introduced into the furnace along with silica, soda, and lime. The organic component is driven off, captured, and treated, and the inorganics are incorporated into the glass.

In situ technology involves insertion of large electrodes into the soil. Graphite is spread on the soil surface between the electrodes to complete the circuit. A negatively pressurized hood is placed over the site to collect any off-gases for later treatment. High voltage is applied across the electrodes to produce temperatures reaching 3,600°C. The vitrification process has shown great promise for treating radioactive and mixed wastes. The radioactive wastes are immobilized in the glass matrix and can then be stored until the radioactivity has decayed to a safe level. In the case of mixed wastes, the vitrification process drives off the nonradioactive components, allowing them to be treated as hazardous wastes, and immobilizes the radioactive component.

1.4 Basic Principles of Waste Management

There are a number of schools of thought that deal with waste management and these differ in their usage between countries or regions. The waste hierarchy remains the cornerstone of most waste minimization strategies. The process is based on the premise that maximum practical benefits from materials should be obtained with the intention of generating the least possible amount of waste. The waste hierarchy refers to the "3Rs" reduce, reuse, and recycle, which classify waste management strategies according to their desirability in terms of waste minimization. The 3Rs in the sequence are meant to be a hierarchy, in order of importance.

The waste hierarchy has taken many forms over the past decade, but the basic concept has remained the cornerstone of most waste minimization strategies. The aim of the system is to extract the maximum practical benefits from products and to generate the minimum amount of waste. Waste management is not a uniform process, and various solutions may be applicable to different communities. For this reason a "fourth R" ("rethink") has been suggested. This has the implied meaning that the basic 3Rs would benefit from further attention and that a thorough effective system of waste management may need an entirely new way of looking at waste. In many EU member states, waste management is the single largest environmental expense. Policy analysis to help improve economic soundness can potentially release significant sums for competing needs (Rasmussen et al. 2005).

Waste source reduction is one such important issue. Source reduction involves efforts to reduce hazardous waste and other materials by modifying industrial production. Source reduction methods involve changes in manufacturing technology, raw material inputs, and product formulation. At times, the term "pollution prevention" may refer to source reduction.

Another method of source reduction is to increase incentives for recycling. Many communities are implementing variable-rate pricing for waste disposal (also known as "pay as you throw"), which has been effective in reducing the size of the municipal waste stream (Reichenbach 2008). Source reduction is typically measured by efficiencies and reductions in the amount of waste. Toxics use reduction is a more controversial approach to source reduction that targets and measures reductions in the upstream use of toxic materials. Toxics use reduction emphasizes the more preventive aspects of source reduction but, owing to its emphasis on toxic chemical inputs, has been opposed more vigorously by chemical manufacturers.

In Europe, toxics use reduction is regulated by the European regulatory framework for the registration, evaluation and authorization of chemicals (REACH) (European Commission 2010). In the mid-1990s, the US EPA considered toxics use reporting or materials accounting as an expansion of public right-to-know on toxic chemical use. The EPA issued an advanced notice of proposed rulemaking in 1996, but toxics use reporting was not adopted.

Extended producer responsibility is a strategy designed to promote the integration of all costs associated with products throughout their life cycle (including end-of-life disposal costs) into the market price of the product. Extended producer responsibility is meant to impose accountability over the entire life cycle of products and packaging introduced to the market. This means that firms which manufacture, import, and/or sell products are required to be responsible for the products after their useful life as well as during manufacture.

The polluter pays principle is a principle where the polluting party pays for the impact caused to the environment. With respect to waste management, this generally refers to the requirement for a waste generator to pay for appropriate disposal of the waste.

1.5 Trends in Waste Management

Current societal issues are forcing the reevaluation and reutilization of otherwise unwanted material often loosely referred to as waste. Waste is an expensive and at times unavoidable result of human activity, and if not managed well can result in dire effects on the environment, quality of life, and public health. Waste and society are interrelated. In wealthier communities the concepts of environmental and product stewardship are also gaining favor. As economic actors, the ecologically conscious consumers in these communities can demand more environmentally friendly products from their retailers, with the resultant relationship between the service sector and environmentalism becoming symbiotic. The degree of environmentalism in a community is also interlinked with disposable income, which may influence growth of the service sector (Clement 2009).

The management and utilization of waste is a means of promoting environmental safety and rational use of resources in a world that is generating large quantities of unwanted material that must be disposed of in a manner that will not adversely impact the environment. Many traditional technologies and policies are not able to cope with current demand (Okonko et al. 2009). In an effort to minimize the effects of waste on the quality of life and the environment, various practices are considered, one such practice being remanufacturing. This involves the return of old products to the factory to be dismantled, thoroughly cleaned, and used again. The proponents of this technology claim that the exact material specification is already known and often requires little extra manufacturing; hence, the old products are a secure source of materials for products the manufacturer plans to sell. In a world where material and energy prices are so volatile, this makes a lot of sense and also means security of supply. Naturally, the environmental benefits accruing from this activity cannot be underestimated, nor can the economic ones.

A common issue persisting within industry is whether a material suitable for reuse or further processing (whether a market currently exists for it or not) is waste or a by-product. To ensure accurate reporting of reduction and recycling targets, there is need for clarity on what can be reduced or recycled. Unclear or ambiguous waste definitions are a common phenomenon throughout the world, leading to courts of justice having to resolve waste governance issues. A clear definition of waste should promote environmental protection through the application of the precautionary principle, with the possibility of discouraging the implementation of the waste hierarchy, because of the bureaucratic processes involved. On the other hand, adoption of a narrow definition of waste will support implementation of the waste hierarchy, but may undermine environmental protection.

Irrespective of the definition adopted, some trade-off between protection and reuse is envisaged. The result has been a paradigm shift towards waste as a resource, and a resultant change in the governance of waste from protection to reuse. The debate on the definition of waste is far from concluded. It is clear, however, that broad definitions of waste create a minefield of regulatory requirements and associated red tape. The approach to viewing material as a renewable resource rather than waste may provide an alternative solution to promoting waste reuse. Regulation of resource use, extended to renewable resources, will favor reuse and recycling initiatives as well as give due regard to virgin resource conservation. It may even lead to the replacement of the waste hierarchy with a resource-based hierarchy. Such a resource hierarchy would typically focus on minimization of the use of virgin resources, followed by waste minimization, and renewable resource reuse, recycling, and energy recovery. This could be considered in conjunction with the protection of environmental and human health, to ensure the least impact through waste recycling and reuse (Oelofse and Godfrey 2008).

1.6 Public Perceptions and Attitude

Ineffective solid waste management practices make an unfavorable impression on foreign investors and tourists. They may result in loss of both investment and revenues from these sources. In terms of physical impacts, pollution causes psychological stress and fear of health risks (Petts and Eduljee 1994). The urgency to tackle and solve issues associated with this ongoing and ever-growing state of affairs is widely recognized by governments all over the world. The management of such issues has to include the application of scientific principles while understanding the mindsets of the communities in which the management processes have to be introduced and implemented.

Progress in the form of industrialization and urbanization has meant that waste generation has to be monitored and managed. Waste management, in general, is a global problem and more so is solid waste management, where the display of trash, garbage, or litter, as it may be referred to in different environments, in urban areas is most unwelcome. The many changes in waste generation in recent times, such as the emergence of electronic waste among others, are highly topical. However, of all types of pollution, solid waste is considered the most visible form of mundane pollution and most of the common forms of disposal have the potential to cause serious damage to the environment if they are not effectively handled. Understanding the social forces driving variation in refuse generation can help further illuminate both the social and the environmental dimensions of waste management in general (Clement 2009).

1.7 Real or Perceived Health Risks and Future Needs

Management of solid waste (mainly landfills and incineration) releases a number of toxic substances, most in small quantities and at extremely low levels. Because of the wide range of pollutants, the different pathways of exposure, long-term low-level exposure, and the potential for synergism among the pollutants, concerns remain about potential health effects, but there are many uncertainties involved in the assessment. It is said that the various published studies suffer from many limitations caused by poor exposure assessment, poor ecological level of analysis, and lack of information on relevant confounders. These issues further confuse and complicate a scenario that is already by its nature highly complex. If anything, it is clear that the design and implementation of future research has to seriously consider and allow for and seek to overcome current limitations (Porta et al. 2009).

Uncertainty regarding waste generation, waste management practices, data on emissions, exposure characterization, and in particular, the health risk associated with the different types of waste management methods is the main cause of the extensive market failure in the management of waste disposal. Several population studies document (scientifically) that the mismanagement of waste disposal can have serious effects on the health and well-being of the population (Guerriero and Cairns 2009). The future of waste management therefore lies in the delicate balancing act of diplomacy, science, availability of information to the public, and above all the good will to want to make a difference.

References

Anon. Warmer Bulletin. J Sustain Waste Manag. 2010;(126):8.
Beede DN, Bloom DE. The economics of municipal solid waste. World Bank Res Obs. 1995;10(2):113–50.
Calvert-Henderson. Waste generation. 2006. http://www.calvert-henderson.com/enviro-waste.htm. Accessed 29 Jan 2010.
Clement M. A basic accounting of variation in municipal solid-waste generation at the county level in Texas, 2006: groundwork for applying metabolic-rift theory to waste generation. Rural Sociol. 2009;74(3):412–29.
DEFRA. EU landfill directive. 2009. http://www.defra.gov.uk/environment/waste/strategy/legislation/landfill/. Accessed 30 Jul 2010.
EEA. About waste and material resources 2010. http://www.eea.europa.eu/themes/waste/about-waste-and-material-resources. Accessed 28 Jan 2010.
EEA. Source apportionment of nitrogen and phosphorus inputs into the aquatic environment. EEA report no 7/2005. Copenhagen: European Environmental Agency; 2005.
Eionet. List of wastes. 2009. http://scp.eionet.europa.eu/definitions/low. Accessed 17 Jan 2010.
Eionet. What is waste? 2010. http://scp.eionet.europa.eu/themes/waste/#introduction. Accessed 28 Nov 2010.
EPA. Municipal solid waste generation, recycling, and disposal in the United States: facts and figures for 2008. Washington, DC: EPA; 2008.
EPA. Resource conservation and recovery act (RCRA). 2009. http://www.epa.gov/oecaerth/civil/rcra/index.html. Accessed 29 Jan 2010.

EPA. Wastes – hazardous wastes – waste types. 2008b. http://www.epa.gov/epawaste/hazard/wastetypes/index.htm. Accessed 29 Jan 2010.

European Commission. Reach – what is REACH? 2010. http://ec.europa.eu/environment/chemicals/reach/reach_intro.htm. Accessed 17 Feb 2010.

European Institutions. Waste management. 2010. http://europa.eu/legislation_summaries/environment/waste_management/index_en.htm. Accessed 28 Jan 2010.

European Parliament and Council. Directive 2008/98/EC of the European parliament and of the council of 19 November 2008 on waste and repealing certain directives Euro-Lex, Brussels; 2008.

Guerriero C, Cairns J. The potential monetary benefits of reclaiming hazardous waste sites in the Campania region: an economic evaluation. 2009. http://www.ehjournal.net/content/8/1/28. Accessed 2 Aug 2010.

Misra N, Abd El-AalBakr A, Niranjan K. Environmental aspects of food processing. In: Brennan JG, editor. Food processing handbook. Weinheim: Wiley-VCH; 2006. p. 606.

Nemerow NJ, Franklin A, Salvato JA, editors. Environmental engineering: environmental health and safety for municipal infrastructure, land use and planning, and industry. 6th ed. Hoboken: Wiley; 2009.

Oelofse S, Godfrey L. Defining waste in South Africa: moving beyond the age of 'waste'. S Afr J Sci. 2008;104:242–8.

Okonko I, Ogun A, Shittu O, Ogunnusi T. Waste utilization as a means of ensuring environmental safety-an overview. Electronic J Environ Agri Food Chem. 2009;8(9):836–55.

Petts J, Eduljee G. Environmental impact assessment for waste treatment and disposal facilities. Chichester: Wiley; 1994.

Porta D, Milani S, Lazzarino A, Perucci CF. Systematic review of epidemiological studies on health effects associated with management of solid waste. 2009. http://www.ehjournal.net/content/8/1/60. Accessed 2 Aug 2010.

Rasmussen C, Vigsø D, Ackerman F, Porter R, Pearce ED. Rethinking the waste hierarchy. Copenhagen: Environmental Assessment Institute; 2005.

Read AD, Philips P, Robinson G. Landfill as a future waste management option in England: the view of landfill operators. Geographic J. 1998;164(1):55–66.

Reichenbach J. Status and prospects of pay-as -you-throw in Europe – a review of pilot research and implementation studies. Waste Manag. 2008;28:2809–14.

Schäfer G, editor. Eurostat pocket books: generation of waste by source, EU-27, 2007/2008 edition. Luxemburg: Eurostat; 2008.

Strange K. Issues in environmental science and technology, vol. 18. London: Royal Society of Chemistry; 2002.

Anon. Toxics release inventory: do communities have a right to know more? III. 2009. http://ncseonline.org/NLE/CRSreports/pesticides/pest-9b.cfm. Accessed 17 Feb 2010.

UNEP. Developing integrated solid waste management plan, Training manual assessment of current waste management systems and gaps, vol. 2. Osaka: UNEP; 2009.

Weimer K. Waste management focusing on climate change. Public lecture at University of Malta, Valletta; 2007.

Chapter 2
Implementation of Hazard Analysis and Critical Control Points System in the Food Industry: Impact on Safety and the Environment

Sueli Cusato, Paula Tavolaro, and Carlos Augusto Fernandes de Oliveira

2.1 Introduction

The present economic situation and global market conditions have led companies to look for ways to increase competitiveness by improving production processes, reducing production costs, and improving product quality. In terms of the food industry, two other factors should also be included: the need to ensure food safety and the need to protect consumers' health. Therefore, the existence of a system that ensures food safety is crucial to preserve a company's image and reputation and to increase local and international market shares.

Food safety has become a common concern worldwide, making public health agencies and governments of several countries look for more efficient ways to monitor production chains (Makiya and Rotondaro 2002).

The hazard analysis and critical control points (HACCP) system is widely recognized as a management tool capable of ensuring food safety. The keyword of the system is "prevention" (Mortimore and Wallace 1998), by means of the identification of possible contaminations before they occur, and of the definition of control measures to maximize food safety in every step of the process (Cullor 1997; Leitão 1993). Compared with traditional methods of inspection and quality control based on the analysis of finished products, HACCP facilitates a stricter control of contaminations (Stevenson 1990).

The HACCP system is recognized as an important tool in the reduction of foodborne diseases (FBDs), and it is a global reference in terms of food safety control. It is recommended by the World Health Organization, the International Commission on Microbiological Specifications for Foods, the *Codex Alimentarius*, and food regulatory agencies in various countries.

S. Cusato • P. Tavolaro • C.A.F. de Oliveira (✉)
Department of Food Engineering, School of Animal Science and Food Engineering,
University of São Paulo, Pirassununga, Brazil
e-mail: carlosaf@usp.br

2.2 The HACCP System

2.2.1 General Principles and Definitions

HACCP is a preventive system for the production of safe food products. It is based on technical and scientific principles applicable to every step of the food production chain, from growing/breeding activities, to production and distribution systems, to the moment the food reaches the final consumer (ICMSF 1991).

HACCP systematic analysis identifies raw materials and processed foods that may contain toxic substances or agents of FBDs, or that are potential sources of contamination. It may also determine the possibility that microorganisms survive or grow during food production, processing, storage, and preparation (ICMSF 1991).

HACCP was developed by Pillsbury Company, after a request from the National Aeronautics and Space Administration in the 1960s, to ensure the safety of foods used in the American space program (Bauman 1990). The system has its own specific concepts and terminology, as follows (Bryan 1993; Silva 1999):

- *Hazard*: unacceptable biological (growth or survival of microorganisms), chemical (pesticides, antibiotics, heavy metals, cleaning products), or physical (pieces of glass, metal, or other materials) contamination, rendering the food unfit for consumption.
- *Severity*: magnitude of the hazard or of the consequences to the health of consumers. Diseases may be classified, in terms of severity, as lethal, chronic, or mild.
- *Risk*: probability that the hazard will occur. Risk levels may be high, moderate, or low, and may vary according to the situation.
- *Critical control point (CCP)*: a place, practice, procedure, or process that may be controlled to prevent, eliminate, or reduce the hazard to acceptable levels.
- *Critical limit*: physical (e.g., time, temperature), chemical (e.g., pH), or biological (e.g., sensorial, microbiological) attribute or value determined for each CCP, which indicates that the operation is controlled.
- *Monitoring*: measurement of time/temperature, pH, or acidity, or visual observation of CCPs in order to assess whether critical limits are met; if they are not met, the CCP is not controlled and corrective actions are necessary.
- *Corrective action*: immediate and specific procedures to be followed whenever critical limits are not met.
- *Verification*: additional tests and/or review of monitoring records in order to confirm whether the HACCP plan is working as designed. Verification may cause some of the steps of the process to be changed in order to ensure food safety.
- *Decision tree*: logical sequence of questions that enable the identification of a raw material, step in the process, or ingredient as a CCP.

HACCP has changed and developed over the years. In 1991, the National Advisory Committee on Microbiological Criteria for Foods published a report determining the basic principles of the system as it is known today (Almeida 1998). Successful implementation of HACCP depends on the understanding and correct application of these principles (Motarjemi and Käferstein 1999), which were described by Mayes and Mortimore (2003), and by Cullor (1997) as follows:

1. Analysis of hazards and identification of preventive measures
2. Identification of CCPs using a decision tree if necessary
3. Definition of critical limits for the preventive measures associated with each CCP
4. Definition of mechanisms for CCP monitoring, and definition of procedures for using these results to adjust and control the process
5. Definition of corrective actions for deviations in critical limits
6. Definition of a recordkeeping procedure for every control
7. Definition of verification procedures

2.2.2 Prerequisites for the Implementation of the HACCP Plan

Before the application of HACCP principles, some "prerequisite programs," such as good manufacturing practices and cleaning procedures, should be established in order to ensure basic hygiene conditions in the processing plant. These prerequisite programs, if correctly implemented, will determine the principles for correct handling of foodstuffs, making HACCP more efficient and easy to manage (Wallace and Williams 2001).

The main prerequisite programs are good manufacturing practices and sanitation standard operating procedures. These programs involve the following aspects: physical structure and maintenance of the premises, water supply, handler health and personal hygiene, pest control, sanitization of premises and equipment, calibration of instruments, quality control of raw material and ingredients, recall procedures, and measures related to consumer complaints (Brasil 1998).

The lack or inadequate implementation of prerequisite programs may lead to more complex HACCP plans, with a greater number of CCPs to be monitored, once hygienic aspects have also been included (Byrne and Bishop 2001). More CCPs means increased difficulty in managing the plan, and affects efficacy in terms of food safety (Roberto et al. 2006).

2.2.3 Steps for HACCP Implementation

2.2.3.1 Preliminary Procedures

Management Commitment, Assembling the HACCP Team, and Technical Training of the Personnel

A basic requirement for the implementation of the HACCP system is related to the staff involved in the program, who should be aware of the characteristics of the system and of the necessary commitment involved with it. The management of the company should be committed to the objectives of the plan and should be aware of the resources that have to be made available. The HACCP team, responsible for creating and implementing the plan, should be multidisciplinary and knowledgeable regarding production, engineering, health, microbiology, and quality assurance issues (SENAI 2000). The team leader should have knowledge of the manufacturing process, leadership skills, and easy access to managers (Mayes 1994; Hajdenwurcell 2002).

The team should also include people involved in daily activities in the company, because they may contribute with information on particularities and limitations of the production process, and their presence may create a sense of commitment to the job.

Employees should be previously trained in good manufacturing and handling practices, as well as in all aspects of HACCP. A continuing education program should be created to enable constant updating (Cezari and Nascimento 1995).

Description of the Product; Creation and Validation of the Flowchart for the Process

The HACCP team should know the food product in detail: microbiological and physical–chemical characteristics, ingredients and formula, packaging materials, specifications for storage and transportation, and retail conditions, besides adequate handling procedures, shelf life, and the type of consumer.

The flowchart should describe all the steps, identify the equipment, and define working conditions (temperature, pressure, etc.). Flowcharts are the basis for the identification of hazards and preventive measures, and they should be periodically validated and adjusted, when necessary, to reflect the real processing conditions (Corlett 1998).

The basic conditions for the application of HACCP principles will have been created after the conclusion of these preliminary stages (Wallace and Williams 2001), as summarized in Fig. 2.1.

2.2.3.2 Application of HACCP Principles

Principle 1: Analysis of the Hazards and Definition of Preventive Measures

The possible physical, chemical, and microbiological contaminations (hazards) should be determined, as well as their respective preventive measures, based on specialized literature, on the knowledge of the raw material, and on the flowchart for the process.

Although the HACCP system was originally developed to ensure food safety and protect the health of consumers, the definition of hazard is generally broader, considering not only factors that are harmless and of no consequence, but also those that cause "loss of quality and economic integrity of the product" and noncompliance

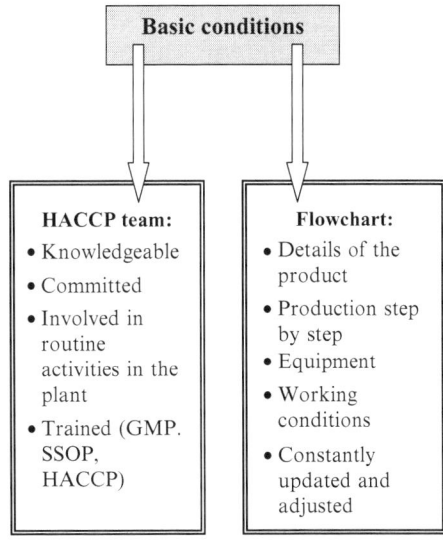

Fig. 2.1 Basic conditions for hazard analysis and critical control points (*HACCP*) implementation in the food industry. *GMP*, good manufacturing practices, *SSOP*, sanitation standard operating procedures

with standards defined by the manufacturer. This broader definition of "hazard," however, may increase the complexity of HACCP, and create a greater number of CCPs (Roberto et al. 2006).

Principle 2: Identification of the CCPs

CCPs are the steps in the process where hazards may be eliminated, prevented, or reduced to acceptable levels are identified in the flowchart by using a decision tree, if necessary.

Principle 3: Definition of Critical Limits

Each CCP should have a critical limit defined in terms of time/temperature, pH, temperature, acidity, etc., in order to ensure the safety of the process. In some cases, safety limits should also be defined, in a way to prevent that critical limits are exceeded. Critical limits may be defined based on specialized literature, present regulations, or the practical expertise of the HACCP team (Cezari and Nascimento 1995).

Principle 4: Definition of Monitoring Procedures

This step involves the definition of controls for each CCP, by means of visual observation, measurements, or laboratory analyses. The frequency with which these controls should be conducted, as well as the person responsible for them, should

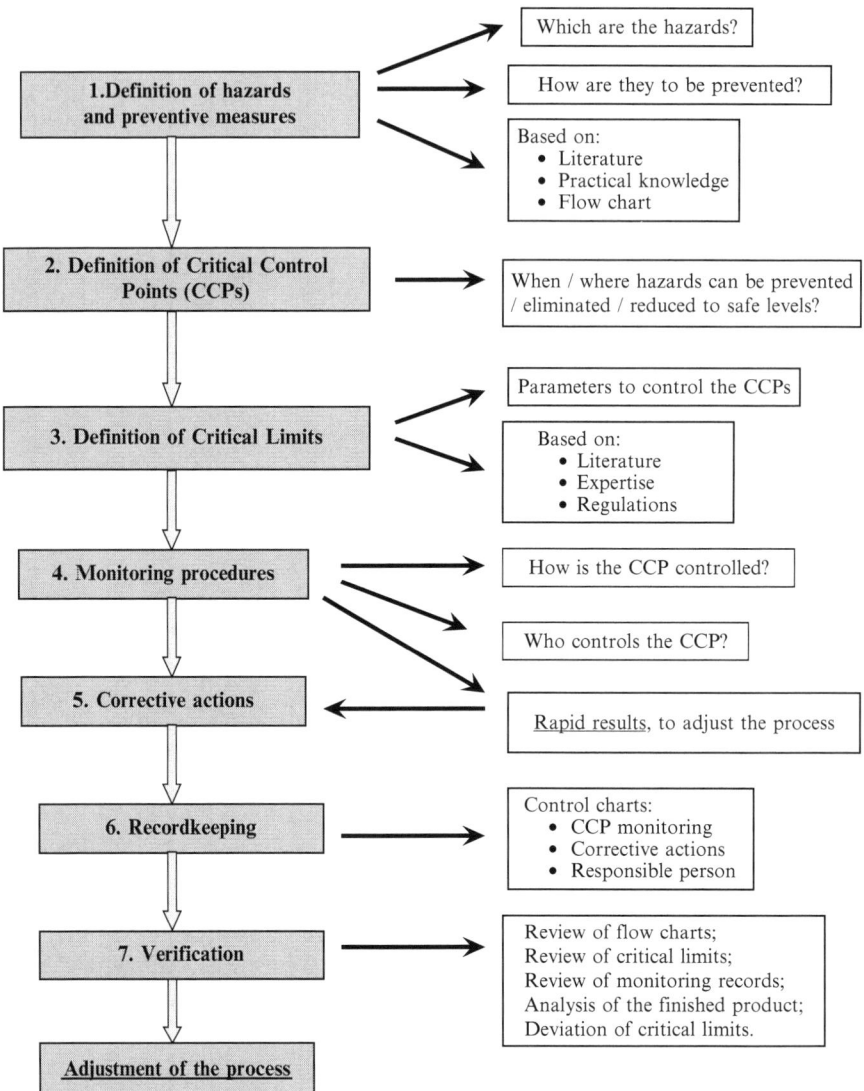

Fig. 2.2 HACCP principles and their application in the food industry

also be defined. The choice of the monitoring procedure should take into account how easy and fast results are obtained to ensure that the process is adjusted without delay, and that the flow of the process is not affected.

Inspection and calibration of the equipment used in CCP monitoring should receive special attention during this stage.

Principle 5: Definition of Corrective Actions

When monitoring shows that critical limits have been exceeded, previously determined corrective actions should be immediately put in place to control the CCP.

Principle 6: Definition of Recordkeeping Procedures

All CCP monitoring procedures should be recorded in control charts, which also have to show the necessary corrective actions. The recordkeeping system should, whenever possible, be integrated in the routine charts of the company to prevent the buildup of time-consuming forms to be completed. Only necessary changes should be made in the charts, such as fields for describing corrective actions and for the signature of the person responsible for the procedure (Mortimore and Wallace 1998).

Principle 7: Definition of Verification Procedures

Verification procedures should be performed periodically to assess whether the HACCP plan is working properly. The following methods of evaluation may be used: review of the flowchart for the process, review of the critical limits, review of CCP monitoring records, laboratory analyses of the finished product, and analysis of deviations in critical limits.

Verification procedures enable adjustments in the HACCP plan, and may ensure the safety of the food. A general overview of the HACCP principles and their application in the food industry can be seen in Fig. 2.2.

2.3 Impact of HACCP in the Food Industry

2.3.1 How the Food Industry Perceives Application of HACCP

HACCP has become an international standard in food safety assurance. Recommended or mandatory use of HACCP is found in the regulations of several countries, and governments, industries, and consumers are showing growing acceptance of the system. The following were the most relevant cases of HACCP adoption (Fermam 2007):

- In 1972, the Food and Drug Administration (FDA) determined the use of HACCP for low-acidity canned foods. Nowadays, the FDA and the US Department of Agriculture require the use of HACCP for fish (since December 1995), poultry, and beef (since July 1996) products. The FDA has required that both US and foreign fruit juice producers use HACCP in their manufacturing processes since January 2001. The same requirement was determined for swine exporters.

- In Brazil, HACCP was made mandatory by the Ministry of Health, in 1994, for all food-handling facilities, by means of *Portaria* 1428 of October 26th, 1993. In 1998, the Ministry of Agriculture determined the use of HACCP in facilities that handle products of animal origin, by means of *Portaria* 46 of February 10th, 1998.
- In the European Union, HACCP is found in Council Directive 93/43/CEE, on the hygiene of food products. This directive was incorporated in the food safety white paper, on January 12th, 2000, and has been periodically revisited and refined with further regulations.
- The Government of Canada, in a joint effort with the fishing industry, introduced in 1993 the Quality Management Program, considered to be the first HACCP-based mandatory inspection program in the world. Canada is moving towards the implementation of a Food Safety Enhancement Program for Agriculture, a system to ensure the safety of all foods, which may further stimulate the adoption of HACCP.

Even in countries where HACCP is not mandatory, training of inspectors is based on this methodology. Some of the reasons for the adoption of HACCP by the food industry are responses to legal requirements, interest in export markets, anticipation of future requirements, and need to lower costs or to increase food safety (Donovan et al. 2001).

Although the system is recognized as an efficient tool in food safety assurance, and despite the efforts of several countries, broader use of HACCP is still prevented by some barriers, described in the following paragraphs.

Studies have shown that HACCP adoption is related to the size of the company and to the market where it is established. Export companies, because of their needs and of their interest in maintaining access to markets, are motivated to meet the standards determined by other countries (Donovan et al. 2001). This is also true for large companies: they have greater financial resources, and personnel with the necessary technical knowledge, making the adoption of the system easier.

The lack of clear understanding of HACCP principles, the implementation process, and the costs/benefits involved is a barrier for the voluntary adoption of the system (Ehiri et al. 1995).

According to Taylor (2003), the implementation of the system has been largely motivated by the requirements of clients, especially in large companies, such as supermarket chains, which demand from their suppliers documented proof of the use of HACCP. Still, according to Taylor (2003) author, for smaller companies whose clients are the final consumers, the greatest pressure for the implementation of the system comes from legal requirements. In countries where regulations are not strict, these companies may not be motivated to adopt HACCP.

Therefore, in small or medium-sized companies, the use of HACCP is still restricted (Taylor 2003; Ehiri et al. 1995). According to Henson et al. (1999), high costs related to the economy of scale and the lack of a clear understanding of the benefits, considered to be limited or of an intangible nature, hinder HACCP adoption. The implementation of the system is still more difficult in companies that operate with small profit margins.

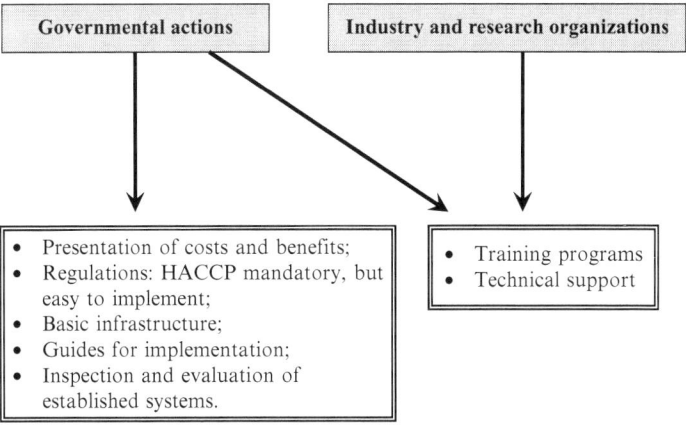

Fig. 2.3 How governments can stimulate HACCP adoption

Other obstacles and difficulties faced by smaller companies are lack of knowledge of the principles of the system, and how they would fit into their reality (methodology), lack of knowledgeable technical personnel (particularly in hazard identification and monitoring), difficulty in recordkeeping, and greater turnover of employees. However, many of these problems stem from the fact that, most of the time, managers and employees are not adequately trained (Motarjemi and Käferstein 1999), making these companies depend on external consultants (Taylor 2003).

Studies carried out by Buchweitz and Salay (2006) in food services in the region of Campinas, Brazil, showed that the lack of information and economic factors are the main reasons for not adopting HACCP. The government and its agencies have a fundamental role in facilitating and stimulating the adoption of HACCP, mainly in small companies. The following aspects should be approached in the process (Suwanrangsi and Keerativiriyaporn 2004), as summarized in Fig. 2.3:

– Demonstration of the benefits of the plan, such as reduction in the number of errors in the manufacturing process, improved company image, and reduction of the costs involved
– Mandatory adoption of HACCP, by means of regulations that make implementation simple
– Creation of training programs for the food industry and for employees of governmental agencies, in a joint effort by the government, research organizations, and the food industry
– Technical support for the companies, also in a joint effort by the government, research organizations, and the food industry
– Implementation of the required basic infrastructure, such as electricity, routes of access (roads), and treated water

- Availability of HACCP implementation guides with all the necessary technical data related to safety standards and regulations, as well as open communication channels with government agencies
- Definition of programs for inspection and evaluation of the systems that have already been implemented according to present regulations

2.3.2 Impact of HACCP on Food Safety

During the past decades, the quest for safety has been challenged by important changes in food production, such as innovations in manufacturing processes, reduced intervals between production and consumption, increased product shelf life, and increased prevalence of some microorganisms (Stevenson 1990; Bauman 1990).

As the food chain became global, FBDs are seen in a new dimension (Motarjemi and Käferstein 1999) and now represent one of the greatest health problems worldwide, affecting millions of people a year (Germano 2003) and leading to significant economic and social consequences (Ruegg 2003; Silva 1999).

Data from the World Health Organization show that, in 2005, 1.8 million people died of gastroenteritis caused by contaminated food and water (World Health Organization 2007). In spite of the technological progress in food production and control, the occurrence of these diseases has recently increased, even in developed countries (Franco and Landgraf 2003).

Food hazards or contamination may come from primary production, still on the farm, from inadequate handling or storage in the food industry, or from errors during preparation at home or in other places where the food is consumed.

Although they have not recently become an issue, FBDs have become increasingly important lately, both in terms of magnitude and in terms of health consequences for the general population. Factors related to the supply chain, demographic situation, lifestyle, health system infrastructure, and the environmental conditions of each country influence the prevalence, increased frequency, and consequences of these diseases (Motarjemi and Käferstein 1999).

When all these facts are taken into account, HACCP is an important tool in modern quality management in the food industry, ensuring the integrity of the product, preventing FBDs, and protecting the health of the consumer (Mortimore and Wallace 1998).

However, HACCP will only become effective when its principles are correctly and broadly applied in all stages of the food production chain. Some of the reasons for the recent increase in FBD frequency all over the world may be failures in implementation or limited application of HACCP, mainly in small companies; lack of knowledge of the final consumer, keeping inadequate food handling practices alive; and low rates of HACCP adoption in developing countries, where most of the FBD outbreaks occur.

2.3.3 Impact of HACCP on the Economy: Cost/Benefit of the System

In general, companies find it difficult to clearly picture the costs and benefits of HACCP (Maldonado et al. 2005). Lack of knowledge of the principles, and of how the plan works, makes it difficult to identify and separate HACCP expenses from production costs (Buchweitz and Salay 2006; Donovan et al. 2001). Therefore, as they are basically interpreted by the perception of the managers, they may be overestimated.

HACCP generally involves high fixed costs related to the creation of the plan, training of the workers, and acquisition of equipment, requiring an economy of scale (Unnevehr and Roberts 1996). Maldonado et al. (2005) emphasized the importance of evaluating the magnitude of costs before the system is implemented. However, this is quite uncommon, as confirmed by Henson et al. (1999), who showed that less than 15% of the companies estimated the costs involved before they began HACCP implementation.

Total relative costs of HACCP involve the sum of all resources made available at the different stages. The technological level of the inividual plant and noncompliance with prerequisite programs contribute to greater costs in the implementation of the system (McAloon 2003; Suwanrangsi 2000). Prerequisite programs determine adequate implementation of good manufacturing practices, and make adoption of the program easier owing to the reduction of the number of CCPs (Bata et al. 2006; Henson et al. 1999). A great number of CCPs make management difficult and make auditing procedures too time-consuming (Wallace and Williams 2001).

In the initial phase of the plan, the main costs are related to the use of external consultants (when required), and to the use of the HACCP team in other positions, different from their routine ones (Bata et al. 2006). In the implementation stage, costs are related to training of employees and adjustment to prerequisite programs and specific HACCP items, such as new equipment, laboratory analyses, and adjustments in the process and in the structure of the plant.

During the maintenance phase, costs are mainly related to time consumed in monitoring CCPs and recording corrective actions (recordkeeping procedures), as well as to hiring people to monitor CCPs (Motarjemi and Käferstein 1999; Roberto et al. 2006; Donovan et al. 2001; Caswell 2000). According to Henson et al. (1999), although difficult to measure, the cost related to the time consumed filling in forms and records is generally greater than expected.

In terms of human resources, lack of trained personnel to develop and implement all aspects of HACCP make most of medium-sized companies use external consultants (Bata et al. 2006), increasing the cost of the system.

In relation to employee training, the following costs should be considered: external costs incurred by the HACCP team, including trips, transportation, meals, and loss in productivity caused by team members being away from regular positions, or when all employees have to be trained, and by a complete interruption in the production cycle (Donovan et al. 2001). Staff training is the basis of the plan and is the key element for the motivation of the team, including plant staff, managers, and supervisors, normally cited as the main obstacles to HACCP implementation in the companies (Henson et al. 1999; Maldonado et al. 2005).

The greater or lesser impact of these elements on total HACCP costs depends, however, on the particular characteristics of each plant (Bata et al. 2006). Implementation of the system may take from some months to several years, and depends on the qualification of the employees, the complexity of the production process (Donovan et al. 2001), the number of CCPs, and the initial condition of the plant.

As for the advantages attributed to the HACCP system, there are several recognized benefits, many of them of an intangible nature or difficult to quantify. The main beneficiary is the consumer, because the system may ensure food safety and lead to the production of higher-quality products (Caswell 2000; Bauman 1995).

Benefits to the public sector are related to the reduction in costs for public health services and sick leaves, besides making it easier for regulatory agencies to monitor processes and products, saving time in audits and decreasing costs in analyses (Donovan et al. 2001; Unnevehr and Roberts 1996).

However, the companies are beneficiaries of most of the advantages of HACCP implementation, by becoming aligned with governmental regulations, and reducing the number of incidents related to the production of unsafe food (Bauman 1995). Economic advantages are related to better control of the process, less reprocessing of products, decrease in raw material and finished product losses, reduction in microbiological counts and consequent increased shelf life of the products, and gains in production efficiency (Henson et al. 1999; Donovan et al. 2001; Maldonado et al. 2005).

Hajdenwurcell (2002) demonstrated other advantages, such as the reduction in the number of laboratory analyses necessary for the finished product, reduction in sampling plans to control the process because of preventive control of CCPs, and reduction in the number of noncompliant products. Hajdenwurcell (2002) also observed that human operational errors may be less frequent owing to better training and greater awareness of the handlers.

After HACCP was implemented in Cargill, McAloon (2003) reported that the system enabled better control of the process, reduced losses and reworks, increased food safety, and improved employee commitment. Besides, McAloon (2003) reported increased productivity and lower production costs. Marthi (2003) showed that when HACCP was implemented in the fishing industry in India, productivity increased owing to fewer interruptions in the production process and to better quality of raw materials.

The use of HACCP increases exporting possibilities, because the system enables harmonization with international trade requirements (Unnevehr and Roberts 1996) and contributes to a positive image of the company, improving consumer confidence and reducing the possibilities of product recall (Ehiri et al. 1995; Motarjemi and Käferstein 1999).

According to Bauman (1995), the high costs of recalls are related to destruction of the products, momentary decreases in sales, and reduction in future sales caused by negative repercussions. Besides, legal actions and financial responsibility should also be considered, as well as costs that are difficult to measure, such as damaged company image and effects on the sales of other products.

In a study among fish-processing industries in Brazil, Donovan et al. 2001) showed that HACCP led to better quality of raw materials owing to greater control of suppliers and, consequently, to final products of higher quality.

The advantages of HACCP related to company image are more difficult to assess. They are, however, undeniable, because the system improves competitiveness and leads to longer permanence in the market, greater consumer confidence, better product/service compliance (Bata et al. 2006), and lower rates of consumer complaints (Motarjemi and Käferstein 1999). In the present, highly competitive market, these gains may make the difference between commercial success and failure.

Reduction in microbiological counts of the products, the ability to attract new clients and to keep existing consumers satisfied were recognized as the greatest benefits of HACCP implementation in dairy factories in the UK (Henson et al. 1999). However, Maldonado et al. (2005) observed that the perception of the benefits by the consumers depended on their awareness of food safety issues.

Khatri and Collins (2007) reported the benefits of HACCP implementation in meat industries in Australia, such as the reduction in losses and reworks of non-compliant products, besides reduction in the number of consumer complaints, improved hygienic conditions of the products, and increased market shares for the companies.

The greater the number of studies that demonstrate the costs and benefits of HACCP to food industries and discuss the elements that make them up, the greater the number of companies that will be motivated to adopt the system (Henson et al. 1999).

2.3.4 HACCP and the Environment

The present integrated economy increasingly demands a more proactive environmental posture from the production sector, making companies reevaluate their competitive strategies. The search for sustainable development demands a review of traditional standards of waste production, manufacturing procedures, and environmental management systems, including practices aiming at waste management and efficient use of nonrenewable resources (Tanimoto et al. 2008).

As new concepts are brought into this discussion, present consumption and production standards must be reviewed and aligned with increasingly clean and sustainable productive processes. "Clean production" involves the use of technologies that enable the use of fewer natural resources, such as water, energy, and raw material, as well as the reduction in waste production and in environmental impacts. Other measures related to production and consumption are also involved in "clean production," such as good operational practices and reduction in losses, adequate storage and discard of residues, redesign of products and production processes, and minimal and efficient use of raw material and energy (Andrade et al. 2001).

Although HACCP was originally conceived to ensure food safety, there are other recognized benefits related to the use of the system, such as reduction in losses during food production. Better trained employees and monitored procedures are responsible for this benefit, because systematic monitoring of some steps of the process leads to immediate responses when critical limits are exceeded, in a way that hazards are controlled without delay, preventing errors and losses during the process. Therefore, fewer failures in the process lead to fewer noncompliant products, that is, fewer products that are rejected and discarded. In the lack of strict control of the process, as proposed in the HACCP system, errors are only identified in the finished product, making reprocessing impossible most of the times, and leading to even greater losses.

Discard of finished product implies added costs for the company and for the environment, mostly related to the necessary treatment of the material before it is discarded, such as the use of energy, water, and chemical products, as well as the cost of the discard process per se. For example, residual waters of food industries, such as dairy or meat plants, contain blood, fat, meat residues, whey and amounts of milk, cheese, yogurt, dairy drinks, and butter. Treatment of these residues involves large amounts of water and produces large volumes of effluent that still have high concentrations of organic material and should be adequately treated before being disposed of into natural water bodies (Chaves 2006). Therefore, HACCP contributes to the reduction of losses in all steps of the process, and has a positive impact on environment conservation.

Packaging material is often discarded together with the products, and it is a waste of natural goods. Although materials such as cardboard, plastic, and cans may be reused after recycling, they are not always recycled and may overload landfills. According to Marinho and Kilperstok (2000), prevention of environmental pollution is a positive attitude that minimizes and may even prevent waste production by means of changes in the types of materials used, or in the production processes.

The use of high-quality raw materials, obtained from reliable companies and stored in adequate conditions, is an indispensable requisite for the quality of the final product (Góes et al. 2001; Ehiri et al. 1995). These issues are approached and foreseen by the HACCP system, as part of the reception of ingredients and raw materials in the food industry, and are important CCPs (Forsythe 2002).

Many of the raw materials delivered to the food industry come directly from primary production (i.e., from farms), where levels of contamination, mainly chemical contamination, may pose serious risks to the health of the consumer, especially in developing countries. Thus, this CCP requires critical limits for the presence of chemical contaminants, ensuring quality control of raw material, and leading to greater environmental awareness and responsibility of the suppliers, by means of controlled and rational use of pesticides and drugs of veterinary use.

Ehiri et al. (1995) and Mortimore and Wallace (1998) showed that auditing suppliers is an important element in monitoring this CCP, because it prevents many problems that would only be identified at the moment of reception of the materials in the food industry, and enables the evaluation of quality standards of the suppliers. In this context, HACCP contributes to stimulating the responsibility of the industries in relation to food safety and quality, and environmental protection.

References

Almeida CR. O Sistema HACCP como instrumento para garantir a inocuidade dos alimentos. Revista Higiene Alimentar. 1998;12(53):12–20.
Andrade JCS, Marinho MMO, Kiperstok A. Uma política nacional de meio ambiente focada na produção limpa: elementos para discussão. Bahia Anal Dados. 2001;10(4):326–32. Salvador.
Bata D et al. Cost of GHP improvement and HACCP adoption of an airline catering company. Food Control. 2006;17(5):414–9. Guildford.
Bauman H. HACCP: concept, development, and application. Food Technol. 1990;44(5):156–9. Champaign.
Bauman HE. The origin and concept of HACCP. In: Pearson AM, Dutson TR, editors. HACCP in meat, poultry and fish processing. London: Chapman & Hall; 1995. p. 1–7.
Brasil MAA. Portaria n.46 de 10 de fevereiro de 1998. Institui o Sistema de Análise de Perigos e Pontos Críticos de Controle – APPCC a ser implantado, gradativamente, nas indústrias de produtos de origem animal sob o regime do serviço de inspeção federal – SIF, de acordo com o manual genérico de procedimentos. Diário Oficial da União, Brasília, 16 Mar 1998. Seção l, p. 24.
Bryan FL. Aplicação do método de análise de risco por pontos críticos de controle, em cozinhas industriais. Revista Higiene Alimentar. 1993;7(25):15–22. São Paulo.
Buchweitz M, Salay E. Analysis of implementation and cost of HACCP system in foodservices industries in the county of Campinas, Brazil. http://www.umass.edu/ne165/haccp1998/buchweitz.html. Accessed 18 Mar 2006.
Byrne DB, Bishop JR. Control of microrganisms in dairy processing: dairy product safety systems. In: Marth EH, Steele JL, editors. Applied dairy microbiology. 2nd ed. New York: Marcel Dekker; 2001.
Caswell JA. Economic approaches to measuring the significance of food safety in international trade. Int J Food Microbiol. 2000;62:261–6. Amsterdam.
Cezari DL, Nascimento ER. Análise de perigos e pontos críticos de controle: manual. Rio de Janeiro: SBCTA; 1995. (Série Qualidade).
Chaves JBP. Contaminação de alimentos: o melhor é preveni-la. Departamento de Tecnologia de alimentos – DTA, Universidade Federal de Viçosa. 2006. http://www.dta.ufv.br/dta/artigos/contal.htm. Accessed 5 Aug 2008.
Corlett Jr DA. HACCP user's manual. Gaithersburg: Aspen Publishers; 1998.
Cullor JS. HACCP (Hazard Analysis Critical Control Points): is it coming to the dairy? J Dairy Sci. 1997;80(12):3449–52. Savoy.
Donovan JA, Caswell JA, Salay E. The effect of stricter foreign regulations on food safety levels in developing countries: a study of Brazil. Rev Agric Econ. 2001;23(1):163–75.
Ehiri JE, Morris GP, McEwen J. Implementation of HACCP in food businesses: the way ahead. Food Control. 1995;6(6):341–5. Guildford.
Fermam RK. HACCP e as Barreiras Técnicas. 2007. http://www.inmetro.gov/barreirastecnicas. Accessed 13 Jul 2008.
Forsythe SJ. Microbiologia da segurança alimentar. Porto Alegre: Artmed; 2002.
Franco BDGM, Landgraf M. Microrganismos patogênicos de importância em alimentos. In: Franco BDGM, Landgraf M, editors. Microbiologia dos alimentos. São Paulo: Atheneu; 2003.
Germano MIS. Treinamento de manipuladores de alimentos: fator de segurança alimentar e promoção da saúde. São Paulo: Varela; 2003.

Góes JAW et al. Capacitação dos manipuladores de alimentos e a qualidade da alimentação servida. Revista Higiene Alimentar. 2001;15(82):20–2. São Paulo.

Hajdenwurcell JR. A experiência da indústria de laticínios na implantação do APPCC: Estudo de caso. Revista Indústria de Laticínios. São Paulo, p. 24–31, July/Aug 2002.

Henson S, Holt G, Northen JC. Cost and benefits of implementing HACCP in the UK dairy processing sector. Food Control. 1999;10:99–106.

International Commission on Microbiological Specifications for Foods – ICMSF. El sistema de análisis de riesgos y puntos críticos: Su aplicación a las industrias de alimentos. Saragossa: Acribia; 1991.

Khatri Y, Collins R. Impact and status of HACCP in Australian meat industry. Br Food J. 2007;109(5):343–54.

Leitão MFF. Análise de Perigos e Pontos Críticos de Controle na Indústria de Alimentos. In: seminário sobre qualidade na indústria de alimentos. Campinas: ITAL; 1993. p. 100–10.

Makiya IK, Rotondaro RG. Integração entre os sistemas GMP /HACCP /ISSO 9000 nas indústrias de alimentos. Revista Higiene Alimentar. 2002;16(99):46–50.

Maldonado ES et al. Cost-benefit analysis of HACCP implementation in the Mexican meat industry. Food Control. 2005;16:375–81.

Marinho M, Kilperstok A. Ecologia Industrial e prevenção da poluição: uma contribuição ao debate regional. Tecbahia. 2000;15(2):47–55.

Marthi B. HACCP implementation: the Indian experience. In: Mayes T, Mortimore S, editors. Making the most of HACCP: learning from others' experience. England: Woodhead; 2003. p. 81–97.

Mayes T. HACCP training. Food Control. 1994;5(3):190–5.

Mayes T, Mortimore S. Making the most of HACCP: learning from others' experience. England: Woodhead; 2003.

McAloon TR. HACCP implementation in the United States. In: Mayes T, Mortimore S, editors. Making the most of HACCP: learning from others' experience. England: Woodhead; 2003. p. 61–80.

Mortimore S, Wallace C. HACCP – a practical approach. Gaithersburg: Aspen; 1998.

Motarjemi Y, Käferstein F. Food safety, hazard analysis and critical control point and the increase in foodborne diseases: a paradox? Food Control. 1999;10:325–33.

Roberto CD, Brandão SCC, da Silva CAB. Cost and investments of implementing and maintaining HACCP in pasteurized milk plant. Food Control. 2006;17(8):599–603.

Ruegg PL. Practical food safety intervention for dairy production. J Dairy Sci. 2003;86(Suppl):E1–9. Savoy.

SENAI. Guia para elaboração do plano APPCC: geral. 2nd ed. Brasília: SENAI/DN; 2000.

Silva JA. As novas perspectivas para o controle sanitário dos alimentos. Revista Higiene alimentar. 1999;13(65):19–25.

Stevenson KE. Implementing HACCP in the food industry. Food Technol. 1990;44(5):179–80.

Suwanrangsi S. HACCP implementation in Thai fisheries industry. Food Control. 2000; 11:377–82.

Suwanrangsi S, Keerativiriyaporn S. How official services foster and enforce the implementation of HACCP by industry and trade. In: Second FAO/WHO Global forum for food safety regulators. Bangkok: WHO; 2004. http://www.foodsafetyforum.org/global2/documents_en.asp. Accessed 13 Jul 2008.

Tanimoto AH, Jesus DS, Santos ARS. Gerenciamento ambiental e simbiose industrial: uma proposta prática para a busca por um desenvolvimento sustentável. http://www.cefetba.br/comunicacao/etc2a10.htm. Accessed 20 Jul 2008.

Taylor E. HACCP and SMEs: problems and opportunities. In: Mayes T, Mortimore S, editors. Making the most of HACCP: learning from others' experience. England: Woodhead; 2003. p. 13–31.

Unnevehr L, Roberts T. Improving cost/benefit analysis for HACCP and microbial food safety: an economist's overview. In: Caswell JA, Cotterill RW, editors. Strategy and policy in the food system: emerging issues. Washington, DC: University of Connecticut/University of Massachusetts; 1996. p. 225–9.

Wallace C, Williams T. Pre-requisites: a help or a hindrance to HACCP? Food Control. 2001;12(4):235–40.

World Health Organization. Food safety and foodborne illness. 2007. http://www.who.int/foodsafety/foodborne_disease/in/. Accessed 1 May 2007.

Chapter 3
Food By-products for Biofuels

Cecilia Hodúr, Zsuzsanna László, and Giovana Tommaso

3.1 Introduction

The rise in global energy usage, together with the disappearance of fossil fuel reserves, has highlighted the importance of developing technologies to harness new and renewable energy sources. In addition to sustainability, climate change is another major issue that has driven the search for clean carbon-neutral fuels.

Biofuels are renewable energy sources, mainly for the transport sector, made from biological sources, biomass. Biofuel can be solid, liquid, or gas produced from any biological carbon source: energy crops and trees, agricultural food and feed plants, agricultural crop wastes, wood wastes and residues, aquatic plants as algae, animal wastes, municipal wastes, and other waste materials.

Biofuels offer the possibility of producing energy without a net increase of carbon in the atmosphere. The energy obtained from plant sources is derived from CO_2 recently captured from the atmosphere. On the other hand, fossil fuels derive energy from long-removed CO_2 stored underground, thus causing a net increase in greenhouse gas concentrations. Biofuel is therefore nearly carbon neutral and less likely to increase atmospheric concentrations of greenhouse gases. The use of biofuels also reduces dependence on petroleum and enhances energy security.

Biofuels include bioethanol, biomethanol, vegetable oils, biodiesel, and biohydrogen. The traditional biofuels are produced from agricultural food crops: bioethanol from corn, wheat, or sugar beet, and biodiesel from rape or sunflower seeds. These food crops require high-quality agricultural land (Neményi et al. 2008).

C. Hodúr (✉) • Zs. László
Faculty of Engineering, University of Szeged, Mars sq. 7, 6724 Szeged, Hungary
e-mails: hodur@mk.u-szeged.hu; zsizsu@sol.cc.u-szeged.hu

G. Tommaso
Department of Food Engineering, School of Animal Science and Food Engineering,
University of São Paulo, Pirassununga, Brazil

On the other hand, biomass can also come from waste plant material. Landfill sites generate gases as the waste decomposes by anaerobic digestion. This waste biomass can be burned and may be a source of renewable energy. Using waste biomass to produce energy can reduce not only the use of fossil fuels, but also the use of food and feed crops, and reduce pollution and waste management problems.

3.2 Biofuels

3.2.1 *Bioalcohols*

The alcohols suitable for motor fuels are methanol (CH_3OH), ethanol (C_2H_5OH), propanol (C_3H_7OH), and butanol (C_4H_9OH), but only ethanol and the methanol are technically and economically suitable for internal combustion engines. The alcohols are usually obtained from biological sources, these are bioalcohols.

Biofuels used for transport may be pure (100%) for dedicated vehicles, or ethanol can be blended with gasoline without the problem of engine modification to produce a blend with up to 15–20% alcohol by volume (E15–E20) IEA (International Energy Agency) 2002. Gasohol is a mixture of 90% gasoline and 10% ethanol or 97% gasoline and 3% methanol. The blend E85 (85% ethanol, 15% gasoline) is an alternative fuel in flexible fuel vehicles, which can be operated on pure gasoline or any ethanol blend with up to 85% ethanol. Gasohol has higher octane number, burns slowly and completely, resulting in reduced air pollutant (NO_x, SO_2) emissions, but it is more volatile, potentially aggravating the photochemical smog problems in large cities with a warm climate.

Anhydrous ethanol can be blended with gasoline without problem. The biologically produced ethanol contains about 5% water; it forms an azeotropic mixture which cannot be purified by distillation. Ethanol containing more than 2% water is not completely miscible with gasoline, but can form an emulsion if a suitable emulsifier (diesohol) is used. Diesohol is a blend of 84.5% diesel fuel, 15% hydrated ethanol (azeotropic ethanol, 96%), and 0.5% emulsifier. The emulsifier consists of a styrene–butadiene copolymer which is soluble in the diesel, and a poly(ethylene oxide)–polystyrene copolymer which is soluble in the alcohol. Using diesohol in engines significantly reduces the visible smoke, carbon dioxide, and particulate emissions and increases the engine thermal efficiency up to 8%.

Ethyl tertiary butyl ether (ETBE) is used as a gasoline additive to reduce carbon monoxide production during the burning of the fuel. There are some benefits from using ETBE – it offers similar air quality as ethanol, does not induce evaporation of gasoline, and does not absorb moisture from the air. ETBE made from bioethanol makes ETBE partially a biofuel, whereas the cheaper methyl tertiary butyl ether made from methanol is usually a fossil fuel (Malca and Freire 2006)

Methanol (also known as wood alcohol) is one of the most industrially important chemicals. It may be used directly as a clean fuel, or as an additive to the gasoline.

It is not as flammable as gasoline. One of the drawbacks of methanol as a fuel is its tendency to corrode some metals, e.g., aluminum. It dissolves the oxide coating that normally protects the aluminum from corrosion.

3.2.1.1 Source of Bioalcohols

Bioethanol is obtained from the conversion of carbon-based feedstock. Feedstocks can be classified into sucrose-containing feedstocks (e.g., sugar beet, sugarcane), starch materials (e.g., wheat, corn), and lignocellulosic biomass (e.g., wood, straw, or grasses). Agricultural feedstocks are considered as renewable, getting energy from the sun by photosynthesis.

The predominant sugar crops are sugarcane (*Saccharum officinarum* L) and sugar beet (*Beta vulgaris* L), with 12–20% sugar content. Less commercial sugar crops are palm (*Phoenix dactylifera* L), sweet sorghum (*Sorghum vulgare* L), and sugar maple (*Acer saccharum* L). Cane sugar comes from countries with warm climates, such as Brazil, India, China, Thailand, Mexico, and Australia. Beet sugar comes from regions with cooler climates, such as northwestern and eastern Europe, northern Japan, and some areas of the USA.

Sugar production causes the formation of different residues with differing properties: Cane molasses can be used in food preparation; however, molasses from sugar beet is unpalatable, and is usually used as industrial fermentation feedstock or animal feed. Dried molasses can be used as a fuel for burning (Blesa et al. 2003).

Corn and wheat are starch-containing materials used for bioethanol production. Corn (*Zea mays* L) is grown in most countries throughout the world. The primary use for maize is as a feed for livestock, forage, silage, or grain. Feed corn (stove) is also being increasingly used for heating. Silage is made by fermentation of chopped green cornstalks. The grain also has many industrial uses, including production of plastics and fabrics. Grains are usually hydrolyzed and enzymatically treated to produce corn syrup, which can be used as a sweetener or can be fermented and distilled to produce grain alcohol (Wang et al. 1999; Kopsahelis et al. 2007).

Starch is found as granules in plant tubers and seed endosperm. Typically, each one contains 70–80% amylopectin molecules and 20–30% smaller amylose molecules. Both components (amylose, amylopectin) consist of polymers of α-D-glucose. Starch can be hydrolyzed to glucose molecules by enzymes, e.g., α-amylase, β-amylase, glucoamylase, and pullulanase (Roy and Nath Gupta 2004; O'Brien and Wang 2008).

The efficient production of ethanol from lignocellulose is thought to be a breakthrough in the fuel market (Kim and Dale 2004; Demirbas 2005). The basic structure of all lignocellulosic material consists of three basic polymers: cellulose, hemicellulose, and lignin (Sjöström 1981). Cellulose consists of long chains with 7,000–15,000 glucose molecules per polymer (Fig. 3.1); hemicellulose (also a polysaccharide) consists of shorter chains, 500–3,000 sugar units. Hemicellulose is a branched polymer, whereas cellulose is unbranched. Lignin is a large, cross-linked, racemic macromolecule with molecular masses in excess of 10,000 u. Different types of lignin have been described depending on the means of extraction.

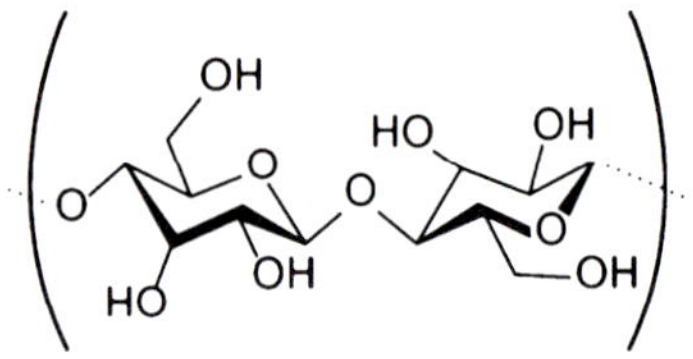

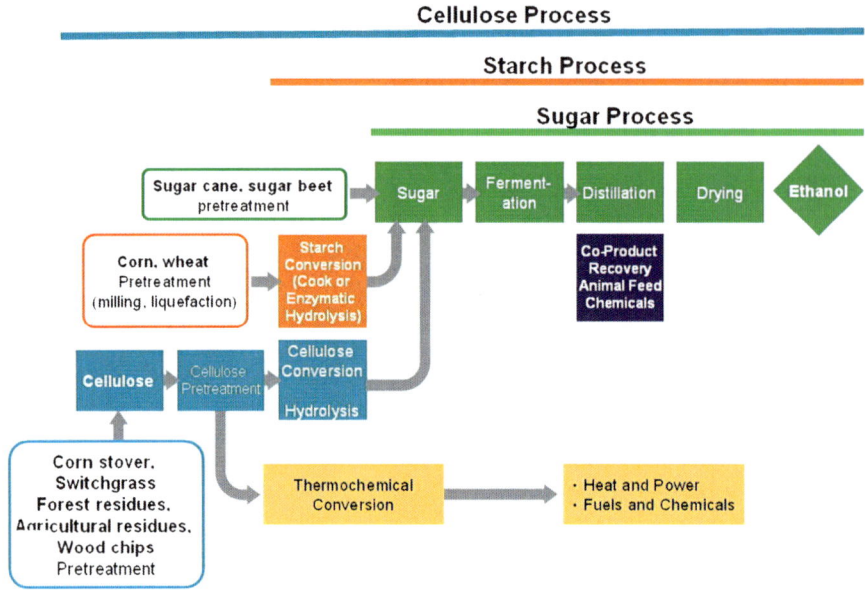

Fig. 3.1 Cellulose, a linear polymer of D-glucose linked by β(1→4)- glycosidic bonds

Fig. 3.2 Ethanol production flowchart. (Adapted from Dale 2007)

3.2.1.2 Short Outline of Bioalcohol Technologies, Process Description

The production of ethanol from sugar involves the following steps: pretreatment of sugar-containing plants, extraction of sugar, fermentation, distillation, dehydration, and denaturing (Gray et al. 2006) (Fig. 3.2.). The common steps independent of the raw material are the following:

- *Fermentation*: In ethanol fermentation from glucose, yeast is added to the mash to ferment and convert the sugars to ethanol and carbon dioxide (Eq. 1):

$$C_6H_{12}O_6 \rightarrow 2C_2H_5OH + 2CO_2 \; (+92.18 \text{ kJ/mol}). \quad (1.1)$$

- Fermentation can be performed in a continuous process, where the fermenting mash flows through several fermenters until it is fully fermented, or in a batch fermentation process, where the mash stays in one fermenter for about 48 h before the distillation process is started.

3 Food By-products for Biofuels

- *Distillation*: The fermented mash (*beer*) contains about 10% alcohol, nonfermentable solids, and yeast cells. The mash is fed to a continuous multicolumn distillation system, where the alcohol leaves the top of the final column at about 96% azeotropic concentration. The mash residue called stillage is removed from the base of the column.
- *Dehydration*: The alcohol passes through a dehydration system where the remaining water is removed. Most ethanol plants use a molecular sieve to capture the water from ethanol.
- *Denaturing*: Ethanol that will be used for fuel is denatured with 2–5% additive, such as gasoline, to make it unfit for human consumption. In the case of starch-containing feedstocks, the process contains more steps to saccharify (to convert starch into sugar) the starch.
- *Milling*: The corn, barley, or wheat passes through hammer mills to grind it into a fine powder called meal.
- *Liquefaction*: The meal is mixed with water and α-amylase, and passes through cookers where starch is liquefied at 120–150°C in a high-temperature stage followed by a lower-temperature (95°C) holding period. The high temperatures reduce bacteria levels in the mash.
- *Saccharification*: The mash is cooled and a secondary enzyme (glucoamylase) is added to convert the liquefied starch to fermentable sugars. The liquefaction and saccharification are called starch hydrolysis.

To produce cellulosic ethanol there are two commonly used methods, the cellulolytic method, where hydrolysis (del Campo 2006) is followed by fermentation (cellulolysis), and gasification, in which synthesis gas is produced that can be converted to ethanol by fermentation or thermochemical catalysis (e.g., the Fischer–Tropsch process).

Cellulolysis contains the following steps before fermentation:

- *Pretreatment*: To make ligocellulosic material (wood or straw) amenable to hydrolysis.
- *Cellulose hydrolysis*: To break down the molecules into sugars.
- *Extraction*: Extraction of the sugar solution from the residual materials (mainly lignin).
- *Gasification*: Complex carbon-based molecules are broken to carbon monoxide, carbon dioxide, and hydrogen. During gasification, instead of breaking the cellulose to sugar molecules, the carbon in the raw material is converted into synthesis gas. This process uses the microorganism *Clostridium ljungdahlii*.
- *Fermentation*: To convert the carbon monoxide, carbon dioxide, and hydrogen into ethanol using *Clostridium ljungdahli*.

Methanol is currently made from natural gas, but can also be made from biomass (Fig. 3.3.) (Demirbas and Gullu 1998) by gasification, partial oxidation with O_2 and H_2O, and catalytic synthesis to produce methanol (Demirbas 2008). Methanol is considerably easier to recover than ethanol, and does not form an azeotrope with water. This is the reason why the more toxic methanol is the preferred alcohol for producing biodiesel.

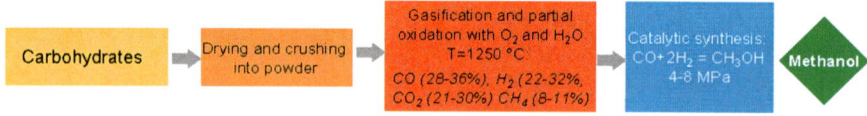

Fig. 3.3 Biomethanol from carbohydrates by gasification and partial oxidation with O_2 and H_2O Demirbas 2008)

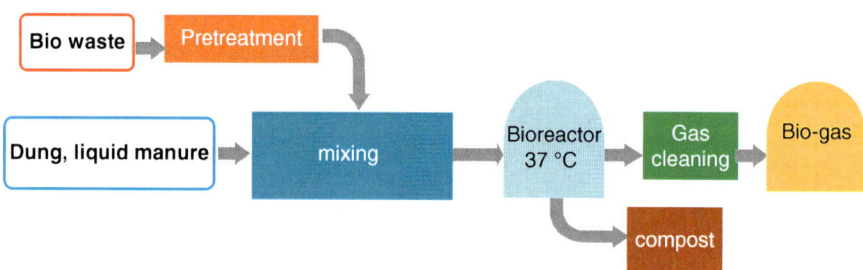

Fig. 3.4 Simplified process of biogas production EC (European Commission) 2004

3.2.2 Biogas

Biogas is an environmental friendly, clean, and cheap fuel (Kapdi et al. 2005). Almost any form of biomass, including sewage sludge, animal wastes, and industrial effluents, can be decomposed through anaerobic digestion into a methane and carbon dioxide mixture called biogas (Fig. 3.4). Anaerobic digestion is a microbially mediated biochemical degradation of complex organic material into simple molecules. Biogas is produced in anaerobic digesters filled with feedstock-like energy crops such as maize silage or biodegradable wastes, including sewage sludge and food waste. The digestion is allowed to continue from 10 days to a few weeks.

Landfill gas is produced by decomposition of organic waste under anaerobic conditions in a landfill. The waste is covered and compressed mechanically, an anaerobic environment is produced, and consequently anaerobic microorganisms thrive. This gas builds up and slowly escapes into the atmosphere. The landfill gases can be captured using appropriate systems and stored for later use. Landfill gas may be hazardous if left uncontrolled, because it is explosive when it mixes with oxygen. The methane contained within biogas is 20 times more potent as a greenhouse gas than carbon dioxide, and it significantly contributes to the effects of global warming. In addition to this, volatile organic compounds contained within landfill gas contribute to the formation of photochemical smog.

The composition of biogas differs depending upon the origin of the anaerobic digestion process. Landfill gas typically has methane concentrations around 50%, whereas advanced waste treatment technologies can produce biogas with 55–75% methane Beszédes et al. (2007).

3.2.2.1 Sources of Biogas

Raw materials may be obtained from a variety of sources – livestock and poultry wastes, night soil, crop residues, food-processing and paper wastes, and materials such as aquatic weeds, water hyacinth, filamentous algae, and seaweed.

Agricultural residues, such as spent straw, hay, cane trash, corn and plant stubble, and bagasse (fibrous residue remaining after sugarcane of sorghum stalks is crushed to extract the juice), are difficult to decompose biochemically; mechanical and/or chemical pretreatments are needed. Mechanical pretreatments consist of size reduction of straw, heat treatment, and/or digestion with strong acids or bases to improve the degradability. Chemical pretreatment methods include bicarbonate treatment, alkaline peroxide treatment, and ammonia treatment. Agricultural residues contain low amounts of nitrogen with 60:90 carbon/nitrogen ratios. The proper carbon/nitrogen ratio for anaerobic digestion is 25:35 (Hills and Roberts 1981); therefore, nitrogen needs to be added to enhance the degradation of agricultural solid residues. The advantages of ammonia treatment is that ammonia will be a source of nitrogen for biodegradation, and no separate wastewater streams are generated from the pretreatment process (Nuntagij et al. 1989; Vigneron et al. 2007; Buendia et al. 2008).

Anaerobic biological treatment of the agricultural solid waste is a process which has received ample attention during the last few years. Conversion to methane provides some energy, and has a beneficial effect on the environment by decreasing the net CO_2 emission to the atmosphere.

Sewage sludge is an unavoidable by-product of wastewater purification processes. Many municipal wastewater treatment plants use anaerobic digestion sludge treatment to stabilize organic matter. Mass reduction, methane production, and improved dewatering properties of the treated sludge are the main features of the process (Tiehm et al. 1997). The disadvantage of the process is the slow degradation of sewage sludge; with a retention time, in conventional digesters, of about 20 days. Anaerobic digestion may be performed under mesophilic and thermophilic conditions. Mesophilic anaerobic digestion of sewage sludge is more widely used than thermophilic digestion, mainly because of the lower energy requirements and higher stability of the process (Gavala et al. 2003). On the other hand, thermophilic anaerobic digestion provides sludge hygienization, increases methane production, and reduces the retention time needed for sludge treatment (Ahring 1994).

3.2.2.2 Short Outline of Biogas Technologies, Process Description

Biogas production systems are relatively simple and can operate at small and large scales in urban or very remote rural locations. Biowastes are pretreated and stored at 70°C to avoid any taint, and then are mixed with substrates. This is the beginning of the hydrolysis. Then the manure is pumped into an anaerobic bioreactor. The fermentation time depends on the quality of the feedstock and the temperature: It is 15–25 days at 30–40°C, but at 50–60°C it is shorter.

The complete anaerobic degradation of an organic substrate to form methane and carbon dioxide takes place in four degradation phases, each of which involves different groups of bacteria:

1. The hydrolysis phase, in which insoluble organic compounds are converted to water-soluble organic products, mainly by extracellular enzymes.
2. The acidification phase, in which larger soluble molecules are degraded by acidogenic bacteria (acidifiers). The products of this process are mainly hydrogen, carbon dioxide, aliphatic acids and alcohols, and small quantities of methyl amines, ammonia, and hydrogen sulfide.
3. The acetogenic phase, in which acetogenic bacteria (acetic acid forming agents) degrade the organic acids and alcohols, except for the chlorine compounds, to form acetic acid, hydrogen, and carbon dioxide.
4. The methanogenic phase, in which the strictly anaerobic, methane-forming bacteria convert acetic acid and chlorine compounds, together with hydrogen, to form biogas (methane and carbon dioxide).

The bacteria of the hydrolysis and acidification phases are facultative anaerobic agents. They are able to withstand fluctuations in environmental conditions without any loss of activity, and they remain active through a pH range from about 3 to 7. The acetogenic and methanogenic bacteria are obligatory anaerobic agents. They are sensitive to oxygen, temperature fluctuations, and low pH values. For substrates containing solids and large molecules that are hard to break down, the hydrolysis phase is the phase that determines the rate of the reaction, and therefore the phase that determines the degree of degradation.

Anaerobic degradation operates more efficiently at two temperatures: 37°C (mesophilic) and 55°C (thermophilic). In mesophilic conditions, the organic components which are easy to decompose are degraded. In thermophilic conditions, other bacteria are able to degrade the difficult-to-decompose components of the organic load.

3.2.3 Lipid Biofuels

Lipid biofuels can also be used as fuels for (diesel) engines. Biodiesel is a general name for alkyl esters from organic feedstock in forms of vegetable oils, animal fats, or aquatic biomass such as algae. Chemically, lipid biofuels are triglyceride molecules, in which three fatty acid groups are attached to one glycerol molecule. Biodiesel is commonly produced by the transesterification of the oils.

Vegetable oils can be used directly as fuels for diesel engines, but may require certain engine modifications to avoid maintenance and performance problems. Vegetable oils from renewable oil seeds can be mixed with gasoline. Gasoline with the label B20 means it contains 20% biofuel.

Vegetable oil (me)ethyl esters, biodiesel, are promising alternative diesel fuels. Biodiesel is environmentally friendly; it is less polluting (e.g., almost sulfur-free) and renewable in contrast to conventional diesel, which is a fossil fuel, use of which may lead to exhaustion (Demirbas 2002).

Biodiesel may contain small but problematic quantities of water: water reduces the heat of combustion of the bulk fuel, resulting in more smoke, harder starting, and less power. Water causes corrosion of vital fuel system components (e.g., fuel pumps, injector pumps, fuel lines) and accelerates the growth of microbial colonies, which can plug up a fuel system. However, the supercritical methanol production method, whereby the transesterification process of oil feedstock and methanol is performed under high temperature and pressure conditions, has been shown to be largely unaffected by the presence of water contamination during the production phase.

Bio-oils are liquid or condensable gaseous fuels made from biomass materials, such as agricultural crops, municipal waters, or agricultural and forestry by-products, via biochemical or thermochemical process, such as pyrolysis (Demirbas 2007). They can substitute conventional fuels in vehicle engines in clear or blended form EC (European Commission) 2004.

3.2.3.1 Sources of Lipid Biofuels

Biodiesel can be made from a range of vegetable oils and animal fats. The most common ones are rapeseed (more than 80%), sunflower, soybean, palm oil, and olive oils. Table 5.2 shows the yields of the most frequently used biodiesel crops.

Palm (*Elaeis*) oil is the most important feedstock in biodiesel production in South Asia, Indonesia, and Malaysia. These regions produce 89% of the world's palm oil. *Elaeis* spp. palm trees have a very large oil yield per hectare. Another advantage is the relatively low production costs.

Rapeseed (*Brassica napus* L.) is the third most important source of vegetable oil in the world, after soybean and palm oil. Rapeseed grows best in mild maritime climates. The leading rapeseed producers are China, India, the European Union, and Canada. Originally, the rapeseed oil was not eatable, but after improvements to the plant developed in Canada, a lot of rapeseed oils (produced from the species *Canola*) are now edible. Natural rapeseed oil contains erucic acid. It is mildly toxic to humans, but it is used as a food additive in small doses. High erucic acid content rapeseed oil is used in lubricants, especially where high heat stability is required. Rapeseed oil is an excellent feedstock for biodiesel production.

Sunflower (*Heliantus annus* L) oil may also be a feedstock for biodiesel. Although sunflower oil yields are lower than rapeseed yields, it is a good choice in countries with a warm and dry climate. However, the sunflower oil used as biofuel cannot compete with diesel, since the growing costs are too high.

Soybean (*Glycine max*) provides one third of all kinds of oils used for biodiesel. The leading soybean producers are the USA and Argentina. Soy contains significant amounts of all the essential amino acids for humans, and so is a good source of protein. Soybeans are the primary ingredient in many processed foods, including dairy product substitutes.

There are suggestions that waste vegetable oil would be the best source to produce biodiesel. Although it has economic benefits, it is even more profitable to convert waste oils into other products, such as soap.

Animal fats (e.g., chicken fat) are limited in supply, and only a small percentage of petroleum usage could be replaced if they were used (Anonymous 2003.

Algae fuel is a biofuel from algae (Cristi 2007). Algae are high-yield, high-cost (30 times more energy per acre than terrestrial crops) feedstocks to produce biofuel. Since the whole organism converts sunlight into oil, algae can produce more oil in an area the size of a two-car garage than an entire football field of soybeans (Yang et al. 2004; Rosenberg et al. 2008).

3.2.3.2 Short Outline of Biodiesel Technologies, Process Description

The basic production methods to obtain fats and oils for raw vegetable oils are rendering, pressure expelling, and solvent extraction (Potter and Hotchkiss 1998). "Rendering" means that fat or oil is heated to melting in different ways, e.g., under a vacuum or by utilizing steam or water. The most common process is mechanical pressing or expelling, a process which presses oil out of seeds. In this case, depending on the type of oil seed and the operation conditions, the oil cake still contains 5–18% residual oil. In large-scale operations the oil is generally removed from cracked seed at low temperature with a nontoxic solvent, e.g., hexane. After the extraction, the solvent is distilled from the oil and recovered for reuse. After extraction, oils have to be purified (generally filtered) to remove insoluble solids and seed residues (Fellows 2000).

Waste vegetable oil processing differs from processing of virgin oils. Because waste oils had been reheated several times during the course of their usage, reheating causes the fatty acid bound to glycerol to break away, and fatty acids become free in the oil. The amount of free fatty acids in the oil has to be reduced, e.g., by esterification during the transesterification process.

Transesterification is the reaction of triglycerides (or other esters) with alcohols to produce alkyl esters and glycerol (Fig. 3.5), generally in the presence of acid or base catalyst. Methanol, owing to its low cost, is the most commonly used alcohol, although other alcohols, e.g., ethanol, produce better fuel characteristic biodiesel. The most common method is the base-catalyzed reaction at low temperature (65°C) and pressure (1.4 bar), resulting in high yield (more than 98%) with minimal side reactions and reaction time.

The main steps of biodiesel processing are the following (Fig. 3.6):

- Pretreatment, where the water and fatty acid content must be controlled, because high levels may cause problems with soap formation.
- Mixing of alcohol and catalyst. The catalyst is typically sodium hydroxide. It is dissolved in the alcohol, and then charged to the reactor and oil is added. Mixing is at 71°C for 1–8 h.
- The resulting biodiesel–glycerol mixture is neutralized, and then the glycerol phase is separated by gravity.

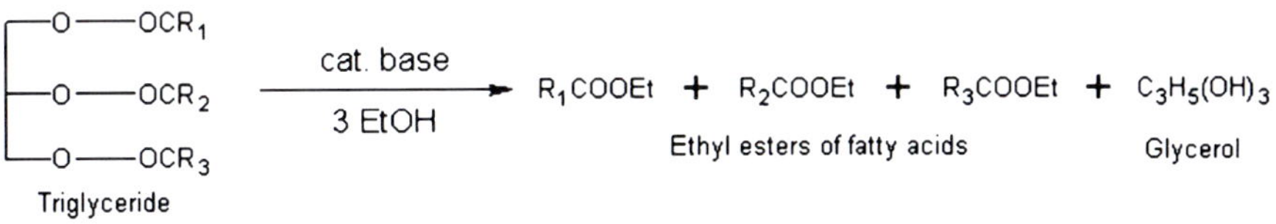

3 Food By-products for Biofuels

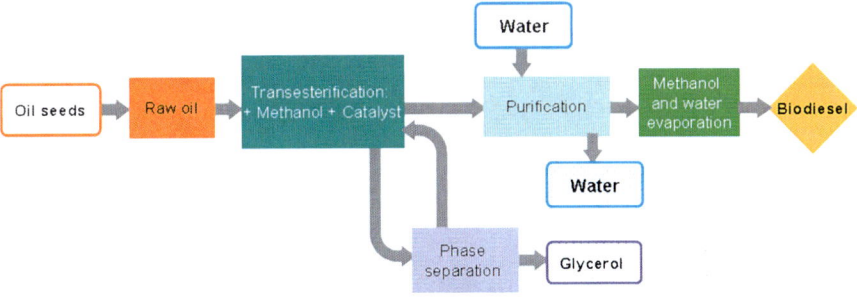

Fig. 3.5 The overall reaction of transesterification

Fig. 3.6 The production of biodiesel

- The excess alcohol in each phase is removed with a flash evaporation process or distillation.
- Biodiesel is purified by gentle washing with warm water to remove residual catalyst and soap, and then dried.

3.3 General Review of Food By-products/Wastes Regarded as Raw Material for Biofuels

The raw materials use in the food industry, such as milk, meat, vegetables, and fruits, are very rich in organic components such as proteins, fats, and carbohydrates. The processes can never result in total conversion; there always remain some residual valuable components that are considered "waste" or the processing waste water itself which may be of highly significant volumes. It is very important to use these components not only because of the utilization of the valuable matter or to minimize environmental fines, but because of the environmental stewardship aspects as well. The potential for utilization is considerable because of the richness of the "raw materials" and the technical development of processing technologies and methods.

Lignocellulosic biomass is renewable, cheap, and readily available with over 10–50 billion tons produced per year at the global level (Sticklen 2006). The sources of lignocellulosic biomass for bioethanol production can be divided into three main categories. One of them is forest residues, which include woods and straws from pulp and paper industries and logging activities. There are also dedicated agricultural crops such as grasses and short rotation crops that are grown for energy purposes. Lastly, there are secondary or tertiary wastes such as municipal solid waste, animal manures, and wastes from food processing industries. All these sources are rich in lignocellulose and have the potential to be used as bioethanol feedstock, rather than using edible crops such as corn and sugarcane as in first-generation bioethanol. Cellulosic ethanol could be more effective and promising as an alternative renewable biofuel than corn ethanol in the long run because it could greatly reduce net greenhouse gas emissions as well as having a higher net fossil fuel displacement potential (Huang et al. 2008).

3.3.1 Wastewater for Biofuels

Biogas is a clean, environmental friendly fuel which may be produced from wastewater by bacterial conversion of organic matter under anaerobic conditions. Raw biogas contains about 55–65% CH_4, 30–45% CO_2, traces of H_2S, and fractions of water vapor. According to Harasimowicz 2007), pure methane has a calorific value of 9,100 kcal/m^3 at 15.5°C and 1 atm.; the calorific value of biogas ranges from 4,800 to 6,900 kcal/m^3. Biogas is of great importance to the food industry since wastewaters from the food-processing industry may represent a high potential for methane production owing to their high organic matter content. Thus, by anaerobic treatment of wastewaters it is possible to achieve not only removal of pollutants of the wastewater streams but also the production of renewable energy in the form of methane (Maya-Altamira et al. 2008). It is thought that anaerobic digestion is the most suitable option for the treatment of high-strength organic effluents (Rajeshwari 2000). The presence of biodegradable components in the effluents coupled with the advantages of anaerobic processes over other treatment methods makes bacterial conversion an attractive option for the treatment of food industry wastewater.

3.3.1.1 Biogas Generation

The anaerobic degradation of organic matter can occur in a liquid medium, as in the treatment of effluents, or in a solid medium, in conventional biodigestors, stabilizing solid residues from the food industry, sometimes increased with animal waste or sludge from aerobic processes of effluent treatment. Methane is formed from a sequence of biochemical reactions characterized by three distinct phases. Figure 3.7 illustrates the anaerobic degradation of organic substances with methane generation.

Initially, the complex organic substances are hydrolyzed to monomers or to small molecules susceptible to microbial absorption. The microorganisms absorb such molecules, using them for energy generation and manufacture of cellular material.

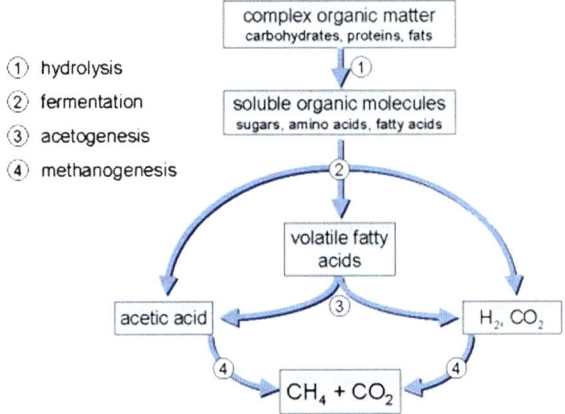

Fig. 3.7 Anaerobic digestion pathways

This stage, called acidogenesis, is a fermentative pathway that produces short-chain fatty acids such as acetic acid, propionic acid, and butyric acid, among others. These acidogenic reactions generate CO_2 and H_2. Of the acids formed, only acetic acid is a substrate for the methanogenic archaea, and because of this, the acetogenic phase is necessary to convert butyric acid and propionic acid into acetic acid. These reactions also generate H_2, are thermodynamically unfavorable, and only occur at very low H_2 partial pressure. The low H_2 partial pressure is possible owing to its consumption by the hydrogenotrophic methanogens that can convert H_2 and CO_2 into CH_4. These methanogenic reactions are responsible for 30% of the methanogenic production in stable reactors. The other 70% of the methane content is generated through the acetoclastic fermentation process in which the methanogenic archaea can ferment acetate to produce CH_4. According to Stams (1997), the key for the stability of the methanogenic production lies precisely in interspecies hydrogen transfer. This process has been described as integral to the symbiosis between certain methane-producing bacteria (methanogens) and nonmethanogenic anaerobes. In this symbiosis, the nonmethanogenic anaerobes degrade the organic substance and produce –among other things – molecular hydrogen (H_2). This hydrogen is then taken up by methanogens and converted to methane via methanogenesis. One important characteristic of interspecies hydrogen transfer is that the H_2 concentration in the microbial environment is very low. Maintaining a low hydrogen concentration is important because the anaerobic fermentative process becomes increasingly thermodynamically unfavorable as the partial pressure of hydrogen increases. So, summarizing, although the methanogenic archaea exploit appropriate conditions to use the H_2 content, acetogenesis occurs, and subsequently acetoclastic methanogenesis.

There are several physical, chemical, and physiological factors in the environment that affect biodegradation of organic compounds, such as the availability of the compounds, the availability of electron donors and acceptors, oxygen concentration, temperature, pH, moisture, salinity, sorption of chemicals to particulate material, and the concentration of the chemicals. Different factors might have different influences according to specific characteristics of the compounds. Anaerobic biodegradability assays are used to establish anaerobic biodegradability to determine the ultimate methane potential of wastes; such assays are also used to determine the rate of biodegradation in general. Methane potential (also called biochemical methane potential) of wastes is defined as the ultimate specific methane production for an indefinite degradation time. In practice, the degradation time is definite and the methane potential is estimated by extrapolation of a methane concentration-time time degradation curve.

3.3.1.2 Biogas Utilization

According to Komiyama et al. (2006), there are two ways of using biogas as an energy source: (1) as a natural gas substitute, to burn it in a boiler or in a stove to obtain heat or to supply it to a gas engine for power generation, and (2) for steam reforming, to obtain hydrogen for fuel cell power generation or to produce synthesis gas ($CO+H_2$), from which dimethyl ether is synthesized, to be used as diesel fuel substitute for farm machines. The merit of steam-reforming biogas, instead of simply using it as a natural gas substitute, lies in its additional freedom to be employed as a renewable energy source. The resulting hydrogen can be used to drive fuel cells, by which electricity and hot water can be produced at higher efficiency than by gas engines or microturbines. One of the major obstacles for steam reforming of biogas is its high sulfur content.

Prior to any kind of utilization, there are three primary compounds that must be removed. First, water vapor in biogas is problematic for compressibility and should be removed prior to storage. Second, biogas typically contains a high percentage of CO_2, which decreases its calorific value. Finally, H_2S is toxic and has corrosive effects on process equipment if it is not removed prior to compression and storage (Strevett et al. 1995). The presence of incombustible gases such as CO_2, H_2S, and water vapor reduces the calorific value of biogas and makes it uneconomical to compress and transport it over long distances. It is therefore necessary to remove these gases before compression (Kapdi et al. 2005). So, for a reasonable utilization of biogas as an alternative energy source, its purification is more than necessary, it is indispensable.

3.3.1.3 Biogas Purification

Improving the heat potential of the produced biogas (when used as fuel), reducing technical problems caused by the presence of H_2S when used in power engines, and lowering CO emissions during the energy production (the parameter strictly related to CO_2 content in the fuel) are the main objectives of biogas purification (Lastella et al. 2002). Several basic mechanisms are involved to achieve selective separation of gas constituents.

Physicochemical methods such as physical adsorption and chemical absorption are commonly used to treat biogas. Physical absorption into a liquid can remove water vapor and H_2S as well as CO_2. Typically, water scrubbing will remove CO_2 and H_2S (Strevett 1995). One of the easiest and cheapest methods involves the use of pressurized water as an absorbent. In this method CO_2 and H_2S are dissolved in water, and are collected at the bottom of the tower. This physical absorption method is perhaps the simplest method for scrubbing biogas (Kapdi et al. 2005). Ethylene glycol scrubbing is required to remove water vapor.

The removal of water vapor and H_2S can be accomplished by physical adsorption on a solid – commonly silica gel or activated carbon. By the proper choice of the adsorbent, the process can remove CO_2, H_2S, moisture, and other impurities either selectively or simultaneously from biogas. H_2S can be adsorbed on activated carbon. The sulfur-containing carbon can then be either replaced with fresh activated carbon or regenerated. The adsorption is a catalytic reaction and carbon acts as a catalyst. Adsorption is generally accomplished at high temperature and pressure. It has good moisture removal capacities, is simple in design, and is easy to operate. But it is a costly process with high pressure drops and high heat requirements (Kapdi et al. 2005).

The purification process can also be done by chemical absorption, which involves formation of reversible chemical bonds between the solute and the solvent. Regeneration of the solvent, therefore, involves breaking of these bonds and, correspondingly, a relatively high energy input. Chemical solvents generally employ either aqueous solution of amines, i.e., monoethanolamine, diethanolamine or triethanolamine, or aqueous solution of alkaline salts, i.e., sodium, potassium, and calcium hydroxides. In H_2S purification with the physical absorption process the consumption of water is very high for absorption of a small amount of H_2S. If chemicals such as sodium hydroxide are added to the water, the absorption process is enhanced. Sodium sulfide or sodium hydrosulfide is formed; these are not regenerated and pose disposal problems. Chemical absorption of H_2S can take place with solutions of an iron salt, such iron chloride. This method is extremely effective in reducing high H_2S levels. The process is based on the formation of insoluble precipitates. $FeCl_3$ can be added directly to the digester slurry (Kapdi et al. 2005).

Another chemical conversion method for the removal of H_2S involves a small amount of oxygen (2–6%) being introduced into the biogas system by using an air pump. As a result, sulfide in the biogas is oxidized to sulfur and the H_2S concentration is lowered. This is a simple and low-cost process. No special chemicals or equipment is required. Depending on the temperature, the reaction time and the place where the air is added, the H_2S concentration can be reduced by 95% to less than 50 ppm. However, care should be taken to avoid overdosing of air, as biogas in air is explosive in the range from 6 to 12%, depending on the methane content (Kapdi et al. 2005).

Chemical solid adsorbents such as quicklime are also capable of removing the three compounds, whereas liquid absorbents, such as ethanolamines, will remove only CO_2 and H_2S (Strevett et al. 1995). Lastella et al. 2000) used Neapolitan Yellow Tuff as a solid adsorbent, which adjusted the CO_2 concentration to the order of 20% and provoked the complete removal of H_2S. Another example is the reaction between

H_2S and iron hydroxides or oxides to form iron sulfide. The biogas is passed through iron oxide pellets to remove H_2S. When the pellets are completely covered with sulfur, they are removed from the tube for regeneration of sulfur; it is a simple exothermic reaction. Also, the dust packing contains a toxic component and the method is sensitive to high water content of biogas.

Membrane separation, cryogenic separation, and chemical conversion can represent an option for biogas treatment as well (Kapdi et al. 2005; Harasimowicz et al. 2007). The principle of the membrane separation process is that some components of the raw gas can be transported through a thin membrane, whereas others are retained. The transportation of each component is driven by the difference in partial pressure over the membrane and is highly dependent on the permeability of the component in the membrane material. For high methane purity, the permeability must be high. A solid membrane constructed from acetate–cellulose polymer in permeable to CO_2 and H_2S up to 20 and 60 times, respectively, more than CH_4. However, a pressure of 25–40 bar is required for the process (Kapadi et al. 2005).

Beyond the physical and chemical methods, the biological treatment of biogas has shown very good results in H_2S and CO_2 removal, as has rinsing the CH_4 content (Table 3.1). Strevett et al. (1995) obtained promising results with a process that effectively removed both CO_2 and H_2S. Through the use of a chemoautotrophic methanogen (*Methanobacterium thermoautotrophicum*), uncoupled methanogenesis techniques, and hollow fiber, CO_2 was converted to CH_4 and H_2S was effectively removed, causing the original CH_4 mass to approximately double. The final purified biogas contained about 96% CH_4, which increased its calorific value from 21 to 35 MJ m^{-3} (Strevett 1995). Komiyama et al. (2006) stated that biodesulfurization and dry desulfurization appear to be comparable in the effectiveness of desulfurization, although the latter may incur a much higher running cost than the former technique. Pagella and Faveri (1999) developed a process for abatement of H_2S based on the combined action of a chemical absorption step and a biological oxidation step exploiting the biocatalytic ability of the bacterium *Thiobacillus ferrooxidans*. In the first stage (chemical step), H_2S was absorbed in a ferric solution. In the second stage (biological step), ferrous ion in the solution was oxidized by the biological action of *Thiobacillus ferrooxidans*. By these means H_2S is definitely converted to sulfur in a closed-reaction cycle. Ramírez-Sáenz (2008) evaluated a biofiltration system for removing H_2S and volatile fatty acids contained in a gaseous stream from an anaerobic digester. The elimination of these compounds allowed the potential use of the biogas while maintaining the methane content throughout the process. The aerobic biodegradation of H_2S was determined in a lava rock biofilter. The overall results suggested that the system used was efficient for the biological removal of H_2S and minor concentrations of volatile fatty acids.

3.3.1.4 Storage

Biogas, containing mainly methane, cannot be stored easily as it does not liquefy under pressure at ambient temperature (the critical temperature and pressure required are 82.5°C and 47.5 bar, respectively). Compressing the biogas reduces the storage requirements, concentrates the energy content, and increases the pressure to the level required to overcome resistance to gas flow. Compression is better in scrubbed biogas.

3.3.2 Solid By-products

Besides wastewater, the solid by-products are also very important and could become a raw material for different kinds of biofuels. The well-known unit operations chopping, sieving, filtration, extraction, distillation, fermentation, etc. could be used for processing by-products although the parameters used might be different. Apart from environmental aspects, the driving force to develop a new processing methods based on "wastes" may be conditioned by economic necessity. The use of by-products of the food industry as a valuable biofuel raw material is well documented.

3.3.2.1 Lignocellulose-Containing By-products

Lignocellulose may be a substrate for the production of value-added products, such as biofuels, biochemicals, biopesticides, and biopromoters, or may be a product itself after biotransformation (e.g., compost, biopulp). In all applications the primary requirement is the hydrolysis of lignocellulose to fermentable sugars by lignocellulolytic enzymes, or appropriate modification of the structure of lignocellulose. Economical and effective lignocellulolytic enzyme complexes containing cellulases, hemicellulases, pectinases, and ligninases may be prepared by simultaneous saccharification and fermentation. (Tengerdy and Szakacs 2003.

The difference between first- and second-generation bioethanol is that in the latter an extra step is required to hydrolyze lignocellulosic biomass. Pretreatment methods are currently used to disrupt the lignocellulosic matter and to remove most of the lignin, thus allowing the cellulases to access the cellulose. Plant genetic engineering can decrease the lignin content and/or need for expensive and harsh pretreatments. Genetic engineering can also be employed to produce microbial ligninases within the biomass crops, so the lignin content of the biomass could be deconstructed during or before bioprocessing

Saska and Ozer (1995) showed that hemicellulose from sugarcane bagasse can be successfully extracted with water. Under the operating conditions of a solid-to-liquid ratio of 1:5, an extraction temperature of 150–170°C, and an extraction time of 15–30 min, 89% of the original amount of xylose was recovered. The major advantages of the water extraction method over the dilute acid pretreatment are less corrosion of equipment, less xylose degradation and thus lower amounts of by-products, including inhibitory compounds, in the extracts, and easier recovery of acid from the hydrolyzate.

Harsh pretreatment conditions could be eliminated for the hydrolysis of sweet sorghum bagasse (Hodúr et al. 2008) compared with the waste wood.

For the simultaneous saccharification and fermentation of the sweet sorghum bagasse 0.67 ml/l cellulase and β-glucosidase enzymes was used at pH 5, 42°C, and with a suspension content of 75 g/l bagasse. In this case the alcohol yield was 61.3% of the optimal yield (Hodúr et al. 2008).

Steam explosion is an effective pretreatment for hemicellulose hydrolysis. In this process, biomass is pretreated by pressurized steam, followed by rapid relief of the pressure, which breaks down the lignocellulosic structure so that the lignin is readily depolymerized and thus the hemicellulose is easily hydrolyzed (Cara et al. 2006). The steam explosion process can result in around 50% insoluble residue of the wood, consisting mainly of cellulose.

Tucker at al. (2003) investigated the combined dilute acid–steam explosion method for biomass treatment. Corn stover was subjected to 1 wt% H_2SO_4 for 70–840 s in a steam explosion reactor at 160, 180, and 190°C. The yields of xylose obtained were 63–77% of thetheoretical yield at 160–180°C, and more than 90% at 190°C.

It is well known that thermochemical pretreatment of lignocelluloses, e.g., dilute acid hydrolysis and steam explosion, can release not only the fermentable pentose and hexose sugars, but also various compounds which are inhibitory to microorganisms and lead to apparent reduction in fermentation yield and productivity. Since the detoxification process can be expensive and constitute a large portion of the whole ethanol production cost, detoxification is a key step and selection of the proper detoxification method becomes very important. For example, one study showed that the detoxification process constituted 22% of the ethanol production cost with willow as feedstock (Sivers et al. 1994).

3.3.2.2 Starch- and/or Sugar-Containing By-products

The main element of juice technologies is squeezing. The juices are very rich in useful components such as vitamins, colorants, sugars, minerals, and amino acids, but the extraction from the tissues is never complete, numerous valuable components remaining in the press-cakes. Classical methods of using these components include fermentation and/or compost preparation. Hodúr et al. (2007) investigated the extraction of valuable components such as sugar, pectin, and colorants. Microwave-assisted extraction was used and they showed that the microwave treatment enhanced not only the pectin yield, but also the biodegradability and the biogas production of the marcs as well.

László et al. (2008) investigated squeezed sweet sorghum juice for ethanol fermentation. They found that the addition of cellulase and glucosidase enzymes considerably enhances the alcohol yield, more than is expected theoretically on the basis of lignocellulose content. In the case of subspecies Monori, the alcohol yield increased from 60% for sugar only to about 85% for sugar and lignocellulose with addition of enzymes.

Because of the high content of starch in pulse food residues, dilute acid pretreatment does not seem to be efficient enough to break down the starch into smaller units of glucose. For this reason, alternative pretreatments, such as enzymatic hydrolysis with amylases, must be considered for subsequent studies.

Harsh pretreatment conditions could be eliminated for agri-food residues (Campo et al. 2006). This is especially appropriate for tomato and red pepper wastes because of the considerable amount of soluble sugars released in the hydrolyzate. Therefore, these residues could be potential feedstocks for ethanol production because of both their low cost and their easy availability. The fact that around 40.29 and 50.20% by weight of soluble sugars were obtained for tomato and red pepper residues after pretreatment without acid must be taken into account when developing new valorization strategies. However, to increase the overall yield of the pretreatment process, an enzymatic hydrolysis must be performed on the solid fraction obtained during the pretreatment step.

Waste mushroom logs have advantages for conversion to fermentable sugar for bioethanol production because of their carbohydrate content and the degradation of lignin (Christopher et al. 2003). Removal of the lignin, which inhibits enzymatic hydrolysis, and some of the hemicellulose from the lignocellulosic biomass provides an accessible surface area for enzyme or chemical reactions for hydrolysis. On the basis of energy consumption, waste mushroom logs are an economically feasible source for bioethanol production. The sugar yield of waste mushroom logs (32.02%) was higher than that of normal wood (24.81%). The high yield of xylose obtained from waste mushroom logs by enzymatic hydrolysis indicates that hemicelluloses were converted to an easily degradable structure by fungal treatment with *Lentinula edodes*. Use of waste mushroom logs in both hydrolysis processes resulted in greater production of glucose and other monosaccharides than did use of normal wood. The faster conversion of waste mushroom logs to fermentable monosaccharides could be due to reduced crystallinity and lignin content following cultivation of *L. edodes* (2003).

The treatment of distillery wastewater, generally known as vinasse, is one of the significant and challenging issues in the industrial production of ethanol. A distillery with a daily ethanol production of 100 m^3 has a vinasse discharge of 1,300 m^3 and a high pollution load, with biological oxygen demand ranging from 30 to 60 g O$_2$/l (Pieper 1990)

The vinasse recycling system results in the buildup of yeast by-products and compounds that inhibit yeast fermentation. This problem can be overcome by recycling a certain percentage of the total vinasse to keep the concentration of undesirable compounds below the level of toxicity that appears in a vinasse with 26% solids content. The recycling of 60% of the generated vinasse is technically feasible and may enhance the ethanol production process with no inhibitory effects. Considering that the amount of water required for the preparation of the fermentation medium in an alcohol-producing plant is about 77% of the total water consumption, the reuse of 60% vinasse reduces the quantity of water required to 46.2% of the total water consumption.

3.3.2.3 Fat/Oil-Containing Byproducts

A variety of oils or fats from animals or vegetables can be used as feedstock for biodiesel production. Nowadays, these technologies are mainly in the research stage, but there are some promising projects.

Waste fats can be converted into biodiesel, according to research by the National Renewable Energy Lab, but there are some problems. Yellow-grease biodiesel (from animal fats) can have as much as 30 ppm sulfur, and hence this is problematic for blending with the 15 ppm sulfur (maximum) gasoline that the US EPA mandated in 2006. Oxidative stability is probably the single largest concern. Low-temperature operability is another issue, especially with animal-fat biodiesel, whereas B20 biodiesel blends cause 2–4% NO_x increases.

In Spain, through a National Technical Centre for Conservation of Fisheries Products (CECOPESCA) and National Association of Fish and Shellfish Canners (ANFACO) project, it is expected that biodiesel will produced from industrial wastewater taken from various stages of the fish canning process. The process for the production of biodiesel is based on collection of fat, which is later purified into oil by centrifugal force and evaporation.

Biodiesel can also be produced from 100% animal fats and fulfills the current EN 14214 for fatty acid methyl esters, as long as state-of-the-art process technology is applied. When chemical properties of various feedstock materials for biodiesel production are compared, the main difference between vegetable oils such as rapeseed oil and animal fats can be found in the diverse fatty acid composition: whereas rapeseed oil and soybean oil have a high content of unsaturated fatty acids, mainly oleic acid and linoleic acid, animal fats such as tallow and lard have a major content of saturated fatty acids (e.g., palmitic acid and stearic acid) (Table 3.2). Biodiesel derived from animal fat (e.g., beef tallow) shows significantly better fuel properties concerning engine performance than does conventional biodiesel (Hilber et al. 2005).

3.3.2.4 Wastewater Sludge for Biofuels

Most branches of the food industry, for instance, the dairy industry, the meat industry, and the cannery industry, have a considerable wastewater output. The problem of pollution is caused not only by the total amount of wastewater production, but also by the high content of organic matter. The sewage sludge is the residue of primary, secondary, or tertiary wastewater technologies. The organic matter content of the sludge is also considerable. Depending on the raw material processed, the sludge may be rich in carbohydrates, lipids, or proteins.

The production of canned maize produces a high volume of wastewater too, with high chemical oxygen demands and biological oxygen demands. After mechanical wastewater treatment, the chemical oxygen demand of the sludge may be more than 100 kg m^{-3} because of the high content of corn starch.

Beszédes et al. (2007) examined the effects of acidic, microwave, and ozone pretreatment on the biogas production and biodegradability of canned maize

Table 3.1 Comparison of composition of reactor and landfill biogas

Matter	Reactor biogas (%)	Landfill biogas content
Methane, CH_4	55–75	54%
Carbon dioxide, CO_2	25–45	42%
Carbon monoxide, CO	0–0.3	–
Nitrogen, N_2	1–5	3.1%
Hydrogen, H_2	0–3	–
Hydrogen sulfide, H_2S	0.1–0.5	88 mg/ml
Oxygen, O_2	Traces	0.8%
Chlorine, Cl_2	–	22 mg/ml
Fluorine, F_2	–	5 mg/ml

Table 3.2 Oil yields of some common crops

Crop	Yield (l oil/ha)
Sunflowers	952
Rapeseed	1,190
Mustard seed	572
Soybean	446
Cotton	325
Olives	1,212
Oil palm	5,950
Algae	7,660

Table 3.3 Parameters of biodegradability after different pretreatments (Beszédes et al. 2007)

Pretreatment	Treatment time (min)	BD_5 (%) [(BOD_5/COD)×100]	Initial biogas production rate (cm^3/g/day)
Untreated	–	26	1.037
Ozone	30	63	3.77
Ozone	60	94	7.40
Ozone/microwave	30+5	96	9.52
Microwave (pH 2)	5	95	25.75

BOD_5 5-day biological oxygen demand, *COD* chemical oxygen demand

production sludge, and the energy balances of the processes were determined when different sludge pretreatments were used. It was found that ozone treatment decreased the chemical oxygen demand, whereas the biological oxygen demand and the biodegradability increased. The combination of microwave and ozone treatment increased the biodegradability relative to ozone treatment alone. The investigation of biogas production showed that all types of pretreatment enhanced methane production: 30 min ozone treatment and 5 min 250-W microwave treatment resulted in a positive energy balance. (Table 3.3.)

Only the short-time ozone or microwave treatment resulted in net energy production. From an energetic aspect, the 30-min ozone pretreatment and the microwave pretreatment at pH 2 were more profitable, because of the high energy demand of the long-time treatment.

The dairy industry has a large-scale technological and cleaning water demand; therefore, the dairy industry has a large amount of wastewater product too. The development of wastewater management technologies has increased the efficiency of removal of pollutant components, but for this reason the mass of produced sewage sludge has also increased. Sludge originating from dairy processing is rich in different type of organic matter and for that reason would be suitable as a raw material for biogas production by applied anaerobic fermentation processes. The dairy sludge is often less degradable by microorganisms because of resistant components. Beszédes et al. (2007) showed that increasing the microwave power caused increased biogas production, but from an energy aspect the higher-level microwave pretreatment was less efficient.

3.4 Biofuel Case Studies and Examples

Whey is a by-product in the cheese making process: Every 100 kg of milk produces about 80–90 kg of liquid whey. Further processing of whey for food is possible but expensive, and sewage treatment is difficult owing to the high organic content of the wastewater. In March 2007, a German dairy products group, Müller Group, began the construction of a plant to produce bioethanol from whey. The group invested €20 million in the plant. According to plans, when maximum capacity is reached, the plant should product ten million liters of ethanol per year. Use of this quantity of ethanol will release an energy equivalent to about 5,079 t of gasoline (or 37,079 barrels), and save approximately 15,450 t of fossil carbon emissions (Murias 2007).

In Aalborg in Denmark, a biogas plant was built in 1997. It receives farm slurry, industrial waste, and organic household waste. The average biogas production is 10,000 m^3/day. The biogas is utilized in a nearby 2-MW heat and power cogenerating unit. The biogas plant is working under "thermophile" conditions at approximately 53°C. The heat and power consumed by the plant are produced by a cogeneration unit, using a minor part of the biogas produced (Energie-Cites 1999).

Waste cooking oil can be used as fuel for diesel engines. Several local projects have been set up to collect used cooking oil from individuals and//or professionals and to use it to fuel public transport. This strategy is all the more environmentally friendly because it reduces the use of fossil fuels, and avoids the disposal of large quantities of oil, which may damage the sewage systems because of their physical properties. In Spain, the city of Valencia ran the ECOBUS project using LIFE funding. On average, some 100 l of waste domestic oil was collected per month. By the end of the project, 800 commercial outlets had between them collected some 800,000 l of used cooking oil. The oil was stored and sent to a transformation plant

3 Food By-products for Biofuels

Table 3.4 Summary of advantages and disadvantages of biofuels

Pros	Cons
Reduce fossil fuel dependence	Use of food as an energy source is unethical; increasing food prices
Do not increase the pollution level, produce less carbon dioxide than other fuels; do not increase global warming	The huge agricultural monocultures have damaging environmental impact
Lower levels of other harmful emissions (carbon monoxide, sulfur dioxide)	Cause other environmental problems (e.g., bioethanol also generates photochemical smog)
Support agricultural development	
Immediately usable with existing infrastructure	The total change to biofuels from fossil fuels is impossible because there is not enough cultivation area
	Much more expensive than fossil fuels

to produce an eco-diesel fuel mix for use by the city's urban buses. During the project, 322,654 l of eco-diesel was used to fuel the Valencia City fleet.

In Austria, the city of Graz has had a similar project since 1999. With the slogan "From the pan to the tank," waste oil is collected from the city's restaurants. It i transformed into biodiesel, and it is used to run 50% of the city's bus fleet. The project was also supported by LIFE, as part of the SMILE project (EC – Directorate General for Energy and Transport 2009).

3.5 Pros and Cons of Biofuels

Recently, the production of biofuels has been on the rise – there are a lot of grants and projects designed to develop new biofuel technologies, and more and more crops are being processed to biofuel. However, the question of whether this is ecologically the best choice remains to be answered (Table 3.4).

References

Ahring BK. Status on science and application of thermophilic anaerobic digestion. Water Sci Technol. 1994;30:241–9.

Anonymous. Waste-fat feedstock only way to boost biodiesel energy efficiency, says top NREL researcher. *Diesel Fuel News*, 10 Nov 2003.

Angelidaki I, Hansen T L, Schmidt J E, Marca E, Mosbæk H, Christensen T H. Method for determination of methane potentials of solid organic waste. Waste Managem. 2004;4:393–400.

Beszédes S, Kertész Sz, László Zs, Zsuzsanna, Szabó G, Hodúr C. (2008): Biogas production of ozone and/or microwave-pretreated canned maize production sludge Ozone Science & Engineering Journal Vol. 31(3) pp. 257–261.

Beszédes S, László Zs, Kertész Sz, Hodúr C, Szabó G, Kiricsi I. The effect of microwave pre-treatment on biogas product of dairy sludge MTA AMB XXXII. Kutatási és Fejlesztési Tanácskozás, Gödöllő, p. 99–104.

Blesa MJ, Miranda JL, Moliner R, Izquierdo MT. Curing temperature effect on smokeless fuel briquettes prepared with molasses and H_3PO_4. Fuel. 2003;82:1669–73.

Buendía IM, Fernández FJ, Villaseñor J, Rodríguez L. Biodegradability of meat industry wastes under anaerobic and aerobic conditions. Water Res. 2008;42:3767–74.

Cara C, Ruiz E, Ballestoros I, Negro MJ, Castro E. Enhanced enzymatic hydrolysis of olive tree wood by steam explosion and alkaline peroxide delignaification. Process Biochem. 2006;41:423–9.

Christopher HV. The molecular composition of lignin in spruce decayed by white-rot fungi (Phanerochaete chrysosporium and Trametes versicolor) using pyrolysis-GC–MS and thermochemolysis with tetramethylammonium hydroxide. International Biodeterioration & Biodegradation. 51:67–75.

Cristi Y. Biodiesel from microalgae. Biotechnol Adv. 2007;25:294–306.

Dale BE. Why cellulosic ethanol is nearer than you may think. NACAA Conference Grand Rapids. 16 July 2007, Michigan.

del Campo I, Alegria I, Zazpe M, Echeverria M, Echverria I. Diluted acid hydrolysis pretreatment of agri-food wastes for bioethanol production. Ind Crops & Prod. 2006;24:214–21.

Demirbas A. Biodiesel from vegetable oils via transesterification in supercritical methanol. Energy Convers Manag. 2002;43:2349–56.

Demirbas A. Bioethanol from cellulosic materials: a renewable motor fuel from biomass. Energ Source. 2005;27:327–37.

Demirbas A. The influence of temperature on the yields of compounds existing in bio-oils obtaining from biomass samples via pyrolysis. Fuel Proc Technol. 2007;88:591–7.

Demirbas A. Biofuel sources, biofuel policy, biofuel economy and global biofuel projections. Energ Convers Manag. 2008;49:2106–16.

Demirbas A, Gullu D. Acetic acid, methanol and acetone from lignocellulosics by pyrolysis. Energy Educ Sci Technol. 1998;2:11–115.

EC – Directorate General for Energy and Transport (2009) SMILE – The gateway to sustainable mobility from http://www.managenergy.net/products/R633.htm.

EC (European Commission). Promoting biofuels in Europe. European Commission, Directorate-General for Energy and Transport, Brussels, 2004. http://europa.eu.int/comm/dgs/energy_transport/index_en.html.

Energie-Cites and Plan-Energi Co. Biogas CHP from: http://www.energie-cites.eu/db/aalborg_139_en.pdf (1999).

IEA (International Energy Agency), 2002, Renewables in global energy supply. An IEA Fact Sheet. Paris, November 2002.

Fellows PJ. Food processing technology. Principles and practice. Boca Raton: CRC.; 2000.

Gavala HN, Yenal U, Skiadas IV, Westermann P, Ahring BK. Mesophilic and thermophilic anaerobic digestion of primary and secondary sludge. Effect of pre-treatment at elevate temperature. Water Res. 2003;37:4561–72.

Gray K. Bioethanol. Curr Opin Chem Biol. 2006;10:41–146.

Gray KA, Zhao L, Emptage M. Bioethanol. Chem Biol. 2006;10:141–6.

Harasimowicz M, Orluk , Zakrzewska-Trznadel G, Chmielewski AG. Application of polyimide membranes for biogas purification and enrichment. J.l of Hazardous Mat. 2007;144:698–702.

Hilber, T., Mittelbach, M., Schmidt, E., 2005, Animal Fats Perform Well in Biodiesel"*5th International Colloquium FUELS in Germany*, January 12–13.

Hills DJ, Roberts DW. Anaerobic digestion of dairy manure and field crop residues. Agric Wastes. 1981;3:179–89.

Hodúr C,. Beszédes S., László Zs., Szabó G., 2007, *Extraction and Biodegradability of Marcs*. CIGR Section VI International Symposium on Food And Agricultural Products: Processing And Innovations Naples, Italy, 24–26 September.

Hodúr, C., Beszédes, S., László, Zs., 2008, Bioethanol production from lignocelluloses containing material. *MTA AMB XXXII. Kutatási és Fejlesztési Tanácskozás, Gödöllő*, 22 Jan 2008. p 94–9.

Huang H-J, Ramaswamy S, Tschirner UW, Ramarao BV. A review of separation technologies in current and future biorefineries. Sep Purif Technol. 2008;62:1–21.

Kapdi SS, Vijay VK, Rajesh SK, Prasad R. Biogas scrubbing, compression and storage: perspective and prospectus in Indian context. Renew Energy. 2005;30:1195–202.

Kim S, Dale BE. Global potential bioethanol production from wasted crops and crop residues. Biomass and Bioenergy. 2004;26:361–375.

Kopsahelis N, Agouridis A, Bekatorou N, Kanellaki M. Comparative study of spent grains and delignified spent grains as yeast supports for alcohol production from molasses. Bioresour Technol. 2007;98:1440–7.

Lastella G, Sharma VK, Testa C, Cornacchia G, Comparato MP. Inclined-plug-flow type reactor for anaerobic digestion of semi-solid waste. Applied Energy. 2000;65:173–185.

Lastella G, Testa C, Cornacchia G, Notornicola M, Voltasio F. Anaerobic digestion of semi-solid organic waste: biogas production and its purification. Energy Conversion and Manag. 2002;43:63–75.

László Z, Beszédes S, Kertész S, Hodúr C, Szabó G, Kiricsi I. Bioethanol from sweet sorghum. Hung Agric Eng. 2007;20:15–7.

Malca J, Freire F. Renewability and life-cycle energy efficiency of bioethanol and bio-ethyl tertiary butyl ether (bioETBE): assessing the implications of allocation. Energy. 2006;31: 3362–80.

Maya-Altamira L, Baun A, Angelidaki I, Schmidt J E. Influence of wastewater characteristics on methane potential in food-processing industry waste waters. Water Research. 2008;42: 2195–2203.

Mirón SA. Comparative evaluation of compact photobioreactors for large scale monoculture of microalgae. J Biotechnol. 1999;70:149–270.

Murias, A, 2007, Fish fat wastes studies for biodiesel production from: http://fis.com/fis/worldnews/worldnews.asp?l=e&ndb=1&id=25625.

Neményi M, Kovács AJ, Lakatos E, Kacz K. Liquid biofuels. Renew Energy. 2008;2:3.

Nuntagij A, Lassus C, Sayag D, André L. Aerobic nitrogen fixation during the biodegradation of lignocellulosic wastes. Biol Waste. 1989;29:43–61.

O'Brien S, Wang Ya-Jane. Susceptibility of annealed starches to hydrolysis by α-amylase and glucoamylase. Carbohydr Polym. 2008;72:597–607.

Pagella C, Faveri D M. H2S gas treatment by iron bioprocess. Chem Eng Sci. 1999;55:2185–2194

Pieper A. Utilization of waste material in alcohol industry. Food Biotechnol. 1990;4:203–4.

Potter NN, Hotchkiss H. Food Science. Gaithersburg: Aspen; 1998.

Rajeshwari KV, Balakrishnan M, Kansal A, Kusum Lata V, Kishore VN. State-of-the-art of anaerobic digestion technology for industrial wastewater treatment. Renew Sustain Energy Rev. 2000;4:135–56.

Ramirez SA, Nascimento MAR, Lora E S, Corrêa PSP, Andrade RV, Rendon MA, Venturini OJ. Biodiesel fuel in diesel micro-turbine engines: Modelling and experimental evaluation Energy. 2008;33:233–240.

Rosenberg JN, Oyler GA, Wilkinson L, Betenbaugh MJ. A green light for engineered algae: redirecting metabolism to fuel a biotechnology revolution. Curr Opin Biotechnol. 2008;19: 430–6.

Roy I, Nath Gupta M. Hydrolysis of starch by a mixture of glucoamylase and pullulanase entrapped individually in calcium alginate beads. Enzyme Microb Technol. 2004;34:26–32.

Saska M, Oser E. Aqueous extraction of sugarcane bagasse hemicellulose and production of xylose syrup. Bioethanol Bioeng. 1995;45:517–23.

Sivers M, Zacchi G, Olsson L, Hahn-Hagerdal B. Cost analysis of ethanol production from willow using recombinant E.coli. Biotechnol Prog. 1994;10:555–60.

Sjöström E. Wood chemistry. Fundamentals and applications. New York: Academic; 1981.

Stam AJM, Oude-Elferink S. Understanding and advancing wastewater treatment. Current Opinion in Biotechnology. 1997;8:328–334.

Sticklen M. Plant genetic engineering to improve biomass characteristics for biofuels. Curr Opin Biotechnol. 2006;17:315–9.

Strevett KA, Vieth RF, Grasso F. Chemo-autotrophic biogas purification for methane enrichment: mechanism and kinetics. The Chem.l Eng. J. and the Biochemical Eng. J. 1995;58:71–79.

Tengerdy RP, Szakacs G. Bioconversion of lignocellulose in solid substrate fermentation. Biochem Eng J. 2003;13:169–79.

Tiehm AN, Kneis U. The use of ultrasound to accelerate the anaerobic digestion of sewage sludge. Water Sci Technol. 1997;36:121–8.

Tucker MP, Kim KH, Newman MM, Nguyen QA. Effects of temperature and moisture on dilute-acid steam explosion pretreatment of corn stover and cellulase enzyme digestibility. Biochem Bioethanol. 2003;105:165–77.

Vigneron V, Ponthieu M, Barina G, Audic J, Duquennoi C, Mazéas L, et al. Nitrate and nitrite injection during municipal solid waste anaerobic biodegradation. Waste Manag. 2007;27:778–91.

Wang S, Thomas KC, Sosulski K, Ingledew WM, Sosulski FW. Grain pearling and very high gravity (VHG) fermentation technologies for fuel alcohol production from rye and triticale. Process Biochem. 1999;34:421–8.

Yang YF, Feng CP, Inamori Y, Maekawa T. Analysis of energy conversion characteristics in liquefaction of algae. Resour Conserv Recycling. 2004;43:21–33.

Chapter 4
Integrated Management Methods for the Treatment and/or Valorization of Olive Mill Wastes

Katerina Stamatelatou, Paraskevi S. Blika, Ioanna Ntaikou, and Gerasimos Lyberatos

4.1 Introduction

The olive tree, originally from Persia and Egypt, has been cultivated for thousands of years (Kapellakis et al. 2008). The Phoenicians and the Greeks contributed to spreading its cultivation in the Mediterranean region, and with the discovery of the new world in the fifteenth century, the olive tree was propagated in areas of the American continent with Mediterranean-like climate conditions. The value of olive oil as a staple, pharmaceutical, and fuel has been greatly appreciated by many since the Roman era (Chazau-Gillig 1994).

More than 30 species of olive trees are known (Siggelakis 1982), but *Olea europea* L. is the only edible species. There are approximately 750 million to 850 million productive olive trees worldwide, occupying a surface area of 7×10^6–8.5×10^6 ha (Molina Alcaide and Nefzaoui 1996; Niaounakis and Halvadakis 2006), and 98% of them are located in the Mediterranean region. The Mediterranean region accounts for not less than 97% of world olive oil production (Al-Malah et al. 2000; WWF 2001), and the European Union countries produce approximately 75% (Fig. 4.1). The three

K. Stamatelatou (✉)
School of Chemical Engineering, National Technical University of Athens, Athens 15780, Greece

Institute of Chemical Engineering and High Temperature Chemical Processes, Patras, Greece
e-mail: astamat@env.duth.gr

P.S. Blika
Department of Chemical Engineering, University of Patras, Vas. Sofias 12, Xanthi 67100, Greece

I. Ntaikou • G. Lyberatos
School of Chemical Engineering, National Technical University of Athens, Athens 15780, Greece

Institute of Chemical Engineering and High Temperature Chemical Processes, Patras, Greece

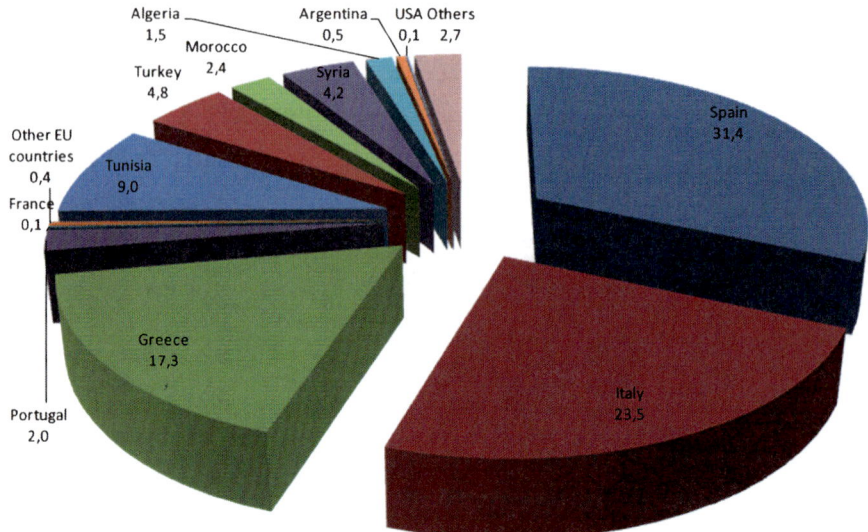

Fig. 4.1 Geographic distribution of world olive oil production (Di Giovacchino 2000)

major olive oil producers worldwide are Spain (890,100 t/year, 36% of the worldwide production), Italy (614,950 t/year, 24% of the world's total), and Greece (402,703 t/year, 17% of the global production), followed by Turkey (168,700 t/year, 4.4% of the total production), Tunisia, and to a lesser extent Portugal, Morocco, Syria, and Algeria (Paraskeva and Diamadopoulos 2006; FAOSTAT 2007).

The process of oil production was initially based on the crushing of olives mixed with hot water, followed by the separation of the oily phase, which is less dense than the aqueous phase. The development of devices that improved fruit crushing and the use of pressure devices that enhanced the extraction process, prior to phase separation, further improved the oil production process. Depending on the extraction technology, there are two main methods used: the traditional and three-phase or two-phase milling.

The traditional method of olive milling is based on application of pressure to the olive fruits. Initially, the olives are ground to form a paste. The olive paste is spread on fiber discs (made of hemp, coconut, or synthetic fiber); these are stacked in a pile and placed in the press. When pressure is applied to the discs (up to 400 atm), the liquid phase percolates. To enhance the speed of percolation as well the subsequent phase separation step, water is added to run down the sides of the discs. The liquid extract is then decanted into a vertical centrifuge to separate the oil from the vegetable water. Proper cleaning of the discs after each extraction is critical to avoid fermentation and production of undesirable compounds that would spoil the oil. The procedure of cleaning, however, is time- and labor-consuming. Production of oil is performed in batches (noncontinuous) and as a result, the paste may be exposed to the oxidizing effects of air and light. On the other hand, this method ruptures the pulp of the drupe and only slightly affects the nut and the skin. As a result, the effect of oil oxidation enzymes present in these organs is reduced. Another advantage is that the water requirement is low and, consequently, loss of polyphenols by washing is reduced, and the exhausted paste, called pomace, is a residue of low moisture and is easy to manage.

In the modern method of olive oil extraction, industrial decanters based on centrifugation are used. The olives are crushed to a fine paste. This paste is then mixed and heated to 27°C (malaxation step) for 30–40 min to enhance the aggregation of the small olive droplets. After malaxation, the paste is pumped to a horizontal decanter. Water is added to facilitate the phase separation into solids (pomace), olive mill wastewater (OMWW), and oil. Further centrifugation in a vertical decanter recovers most of the residue oil contained in the OMWW fraction. Pomace can be further processed to extract its residue oil through extraction with hexane or another solvent. The oil produced is called pomace olive oil.

In this type of olive processing, the so-called three-phase oil extraction, large quantities of OMWW, containing polyphenols washed off from the oil, are produced. In an attempt to solve this problem, two-phase oil decanters have been manufactured. This type of decanter uses less added water, at the expense of its extraction efficiency. The olive paste is separated into two phases: oil and wet pomace. This two-phase decanter, instead of having three exits (for oil, water, and solids) has only two. The water is removed with the pomace, resulting in a rather semisolid residue (two-phase olive mill pomace) that is much harder to process. In fact, the performance of such a process in the existing pomace oil extraction plants to render the pomace dry enough to be processed is highly energy dependent. Therefore, in the attempt to reduce waste quantities, a new type of waste is produced that is harder to manage. Another mode of oil extraction using the three-phase decanter but requiring lower quantities of water is called two and a half phase extraction.

Other oil extraction techniques have also been developed, such as the Sinolea method (Boskou 2006) or combinations of the traditional and modern decanter-based methods, but these have limited acceptance so far owing to low efficiency or low productivity. Figure 4.2 shows the extent of the most widespread olive oil extraction technologies.

4.1.1 Waste Production from Olive Oil Extraction

During the three olive oil extraction process types, the following categories of wastes are produced:

- Solid: leaves and twigs (used as feed for goats and bulking material for compost) and pomace (crushed olive stones and residues of the fruit flesh – olive pulp).
- Liquid: added water, olive juice (vegetation water), unrecoverable oil, and fine olive pulp. This portion constitutes the OMWW.
- Gaseous: fumes from malaxation and gases from the burner exhaust.

The OMWW (also called *aqua reflue* in Italy, *alpechin* in Spain, *katsigaros* or *liozoumi* in Greece, and *zebar* in Arab countries) is produced in large quantities; the annual world OMWW production is estimated to be between 7×10^6 m^3 and more

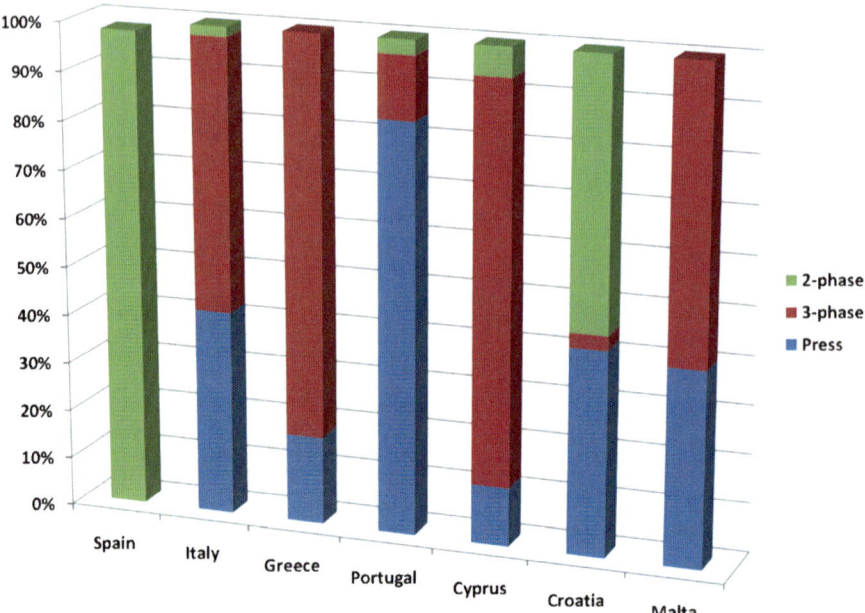

Fig. 4.2 Technologies used by European olive oil mills (IMPEL 2003)

than 30×10^6 m³. Generally, processing 100 kg of olives results in the production of 35 kg of pomace, 55–200 L of OMWW (depending on the extraction process), and 5 kg of leaves. According to Scioli and Vallaro (1997), about 2×10^6 m³ OMWW is generated in Italy annually. Greece generates $1 \times 10^6 – 1.5 \times 10^6$ m³ OMWW annually (Iconomou et al. 2000; Papafotiou et al. 2004), Tunisia 600,000–700,000 m³ OMWW (Ammar et al. 2004; Kallel et al. 2008), and Turkey around 1×10^6 m³ OMWW (Eroğlu et al. 2004). In Spain, where the two-phase olive oil extraction is applied, around 2×10^6 t per year of two-phase olive mill pomace or alpeorujo are produced by almost 2000 Spanish olive oil factories, most of them located in Andalusia (Borja et al. 2006c).

The main characteristics of OMWW are its high chemical oxygen demand (COD) concentration (45–220 mg/L), low pH (4–5), high suspended solids concentration (up to 50 g/L), and other recalcitrant organic compounds, such as water-soluble phenols and polyphenols originating from the olives (Azbar et al. 2004; Davies et al. 2004). These compounds are beneficial if recovered but inhibitory or even toxic to biological OMWW treatment. The characteristics of OMWW differ (Table 4.1) as a result of the different olive varieties and ripeness, the harvesting time, the climate conditions, the cultivation of soil, the presence of pesticides and fertilizers, and the extraction procedure used (Ammar et al. 2005; Borja et al. 2006b). A typical OMWW composition by weight is 83–94% water, 4–16% organic compounds, and 0.4–2.5% mineral salts. The mineral fraction typically contains 47% potassium salts and 7% sodium salts. The main organic constituents are oils (1–14%), polysaccharides (13–53%), proteins (8–16%), organic acids (3–10%), polyalcohols (3–10%), and 2–15% phenolic

Table 4.1 Olive mill wastewater characteristics

Property	Azbar et al. (2004)	Niaounakis and Halvadakis (2006)	Borsani and Ferrando (1996)	Paredes et al. (1999)	Sierra et al. (2001)	Galiatsatou et al. (2002)	Eroğlu et al. (2004)	Al-Malah et al. (2000)
pH	3–5.9	4–6		4.8–5.5	4.5–6	4.9–6.5	4.86	4.52
BOD	23–100 g/L	35–110 g/L			35–100 g/L	15–120 g/L	17.88 g/L	13.2 g/L
COD	40–220 g/L	40–220 g/L			40–195 g/L	30–150 g/L	72.20 g/L	320 g/L
Carbohydrates			2–8%	3.37–32.91%		2–8 g/L		
Polyphenols	0.002–80 g/L	0.5–24 g/L		1.32–3.99%	3–24 g/L	1.5–2.4 g/L	0.13 g/L	3.12 g/L
Fats, oils	1–23 g/L		0.03–1%	0.55–11.4%	0.3–23 g/L	1.3 g/L		
Total solids	1–102.5 g/L						42.24 g/L	
Suspended solids							3.48 g/L	2.17 g/L
N	0.3–1.2 g/L			0.58–1.13%	5–15 g/L	0.5–2%		
K		4 g/L	0.87% K_2O	3.30–6.94%	2.7–7.2 g/L		7.81 g/L	
P			0.22% P_2O_5	0.06–0.32%	0.3–1.1 g/L			
Ca				0.32–0.53%	0.12–0.75 g/L		0.55 g/L	
Na				0.04–0.48%	0.04–0.90 g/L		0.41 g/L	
Mg				0.06–0.22%	0.10–0.40 g/L		0.28 g/L	

BOD biological oxygen demand, *COD* chemical oxygen demand

compounds corresponding to a concentration of 3–10 g/L (Andreoni et al. 1986; Cabrera et al. 1996). Storage also plays an important role in the OMWW composition, since physicochemical as well as aerobic and anaerobic microbial processes take place naturally, resulting in emissions of volatiles, an increase of the acidity, and precipitation of suspended solids.

4.1.2 Olive Oil Wastewater Management Technologies

OMWW management remains an unresolved problem mainly because of the high cost caused by:

- The scattered location of small to medium capacity production units. The cost of any technological solution of acceptable efficiency is prohibitive to each production unit. The need to develop centralized management plants involves the transfer of the OMWW, raising the treatment costs. Moreover, the proper location of one or more centralized plants is another important issue.
- The construction and operating cost of treatment plants based on a single or a combination of methods aiming to reduce the organic load and the phytotoxic load of the wastewater to acceptable levels. To decrease the technology cost, recovery or production of energy and/or added-value materials is also considered as part of the integrated management scheme. Another practice is to apply low-cost practices such as irrigation, wetlands, and lagooning but land suitability and availability is a critical factor in this case.
- The seasonal production of OMWW, usually from November to February. A treatment plant should operate throughout the year for it to be economically viable and profitable. Moreover, intermittent operation would cause reduced efficiency of microbial-based technologies or would be cost-prohibitive in the case of incineration technologies. The possibility of co-treating other agro-industrial wastes with OMWW would secure the continuous year-round operation of these plants.

Numerous technological methods have been developed for OMWW treatment. These methods are based on physicochemical or biological processes and are reviewed in the following sections. The potential of using OMWW as a material to recover energy and/or added-value products or even to produce profitable materials is also considered.

4.2 Physicochemical Treatment

4.2.1 Simple Physical Processes

Processes such as settling, centrifugation, filtering, and flotation result in the separation of the liquid bulk from the particulate matter, which contains the highest organic portion of the wastewater.

Settling (or sedimentation) is a natural process based on the effect of gravity on particles that have different density, such as solid and liquid particles. As a result, when leaving the OMWW at rest, the solids tend to aggregate into sludge at the bottom of a tank, thus becoming separated from the main bulk of the wastewater. Georgacakis and Dalis (1993) found that, the supernatant and the sludge streams obtained after a 10-day settling corresponded to 68% and 32%, respectively, of the total volume, and contained 22 and 162.4 g/L COD, respectively. Further treatment of both streams is necessary, using methods that are applicable to liquid- and sludge-type wastewaters.

Centrifugation of OMWW results in an extra phase – an oily phase formed on the surface, where a second quality oil can be recovered up to 30–50% (Mitrakas et al. 1996). Owing to oil recovery and more efficient solid separation, centrifugation results in higher COD (70%) removal from the bulk volume of the wastewater. Addition of sulfuric acid to lower the pH to 2 enhances oil recovery after centrifugation (47%), resulting in high COD removal (67.8%) and a smaller (but more cohesive) volume of sludge (15%) with less water (80%). However the quality of oil is low in this case because of hydrolysis. Addition of lime to precipitate the calcium salts of fatty acids also improves COD removal (83%) but reduces oil recovery (12%), and the sludge produced becomes bulky. In this case too, the OMWW after treatment with centrifugation contains 50–70 g COD/L, which necessitates further treatment.

Filtration is an alternative to sedimentation. Application of pressure, forcing the OMWW to pass through filters, leads to removal of suspended and colloidal solids (Velioğlu et al. 1987). The main advantage of filtering is that low-cost materials such as straw, turf, and sand can be used for making filters with different porosities. However, the high solid content of OMWW results in clogging of the filters, which is the main disadvantage of this process. This disadvantage can be relieved by setting up appropriate configurations where more than one filter is used and combining filtration with processes such as sedimentation, centrifugation, and membrane or resin treatment.

Flotation is another method for removing suspended solids from the OMWW. It involves the addition of a gas such as air in small bubbles, which adhere or become trapped in the particulate mass, decreasing its specific gravity, making it buoyant. In this way, the suspended solids (containing small quantities of oil, 0.2%) become aggregated in the surface and are collected. The efficiency of this method is based on the formation of tiny gas bubbles in the wastewater. This can be achieved through the dissolved-air flotation method or pressure flotation, in which gas (usually air) is pressured and then the pressure is relieved. During pressure relief, fine gas bubbles are formed. According to Mitrakas et al. (1996), the dissolved-air flotation method is not as efficient as centrifugation, and moreover, the extra step of oil recovery from the floated mass renders this technique more expensive.

4.2.2 Simple Chemical Processes

A series of treatment methods are based on the chemical characteristics of the OMWW such as the colloidal matter and the chemical nature of certain compounds contained in the OMWW. Specifically, methods such as coagulation/flocculation and chemical oxidation result in COD removal, whereas adsorption and ion-exchange application leads to selective removal of certain classes of compounds from the OMWW.

Flocculation is a process of contact and adhesion whereby the particles of a dispersion form larger clusters. "Flocculation" is synonymous with "agglomeration" and "coagulation," although some researchers prefer to distinguish the two terms; flocculation results in the formation of loosely structured flocs, whereas coagulation results in the formation of more compact aggregates that can be easily separated. OMWW can be processed through flocculation since it contains colloids (Everett 1972). The negative surface of colloids can be neutralized by adding hydrogen ions (e.g., H_2SO_4, HCl, HNO_3) or calcium cations [e.g., $CaCO_3$, $Ca(OH)_2$, CaO]. In this way, the colloids are destabilized and collapse, forming flocs. The flocculants that have been studied include lime, ferric chloride ($FeCl_3$), ferric sulfate [$Fe_2(SO_4)_3$], aluminum sulfate [$Al_2(SO_4)_3$], sodium silicate (Na_2SiO_3), and various polyelectrolytes.

Lime stabilization is currently applied as a minimal pretreatment procedure for the removal of organic matter (Grappelli et al. 1987; Lolos et al. 1994). Lime is a mixture of minerals, with the predominant compounds being the oxides, hydroxides and carbonates of calcium. Calcium oxide when hydrated is converted to calcium hydroxide. Lime treatment is a low-cost and effective method for reducing the amounts of numerous pollutants in OMWW such as suspended solids, COD, fats, and polyphenols. Aktas et al. (2001) added lime until pH 12 and reported removal of these compounds in a range from 63% to 95%. Tsonis et al. (1987) found that adding lime to increase the pH to 11 resulted in 55% COD removal and 70% color removal. However, a hydrated lime, $Ca(OH)_2$ at a concentration of 10 g/L can precipitate more than 50% of the initial COD and remove 50% of the initial color within 12 h (Zouari 1998). The main disadvantages of lime treatment are the increase of the effluent pH and hardness and the large volumes of sludge produced. Although lime decreases the microbial activity, this is temporal, since calcium makes organic matter more amenable to biodegradation. Therefore, further treatment of the liquid effluent as well as the sludge derived from OMWW after lime treatment is necessary.

The other inorganic flocculants such as the ferric compounds and sodium silicate are not efficient; in the presence of water, ferric chloride and ferric sulfate are converted to iron hydroxide, which forms flakes upon which the colloids coagulate and precipitate. For the flakes to form, much energy and water are needed, which makes this process inexpedient. Sodium silicate, on the other hand, causes the colloids to precipitate, forming a gel-like sludge which is difficult to manipulate. Aluminum sulfate has been combined with sulfuric acid and lime in a multistep process during which the initial COD of 240 g/L was reduced to 18 g/L in the effluent, whereas the COD removed ended up in the sludge (Fortunato 1988).

Other types of flocculants are the polyelectrolytes. Polyelectrolytes are water-soluble polymers which have portions of their subunits charged. However, nonionic water-soluble polymers are also called polyelectrolytes by researchers. Generally, the organic flocculants are more expensive than the inorganic flocculants, but the dose required is much lower and the sludge produced is much less. Moreover, the pH is not changed during flocculation with polyelectrolytes. Both anionic and cationic polyelectrolytes have been used for removal of solids from OMWW. Applying a dose of 1% w/v has proved sufficient to remove the total suspended solids completely and to reduce the organic content in the aqueous phase (Sarika et al. 2005). Polyelectrolytes such as polyacrylamide and polyamine can bring about a reduction of the 5-day biological oxygen demand by up to 50% in a serial step process (Gamero et al. 1982, 1983). Chitosan, a water-soluble cationic polysaccharide made from chitin, has been used together with a water-soluble calcium compound (such as calcium hypochlorite or calcium nitrate) or organic polymer materials (such as Prastol™ and/or Zetag™) to form larger and more solid flakes, resulting in 35% purification of OMWW (Knudsen and Larsen 1992). Chitosan is a natural product since its precursor, chitin, is derived from fishery wastes. The sludge produced by its flocculent effect has nutritional value and can be used as fertilizer or feed.

4.2.3 Adsorption/Ion Exchange

Adsorption is a process through which a compound accumulates on the surface of a solid matrix. If diffusion of the compound into the matrix of the complex takes place, then "absorption" is a more appropriate term. Sorption involves both adsorption and absorption. The mechanism of binding may be physical (weak van der Waals forces) or chemical (chemisorption, covalent bonding). Ion exchange is another mechanism for transferring ionic compounds selectively from the liquid phase onto a solid substrate, through exchange of ions between the compound and the solid complex. Available materials for ion exchange are chelate-producing, semiacid cation exchangers. Regeneration of this material is feasible with strong/fully dissociated inorganic acids, such as sulfuric acid. If phenols and polyphenols are to be removed, a semiacid anion exchanger is more appropriate. This exchanger can be regenerated with methanol.

Compounds that can be removed from OMWW via adsorption and ion exchange are coloring substances (mainly tannic acids), hardly or nonbiodegradable pollutants, and bactericidal or inhibiting compounds (phenols, polyphenols, long-chain fatty acids, etc.). The adsorption can be enhanced by applying centrifugation and/or filtration as preceding steps, so a large portion of solids is initially removed.

Activated carbon is the most widely studied adsorbent because of its efficiency. The high efficiency of activated carbon is due to its high specific surface area (500–1,500 m^2/g) and its high adsorptive capacity, but unfortunately it cannot be reused. However, its calorific value is very high, so it can be incinerated without problems. It has been estimated that between 60% and 80% of the organic constituents from OMWW can be adsorbed on activated carbon (Improlive).

Azzam et al. (2004) reported that the maximum adsorption capacity of activated carbon could be reached in less than 4 h in a wide range of concentrations (3–24 g/L), resulting in removal of phenolics and organic matter of up to 94% and 83%, respectively. Similar results were obtained by El-Shafey et al. (2005), who also applied treatment with activated carbon after coagulation and filtering. Sabbah et al. (2004) also applied treatment with activated carbon to reduce the amount of phenolics by up to 95%. In this work, adsorption was applied prior to anaerobic digestion and resulted in enhancement of the biological process because of the elimination of an inhibiting factor such as the phenolics.

There are two types of activated carbon: powdered and granular. Powdered activated carbon is mainly produced by the partial incineration of waste liquors from paper manufacture, and is difficult to remove from the treated effluent. Moreover, there is hardly any efficient method for regenerating this material after recovery from the effluent. In conclusion, although the cost of its production is low, the operating cost is high, since it cannot be reused and its disposal requires an additional cost. Granular activated carbon, on the other hand, such as that produced from coal, is an expensive adsorbent because of the multistep process applied to make a product of uniform size and acceptable hardness. Activated carbon can also be produced from the solid wastes originating from olive oil processing (see Sect. 4.4.4)

Bentonite and other clay minerals are natural ion exchangers and are used sometimes as low-cost adsorbents. The cations deposited on the surface of the mineral can be replaced by other cations, found in OMWW in the form of polluting particles. Bentonite has a high specific surface area (owing to the microscopic voids in its interior) and its adsorption capacity is also remarkable. Combination of lime [in the form of $Ca(OH)_2$] and bentonite results in adsorption of lipids and other inhibitory compounds found in OMWW, rendering it more amenable to anaerobic digestion according to Beccari et al. (2002)

Activated clay, as cited by Al-Malah et al. (2000), is a new low-cost adsorbent, which has been tested for treating OMWW following settling, centrifugation, or filtration. The maximum adsorption was achieved in less than 4 h at the concentrations of clay tested and resulted in removal of 81% and 71% in the case of phenols and total organic matter, respectively.

Specific resins can also be used for removing organic compounds from OMWW. Zouari (1998) studied DUOLITE™ XAD™ 761. DUOLITE™ XAD™ 761 is an aromatic resin, capable of adsorbing protein, iron complexes, tannins, hydroxymethyl furfural, and other ingredients responsible for off-flavors (Duolite XAD 761 technical sheets from Rohm and Haas Co). In the case of OMWW, more than 50% of the chromophores responsible for the dark color were removed. Resins are specially used for recovering added-value products from OMWW (Agalias et al. 2007)

4.2.4 Chemical Oxidation

4.2.4.1 Advanced Oxidation Processes

The organic content of the OMWW can be oxidized using an appropriate oxidizing agent or a mixture of oxidizing agents, such as oxygen or oxygen derivatives, chlorine or chlorinated derivatives and potassium permanganate.

Oxygen derivatives such ozone and hydrogen peroxide have been widely used for the oxidation of organic compounds. Ozone is selectively reactive on the double bonds of a compound. It is expected that toxic compounds which include a double bond in their molecule (e.g., phenols, unsaturated fatty acids) will be broken down to smaller units with (hopefully) less toxic properties, making the wastewater more amenable to anaerobic digestion. On the other hand, readily biodegradable organic compounds, such as sugars and proteins, will remain intact upon treatment with ozone.

Improvement of the conventional oxidation processes can be achieved by combination of oxidants as well as combinations of oxidants with UV radiation. The resultant oxidation systems are known as advanced oxidation processes (AOPs). The characteristic feature of AOPs is the production of the highly oxidative hydroxyl radical (HO·) at ambient temperature through photochemical and nonphotochemical pathways.

Common AOPs are ozone and/or hydrogen peroxide combined with UV. The disadvantage of ozone treatment is that mass transfer of ozone in OMWW is limited. Moreover, the solar UV is not usable in either system, since the required UV energy is not available in the solar spectrum. A semiconductor such as titanium dioxide (TiO_2), when in contact with water, can absorb solar UV radiation and produce hydroxyl radicals. This process is called photocatalysis and is particularly interesting since it uses solar radiation for producing the oxidative radicals. During photocatalytic oxidation, several reactions occur resulting in the production of hydroxyl and oxygen radicals. First, UV light is absorbed by the semiconductor (usually titanium dioxide) and excitation of electrons occurs from the valence to the conduction band. The electron transfer causes "holes" to appear on the surface of the semiconductor. The excited electrons as well as the electron holes are strong oxidizing agents. Specifically, the holes react with water to produce hydroxyl ions and, further, hydroxyl radicals. The electrons produced react with oxygen and convert it to oxygen radical (O_2). Photocatalysis has significant advantages over the other AOPs: the degradation of the organic pollutants is complete (mineralization), fast, cheap, as it uses solar light, and occurs at ambient temperature and pressure.

Another oxidative process, called the Fenton process, combines chemical oxidation and coagulation of organic compounds. The Fenton reagent is composed of a mixture of hydrogen peroxide (H_2O_2) and ferrous sulfate ($FeSO_4$). As in photocatalysis, reactive oxidizing agents are produced. The mechanism of the reactions occurring in Fenton systems is rather complex, but can be outlined by the following reactions:

$$Fe^{2+} + H_2O_2 \rightarrow Fe^{3+} + HO^- + HO^·$$
$$HO^· + Fe^{2+} \rightarrow Fe^{3+} + HO^-$$
$$HO^· + RH \rightarrow H_2O + R^·$$
$$R^· + Fe^{3+} \rightarrow R^+ + Fe^{2+}$$

The first two reactions are the initiating and terminating reactions respectively, whereas the last two are propagation reactions, regenerating the ferrous ion to be oxidized through the first reaction again. The hydroxyl radical can attack and break down the organic compounds (RH) through the last two reactions. As other side reactions also occur at the same time, the increase in peroxide concentration does not always lead to an increase in the efficiency of the oxidation. However, in the absence of dissolved oxygen, complexes of the organic radicals are formed, slowing down the degradation process. Moreover, the concentration of the ferrous ion should be kept low in order to discourage the termination reaction.

The other important aspect of the Fenton process is the coagulation it causes. Ferric complexes are formed as the result of the reaction of ferric ions with hydroxyl ions. In a pH range from 3.5 to 7, the ferric complexes polymerize, interact with the colloidal suspended solids, and form insoluble aggregates of ferric oxide hydrates. In this way, precipitation takes place, but at a slow rate. The rate can be increased significantly by the addition of polyelectrolytes. In the presence of calcium hydroxide, the coagulation nuclei are formed from the insoluble ferrous hydroxide and calcium sulfate.

The use of model compounds in studying the efficiency of oxidative processes is a very common practice, since it allows a more complete analytical characterization of the reaction medium that would elucidate the reaction mechanisms and pathways (Mantzavinos and Kalogerakis 2005). The selection of the model compounds is based on their abundance in OMWW. They can be classified as cinnamic acid derivatives (coumaric, caffeic, and ferulic acids), benzoic acid derivatives (vanillic, gallic, veratric, syringic, protocatechuic, and hydroxybenzoic acids), and as tyrosol and related compounds (hydroxytyrosol and hydroxyphenylacetic acid). A number of studies have focused on the application of AOPs on model compounds (Miranda et al. 2000, 2001, 2002; Amat et al. 2003; Gernjak et al. 2003b; Mantzavinos 2003a, b). A common conclusion of all these studies is that the number, type, and position of the functional groups attached to the aromatic ring contribute to the reactivity of the phenolic compound (Mantzavinos and Kalogerakis 2005). In the case of both cinnamic acid (Miranda et al. 2000; Mantzavinos 2003a) and benzoic acid (Miranda et al. 2001; Gernjak et al. 2003b; Mantzavinos 2003b) derivatives subject to various AOPs (i.e., photocatalysis, Fenton process, photo-Fenton process), their reactivity was found to increase with increasing number of hydroxy groups and decreasing number of methoxy groups.

The effect of combining ozone or hydrogen peroxide with UV has been tested on four phenols (caffeic, *p*-coumaric, syringic, and vanillic acids) and compared with the individual processes, such as single photolysis and ozonation (Benítez et al. 1995, 1996a, 1997a, 1998). The ozone/UV system resulted in a lower degree of mineralization of phenols but faster kinetics, in contradiction to the H_2O_2/UV system.

Other studies performed on real OMWW have shown that ozone has limited effectiveness on highly concentrated OMWW since it is scarcely soluble in water. The extent of COD removal achieved during OMWW ozonation is up to 20–30% (Benitez et al. 1997b, 1999; Andreozzi et al. 1998; Beltrán-Heredia et al. 2001), but it can be used as a pretreatment method. Ozone is particularly efficient in total phenolics removal (Bettazzi et al. 2007): up to 91% with an initial pH of 12. A portion of the COD was also removed (up to 19%). Specific tests conducted on gallic and *p*-coumaric synthetic media confirmed the efficiency of ozone in degrading phenolic compounds. Biological tests, under aerobic or anaerobic conditions, indicated a significant increase of the biodegradability of treated OMWW samples. Specifically, respirometric tests showed an increase in the biological oxygen demand by about 20%, and anaerobic batch tests yielded a methane production up to 8 times higher. The effect of ozone as a function of pH has also been studied by several researchers (Benitez et al. 2005; Monteagudo et al. 2005; Saroj et al. 2005). Gernjak et al. (2003a) applied various AOPs such as solar light photocatalysis over titanium dioxide or solar light coupled with the Fenton reaction on OMWW in pilot-scale photoreactors. To enhance the process efficiency, a pretreatment step such as flocculation and/or decanting was employed to remove suspended solids as they obstructed light from entering the liquid. Rivas et al. (2001a) applied the Fenton reaction on OMWW and reported a COD removal of 80–90% at hydraulic retention times ranging between 1 and 8 h from an influent COD of 15 g/L. However, the cost of this process is very high and the operating conditions need to be optimized.

Pretreatment with activated carbon has been shown to facilitate the catalytic decomposition of the chlorinated compounds besides the adsorption of the derivatives and other organic substances. On the basis of this concept, a treatment procedure has been tested involving oxidation with sodium hypochlorite (NaOCl), decanting, filtration, activated carbon catalysis, and aeration–ozonation (Bellido 1987, 1989a, b; González-López et al. 1994). This procedure has been tested on a pilot scale: automation was incorporated in the various oxidation steps, lowering the required operating time significantly.

4.2.4.2 Wet Oxidation

Wet (air) oxidation involves the oxidation of the organic matter in water in the presence of oxygen (or air). The reaction takes place at a temperature above the water boiling point (100°C), but below the critical point (374°C). The common temperature range is 120–330°C. The system should also be kept under pressure (10–220 bar) to avoid excessive evaporation. A major drawback of this process is the high installation and operating cost because of the severe conditions required, mainly the high pressure and the high corrosion rates (as a result of the corroding conditions). The cost associated with high temperatures is not of major concern as the process becomes thermally self-sustained because of the typical organic load of OMWW. Another disadvantage of this technology is the low quality of the gaseous emissions.

In various studies, chemical oxidation of OMWW has occurred under supercritical conditions (500°C, 250 bar). Rivas et al. (2001b) found that 98% phenol removal and 99.9% COD removal were obtained from a diluted OMWW influent (initial COD 3.3 g/L) after treatment for 1–3 min. Subcritical conditions have also been tested; Rivas et al. (2005) found that 50% COD removal and 80% phenol removal could be achieved from a diluted OMWW effluent (initial total organic content 4.9 g/L) after 6 h of treatment with 2 g/L CuO/C or Pt/Al_2O_3 catalysts with or without H_2O_2 at 180°C and 70-bar pressure. Gomes et al. (2007) tested other catalysts such as platinum–carbon and iridium–carbon that showed a high efficiency in the wet air oxidation of OMWW, which had already been centrifuged. The COD was reduced to 2.8 g/L at 200°C and 6.9-bar oxygen partial pressure.

Minh et al. (2006, 2007a, b) performed catalytic wet air oxidation experiments on individual phenolic compounds on platinum- and ruthenium-supported titanium oxide and zirconium oxide catalysts in batch or fixed-bed reactors at 140°C and 50-bar total air pressure. Ruthenium-supported catalysts were stable under these conditions. Tests on real OMWW using platinum and ruthenium catalysts in batch and continuous reactors at 190°C and 70-bar air operating pressure were also conducted by Minh et al. (2008). They performed anaerobic digestion tests on the oxidized effluent and found that the biodegradability of the oxidized waste was enhanced.

4.2.4.3 Electrochemical Oxidation

Electrochemical oxidation involves the application of an electrical current to enhance the oxidation process. The electrooxidative processes (taking place at the anode of the electrolytic cell) for the oxidation of recalcitrant organic substances contained in several industrial wastewaters, landfill leachate, and domestic sewage have been extensively studied. The parameters affecting the efficiency of this technology are the electrode material, the electrolyte, and the design of the electrolytic cell (membrane, recycling, two- or three-dimensional electrodes, etc.). The electrodes used are coated with SnO_2, PbO_2, Pt, or Pt–Ti for the anode and steel for the cathode. The electrolytes used are NaCl, Na_2SO_4, and H_2SO_4 in concentrations ranging from 0.2 to 2 N. The above-mentioned parameters were varied in several studies aiming to maximize the COD removal (Vigo et al. 1983a; Israilides et al. 1997; Polcaro et al. 2002)

Longhi et al. (2001) showed that it is possible to reduce the concentration of the phenolic components without decreasing the total organic content of OMWW to a great extent. In this concept, electrochemical oxidation could be used as a pretreatment method for rendering the wastewater more amenable to anaerobic digestion. The biotoxicity of the electrochemically oxidized OMWW was investigated by Giannis et al. (2007), who used a titanium–tantalum–platinum–iridium electrode and 3% NaCl as the electrolyte. COD removal reached 70% after 8 h of treatment, and color and turbidity were completely removed. However, bioassays with *Daphnia magna* and *Artemia salina* indicated that the ecotoxicity of the treated wastewater remained unchanged, possibly due to the formation of chlorinated by-products. They also used $FeCl_3$ with NaCl, and this combination resulted in the electrocoagulation of OMWW and its separation into a supernatant phase and a settled solids phase. Under optimal conditions (2% Na_2SO_4 plus 1% $FeCl_3$; 24 V), the removal efficiency of COD reached 85.5% in the supernatant phase. Electrochemical oxidation can also be combined with the Fenton reagent (the electro-Fenton process). The electro-Fenton process has been applied on OMWW as well as model compounds (Bellakhal et al. 2006; Khoufi et al. 2004).

4.2.5 Membrane Technologies

These technologies achieve the separation of compounds that are in the same phase, i.e., all components are in solution based on molecule size. The membrane methods that are applicable in the treatment of OMWW are microfiltration, ultrafiltration, nanofiltration, and reverse osmosis.

Ultrafiltration and nanofiltration are pressure-driven processes used to separate and concentrate macromolecules and colloids from wastewater, respectively (molecular size in the range 0.1–0.01 mm). The wastewater is forced to pass through a perforated membrane of a specific pore size. All molecules smaller than the membrane pore size pass through the membrane, whereas larger molecules are retained and concentrated on the feeding side of the membrane. The reverse osmosis system uses a semipermeable membrane to remove pure water from an aqueous solution, that is, the water moves not to equalize but to maximize the osmotic pressure between the water and the aqueous solution. Reverse osmosis succeeds in separating molecules in the ionic range of less than 0.001 mm. Ultrafiltration is used as a pretreatment step before nanofiltration and/or reverse osmosis.

The constituents of a solution can be fractionated using a membrane with a limiting pore size, or by installing a series of membranes with successively smaller pores. In the membrane separation process, the liquid to be treated circulates with flow parallel to the filtering surface, thus creating sufficient turbulence to avoid fouling. It is for this reason that cross-flow filtration maintains its efficiency for long periods (Paraskeva et al. 2007a).

Membrane separation technologies for OMWW treatment have been studied by several researchers (Pompei and Codovilli 1974; Carrieri 1978; Vigo et al. 1981, 1983a, b; Canepa et al. 1987, 1988; Paraskeva et al. 2007a; Russo 2007). Turano et al. (2002) proposed an integrated centrifugation–ultrafiltration system for the reduction of pollution caused by wastewaters and selective separation of useful products, such as fats, sugars, and polyphenols. The combination of these two processes resulted in a 90% reduction of COD values. Canepa et al. (1988) proposed a combined system of membrane processes to treat OMWW and showed that the cost is affordable for olive mill enterprises. Similarly, Paraskeva et al. (2007b) found that the ultrafiltration process results in the effective separation of high molecular weight constituents and any suspended solid particles and/or aggregates. In the subsequent steps, the nanofiltration and/or reverse osmosis processes yielded effective separation of the largest part of polyphenols contained in the OMWW. These "toxic" fractions may be used for their phytotoxic properties, and if they have legal acceptance to be exploited commercially, the proposed method is economically viable.

The permeate of the reverse osmosis process is rich in metal ions which may be beneficial for irrigation. Russo (2007) applied microfiltration and utrafiltration processes on OMWW without prior centrifugation, using ceramic and polymeric membranes, leading to recovery of phenols ranging from 31% to 75%.

Another separation technology, emulsion liquid membrane separation, was proposed by Li (1968) for a variety of separation operations, including removal of metal ions and phenols from OMWW (Boyadzhiev and Bezenshek 1983; Zhang et al. 1987; Draxler et al. 1988; Correia and Carvalho 2001, 2003). Emulsion liquid membrane separation is a three-phase dispersion system in which a primary emulsion is dispersed in a continuous phase (such as the OMWW) forming droplets. The solutes move through the liquid membrane from the external environment and are concentrated in the droplet interior. After permeation, the emulsion is separated from the continuous phase, and afterwards, high voltage is applied to split the emulsion.

The liquid membrane consists of a diluent, a surfactant to stabilize the emulsion, and a carrier reagent (extractant) in the case of separation of solutes by chemical reaction. The extraction of phenol with emulsion liquid membrane separation is usually performed using membranes with aliphatic diluents. The application and efficiency of emulsion liquid membrane separation have been studied in the case of the extraction of phenol and *o*-cresol using sulfuric acid salts of trioctylamine as a carrier (Dobre et al. 1999), as well as the extraction of tyrosol with, and without, extractant (Reis et al. 2006). However, this technique has not been extensively studied so far.

4.2.6 Thermal Processes

The thermal processes aim at reducing the volume of OMWW, either by reducing the water content (evaporation, distillation, drying) or by combusting/pyrolyzing the organic matter. The products of these processes (concentrated paste or ash and gaseous emissions) require further management though. A potential advantage may be the recovery of energy from combustion.

4.2.6.1 Evaporation/Distillation

Evaporation of OMWW takes place in evaporation ponds where it is left to be condensed by the heat of the sun or, alternatively, evaporation can be accelerated by providing heat in industrial evaporators.

Natural evaporation requires large lagoons (artificial evaporation ponds or storage lakes). The OMWW is left to dry in these lagoons for a long period of time (7–8 months), usually the time interval between the end of the olive milling period and the beginning of the subsequent one. Because of the long storage time, biodegradation also occurs. This technique was one of the first processes to be used for the treatment of OMWW and is applied in many Mediterranean countries (Leon-Cabello and Fiestas Ros de Ursinos 1981; Shammas 1984; Balice et al. 1986; Shabou et al. 2008). COD removal is from 20–30% to 75–80% after 2–4 months. Michelakis et al. (1999) recorded the evaporation rate of OMWW in a time period ranging from January to August. They found that the evaporation rate of OMWW is enhanced because of its dark color, which increases the absorption of solar light and, consequently, the OMWW temperature. On the other hand, the oily layer on the surface decreases the evaporation rate. Rainfall also negatively affects the evaporation rate. The sludge formed on the bottom of the pond should be regularly removed. Duarte and Neto (1996) proposed a device consisting of inclined plates for the separation of the solid and liquid phases.

The investment and operating costs for evaporation in lagoons are low; however, there are serious drawbacks. One of them is that large areas of land are required. The soil must be covered by an impervious layer to avoid leakage to the water-receiving body. An appropriate load of OMWW is 1–2.5 m^3 OMWW/m^2 pond (Escolano Bueno 1975) or 2 t OMWW/m^3 pond (Kasirga 1988; Azbar et al. 2004). Jarboui et al. (2008) studied the fate and impact of OMWW on clay soils used for evaporation ponds for 8 years in the locality of Agareb (Tunisia). They found that the inclusion of sand layers between the clay layers increases the infiltration of OMWW. In addition, the discharge of OMWW at the site affects the clay's properties and increases its permeability. Therefore, the clay layers should be well compacted to avoid the pollution of groundwater, especially by the phytotoxic and antibacterial phenolic compounds contained in the OMWW (Arambarri and Cabrera 1986; Sierra et al. 2001).

Other problems are the odors produced, the insect growth, deposition of silt, leakages, etc. (Rosa and Vieira 1995). The release of emission gases, namely, volatile phenols and sulfur dioxide, pollutes the environment (Gianfranco et al. 2003). The odors are associated with the prevalence of anaerobic conditions, mainly in large ponds.

For these reasons, the design of evaporation ponds should be based on parameters such as the volume of OMWW to be evaporated, the local climatic conditions, the ground hydrological features, the depth of the aquifer, and the proximity of inhabited areas. Le Verge and Bories (2004) performed a cost analysis for constructing evaporation ponds.

Evaporation can be assisted by supplying heat in industrial evaporators and distillation devices. Distillation offers the advantage of separating the components of the volatile steam produced during the evaporation. The heat required to raise the temperature to the vaporization level of water is quite high (100 kWh/m^3) and this makes the process unfavorable. Distillation processes, such as vacuum, multiple effect (to reduce energy requirements), and flash evaporation (Rozzi and Malpei 1996) have been tested on OMWW in pilot plants in Italy and Spain (Ballester et al. 1991). The distillate obtained from the single or multiple effect evaporation (the most commonly applied evaporation process) has a COD corresponding to less 5–7% of the initial COD. A concentrated mass of 30–50% water content is left at the bottom, corresponding to 10–20% of the liquid feed. The vapor can be either condensed and subjected to further treatment, or distilled to obtain an alcoholic mixture.

The energy consumption of evaporation can be reduced and the efficiency of evaporation can be increased if an appropriate pretreatment step is applied to the OMWW (Azbar et al. 2004), such as centrifugal separation, chemical precipitation, filtration, or increase of the pH (to keep the volatile organic compounds in the solid fraction during evaporation).

Solar energy could be also used to reduce the cost; Potoglou et al. (2003) investigated the efficiency of a solar distillation device on OMWW with promising results.

4.2.6.2 Drying

Further reduction of the moisture contained in the semisolid residues obtained from the various treatment methods is feasible through drying. In addition, drying of the two-phase olive mill waste is also necessary to make use of this by-product. The types of drying applied are contact, convection, and radiation drying processes. In conventional drying, hot gases pass through the olive cake and vaporize the water which is conveyed by the hot gas flow. Typical driers of this type are drum driers and fluidized-bed driers. After drying, the olive cake is deoiled with an organic solvent (hexane) and then can be incinerated for energy production, reused in agriculture, or land-filled. However, in the case of two-phase OMWW (pomace), which has high moisture and sugar content, serious problems affect the drying process taking place in driers designed for the three-phase oil seed. Moreover, the sugars tend to polymerize into a sticky material, causing operational problems. For these reasons, new driers have been designed and the operating conditions determined for this type of waste (Arjona et al. 1999).

4.2.6.3 Combustion

Combustion or incineration involves the thermal oxidation of the organic matter into ash, gas emissions (flue gas and suspended particulates), and heat. The gas emissions should be cleaned prior to dispersion in the atmosphere. Combustion of OMWW is not feasible unless pretreatment methods concentrating and drying the organic material are applied first.

Arpino and Carola (1978) described the efficiency of a combustion plant treating 20 m^3 OMWW/day at a combustion temperature of 800°C, by ejecting smoke at 400°C at a speed of 4 m/s. The fuel was oil but other fuels could also be used, such as exhausted olive stones, reducing the cost (Baccioni 1981; Vitolo et al. 1999). Amirante and Mongelli (1982) reported on another combustion plant and studied its efficiency in terms of heat recovery and air pollution, which are the main drawbacks of this technology. Another problem associated with the operating cost of incinerators is the seasonal production of OMWW. More fuels should be fed to such plants in order to secure the continuous operation of such a plant.

Incineration taking place in the presence of a celliform catalyst made of platinum, iridium, etc. is another option (Papaioannou 1988). Although lower temperatures are required for combustion (350°C), the high cost of the catalysts and their short lifetime because of poisoning from phosphorus make the application of this method problematic.

4.2.6.4 Pyrolysis

Pyrolysis is the thermal destruction of the organic matter in the absence of oxygen. It is applied on concentrated OMWW (Petarca et al. 1997) or olive cake. The problems associated with the melting of salts present in the OMWW at high concentrations and the formation of scales on the pipe surface have been mentioned in several studies (Di Giacomo et al. 1989; Di Giacomo 1990).

The main by-product of pyrolysis is charcoal, the commercial value of which contributes to reducing the operating costs. The result of pyrolysis and subsequent activation of a mixture of OMWW and fly ash from coal combustion is an adsorbent material (Rovatti et al. 1992). Pyrolysis can also be assisted by microwave power.

4.3 Biological Treatment

4.3.1 *Anaerobic Digestion*

Anaerobic digestion is a major biological process applied for the stabilization of high-organic-load wastes, since anaerobic microorganisms are able to convert the organic matter, producing small amounts of excess sludge (in contrast to the aerobic microorganisms) and an energetically rich gas mixture (biogas). It is a complex process performed by numerous microorganisms and involves four major steps:

1. Disintegration/hydrolysis. The disintegration of the organic agglomerates and the breaking down (hydrolysis) of polymeric organic molecules (cellulose, proteins, lipids) into smaller units by extracellular enzymes are the first steps in anaerobic digestion. The OMWW contains particulate matter consisting of lignocellulosic complexes, and its disintegration/hydrolysis is very slow. The lignin content of OMWW is correlated with the nonbiodegradable fraction under anaerobic conditions. Appropriate pretreatment methods aiming at disintegrating and/or hydrolyzing lignocellulosic matter (alkali, heat treatment, enzymatic treatment, ultrafiltration, centrifugation, dilution, acidification, aerobic pretreatment; Boari et al. 1984; Hamdi and Ellouz 1993; Hamdi 1996) enhance the efficiency of this step and consequently of the whole process.
2. Acidogenesis. The small soluble molecules (sugars, amino acids, etc.) derived from the first step are taken up by a diverse group of microorganisms to produce

a mixture of acids (acetic, propionic, butyric, valeric, lactic, succinic, formic), alcohols, hydrogen, and carbon dioxide.

3. Acetogenesis. The alcohol and acid mixture is further converted to acetic acid and hydrogen. This step can occur at low hydrogen partial pressure (10^{-6} atm); otherwise, it is thermodynamically unfavorable, resulting in acid accumulation and a pH decrease to inhibitory levels (under 6). Hydrogen should be consumed as soon as it is produced (intermediate hydrogen transfer) by bacteria grown on carbon dioxide and hydrogen to produce either acetic acid (homoacetogens) or methane (hydrogen-utilizing methanogens)
4. Methanogenesis. The final step in anaerobic digestion is performed by acetoclastic methanogens (belonging to a special group of microorganisms called archaea) and hydrogen-utilizing methanogens. Methane production occurs mostly via acetate conversion (70%). Methane is produced along with carbon dioxide and contains the chemical energy of the parent compounds initially present in OMWW. The percentage of methane in the biogas from OMWW ranges between 60% and 70%.

The temperature, the pH, and the chemical composition of OMWW influence the anaerobic digestion process to a great extent. Anaerobic biodegradation can take place under psychrophilic (above 20°C), mesophilic (25–40°C), or thermophilic (50–65°C) conditions. Biodegradation under thermophilic conditions allows higher loading rates and destruction of pathogens. However, thermophilic microorganisms are more sensitive to variations in operating conditions and toxicants. In addition, operation at a higher temperature implies reduced energy efficiency.

The OMWW should be supplemented with bicarbonate alkalinity (to buffer the potentially decreasing pH) and a nitrogen compound (as a nutrient source) so that viable environmental conditions for the microorganisms can be sustained (Rozzi et al. 1988, 1994). The constituents of OMWW having an adverse effect on anaerobic microorganisms are:

- The phenolic compounds (Boari et al. 1984; Hamdi 1991a, b, 1993a; Beccari et al. 1998), found at concentrations of 0.3–24 g/L (Fedorak et al. 1984; Sierra et al. 2001; Niaounakis and Halvadakis 2006)
- The long-chain fatty acids derived from the hydrolysis of lipids. Anaerobic microorganisms are irreversibly inhibited at concentrations of 0.5 g/L in the case of oleate and 1.0 g/L in the case of stearate (Angelidaki and Ahring 1992)
- Mineral ions, especially K^+ and Na^+. Sodium causes moderate inhibition at concentrations between 3.5 and 5.5 g/L and strong inhibition at concentrations above 8.8 g/L. On the other hand, potassium seems to be more toxic, since inhibition is induced from 400 mg/L, especially in thermophilic conditions (Kugelman and McCarthy 1964; Chen et al. 2008). However, according to Mignone (2005), potassium causes moderate inhibition at concentrations between 2.5 and 4.5 g/L and strong inhibition at concentrations above 12 g/L.

It has been found that proper acclimation of the anaerobic biomass may increase the levels of microorganism tolerance towards these inhibitory constituents (Beltrán et al. 2001; Tan et al. 2006; Akram and Stuckey 2008). Therefore, the starting up of bioreactors is of major importance and takes a long time (2–6 months) (Beccari et al. 1996; Forster-Carneiro et al. 2004; Yacob et al. 2005; Angelidaki et al. 2006). The seasonal nature of OMWW results in halting and restarting the operation during the year. Tsonis (1991) found that less than 30 days is required for the bioreactor to start up following prolonged periods of time under nonfeeding conditions.

Moreover, the application of pretreatment methods aiming to reduce the amount of phenolic compounds and long-chain fatty acids has been proposed to precede the anaerobic digestion in order to enhance the process efficiency. These methods may be physicochemical, as already reviewed in Sect. 4.2 (Georgacakis and Dalis 1993; Hamdi 1993a, b; Tsonis and Grigoropoulos 1993; Chakchouk et al. 1994), or biological, mainly based on the ability of certain fungi to biodegrade phenolic compounds (Hamdi et al. 1992; Ranalli 1992; Boari et al. 1993; Gharsallah 1994; Borja et al. 1995; Yesilada et al. 1995; Beccari et al. 1998). Another practice to decrease the toxicant concentration is to dilute the OMWW with tap water or treated effluent or wastewater devoid of contaminants such as cheese whey and manure. This last option, namely, codigestion, is the most favorable one, since mixing different types of wastewaters results in a mixture more balanced in nutrients and alkalinity. Moreover, codigestion may secure the operation of the anaerobic digestion throughout the year, given the seasonal nature of the various agro-industrial wastewaters (Carrieri et al. 1986, 1992; Angelidaki and Ahring 1996, 1997a, b; Lyberatos et al. 1997; Gavala et al. 1999; Angelidaki et al. 1997, 2002; Marques et al. 1998). Blika et al. (2009) found that in general a 1:2 dilution of the wastewater is required to secure safe operation. This is probably attributed to the reduction of the various inhibitory compounds brought about upon dilution.

Besides the application of appropriate pretreatment methods, the design and operation optimization of anaerobic bioreactors contributes to an increase of the conversion rates of OMWW. Particularly, the bioreactors minimizing the biomass loss through the effluent by permitting the accumulation of biomass in the interior in the form of granules, biofilm, or flocs, called high-rate bioreactors, can be subjected to high organic loading rates of 12–18 g/L/day, resulting in high COD removals of 70–90% (Table 4.2).

The anaerobic digestion of OMWW has also been studied in a two-stage process. Georgacakis and Dalis (1993) studied the anaerobic digestion of a supernatant phase and the sludge coming from the settling of OMWW for 10 days. Two different types of anaerobic digesters were used, a fixed-bed type for the supernatant and a plug-flow type for the sludge. COD reduction reached 94.02%. Dalis et al. (1996) evaluated the anaerobic biodegradation of OMWW in a two-stage pilot plant with an upflow type and an anaerobic filter (fixed-bed-type reactor) working in series. The organic loading rate ranged between 2.8 and 12.7 g COD/L/day, resulting in 83% COD reduction in the first stage and a further 8% COD reduction in the second stage.

The two-stage process can also be applied with the prospect of producing hydrogen in the first stage (operating at low hydraulic retention times of 1-day order of magnitude) and methane in the second stage (Eroğlu et al. 2006; Koutrouli et al. 2009). Hydrogen production from OMMW (Eroğlu et al. 2006; Ntaikou et al. 2008) and olive pulp (Gavala et al. 2005, 2006; Koutrouli et al. 2009) is possible

Table 4.2 Overview of the operating conditions and performance of various anaerobic bioreactor configurations treating olive mill wastewater

Reactor	COD in (g/L)	HRT (days)	OLR (g COD/L/day)	COD removal (%)	References
UASBR	5–19	1	5–18	75 (soluble)	Ubay and Ozturk (1997)
	15–22	0.83–2		70 (soluble)	
UASBR	10–60		12–18	70–75	Beccari et al. (1996)
UASBR	40	5	8	80–85	Sabbah et al. (2004)
UASBR	13–18		16–21.5	70	Boari et al. (1984)
ABR	1–5	2		90–95	Uyanik (2001)
ABR					Khabbaz et al. (2004)
CSTR	89.5–95	20	1.7–4.75	60–70	Gelegenis et al. (2007)
PABR	30–46	3.75–17.5	2.6–12.3	60–90	Stamatelatou et al. (2009)
Hybrid (combination of filter and UASBR)		0.20–1.02	8	>89	Borja-Padilla et al. (1996a)
			17	76	

UASBR upflow anaerobic sludge blanket reactor, *ABR* anaerobic baffled reactor, *CSTR* continuous stirred tank reactor, *PABR* periodic anaerobic baffled reactor, *HRT* hydraulic retention time, *OLR* organic loading rate

because of the high soluble carbohydrate content of these wastes. Hydrogen can be accumulated in the acidogenic phase of the anaerobic digestion process, as long as the methanogenic or other bacteria able to consume hydrogen are absent. Hydrogen production is favored at pH ranging from 5 to 6 and is inhibited by the high hydrogen partial pressure (product inhibition). In the study of Ntaikou et al. (2008), a continuous stirred tank reactor fed with diluted OMWW without any kind of pretreatment was operated at different hydraulic retention times, leading to a maximum yield of 1.3 L H_2 per liter of raw OMWW, and the COD reduction was negligible. Photofermentative production following a dark fermentation step yielded 29 L H_2 per liter of OMWW, but it required large surface areas for the photobiological step.

The olive pulp coming from the two-phase olive mills is considered difficult to treat through anaerobic digestion (Niaounakis and Halvadakis 2006) owing to the high solid content of the olive pulp and the slow startup procedure. However, several studies have focused on the application of this technology to olive pulp and have shown that it is a stable process over a wide range of operating conditions (Borja et al. 2003a, 2006a; Kalfas et al. 2006)

4.3.2 Aerobic Treatment

During aerobic treatment, the microorganisms growing utilize oxygen to degrade the organic compounds contained in the wastewater. This is a more robust process than anaerobic treatment since aerobic microorganisms are more energy efficient and not as sensitive as anaerobic microorganisms to various inhibitory agents. In addition, they convert almost completely the fed organics into carbon dioxide and water (complete mineralization). Finally, many organic compounds that are nondegradable under anaerobic conditions may well be degraded under aerobic conditions. However, the inhibitory effect of phenols and especially tyrosol on aerobic bacteria has been demonstrated (Ragazzi and Veronese 1982; Olori et al. 1990). Generally, aerobic microorganisms grow at higher rates than anaerobic microorganisms. As a result, the degradation of wastewaters of high COD concentration is accompanied by a high biomass production (sludge) and demands excessive oxygen (or air supply). The cost of aerobic processing is regarded as prohibitive in the case of OMWW because of the sludge production and air demand.

The activated sludge process (consisting of aeration and sedimentation tanks) is commonly applied for municipal sewage treatment. OMWW cannot be treated in this type of process (Di Giovacchino et al. 1988; Mascolo et al. 1990) without proper dilution with easily biodegradable wastewaters such as municipal sewage. The inability to maintain a reasonably high dissolved oxygen concentration in the aeration tank (which is a consequence of the high COD concentration of OMWW) makes the application of this process infeasible. Tsonis (1997) studied the extent to which the anoxic step for nitrogen removal may receive OMWW as an external nonnitrogenous carbon source and found that OMWW could be added in a ratio of 1:1,000–1:2,000 (volume of OMWW per volume of municipal type wastewater). In another study, Tsonis (1998) found that the tolerable amount of low-strength waste streams which could be directly discharged in municipal sewage treatment plants corresponds to the processing of 1,500 t of olives per 1,000 equivalent population capacity of the sewage treatment plant. The low-strength waste streams originating from the three-phase olive milling include the olive washing and oil separation steps and account for 40% of the OMWW organic load.

Other types of aerobic systems are based on the formation of biofilm on an inert supporting material, aiming at increasing the degradation rate. Tziotzios et al. (2007) developed a packed-bed bioreactor filled in with rippled plastic hollow tubes onto which biomass able to degrade phenolics was attached. The origin of the biomass was the olive pulp and its indigenous bacteria. The biomass had been acclimated on a synthetic phenolic feeding medium prior to feeding the OMWW. Since the wastewater trickled through the filter, the dissolved oxygen was adequate without the need to provide external aeration. The filter achieved 60% phenolic removal and 70% COD removal within an operating cycle of 3–5 days.

The rotating biological contactor is another biofilm reactor, in which the support media are slowly rotating discs that are partially submerged in a tank with wastewater. Oxygen is transferred to the wastewater and, moreover, the microorganisms come into direct contact periodically with oxygen during the contactor rotation. The rotating biological contactor technology has been tested on OMWW with success after some technical improvements (Siskos 1999).

Aerobic processes based on pure cultures of fungi or bacteria have been developed to detoxify OMWW and improve the biodegradability in anaerobic digestion. Filamentous fungi are particularly efficient for this cause (Zervakis and Balis 1996). Lignin and phenolic compounds can be degraded mainly by the white rot fungi producing phenol oxidases and peroxidases as well as laccases. The main advantage of these enzymes is that they are not substrate-specific, resulting in their catalyzing the degradation of a large number of aromatic pollutants found in OMWW. However, the use of fungi on a large scale is difficult compared with bacteria, since the activity of fungi is influenced by the presence of other microorganisms already contained in the OMWW.

Numerous studies have been conducted regarding the detoxification of OMWW by fungi (Table 4.3).

The use of pure lignolytic enzymes, free or immobilized, has also been investigated (Sayadi and Ellouz 1995; Hamdi 1996; D'Annibale et al. 1998, 2000). Enzymes are susceptible to thermal and pH denaturation, proteolysis, and inactivation. By enzyme immobilization, the process is more stable and enzyme reuse is allowed.

Aerobic bacteria of the *Azotobacter* genus are also capable of degrading phenolic compounds (Hardisson et al. 1969; Rubia et al. 1987; Moreno et al. 1990; García-Barrionuevo et al. 1992; Balis et al. 1996; Papadelli et al. 1996; Chatzipavlidis et al.

Table 4.3 Some fungal species used to detoxify olive mill wastewater

Species	References
Aspergillus niger	Hamdi (1991a); Hamdi et al. (1991a, b); Hamdi and Ellouz (1992a, b); Hamdi and García (1993); García-García et al. (2000); Cereti et al. (2004); Crognale et al. (2006)
Aspergillus terreus	Martínez-Nieto et al. (1993); Borja-Padilla and González (1994); Borja-Padilla et al. (1995d, e, 1998b); García-García et al. (2000)
Coriolus versicolor	Yesilada and Fiskin (1996); Yesilada et al. (1998)
Funalia trogii	Yesilada et al. 1995, 1998
Geotrichum candidum	Borja-Padilla et al. (1992b, e, h, 1995c, 1998b); Martín-Martín et al. (1993); Assas et al. (2000, 2001); Fadil et al. (2003); Ayed et al. (2005); Aissam et al. (2007)
Lentinus edodes	Vinciguerra et al. (1993, 1995) D'Annibale et al. (1998, 2000); García-García et al. (2000); Reverberi et al. (2004); D'Annibale et al. (2004); Tomati et al. (2004)
Phanerochaete chrysosporium	Sayadi and Ellouz (1992, 1995); Gharsallah et al. (1998); García-García et al. (2000); Kissi et al. (2001); Dias et al. (2004); Dhouib et al. (2005, 2006); Ahmadi et al. (2006); Mebirouk et al. (2006)
Phanerochaete flavido-alba	Pérez et al. (1998); Hamman et al. (1999); Blánquez et al. (2002); Rubia et al. (2008)
Pleurotus ostreatus	Sanjust et al. (1991); Flouri et al. (1996); Zervakis et al. (1996); Setti et al. (1998); Kissi et al. (2001); Aggelis et al. (2003); Olivieri et al. (2006)

1996; Ehaliotis et al. 1999; Piperidou et al. 2000). Azotobacters fix atmospheric N_2, and they use phenolic compounds as a carbon and energy source. Two aerobic bacterial strains, a chlorophenol-degrading bacterium (*Ralstonia* sp. LD35) and *Pseudomonas putida* DSM1868, capable of metabolizing 4-methoxybenzoic acid have been tested individually as well as in co-culture (Di Gioia et al. 2001a, b, 2002; Bertin et al. 2001). Their activity was complementary and quite effective towards several aromatic molecules present in OMWW. The olive pulp from two-phase olive mills is an ideal source of aromatic-compound-degrading aerobic bacteria (Jones et al. 2000).

Another type of microorganism able to detoxify OMWW is microalgae. Microalgae grow on mainly sugars and salts, and the biomass produced can be used for animal feeding (Sánchez-Villasclaras et al. 1996).

4.3.3 Phytoremediation

Phytoremediation (phytodepuration) technology is based on the capacity of plants and their associated microorganisms (mycorrhizal fungi and bacteria) to remove pollutants from contaminated water or soil. Wetlands (natural or constructed) are phytoremediating systems that use hydrophytes such as reeds and other marshal plants and are regarded as a low-cost treatment method. Two types of constructed wetlands are currently used for wastewater treatment: free-water surface and subsurface flow systems:

1. Surface flow wetlands are densely vegetated by plants and typically have water depths less than 0.4 m. Open water areas can be incorporated into a design to enable the optimization of hydraulics and wildlife habitat enhancement.
2. Subsurface flow wetlands use a bed of soil or gravel as a substrate for the growth of rooted emergent wetland plants. Pretreated wastewater flows by gravity, horizontally or vertically, through the bed substrate, where it contacts a mixture of facultative microbes living in association with the substrate and plant roots. The bed depth in subsurface flow wetlands is typically between 0.6 and 1.0 m, and the bottom of the bed is inclined to minimize water flow overland. Oxygen may enter the bed substrate by direct atmospheric diffusion and through the roots of the plant, resulting in a mixture of aerobic and anaerobic zones. Most of the saturated bed is anoxic or anaerobic under most wastewater design loadings (Haberl and Langergraber 2002).

The processes taking place in wetlands are both physicochemical and biological: settling of suspended particulate matter, filtration and chemical precipitation through contact of the wastewater with the substrate and litter, chemical transformation, adsorption and ion exchange on the surfaces of plants, substrate, sediment, and litter, uptake and degradation of pollutants and nutrients by microorganisms and plants, and predation and natural die-off of pathogens.

Phytoremediation technology has been applied for OMWW treatment too (Skerratt and Ammar 1999; Santori and Cicalini 2002; Cicalini et al. 2002; Bubba et al. 2004); Kapellakis et al. (2004) applied OMWW to constructed wetlands planted with *Phragmites australis* and reported relatively high dry weight biomass production, ranging from 66.67 t/ha to 74.79 t/ha. It was found that applying 1,500 kg COD/ha/d OMWW resulted in a COD reduction of 74% and total Kjeldahl nitrogen removal of 86%. In this case, effluent recirculation was applied, and 79 and 93% removal efficiencies were observed, respectively.

A phytoremediation system has been constructed in an absorbing tank, consisting of a section for draining OMWW and a section for growing plants (Santori and Cicalini 2002). The absorbing tank receives OMWW during the olive milling season (November to February). The waste flows to the drainage layer, where it undergoes biodegradation; the trees coexist with the waste in a state of a vegetative rest during winter, and after resumption of the vegetative functions of the trees, the wastewater is taken up and further degraded by this complex natural system. Trees with high adaptability and good physiological growth in the presence of OMWW in these wetlands belong to the following families: Salicaceae, Pinaceae, Fagaceae, and Cupressaceae. The more efficient genera for the process of phytoremediation are *Pinus*, *Quercus*, and *Cupressus*.

Wetland technology is a low-cost treatment method for OMWW management and it can be optimized in terms of wetland design, plant selection, and OMWW loadings. However, it is a land-intensive technology.

4.3.4 Irrigation of Agricultural Land

Irrigation is the process of spreading OMWW across the land and, especially in the olive groves themselves, to provide the soil with nutrients and water. It is the oldest practice for the disposal of OMWW (Cabrera et al. 1996). Controlled land application of OMWW generally enhances soil fertility (Morisot 1979; Morisot and Tournier 1986; Fiestas Ros de Ursinos 1986b; Marsilio et al. 1990; Levi-Minzi et al. 1992; Cabrera et al. 1996; Cox et al. 1997; Ben Rouina et al. 1999; Rinaldi et al. 2003). The organic matter, nitrogen, phosphorus, and exchangeable potassium are enriched through OMWW application.

However, parameters such as the nature of the soil, the water table, the type of irrigated crop, and the soil moisture content should be always taken into account (Kapellakis et al. 2008). The soil should have adequate porosity, permeability, and hydraulic conductivity, thus allowing infiltration of OMWW, and avoiding stagnancy and runoff. Moreover, a deep water table protected by an impervious soil layer is required to prevent groundwater pollution. Climatic conditions are also critical: low rainfall and high evaporation have a positive effect. Calcareous alkaline pH soils are very effective in reducing organic and inorganic pollution of OMWW (Cabrera et al. 1995; Garcia-Ortiz et al. 1999), but acid soils and sands, which are poor in bases, change their structure and become nutritionally imbalanced when OMWW is applied.

The determination of the appropriate OMWW dose on land is a very crucial factor, since it is influenced by all the above-mentioned parameters. Normally, applying OMWW doses of less than 1,000 m^3/ha causes no problems (Garcia-Ortiz et al. 1999). In Italy, irrigation with OMWW is common practice. The recycling of the total yearly Italian production of OMWW (about 1,600,000 m^3) on the soil, as fertilizer and irrigation, requires only 2.5–3% of the total Italian surface cultivated with olive trees. In fact, in Italy about one million hectares is cultivated with olive trees but to spread 1,600,000 m^3 OMWW requires approximately 30,000 ha. Currently it is possible to spread 50–80 m^3/ha as Italian law 574/1996 permits (Niaounakis and Halvadakis 2006).

The continuous OMWW application may lead to an accumulation of organic components such as lipids (Gonzalez-Vila et al. 1995; Lopez et al. 1996; Cox et al. 1997). Uncontrolled OMWW application results in increased C/N ratio and salinity of the soil due to potassium and sodium replacement of soil cations (Paredes et al. 1987; Sierra et al. 2001).

The plant growth and physiological processes are generally also positively affected by OMWW (Ben Rouina et al. 1999; Paredes et al. 1999; Cereti et al. 2004). Some phytotoxic effects are temporal (Wang et al. 1967), and are directly correlated with the volume of OMWW applied and the type of crop (D'Annibale et al. 2004). The impact of the application of OMWW on several plants, such as the olive tree (Marsilio et al. 1990; Andrich et al. 1992; Ben Rouina et al. 1999), tomato and chicory seeds (Komilis et al. 2005), and grapevines (Di Giovacchino et al. 1996), has been tested.

In general, land spreading of OMWW is not universally applicable and it is limited to cases where soil of low permeability is near the olive mills.

4.3.5 Composting

Composting is an aerobic process of biodegradation of a heterogeneous solid organic substrate into carbon dioxide, water, mineral salts, and a stabilized organic material containing humus-like substances called compost that can be used as a fertilizer. Composting takes place in open air pile systems called windrows or closed systems that require air supply. As a result of the decomposition of organic matter, heat is released, raising the compost pile temperature to thermophilic levels (60–70°C), where pathogens are destroyed. Composting of olive mill wastes, alone or with other solid wastes (co-composting), has been extensively studied (Paredes et al. 1996a; Papadimitriou et al. 1997; Paredes 1998; Vlyssides et al. 1999; Filippi et al. 2002). Olive cake composting with OMWW has also been studied (Mari et al. 2003, 2005; Hachicha et al. 2006, 2008; Sellami et al. 2008; Ntougias et al. 2008).

Manure is another typical waste to be composted with OMWW (Cayuela et al. 2004). Especially, OMWW coming from two-phase olive mills (pomace) is often mixed with manure in order to be composted, as studied by Sciancalepore et al. (1994, 1995, 1996). In particular, a mixture of crude olive cake, two-phase pomace, and olive leaves inoculated with cow manure resulted in a compost product of high quality after 6 months of composting. Roig et al. (2004) found that addition of 0.5% sulfur (dry weight basis) and adjustment of the moisture content to 40% in a mixture of olive pulp and sheep manure was sufficient to decrease the compost pH by 1.1 units without increasing the electrical conductivity to levels that could reduce the agricultural value of the compost. Municipal biosolids can also be added as a cosubstrate along with cattle manure in olive pulp to form a mixture to be vermicomposted, that is, to be composted with the aid of earthworms (Nogales et al. 1998, 1999; Sainz et al. 2000). Moreover, olive mill waste residues from the sedimentation, flocculation, filtering, etc. can be further treated through composting (Negro Alvarez and Solano 1996).

Although problems of odor emission and leachate production requiring treatment do exist, composting is applied to treat the solid residues of olive mill wastes in Italy (10%), Spain (3%), and Greece (1%) (INASOOP 2003). In the case of olive pomace (produced in two-phase olive mills), the transportation to a composting site is a major technical obstacle.

4.4 Valorization of Olive Mill Wastewater

Certain organic compounds contained in OMWW, i.e., sugars, tannins, polyphenols, polyalcohols, pectins, and lipids, may become an important source for the extraction or generation of high-added-value products. Such products can be classified in three distinct categories according to the method used for their generation:

1. Polyphenols and pectins can be directly recovered from OMWW using either specific solvents or membrane techniques.
2. Enzymes are produced via microbial activity of selective microorganisms during bioconversion of compounds such as polyphenols and lipids which act as promoters of enzymatic expression.
3. Biofuels, bioplastics, biosurfactants, etc. are produced via the microbial conversion of compounds such as sugars, fatty acids, and lipids. Some biofuels are produced via the physicochemical conversion of the organic matter of the solid or semisolid types of olive mill wastes.

Other products that can be derived from the processing of olive mill wastes are compost (see Sect. 4.3.5) and activated carbon.

4.4.1 Recovery of Phenolic Compounds and Pectins

Polyphenols from plant sources are interesting natural antioxidants, showing in vitro higher effectiveness than vitamins E and C on a molar basis, and can be used for enhancing food properties, for nutritional purposes and for preservation (Rice-Evans et al. 1997; Antolovich et al. 2007). Among various plants, the olive tree (*Olea europaea* L.) has been recognized as one of the major source of phenols (Blekas et al. 2002; Visioli and Gali 2002), containing them in the fruit, leaves, and oil. The olive leaf, for example, is a well-known source of biophenols that is marketed under multiple trade names as a nutraceutical. The biophenolic fraction of olive oil comprises only 2% of the total phenolic content of the olive fruits, with the remaining 98% being lost in OMWW (Rodis et al. 2002). OMWW biophenols either may naturally occur in the fruits ending up in the waste or may be generated via conversion of precursor substances during oil processing. The reported biological activities of major biophenols detected in OMWW are shown in Table 4.4 (Obied et al. 2005).

Different methods have been proposed for the recovery of phenols from OMWW based on the chemical and physical properties of the phenols. These processes require removal of the solids from the wastewater to enhance the recovery efficiency and include extraction techniques with selective organic solvents (either directly or via prior attachment to surfactants and sorption in resins) as well as membrane filtration techniques.

Direct liquid–liquid extraction with ethyl acetate as solvent has led to a recovery of phenols of up to 90% from OMWW (Khoufi et al. 2008). A drawback of this method is the large quantities of solvent needed: the required ratio of wastewater to solvent is 1:1. However, recovery of the solvent via evaporation and reuse may reduce the amount of wasted solvent. An alternative approach is to apply a two-step process, according to which the OMWW is first passed through a column packed with resins

Table 4.4 Major groups of phenols detected in olive mill wastewater and their biological activities

Bioactivity	Phenol								
	Hydroxytyrosol	Tyrosol	Oleuropein	Caffeic acid	Vanillic acid	Verbascoside	p-Coumaric acid	Catechol	Rutin
Antioxidant	•	•	•	•	•	•	•	•	•
Cardioprotective	•	•	•						•
Antiatherogenic	•	•	•	•		•			•
Cardioactive	•	•				•			
Chemopreventive	•			•		•	•		•
Chemoprotective				•		•			
Antimicrobial	•	•	•	•	•				
Anti-inflammatory	•			•		•	•		•
Skin bleaching	•								
Hypoglycemic			•						
Antihypertensive			•						
Anti-inflammatory			•						
Cytostatic			•						
Molluscicidal			•						
Endocrinal activity			•						
Enzyme modulation				•					
Antidepressive-like activity						•			
Antihypertensive						•			
Sedative									
Phytotoxic								•	
Anticancer								•	

onto which phenols are absorbed, and subsequently phenols are recovered using organic solvents and the resins are regenerated and reused. Agalias et al. (2007) developed such a process which is less solvent demanding on a pilot scale. The adsorbent resins used were XAD16 and XAD7HP and the solvent used for the regeneration of the resins and recovery of the phenols was a mixture of ethanol and 2-propanol (50:50). Cloud point extraction using surfactants such as Genepol X-080 (Gortzi et al. 2008) and Triton X-114 (Katsoyannos et al. 2006) has also been investigated. In this method, a surfactant solution is heated to its cloud point temperature, i.e., the temperature at which the solution will become turbid because of incomplete solubilization, and the substance to be recovered is entrapped in the micelle of the surfactant. In such procedures, however, fats have to be removed from the OMWW first, in order to enhance phenol recovery. In the case of Triton X-114, the recovery of phenols reached 60% when the surfactant was used at a concentration of 6% in the wastewater.

Membrane processes are also reported as a potential technology for selective fractionation and total recovery of phenols from OMWW. Membrane separation was discussed in Sect. 4.2.5 as a treatment method with the significant advantage of utilizing the fractions obtained for their phenolic content.

Pectins are complex polysaccharides that naturally occur in higher plants and are widely used as gelling agents, stabilizers, and emulsifiers in the food industry. Commercial pectins are available from apple pomace and citrus peels. The availability of other pectic sources, especially waste materials, is always being explored. The olive pulp cell walls are known to contain about one third of arabinose-rich pectic polysaccharide (Coimbra et al. 1994). Any waste stream containing pieces of the olive fruit flesh can be potentially used as a raw material for extraction of pectins. The two-phase olive mill pomace is an appropriate waste for this cause, and it has been studied for its pectin content and quality (Cardoso et al. 2003). It was shown that the olive pectin extract from olive wastes had useful properties for practical applications.

4.4.2 Production of Enzymes

OMWW has been considered as a promising substrate for production of enzymes. Interest is mainly focused on the production of lipases and lignin-oxidizing enzymes, but the potential use for production of pectic enzymes has also been reported.

Lipases can be used as biocatalysts in many biotechnological applications, such as synthesis of polymeric compounds and therapeutics, production of agrochemicals such as herbicides, and for the synthesis of flavors and fragrance (Jaeger and Eggert 2002). They can be produced in significant amounts from OMWW by selective bacterial and yeast strains during fermentation of the carbohydrates and lipids of the wastewater towards metabolic products, such as ethanol and methanol (Scioli and Vallaro 1997). Owing to the consumption of carbohydrates and lipids and moreover reduction of the amount of phenolics, a high reduction of the total COD was observed, reaching 80% (Scioli and Vallaro 1997). The production of lipase (EC 1.3.1.3.)[1] from OMWW is reported to differ significantly according to the type and strain of microorganism used, as well as the operating conditions and addition of nutrients and trace elements. It has been found that lipase activity is connected not only to the presence of fats as expected (Novotny et al. 1988; Ota et al. 1990; Hadeball 1991; Lie et al. 1991), but also to the presence of ethanol or Cu^{2+} and Zn^{2+} ions (Shabtai and Daya-Mishne 1992; Smeltzer et al. 1992). D' Annibale et al. (2006a, b) investigated the potential production of lipases from undiluted OMWW using 12 fungal strains belonging to well-known lipolytic species, such as *Aspergillus oryzae*, *Aspergillus niger*, *Candida cylindracea*, *Geotrichum candidum*, *Penicillium citrinum*, *Rhizopus arrhizus*, and *Rhizopus oryzae*. It was shown that all strains were capable of growing in OMWW and produce lipase, but *C. cylindracea* was the most promising in the conditions tested. The production of enzyme can be stimulated by the type of nitrogen source used, the addition of yeast extract to the medium, pH values, and agitation mode and rate (D'Annibale et al. 2006a).

OMWW has also been considered as a possible substrate for production of lignin-oxidizing enzymes. The three main enzymes involved in this process are laccases (EC 1.10.3.2), lignin peroxidases (EC 1.11.1.14), and manganese-dependent peroxi-dases (EC 1.11.1.13). These enzymes are of commercial interest, since they can be successfully used for the delignification of lignocellulosic biomass, thus enhancing the exposure of carbohydrates to fermentative organisms for subsequent biofuel production (Elander and Hsu 1995), in the paper industry (Widsten and Kandelbauer 2008), for wastewater decolorization (Mohorcic et al. 2006; Rodríguez Pérez et al. 2008), etc. Lignin-oxidizing enzymes are mainly produced by white rot fungi. Fenice et al. (2003) investigated the production of lacasse and manganese-dependent peroxidases from OMWW using the white rot fungus *Panus tigrinus* via both submerged and solid-state fermentation. In the first case, twofold-diluted OMWW was used in stirred tank and air lift reactors. For the solid-state fermentation, OMWW was used to moisturize maize stalks, on which the fungus was grown, but the enzymatic activities observed were lower. The recovery of the enzymes produced can be achieved in microfiltration or ultrafiltration membrane units at relatively low pressures.

Pectinase is an enzyme that breaks down pectins, which are structural polymers of plant cells. It is commonly used in industrial processes (such as in fruit juice extraction) involving the degradation of plant materials. The production of pectinase by OMWW was reported by Federici et al. (1988) where the pretreated OMWW was fermented in batch fermentors using the yeast *Cryptococcus albidus* var. *albidus* IMAT 4735. A maximum enzymatic activity of 280 U/L was observed after 36 h of cultivation. It was further shown that the enzyme was an endopolygalacturonase having a large spectrum of activity on pectin with different degrees of methylation. Further experiments with the purified enzyme showed that when used as additive in olive paste in the malaxation phase it can lead to an approximately 10% increase in the olive oil yield (Petruccioli et al. 1988).

---

[1] Enzyme nomenclature database: http://www.expasy.org/enzyme/

4.4.3 Generation of Biofuels and Bioproducts

The impending shortage of energy resources and degradation of the environment owing to the heedless use of fossil fuels have directed scientific interest to biofuels as well as biodegradable products. Biofuels are defined as fuel products obtained from biological carbon sources, usually photosynthetic plants, via biological or physicochemical methods. Biofuels produced from olive mill wastes include solid (wood pomace, charcoal), liquid (bioethanol, biodiesel), and gaseous (hydrogen, biogas) materials. Other products, such as bioplastics, biosurfactants, and activated carbon, can be derived using the olive mill wastes as raw material.

4.4.3.1 Biofuels

Wood pomace is the solid residue after pomace oil extraction and is usually utilized by olive mills for the production of the required thermal energy (see Sect. 4.2.6.2). Charcoal is another biofuel produced via pyrolysis of concentrated OMWW or olive cake (see Sect. 4.2.6.3). The production of the other fuels is based on the exploitation of mainly carbohydrates and lipids contained in the olive mill wastes. Biogas is the final product of anaerobic digestion. Anaerobic digestion is a major biological treatment technology aiming to reduce the organic content of the waste and convert it to a methane-rich gaseous mixture. The application of this process was thoroughly reviewed in Sect. 4.3.1. Biohydrogen and alcohols are generated via fermentative conversion of carbohydrates. Even though technological achievements in the field of both abovementioned biofuels are quite advanced and numerous wastes have been tested for their generation, very few studies concerning OMWW exploitation in those directions have been reported. Hydrogen is a major intermediate product in anaerobic digestion of olive mill wastes and it can be accumulated by the mixed acidogenic population under certain conditions, as already mentioned in Sect. 4.3.1. On the other hand, alcohol production in high yields can be achieved through the utilization of specific strains of microbes belonging to yeasts or bacteria. Transformation of sugars to ethanol and recovery of the alcohol by distillation has been extensively studied. Oliveira (1974) studied the effect of wine yeasts of *Saccharomyces cerevisiae*, bread yeast and *Candida utilis* and the natural fauna on ethanol production from OMWW and they found that the amount of alcohol produced was in the range 0.5–0.57% (w/v). The inhibitory effect of phenolics on yeasts has been confirmed (Bambalov et al. 1989), and the application of pretreatment methods for degrading phenols (by white rot fungi) substantially increases ethanol yield (Massadeh and Modallal 2008). Bioethanol production from two-phase olive mill pomace has also been investigated (Georgieva and Ahring 2007). In this study, enzymatic hydrolysis increased the concentration of soluble sugars by 75% and a maximum ethanol production was achieved (11.2 g/L) with productivity of 2.1 g/L/h. Ethanol yields were in the range 0.49–0.51 g/g.

Biodiesel is another biofuel which is produced via chemical treatment of lipids that are accumulated in the microbial cell of oleaginous microorganisms when they are grown on different carbon sources, among which are oils (Aggelis and Sourdis 1997) and carbohydrates (Alvarez and Steinbuchel 2003). Although attempts at biodiesel production via gasification of olive cake (Demirbas et al. 2000) and chemical conversion of used olive oil (Dorado et al. 2004) have been made with promising results, OMWW does not seem to have caught the attention of researchers, even though its content of residual oil and carbohydrates could possibly support the growth of oleaginous microorganisms.

4.4.3.2 Biopolymers

Biopolymers are polymers produced by living organisms. Microbial polymers present a greater range of structures and properties than those obtained from seaweeds and plants. As a result, they can be used in different applications. OMWW has been considered as a potential substrate for production of biopolymers via microbial activity, and especially for production of polyhydroxyalkanoates (PHAs) and extracellular polymers (EPS).

PHAs are biodegradable polyesters produced by numerous bacteria and also by several plants and animals. However, only bacteria can synthesize PHAs with high molecular weights and accumulate them as intracellular storage reserves of carbon and energy under stress conditions in the form of inclusion bodies (Luengo et al. 2003). Poly(3-hydroxybutyrate) (PHB) and poly(3-hydroxyvalerate) are the most well known PHAs, and attract high interest because they have properties similar to those of conventional plastics such as polypropylene and polyethylene (Howells 1982), thus being promising candidates for the replacement of fossil-fuel-based thermoplastics in several applications. Production of PHAs from OMWW has been investigated via both pure and mixed cultures. In terms of pure cultures, research has been focused on the use of *Azotobacter chroococcum*, which can produce PHAs, mainly PHB, from untreated OMWW to which a source of nitrogen has been added (Martinez-Toledo et al. 1995). Using the same strain, Gonzalez-Lopez et al. (1996) reported that the type of nitrogen source used can have a significant effect on the accumulation capacity of the bacterium. Final PHB yields obtained from experiments with ammonium acetate, ammonium chloride, ammonium sulfate, and gaseous nitrogen were 59.4%, 50.2%, 50.6%, and 33.7% respectively, measured as a percentage of dry bioplastic weight per dry microbial biomass weight. Pozo et al. (2002) showed that other parameters such as the oxygen supply and the dilution rate of OMWW can further affect the yield of PHAs from *A. chroococcum*. In that study, the maximum yield obtained was 80% PHA per dry biomass when the oxygen supply was decreased and a medium with 60% OMWW was used. Enriched mixed cultures have also been used for production of PHAs from OMWW. The enrichment method was either the "feast and famine" method, during which cultures are subjected to growth periods with external substrate accessibility (feast) and unavailability (famine) (Dionisi et al. 2005), or the successive limitation of nitrogen and carbon sources (Ntaikou et al. 2008). In both studies the OMWW used was pretreated with acidogenic microbes so as to convert the sugars contained into fatty acids, and the predominant bioplastic detected was PHB.

EPS are carbohydrate biopolymers that might be a valid alternative to plant and algal products, considering that their properties are almost identical to those of the currently used gums, which have several industrial applications (Sutherland 1996). OMWW has been considered an interesting substrate for EPS production. This is mainly because it has a high C/N ratio (Alburquerque et al. 2004; Baddi et al. 2004), which could limit cellular growth but favor or stimulate EPS production. OMWW has been used as an organic substrate for the production of pullulan by fermentation (Israelides et al. 1994). Furthermore, Lopez and Ramos-Cormenzana (1996) and Lopez et al. (2001) described xanthan production from OMWW, but the process must be optimized depending on the type and quality of the OMWW used. EPS production by different species of *Azotobacter* (Fiorelli et al. 1996) from OMWW has also been shown. The antimicrobial properties of OMWW have a negative impact in this case too (Tomati et al. 1996) and require dilution of the wastewater (more than 1:1) for it to be used as a culture medium.

4.4.3.3 Biosurfactants

Biosurfactants are surface-active substances synthesized by living cells, such as yeasts, bacteria, and fungi, and their structure includes a hydrophilic moiety composed of amino acids or peptides, anions or cations, or monosaccharides, disaccharides, or polysaccharides. The hydrophobic portion is often made up of saturated, unsaturated, or hydroxylated fatty acids (Georgiou et al. 1992), or is composed of amphiphilic or hydrophobic peptides. Biosurfactants are predominantly synthesized via hydrocarbon uptake by hydrocarbon-degrading microorganisms. Some biosurfactants, however, have been reported to be produced on water-soluble compounds, such as glucose, sucrose, glycerol, and ethanol (Cooper and Goldenberg 1987; Passeri et al. 1992; Hommel and Huse 1993). They find applications in an extremely wide variety of industrial processes, such as emulsification for emulsion polymerization, foaming for food processing, manufacture of detergents for household and industrial cleaning, wetting and phase dispersion for cosmetics and textiles, and solubilization for agrochemicals (Lin 1996).

The suitability of OMWW for biosurfactant production was first investigated by Mercade et al. (1993). In this study, several strains of *Pseudomonas* sp. were proved to be capable of using OMWW as the sole carbon source and of accumulating rhamnolipids. In all cases, the dilution of OMWW as well as the addition of $NaNO_3$ as the nitrogen source was necessary. Conversion yields of 0.058 g rhamnolipid per gram of substrate were achieved, and the COD was reduced by approximately 50% within 72 h. An improved process for rhamnolipid production was developed by studying further the growth of *Pseudomonas aeruginosa* in a stirred tank fermentor (Mercade and Manresa 1994).

4.4.4 Production of Activated Carbon

Activated carbon has been prepared from olive stones and solvent-extracted olive pulp via carbonization at 850°C and physical activation either with CO_2 or with steam at 800°C (Mameri et al. 2000; Galiatsatou et al. 2001, 2002) and treatment with KOH and H_3PO_4 followed by activation with CO_2 at 840°C for different periods of time (Moreno-Castilla et al. 2001). Olive cake has been used by Cimino et al. 2005, to produce activated carbon via acid treatment prior or after the carbonisation effect. Activated carbon can be also produced from olive husks (Michailof et al. 2008).

Alternatively, OMWW has been mixed with the residual fly ash produced by coal combustion in thermoelectric power plants in order to increase the adsorptive capacity of fly ash by means of a new method of aggregation (Rovatti et al. 1992). The product of the aggregation was subjected to a pyrolysis process (see Sect. 4.2.6.3) which resulted in the formation of an oily liquid fraction of high calorific value, a hydrogen-rich gaseous mixture, and a solid carbonaceous matrix dry residue. The solid residue, after activation, could be used as an adsorbent material, and its adsorptive capacity for organic vapors was evaluated in a fixed-bed bench-scale plant, using toluene vapors as the adsorbed gas.

Activated carbon can be used as an absorbent in liquid-phase and gas-phase systems (for the removal of phenols from OMWW or herbicides from water; El-Sheikh et al. 2004), in catalysts, or as a support for catalysts. The role of activated carbon in the OMWW treatment was reviewed in Sect. 4.2.3.

4.5 Conclusions

Olive mill wastes have been the subject of numerous studies with the aim of making use of them and/or rendering them safe for disposal in the environment. It is obvious that there is not a single process that can achieve these goals. The integrated management systems that have already been or can be further developed should combine both physicochemical and biological processes to achieve one or more of the following (ranked from the highest to the lowest value of economic profits from the final products):

- Extraction or production of high-added-value products (phenols, pectines, enzymes, biopolymers, biosurfactants)
- Production of biofuels (hydrogen, ethanol, biogas, biodiesel)
- Production of solid fuels, activated carbon
- Production of compost
- Disposal in land or phytoremediation systems
- Reduction of the organic matter content to acceptable levels for disposal in water-receiving bodies.

The potential of production of high-added-value products from OMWW does not guarantee the viability of the process if the products do not meet the required standards that would make them commercially profitable. Further research is needed to increase the yield, the productivity, and the purity of the final product. Moreover, it is imperative that the integrated management schemes be tested in pilot-scale or full-scale units for long periods in order to test them under real conditions, where the variability of the influent waste, the operational problems of the different individual processes, the coupling of the individual processes of different timescales in a continuous mode of operation, the market availability for the various products, etc. could affect the effectiveness as well as the economic feasibility of the process. Therefore, apart from any scientific and/or technological advances that would improve the OMWW processing, the success of a technological and feasible solution lies at a local and central governmental level, where decisions would be taken to fund and support the application of integrated management schemes in real-scale and real-time operations.

References

Agalias A, Magiatis P, Skaltsounis A-L, Mikros E, Tsarbopoulos A, Gikas E, et al. A new process for the management of olive oil mill waste water and recovery of natural antioxidants. J Agric Food Chem. 2007;55(7):2671–6, http://www.dow.com/products/product_detail.page?product=1120243.

Aggelis G, Sourdis J. Prediction of lipid accumulation-degradation in oleaginous micro-organisms growing on vegetable oils. Int J G. 1997;72(2):159–65.

Aggelis G, Iconomou D, Christou M, Bokas D, Kotzailias S, Christou G, et al. Phenolic removal in a model olive oil mill wastewater using *Pleurotus ostreatus* in bioreactor cultures and biological evaluation of the process. Water Res. 2003;37(16):3897–904.

Ahmadi M, Vahabzadeh F, Bonakdarpour B, Mehranian M, Mofarrah E. Phenolic removal in olive oil mill wastewater using loofah-immobilized *Phanerochaete chrysosporium*. World J Microb Biot. 2006;22(2):119–27.

Aissam H, Penninckx M-J, Benlemlih M. Reduction of phenolics content and COD in olive oil mill wastewaters by indigenous yeasts and fungi. World J Microb Biot. 2007;23(9):1203–8.

Akram A, Stuckey D-C. Biomass acclimatisation and adaptation during start-up of a submerged anaerobic membrane bioreactor (SAMBR). Environ Technol. 2008;29(10):1053–65.

Aktas E-S, Imre S, Ersoy L. Characterization and lime treatment of olive mill wastewater. Water Res. 2001;35(9):2336–40.

Alburquerque J-A, Gonzalvez J, Garcia D, Cegarra J. Agrochemical characterisation of alperujo, a solid by-product of the two-phase centrifugation method for olive oil extraction. Bioresour Technol. 2004;91(2):195–200.

Al-Malah K, Azzam M-O-J, Abu-Lail N-I. Olive mills effluent (OME) wastewater post-treatment using activated clay. Sep Purif Technol. 2000;20(2):225–34.

Alvarez H-M, Steinbuchel A. Triacylglycerols in prokaryotic microorganisms. Appl Microbiol Biotechnol. 2003;60(4):367–76.

Amat A-M, Arques A, Beneyto H, García A, Miranda M-A, Seguí S. Ozonisation coupled with biological degradation for treatment of phenolic pollutants: a mechanistically based study. Chemosphere. 2003;53:79–86.

Amirante P, Mongelli G-L. Prove sperimentali di trattamento delle acque reflue di un oleificio con un impianto di incenerimento. Rivista Italiana Delle Sostanze Grasse. 1982;59(LIX 6):295–300.

Ammar E, Mekki H, Ueno S, Ben Zina M. Traitement de l'effluent des huileries d'olive et son intégration dans des briques de construction. Ann Chimie Sci Materiaux. 2004;29:33–46.

Ammar E, Nasri M, Medhioub K. Isolation of Enterobacteria able to degrade simple aromatic compounds from the wastewater of olive oil extraction. World J Microb Biot. 2005;21:253–9.

Andreoni V, Ferrari A, Ranaldi G, Sorlini C. The influence of some phenolic acids present in oil mill wastes on microbic groups for methanogenesis. Proceedings of the international symposium on olive by-products valorization; Seville, Spain: FAO, UNDP (Food and Agriculture Organization of the United Nations); 1986. p.11–2.

Andreozzi R, Longo G, Majone M, Modesti G. Integrated treatment of olive oil mill effluents (OME): study of ozonation coupled with anaerobic digestion. Water Res. 1998;32:2357–64.

Andrich G, Balzini S, Zinnai A, Silvestri S, Fiorentini R. Effect of olive oil wastewater irrigation on olive plant products. Agr Med. 1992;122:97–100.

Angelidaki I, Ahring B-K. Codigestion of olive oil mill wastewaters with manure, household waste or sewage sludge. Biodegradation. 1997;8(4):221–6.

Angelidaki I, Ahring B-K. Codigestion of olive oil mill wastewaters together with manure, household waste or sewage sludge, biomass energy environment. In: Chartier P, editor. Proceedings of the 9th, European bioenergy conference. Oxford: Elsevier; 1996. vol 1 p.290–5.

Angelidaki I, Ahring B-K. Codigestion of olive oil mill wastewaters together with manure, household waste or sewage sludge, global environmental biotechnology. In: Wise DL, editor. Proceedings of the 3rd international symposium international society of environmental biotechnology. Dordrecht: Kluwer; 1997b. p.173–80.

Angelidaki I, Ahring BK. Effects of free long-chain fatty acids on thermophilic anaerobic digestion. Applied Microbiology and Biotechnology. 1992;37(6):808–12.

Angelidaki I, Ellegaard L, Ahring B-K. Modelling anaerobic codigestion of manure with olive oil mill effluent. Proceedings of the 8th IAWQ International Conference on Anaerobic Digestion; 1997 May 25–29; Sendai, Japan. Water Sci Tech. 1997;36(6–7):263–70.

Angelidaki I, Ahring B-K, Deng H, Schmidt J-E. Anaerobic digestion of olive oil mill effluents together with swine manure in UASB reactors. Water Sci Technol. 2002;45(10):213–8.

Angelidaki I, Chen X, Cui J, Kaparaju P, Ellegaard L. Thermophilic anaerobic digestion of source-sorted organic fraction of household municipal solid waste: start-up procedure for continuously stirred tank reactor. Water Res. 2006;40:2621–8.

Antolovich M, Prenzler P, Robards K, Ryan D. Sample preparation in the determination of phenolic compounds in fruits. Analyst. 2007;125(5):989–1009.

Arambarri P, Cabrera F. Vegetation waters as spreading agents of water contamination by heavy metals. In: International symposium on olive by-products valorisation. Seville: FAO, UNDP; 1986. p.145–6.

Arjona R, García A, de Ollero Castro P. The drying of alpeorujo, a waste product of the olive oil mill industry. J Food Eng. 1999;41(3):229–34.

Arpino A, Carola C. Lo smaltimento delle acque di vegetazione provenienti dagli impianti di estrazione dell'olio di oliva. Nota II: l'incenerimento delle acque di vegetazione. Rivista Italiana Delle Sostanze Grasse. 1978;55(LV)(1):24–8.

Assas N, Marouani L, Hamdi M. Scale down and optimization of olive mill wastewaters decolorization by *Geotrichum candidum*. Bioprocess Eng. 2000;22(6):503–7.

Assas N, Ayed L, Marouani L, Hamdi M. Decolorization of fresh and stored-black olive mill wastewaters by *Geotrichum candidum*. Process Biochem. 2001;38(3):361–5.

Ayed L, Assas N, Sayadi S, Hamdi M. Involvement of lignin peroxidase in the decolourization of black olive mill wastewaters by *Geotrichum candidum*. Lett Appl Microbiol. 2005;40(1):7–11.

Azbar N, Bayram A, Filibeli A, Muezzinoglu A, Sengul F, Ozer A. A review of waste management options in olive oil production. Crit Rev Env Sci Tec. 2004;34(3):209–47.

Azzam M-O-J, Al-Malah K-I, Abu-Lail N-I. Dynamic post-treatment response of olive mill effluent wastewater using activated carbon. J Environ Sci Health A Tox Hazard Subst Environ Eng. 2004;39A(1):269–80.

Baccioni L. Recycling of water and combustion of sediments: a solution for purification of water in oil mills. Rivista Italiana Delle Sostanze Grasse. 1981;58(LVIII 1):34–7.

Baddi G-A, Alburquerque J-A, Gonzalvez J, Cegarra J, Hafidi M. Chemical and spectroscopic analyses of organic matter transformations during composting of olive mill wastes. Int Biodeter Biodegr. 2004;54(1):39–44.

Balice V, Carrieri C, Cera O, Di Fazio A. Natural biodegradation in olive mill effluents stored in open basins. Proceedings of international symposium on olive by-products valorization. Seville, Spain: 4–7 Mar 1986.

Balis C, Chatzipavlidis I, Flouri F. Olive mill waste as a substrate for nitrogen fixation. In: Proceedings of olive oil processes and by products recycling. 1995 Sep 10–13; Granada, Spain. Int Biodete Biodegr. 1996;38(3–4):169–78.

Ballester D-L, García V-A-F, Pérez A-J-P. ES2021191 A Instalación para la depuración integral del alpechín. 16 Oct 1991.

Bambalov G, Israilides C-J, Tanchev S. Alcohol fermentation in olive oil extraction effluents. Biol Waste. 1989;27(1):71–5.

Beccari M, Bonemazzi F, Majone M, Riccardi C. Interaction between acidogenesis and methanogenesis in the anaerobic treatment of olive oil mill effluents. Water Res. 1996;30(1):183–9.

Beccari M, Majone M, Torrisi L. Two reactor system with partial phase separation for anaerobic treatment of olive oil mill effluents. Proceedings of 19th biennial conference of the international association on water quality. 1998 Jun 21–26; Vancouver, Canada. Part 4 (of 9). Water Sci Technol. 1998;38(4–5):53–60.

Beccari M, Majone M, Carucci G, Lanz A-M, Petrangeli P-M. Removal of molecular weight fractions of COD and phenolic compounds in an integrated treatment of olive oil mill effluent. Biodegradation. 2002;13(6):401–10.

Bellakhal N, Oturan M-A, Oturan N, Dachraoui M. Olive oil mill wastewater treatment by the electro-Fenton process. Environ Chem. 2006;3(5):345–9.

Bellido E. Consideración en torno a la contaminación por alpechín. Diseño de una instalación para realizar un proceso industrial de depuración y prímenos resultados obtenidos a escala gemí-industrial Encuentro Euroamericano sobre Innov. Tecnológica, Medio Ambiente y Desarrollo. Seville; 1987.

Bellido E. Un nuevo concepto de la depuración de las aguas residuales de las almazaras, Reunión Internacional sobre Innovación Tecnológica. Madrid (in Spanish); 1989a.

Bellido E. Contaminación por alpechín. Un nuevo tratamiento para disminuir su impacto en diferentes ecosistemas. Ph.D. thesis. University of Córdoba; 1989b.

Beltrán F-J, García-Araya J-F, Navarrete V, Gimeno O. Treatment of wastewater from table olive industries: quantum yield of photolytic processes. B Environ Contam Tox. 2001;67(2):195–201.

Beltrán-Heredia J, Torregrosa J, García J, Domínguez J-R, Tierno J-C. Degradation of olive mill wastewater by the combination of Fenton's reagent and ozonation processes with an aerobic biological treatment. Water Sci Technol. 2001;44:103–8.

Ben Rouina B, Taamallah H, Ammar E. Vegetation water used as a fertilizer on young olive plants. In: Metzidakis I-T, Voyiatzis D-G, editors. Proceedings of the 3 rd ISHS symposium on olive growing. Acta Hortic. 1999;1:474–80.

Benitez F-J, Real F-J, Acero J-L, Leal A-I, Garcia C. Gallic acid degradation in aqueous solutions by UV/H_2O_2 treatment, Fenton's reagent and the photo-Fenton system. J Hazard Mater. 2005;126:31–9.

Benítez F-J, Beltrán Heredia A-J, Acero J-L, González T. Oxidation of vanillic acid as a model of polyphenolic compound present in olive oil wastewaters. II. Photochemical oxidation and combined ozone-UV oxidation. Toxicol Environ Chem. 1995;47(3/4):141–53.

Benítez F-J, Beltrán Heredia A-J, Acero J-L. Oxidation of vanillic acid as a model of polyphenolic compounds in olive oil wastewaters. III. Combined UV radiation hydrogen peroxide oxidation. Toxicol Environ Chem. 1996;56(1–4):199–210.

Benítez F-J, Beltrán Heredia A-J, Acero J-L, Pinilla M-L. Simultaneous photodegradation and ozonation plus UV radiation of phenolic acids major pollutants in agro industrial wastewaters. J Chem Technol Biotechnol. 1997a;70(3):253–60.

Benítez F-J, Beltrán Heredia A-J, Torregrosa J, Acero J-L. Improvement of the anaerobic biodegradation of olive mill wastewaters by prior ozonation pretreatment. Bioprocess Eng. 1997b;17:169–75.

Benítez F-J, Beltrán Heredia A-J, González T, Real F. Kinetics of the elimination of vanillin by UV radiation catalyzed with hydrogen peroxide. Fresen Environ Bull. 1998;7(11/12):726–33.

Benítez F-J, Beltrán Heredia A-J, Torregrosa J, Acero J-L. Treatment of olive mill wastewaters by ozonation, aerobic degradation and the combination of both treatments. J Chem Technol Biotechnol. 1999;74:639–46.

Bertin L, Majone M, Di Gioia D, Fava F. An aerobic fixed-phase biofilm reactor system for the degradation of the low-molecular weight aromatic compounds occurring in the effluents of anaerobic digestors treating olive mill wastewaters. Biotech. 2001;87(2):161–77.

Bettazzi E, Caretti C, Caffaz S, Azzari E, Lubello C. Oxidative processes for olive mill wastewater treatment. Water Sci Technol. 2007;55(10):79–87.

Blánquez P, Caminal G, Sarra M, Vicent M-T, Gabarrell X. Olive oil mill waste waters decoloration and detoxification in a bioreactor by the white rot fungus *Phanerochaete flavido-alba*. Biotechnol Prog. 2002;18(3):660–2.

Blekas G, Psomiadou E, Tsimidou M, Boskou D. On the importance of total polar phenols to monitor the stability of Greek virgin olive oil. Eur J Lipid Sci Technol. 2002;104(6):340–6.

Blika PS, Stamatelatou K, Kornaros M, Lyberatos G. Anaerobic digestion of olive mill wastewater. Global Nest J. 2009;11(3):364–72.

Boari G, Brunetti A, Passino R, Rozzi A. Anaerobic digestion of olive oil mill wastewaters. Agric Wastes. 1984;10(3):161–75.

Boari G, Mancini I-M, Trulli E. Anaerobic digestion of olive mill effluent pretreated and stored in municipal solid waste sanitary landfills. Water Sci Technol. 1993;28(2):27–34.

Borja R, Martin A, Alanso V, Garcia I, Banks C-J. Influence of different aerobic pretreatments on the kinetics of anaerobic digestion of olive mill wastewater. Water Res. 1995;29(2):489–95.

Borja R, Rincon B, Raposo F, Alba J, Martin A. Kinetics of mesophilic anaerobic digestion of the two-phase olive mill solid waste. Biochem Eng J. 2003;15(2):139–45.

Borja R, Raposo F, Rincon B. Anaerobic digestion of two-phase olive mill solid wastes. Tulln, Austria: CROPGEN; 2006a.

Borja R, Sánchez E, Raposo F, Rincón B, Jiménez A-M, Martín A. Study of the natural biodegradation of two-phase olive mill solid waste during its storage in an evaporation pond. Waste Manag. 2006b;26:477–86.

Borja R, Rincon B, Raposo F. Review: Anaerobic biodegradation of two-phase olive mill solid wastes and liquid effluents: kinetic studies and process performance. J Chem Technol Biotechnol. 2006c;81:1450–62.

Borja-Padilla R, González A-E. Comparison of anaerobic filter and anaerobic contact process for olive mill wastewater previously fermented with *Geotrichum candidum*. Process Biochem. 1994;29(2):139–44.

Borja-Padilla R, Alba-Mendoza J, González Berecca A. Kinetic study of the anaerobic digestion of olive mill wastewater, obtained from oil extraction using Olivex, and previously biotreated with *Geotrichum candium*. Grasas Aceites. 1992a;43(4):219–25.

Borja-Padilla R, Martín-Martín A, Durán Barrantes M-M. Kinetic study of the biomethanation of olive mill wastewater previously subjected to aerobic treatment with *Geotrichum candidum*. Grasas Aceites. 1992b;43(2):82–6.

Borja-Padilla R, Martín-Martín A, Maestro-Durán R, Alba-Mendoza J, Ros F, de Ursinos J-A. Enhancement of the anaerobic digestion of olive mill wastewater by the removal of phenolic inhibitors. Process Biochem. 1992c;27(4):231–7.

Borja-Padilla R, Alba-Mendoza J, Garrido-Hoyos S-E, Martínez L, García-Pareja M-P, Incerti C, et al. Comparative study of anaerobic digestion of olive mill wastewater (OMW) and OMW previously fermented with *Aspergillus terreus*. Bioprocess Eng. 1995a;13(6):317–22.

Borja-Padilla R, Alba-Mendoza J, Garrido-Hoyos S-E, Martínez L, García-Pareja M-P, Monteoliva-Sánchez M, et al. Effect of aerobic pre-treatment with *Aspergillus terreus* on the anaerobic digestion of olive mill wastewater. Biotechnol Appl Biochem. 1995b;22(2):233–46.

Borja-Padilla R, Martín-Martín A, Alonso V, García I, Banks C-J. Influence of different aerobic pretreatments on the kinetics of anaerobic digestion of olive mill wastewater. Water Res. 1995c;29(2):489–95.

Borja-Padilla R, Alba-Mendoza J, Banks C-J. Anaerobic digestion of wash waters derived from the purification of virgin olive oil using a hybrid reactor combining a filter and a sludge blanket. Process Biochem. 1996;31(3):219–24.

Borja-Padilla R, Alba-Mendoza J, Mancha A, Martín-Martín A, Alonso V, Sánchez E. Comparative effect of different aerobic pretreatments on the kinetics and macroenergetic parameters of anaerobic digestion of olive mill wastewater in continuous mode. Bioprocess Eng. 1998;18(2):127–34.

Borsani R, Ferrando B. Ultrafiltration plant for olive vegetation waters bypolymeric membrane batteries. Desalination. 1996;108:281–6.

Boskou D. Olive oil: chemistry and technology, Contributor American Oil Chemists Society, AOCS Press; 2006.

Boyadzhiev L, Bezenshek E. Carrier mediated extraction: application of double emulsion technique for mercury removal from waste water. J Membrane Sci. 1983;14:13–8.

Bubba M, del Checchini L, Pifferi C, Zanieri L, Lepri L. Olive mill wastewater treatment by a pilot-scale subsurface horizontal flow (SSF-h) constructed wetland. Ann Chim. 2004;94(12):875–87.

Cabrera F, López R, Martínez-Bordiú A, Dupuy de Lome E, Murillo J-M. Land treatment of olive oil mill waste water. Proceedings of Olive Oil Processes and By-Products Recycling; 1995 10–13 Sep; Granada, Spain. Int Biodeter Biodegr. 1995;38(3–4):215–25.

Cabrera F, Lopez R, Martinez-Bordiu A, Dupuy de Lome E, Murillo J-M. Land treatment of olive oil mill wastewater. Int Biodeter Biodegr. 1996;38(3–4):215–25.

Canepa P, Marignetti N, Gagliardi A. Trattamento delle acque di vegetazione mediante processi a membrana. Inquinamento. 1987;5:73–7.

Canepa P, Marignetti N, Rognoni U, Calgari S. Olive mills wastewater treatment by combined membrane processes. Water Res. 1988;22(12):1491–4.

Cardoso S-M, Coimbra M-A, Lopes-da-Silva J-A. Calcium-mediated gelation of an olive pomace pectic extract. Carbohydr Polym. 2003;52:125–33.

Carrieri C. Ultrafiltration of vegetation waters from olive oil extraction plants; preliminary experiences. Olii grassi derivati. 1978;14:29–31.

Carrieri C, Balice V, Rozzi A, Santori M. 1986. Anaerobic treatment of olive mill effluents mixed with sewage sludge. Preliminary results. Proceedings of International Symposium on Olive By-Products Valorization. Seville, Spain; 4–7 Mar 1986.

Carrieri C, Di Pinto A-C, Rozzi A. Anaerobic co-digestion of sewage sludge and concentrated soluble wastewaters. Proceedings of 2nd IAWQ International Symposium on Waste Management Problems in Agro Industries; 1992 Sep 23–25; Istanbul, Turkey. Water Sci Tech. 1992;28(2):187–97.

Cayuela M-L, Bernal M-P, Roig A. Composting olive mill wastes and sheep manure for orchard use. Compost Sci Util. 2004;12(2):130–6.

Cereti C-F, Rossini F, Federici F, Quaratino D, Vassilev N, Fenice M. Reuse of microbially treated olive mill wastewater as fertiliser for wheat (Triticum durum Desf.). Bioresour Technol. 2004;91:135–40.

Chakchouk M, Hamdi M, Foussard J-N, Debellefontaine H. Complete treatment of olive mill wastewaters by a wet air oxidation process coupled with a biological step. Environ Technol. 1994;15(4):323–32.

Chatzipavlidis I, Antonakou M, Demou D, Flouri F, Balis C. Bio-fertilization of olive oil mills liquid wastes. The pilot plant in Messinia, Greece. In: Proceedings of Olive Oil Processes and By Products Recycling; 1995 Sep 10–13; Granada, Spain. Int Biodeter Biodegr. 1996;38(3–4):183–7.

Chazau-Gillig S. The civilisation of olive trees and cereals. Olivae. 1994;53:14.

Chen Y, Cheng J-J, Creamer K-S. Inhibition of anaerobic digestion process: a review. Bioresour Technol. 2008;99:4044–64.

Cicalini A-R, Santori F, Bànné E-T-G, Gigliotti G, Businelli M, Pasqualetti M, Rambelli A, D'Annibale A, Moscatelli M-C, Marinari S, Grego S. Induction of chemical, biochemical and biological diversity by olive mill wastewaters in the rhizosphere of perennial plants, InterCOST workshop on soil-microbe-root interactions: maximising phytoremediation/bioremediation. 2002.

Cimino G, Cappello R-M, Caristi C, Toscano G. Characterization of carbon from olive cake by sorption of wastewater pollutants. Chemosphere. 2005;61:947–55.

Coimbra M-A, Waldron K-W, Selvendran R-R. Isolation and characterisation of cell wall polymers from olive pulp (*Olea europeae* L.). Carbohydr Res. 1994;252:245–62.
Cooper D-G, Goldenberg B-G. Surface-active agents from two *Bacillus* species. Appl Environ Microbiol. 1987;53(2):224–9.
Correia P-F-M-M, Carvalho J-M-R. A comparison of models for 2-chlorophenol recovery from aqueous solutions by emulsion liquid membranes. Chem Eng Sci. 2001;56:5317–25.
Correia P-F-M-M, Carvalho J-M-R. Recovery of phenol from phenolic resin plant effluents by emulsion liquid membranes. J Membrane Sci. 2003;225:41–9.
Cox L, Celis R, Hermosin M-C, Becker A, Cornejo J. Porosity and herbicide leaching in soils amended with olive-oil wastewater. Agr Ecosyst Environ. 1997;65:151–61.
Crognale S, D'Annibale A, Federici F, Fenice M, Quaratino D, Petruccioli M. Olive oil mill wastewater valorisation by fungi. J Chem Technol Biotechnol. 2006;81(9):1547–55.
D'Annibale A, Crestini C, Vinciguerra V, Sermanni G-G. The biodegradation of recalcitrant effluents from an olive mill by a white rot fungus. J Biotech. 1998;61(3):209–18.
D'Annibale A, Stazi S-R, Vinciguerra V, Sermanni G-G. Oxirane-immobilized *Lentinula edodes* laccase: stability and phenolics removal efficiency in olive mill wastewater. J Biotech. 2000;77:265–73.
D'Annibale A, Casa R, Pieruccetti F, Ricci M, Marabottini R. *Lentinula edodes* removes phenols from olive-mill wastewater: impact on durum wheat (Triticum durum Desf.) germinability. Chemosphere. 2004;54(7):887–94.
D'Annibale A, Brozzoli V, Crognale S, Gallo A-M, Federici F, Petruccioli M. Optimisation by response surface methodology of fungal lipase production on olive mill wastewater. J Chem Technol Biotechnol. 2006a;81(9):1586–93.
D'Annibale A, Sermanni G-G, Federici F, Petruccioli M. Olive-mill wastewaters: a promising substrate for microbial lipase production. Bioresour Technol. 2006b;97(15):1828–33.
Dalis D, Anagnostidis K, López A, Letsiou I, Hartmann L. Anaerobic digestion of total raw olive oil wastewater in a two stage pilot plant (up flow and fixed bed bioreactors). Bioresour Technol. 1996;57(3):237–43.
Davies L-C, Novais J-M, Martins-Dias S. Detoxification of olive mill wastewater using superabsorbent polymers. Environ Technol. 2004;25(1):89–100.
Fiestas Ros de Ursinos J-A. Possibilities of using olive mill wastewater (alpechín) as a fertilizer. Proceedings of the International Symposium on Olive By-Products valorization, Seville, Spain: FAO, UNDP (Food and Agriculture Organization of the United Nations); 1986b. p.321–30.
Demirbas A, Caglar A, Akdeniz F, Gullu D. Conversion of olive husk to liquid fuel by pyrolysis and catalytic liquefaction. Energ Source. 2000;22(7):631–9.
Dhouib A, Aloui F, Hamad N, Sayadi S. Complete detoxification of olive mill wastewaters by integrated treatment using the white rot fungus *Phanerochaete chrysosporium* followed by anaerobic digestion and ultrafiltration. Biotechnology. 2005;4(2):153–62.
Dhouib A, Aloui F, Hamad N, Sayadi S. Pilot-plant treatment of olive mill wastewaters by *Phanerochaete chrysosporium* coupled to anaerobic digestion and ultrafiltration. Process Biochem. 2006;41(1):159–67.
Di Giacomo G. Research Report–ERSA. Apr 1990.
Di Giacomo G, Bonfitto E, Brunetti N, Del Re G, Jacoboni S. Pyrolysis of exhausted olive oil husks coupled with two-stages thermal decomposition of aqueous olive oil mills effluents. In: Ferrero GL, Maniatis K, Buenkes A, Bridgwater AV, editors. Pyrolysis and gasification. New York: Elsevier; 1989. p. 586–90.
Di Gioia D, Bertin L, Fava F, Marchetti L. Biodegradation of hydroxylated and methoxylated benzoic, phenylacetic and phenylpropenoic acids present in olive mill wastewaters by two bacterial strains. Res Microbiol. 2001a;152(1):83–93.
Di Gioia D, Bertin L, Fava F, Marchetti L. Biodegradation of synthetic and naturally occurring mixtures of mono-cyclic aromatic compounds present in olive mill wastewaters by two aerobic bacteria. Appl Microbiol Biotechnol. 2001b;55(5):619–26.

Di Gioia D, Barberio C, Spagnesi S, Marchetti L, Fava F. Characterization of four olive mill wastewater indigenous bacterial strains capable of aerobically degrading hydroxylated and methoxylated monocyclic aromatic compounds. Arch Microbiol. 2002;178(3):208–17.

Di Giovacchino L, Mascolo A, Seghetti L. Sulle caratteristiche delle acque di vegetazione delle olive. Nota II. Rivista Italiana Delle Sostanze Grasse. 1988;65(LXV):481–8.

Di Giovacchino L, Basti C, Costantini N, Ferrante M-L, Angelis A de. Risultati di esperienze pluriennali di spargimento di acque di vegetazione delle olive sul terreno agrario Atti Convegno L'utilizzo dei residui dei frantoi oleari, Viterbo: 1996. pp.35–42.

Di Giovacchino L. In: Hardwood, J. and Aparicio, R. (eds.). Technological aspects. Handbook of olive oil. Analysis and properties. Aspen Publications, Maryland; 2000.

Dias A-A, Bezerra R-M, Pereira A-N. Activity and elution profile of laccase during biological decolorization and dephenolization of olive mill wastewater. Bioresour Technol. 2004; 92(1):7–13.

Dionisi D, Carucci G, Petrangeli-Papini M, Riccardi C, Majone M, Carrasco F. Olive oil mill effluents as a feedstock for production of biodegradable polymers. Water Res. 2005;39(10):2076–84.

Dobre T, Guzun-Stoica A, Floarea O. Reactive extraction using sulphuric acid salts of trioctylamine. Chem Eng Sci. 1999;54:1559–63.

Dorado M-P, Ballesteros E, Mittelbach M, Lopez-Aparicio F-J. Kinetic parameters affecting the alkali-catalyzed transesterification process of used olive oil. Energ Fuel. 2004;18(5):1457–62.

Draxler J, Fürst W, Marr R. Separation of metal species by emulsion liquid membranes. J Membrane Sci. 1988;38:281–93.

Duarte E-A, Neto I. Evaporation phenomenon as a waste management technology. Water Sci Technol. 1996;33(8):53–61.

Ehaliotis C, Papadopoulou K, Kotsou M, Mari I, Balis C. Adaptation and population dynamics of *Azotobacter vinelandii* during aerobic biological treatment of olivemill wastewater. FEMS Microbiol Ecol. 1999;30(4):301–11.

Elander R-T, Hsu T. Processing and economic impacts of biomass delignification for ethanol production. Appl Biochem Biotechnol. 1995;51–52(1):463–78.

El-Shafey E-I, Correia P-F-M, de Carvalho J-M-R. An integrated process of olive mill wastewater treatment. Separ Sci Tech. 2005;40:2841–69.

El-Sheikh A-H, Newman A-P, Al-Daffaee H-K, Phull S, Cresswell N. Characterization of activated carbon prepared from a single cultivar of Jordanian olive stones by chemical and physicochemical techniques. J Anal Appl Pyrol. 2004;71(1):151–64.

Eroğlu E, Gündüz U, Yücel M, Türker L, Eroğlu I. Photobiological hydrogen production by using olive mill wastewater as a sole substrate source. Int J Hydrogen Energ. 2004;29(2):163–71.

Eroğlu E, Eroğlu I, Gondoz U, Torker L, Yocel M. Biological hydrogen production from olive mill wastewater with two-stage processes. Int J Hydrogen Energ. 2006;31:1527–35.

Escolano Bueno A. Tests on removal of waste liquid from olive oil extraction (alpechín) by disposal in ponds or lagoons for percolation and evaporation. Grasas Aceites. 1975;26(6): 387–96.

Everett D-H. Definitions, terminology and symbols in colloid and surface chemistry. Pure Appl Chem. 1972;31(4):579–638.

Fadil K, Chahlaoui A, Ouahbi A, Zaid A, Borja-Padilla R. Aerobic biodegradation and detoxification of wastewaters from the olive mill industry. Int Biodeter Biodegr. 2003;51(1):37–41.

FAOSTAT: Food and Agricultural Organization of the United Nations Statistics Division. 2007. ProdSTAT: crops: olive oil (19 April 2007), based on 2005 data, http://faostat.fao.org/site/567/default.aspxS.

Federici F, Montedoro G, Servili M, Petruccioli M. Pectic enzyme production by cryptococcus albidus var. albidus on olive vegetation waters enriched with sunflower calathide meal. Biol Waste. 1988;25(4):291–301.

Fedorak P-M, Hrudey S-E. The effects of phenol and some alkyl phenolics on batch anaerobic methanogenesis. Water Res. 1984;18(3):361–7.

Fenice M, Sermanni G-G, Federici F, D'Annibale A. Submerged and solid-state production of laccase and Mn-peroxidase by *Panus tigrinus* on olive mill wastewater-based media. J Biotechnol. 2003;100(1):77–85.

Filippi C, Bedini S, Levi-Minzi R, Cardelli R, Saviozzi A. Co-composting of olive oil mill by-products: chemical and microbiological evaluations. Compost Sci Util Sample Doc. 2002;10(1):63–71.

Fiorelli F, Pasetti L, Galli E. Fertility-promoting metabolites produced by *Azotobacter vinelandii* grown on olive-mill wastewaters. Int Biodeter Biodegr. 1996;38(3–4):165–7.

Flouri F, Sotirchos D, Ioannidou S, Balis C. Decolorization of olive oil mill liquid wastes by chemical and biological means. In: Proceedings of Olive Oil Processes and By Products Recycling; 1995 Sep 10–13; Granada, Spain. Int Biodeter Biodegr. 1996:38(3–4):189–92.

Forster-Carneiro T, Fernández LA, Pérez M, Romero LI, Álvarez CJ. Optimization of sebac start up phase of municipal solid waste anaerobic digestion. Chem Biochem Eng Q. 2004;18(4):429–39.

Fortunato V. Trattamento chimico-fisico delle acque discarico dei frantoi oleari e o delle acque di vegetazione delle olive. IT1191528 A 23 Mar 1988.

Galiatsatou P, Metaxas M, Kasselouri-Rigopoulou V. Mesoporous activated carbon from agricultrural by-products. Microchimica Acta. 2001;136(3/4):147–52.

Galiatsatou P, Metaxas M, Arapoglou D, Kasselouri-Rigopoulou V. Treatment of olive mill waste water with activated carbons from agricultural by-products. Waste Manag. 2002;22(7):803–12.

Gamero D-R-N, Vilches D-J-M-P, Alba-Mendoza J. Procedimiento de depuración por separación-recuperación total de sólidos en suspensión y aceite contenido en alpechines. ES820395 A. 16 Jan 1982.

Gamero D-R-N, Vilches D-J-M-P, Alba-Mendoza J. Mejoras introducidas en la patente principal no 497902 por Procedimiento de depuración por separación-recuperación total de sólidos en suspensión y aceite contenido en alpechines. ES8307286 A. 16 Oct 1983.

García-Barrionuevo A, Moreno E, Quevedo-Sarmiento J, González-López J, Ramos-Cormenzana A. Effect of wastewater from olive oil mill (alpechín) on *Azotobacter chroococcum* nitrogen fixation in soil. Soil Biol Biochem. 1992;24(3):281–3.

García-García I, Jiménez-Pena P-R, Bonilla-Venceslada J-L, Martín-Martín A, Martín-Santos M-A, Ramos-Gomez E. Removal of phenol compounds from olive mill wastewater using *Phanerochaete chrysosporium, Aspergillus niger, Aspergillus terreus and Geotrichum candidum*. Process Biochem. 2000;35(8):751–8.

Garcia-Ortiz A, Beltran G, Gonzalez P, Ordonez R, Giraldez J-V. Vegetation water (alpechin) application effects on soils and plants. Acta Horticulturae. 1999;474:749–52.

Gavala H-N, Skiadas I-V, Lyberatos G. On the performance of a centralised digestion facility receiving seasonal agroindustrial wastewaters. Water Sci Technol. 1999;40(1):339–46.

Gavala H-N, Skiadas I-V, Ahring B-K, Lyberatos G. Potential for biohydrogen and methane production from olive pulp. Water Sci Technol. 2005;52(1–2):209–15.

Gavala H-N, Skiadas I-V, Ahring B-K, Lyberatos G. Thermophilic anaerobic fermentation of olive pulp for hydrogen and methane production: modelling of the anaerobic digestion process. Water Sci Technol. 2006;53(8):271–9.

Gelegenis J, Georgakakis D, Angelidaki I, Christopoulou N, Goumenaki M. Optimization of biogas production from olive-oil mill wastewater, by codigesting with diluted poultry-manure. Appl Ener. 2007;84:646–63.

Georgacakis D, Dalis D. Controlled anaerobic digestion of settled olive oil wastewater. Bioresour Technol. 1993;46(3):221–6.

Georgieva T-I, Ahring B-K. Potential of agroindustrial waste from olive oil industry for fuel ethanol production. Biotechnol J. 2007;2(12):1547–55.

Georgiou G, Lin S-C, Sharma M-M. Surface-active compounds from microorganisms. Biotechnology. 1992;10(1):60–5.

Gernjak W, Krutzler T, Glaser A, Malato S, Caceres J, Bauer R. Photo-Fenton treatment of water containing natural phenolic pollutants. Chemosphere. 2003a;50:71–8.

Gernjak W, Maldonado M-I, Malato S, Caceres J, Krutzler T, Glaser A. Degradation of polyphenolic content of olive mill wastewater (OMW) by solar photocatalysis. In: Vogelpohl A, editor. 3 rd International conference on oxidation technologies for water and wastewater treatment. 2003a. p.879–4.

Gharsallah N. Influence of dilution and phase separation on the anaerobic digestion of olive mill wastewaters. Bioprocess Eng. 1994;10:29–34.

Gharsallah N, Labat M, Aloui F, Sayadi S. The effect of Phanerochaete chrysosporium pretreatment of olive mill waste waters on anaerobic digestion. Proceedings of the 4th International Symposium of the International Society for Environmental Biotechnology; 20–25 June 1998; Belfast, UK. Resour Conser Recy. 1998;27(1–2):187–92.

Gianfranco R, Michele R, Introna M. Volatilisation of substances after spreading olive oil wastewater on the soil in a Mediterranean environment. Agr Ecosyst Environ. 2003;96:49–58.

Giannis A, Kalaitzakis M, Diamadopoulos E. Electrochemical treatment of olive mill wastewater. J Chem Technol Biotechnol. 2007;82:663–71.

Gomes H-T, Figueiredo J-L, Faria J-L. Catalytic wet air oxidation of olive mill wastewater. Catalysis Today. 2007;124(3–4):254–9.

Gonzalez-Lopez J, Pozo C, Martinez-Toledo M-V, Rodelas B, Salmeron V. Production of polyhydroxyalkanoates by *Azotobacter chroococcum* H23 in wastewater from olive oil mills (Alpechin). Int Biodeter Biodegr. 1996;38(3–4):271–6.

González-López J, Bellido E, Benítez C. Reduction of total polyphenols in olive mill wastewater by physico chemical purification. J Environ Sci Heal A. 1994;29(5):851–65.

Gonzalez-Vila F-J, Verdejo T, Del Rio J-C, Martin F. Accumulation of hydrophobic compounds in the soil lipidic and humic fractions as result of a long term land treatment with olive oil mill effluents (alpechin). Chemosphere. 1995;31(7):3681–6.

Gortzi O, Lalas S, Chatzilazarou A, Katsoyannos E, Papaconstandinou S, Dourtoglou E. Recovery of natural antioxidants from olive mill wastewater using Genapol-X080. J Am Oil Chem Soc. 2008;85(2):133–40.

Grappelli A, Galli E, Palma G, Tomati U. Procedimento per la depurazione di reflui vegetali agricoli, in partico lare acque di vegetazione. GR870652 A. 30 Jun 1987.

Haberl R, Langergraber G. Constructed Wetland Technology, InterCOST Workshop on soil-microbe-root interactions: maximising phytoremediation/bioremediation. 2002.

Hachicha S, Chtourou M, Medhioub K, Ammar E. Compost of poultry manure and olive mill wastes as an alternative fertilizer. Agron Sust Dev. 2006;26(2):135–42.

Hachicha S, Sallemi F, Medhioub K, Hachicha R, Ammar E. Quality assessment of composts prepared with olive mill wastewater and agricultural wastes. Waste Manag. 2008;28(12):2593–603.

Hadeball W. Production of lipase by *Yarrowia lipolytica*. Lipase from yeasts. Acta Biotechnologica. 1991;2:159–67.

Hamdi M. Effects of agitation and pretreatment on the batch anaerobic digestion of olive mill wastewater. Bioresour Technol. 1991;36(2):173–8.

Hamdi M. Nouvelle conception d'un procédé de dépollution biologique des margines, effluents liquides de l'extraction de l'huile d'olive. Biologie cellulaire et Microbiologie, Thèse doctorat. France: Université de Provence; 1991b.

Hamdi M. Future prospects and constraints of olive oil mill wastewaters and treatment. A review. Bioprocess Eng. 1993a;8(5–6):209–14.

Hamdi M. Thermoacidic precipitation of darkly coloured polyphenols of olive mill wastewaters. Environ Technol. 1993b;14:495–500.

Hamdi M. Anaerobic digestion of olive mill wastewaters. Process Biochem (Oxford). 1996;31(2):105–10.

Hamdi M, Ellouz R. Bubble column fermentation of olive mill wastewater by *Aspergillus niger*. J Chem Technol Biotechnol. 1992a;54(4):331–5.

Hamdi M, Ellouz R. Use of *Aspergillus niger* to improve filtration of olive mill wastewaters. J Chem Technol Biotechnol. 1992b;53(2):195–200.

Hamdi M, Ellouz R. Treatment of detoxified olive mill wastewaters by anaerobic filter and aerobic fluidized bed process. Environ Technol. 1993;14(2):183–8.
Hamdi M, García J-L. Anaerobic digestion of olive mill wastewaters after detoxification by prior culture of *Aspergillus niger*. Process Biochem. 1993;28(3):155–9.
Hamdi M, Bou Hamed H, Ellouz R. Optimization of the fermentation of olive mill waste-waters by *Aspergillus niger*. Appl Microbiol Biotechnol. 1991a;36(2):285–8.
Hamdi M, Khadir A, García J-L. The use of *Aspergillus niger* for the bioconversion of olive mill wastewaters. Appl Microbiol Biotechnol. 1991b;34(6):828–31.
Hamdi M, Garcia J-I, Ellouz R. Integrated biological process olive oil mill wastewater treatment. Bioprocess Eng. 1992;8:79–84.
Hamman O, de la Ben Rubia T, Martínez J. Decolorization of olive mill wastewaters by *Phangerochaete flavido-alba*. Environ Toxicol Chem. 1999;18(11):2410–5.
Hardisson C, Sala J-M, Stainer R-Y. Pathways for the oxidation of aromatic compounds by Azotobacter. J Genet Microbiol. 1969;59:1–11.
Hommel R-K, Huse K. Regulation of sophorose lipid production by *Candida* (torulopsis) *apicola*. Biotechnol Lett. 1993;15(8):853–8.
Howells E-R. Opportunities in biotechnology for the chemical industry. Chem Ind. 1982; 8:508–11.
Iconomou D, Diamantitis G, Zanganas P, Theochari I, Israilides K, Kouloumbis P. Reduction of phenolic concentration in olive-oil mill wastewater by biotechnological means. In: Tsihrintzis V-A, Korfiatis G-P, Katsifarakis K-L, Demetracopoulos A-C, editors. Proceedings of the International Conference on Production and Restoration of the Environment V. Thassos, Greece, 2000. Thessaloniki: Publisher Bouris I; 2000. p.569–72.
INASOOP, COLL-CT-2003-500467. Integrated approach to sustainable olive oil and table olives production; 2003.
IMPEL (European union network for the implementation and enforcement of environmental law). IMPEL OLIVE OIL PROJECT. 2003. Report.
Israelides C-J, Scalon B, Smith A, Harding S-E, Jumel K. Characterization of pullulans produced from agro-industrial wastes. Carbohyd Polym. 1994;25:203–9.
Israilides C-J, Vlyssides A-G, Mourafeti V-N, Karvouni G. Olive oil wastewater treatment with the use of an electrolysis system. Bioresour Technol. 1997;61(2):163–70.
Jaeger K-E, Eggert T. Lipases for biotechnology. Curr Opin Biotechnol. 2002;13(4):390–7.
Jarboui R, Sellami F, Kharroubi A, Gharsallah N, Ammar E. Olive mill wastewater stabilization in open-air ponds: impact on clay–sandy soil. Bioresour Technol. 2008;99:7699–708.
Jones C-E, Murphy P-J, Russell N-J. Diversity and osmoregulatory responses of bacteria isolated from two-phase olive oil extraction waste products. W J Microb Biot. 2000;16:555–61.
Kalfas H, Skiadas I-V, Gavala H-N, Stamatelatou K, Lyberatos G. Application of ADM1 for the simulation of anaerobic digestion of olive pulp under mesophilic and thermophilic conditions. Water Sci Technol. 2006;54(4):149–56.
Kallel M, Belaid C, Boussahel R, Ksibi M, Montiel A, Elleuch, B. Olive mill wastewater degradation by Fenton oxidation with zero-valent iron and hydrogen peroxide, J Hazard Mater. 2008. Article in Press.
Kapellakis I-E, Tsagarakis K-P, Angelakis A-N. The use of constructed wetlands on olive mill wastewater treatment. In: Proceedings of the 9th International IWA Specialist Conference on Wetlands Systems for Water Pollution Control, 2004 Sep 27–30; Palais Des Papes, Avignon, France. vol 1 p.113–22.
Kapellakis E, Tsagarakis K-P, Crowther J-C. Olive oil history, production and by-product management. Rev Environ Sci Biot. 2008;7:1–26.
Kasirga E. Treatment of olive oil industry wastewaters by anaerobic stabilization method and development of kinetic model, Unpublished PhD thesis. Dokuz Eylul University, Graduate School of Natural and Applied Sciences, Izmir, Turkey (in Turkish). 1988.
Katsoyannos E, Chatzilazarou A, Gortzi O, Lalas S, Konteles S, Tataridis P. Application of cloud point extraction using surfactants in the isolation of physical antioxidants (phenols) from olive mill wastewater. Fresen Environ Bull. 2006;15(9 B):1122–5.

Khabbaz M-S, Vossoughi M, Shakeri M. Performance of an anaerobic baffled reactor for olive mill oil wastewater treatment. In: Proceedings of the 39th Central Canadian Symposium on Water Quality Research. Burlington, Ontario, Canada, 9–10 Feb 2004.

Khoufi S, Aouissaoui H, Penninckx M, Sayadi S. Application of electro-Fenton oxidation for the detoxification of olive mill wastewater phenolic compounds. Water Sci Technol. 2004;49(4):97–102.

Khoufi S, Aloui F, Sayadi S. Extraction of antioxidants from olive mill wastewater and electrocoagulation of exhausted fraction to reduce its toxicity on anaerobic digestion. J Hazard Mater. 2008;151(2–3):531–9.

Kissi M, Mountadar M, Assobhei O, Gargiulo E, Palmieri G, Giardina P, et al. Roles of two white rot basidiomycete fungi in decolorisation and detoxification of olive mill wastewater. Appl Microbiol Biotechnol. 2001;57(1–2):221–6.

Knudsen C-H, Larsen S-T. Process and plant for purification of agricultural waste material. WO921120 A. 09 Jul 1992.

Komilis D-P, Karatzas E, Halvadakis C-P. The effect of olive mill wastewater on seed germination after various pretreatment techniques. J Environ Manage. 2005;74(4):339–48.

Koutrouli E-C, Kalfas H, Gavala H-N, Skiadas I-V, Stamatelatou K, Lyberatos G. Hydrogen and methane production through two-stage mesophilic anaerobic digestion of olive pulp. Bioresour Technol. 2009;100:3718–23.

Kugelman I-J, McCarthy P-L. Cation toxicity and stimulation in anaerobic waste treatment. J Water Pollut Cont Fed. 1964;37:97–116.

Le Verge S, Bories A. Les basins d'e´vaporation naturelle des margines, Le Nouvel Olivier (OCL). 2004:5–10.

Leon-Cabello R. Fiestas Ros de Ursinos J-A. Evaluacion de alpechins en balsas de evaporacion. In: Proceedings de Congreso sobre biotechnologias de bajo costo para la depuracion de aguas residuales, Madrid: FAO; 18–20 Nov1981.

Levi-Minzi R, Saviozzi A, Riffaldi R, Falzo L. Land application of vegetable water: effects on soil properties. Olivae. 1992;40:20–5.

Li N-N. Separating hydrocarbons with liquid membranes, U.S. Patent 3,410,794. 1968.

Lie E, Persson A, Molin G. Screening for lipase-producing microorganisms with a continuous cultivation system. Appl Microbiol Biotechnol. 1991;35(1):19–20.

Lin S-C. Biosurfactants:recent advances. J Chem Technol Biotechnol. 1996;66(2):109–20.

Lolos G, Skordilis A, Parissakis G. Polluting characteristics and lime precipitation of olive mill wastewater. J Environ Sci Heal A. 1994;29(7):1349–56.

Longhi P, Vodopivec B, Fiori G. Electrochemical treatment of olive oil mill wastewater. Ann Chim. 2001;91(3–4):169–74.

Lopez M-J, Ramos-Cormenzana A. Xanthan production from olive-mill wastewaters. Int Biodeter Biodegr. 1996;38(3–4):263–70.

Lopez R, Martinez-Bordiu A, Dupuy de Lome E, Cabrera F, Sanchez M-C. Soil properties after application of olive mill wastewater. Fresen Environ Bull. 1996;5:49–54.

Lopez M-J, Moreno J, Ramos-Cormenzana A. The effect of olive mill wastewaters variability on xanthan production. J Appl Microbiol. 2001;90(5):829–35.

Luengo J-M, Garcia B, Sandoval A, Naharro G, Olivera E-R. Bioplastic from microorganisms. Curr Opin Microbiol. 2003;6(3):251–60.

Lyberatos G, Gavala H-N, Stamatelatou A. An integrated approach for management of agricultural industries wastewaters, nonlinear analysis, theory. Meth Appl. 1997;30(4):2341–51.

Mameri N, Aioueche F, Belhocine D, Grib H, Lounici H, Piron D-L, et al. Preparation of activated carbon from olive solid residue. J Chem Technol Biotechnol. 2000;75(7):625–31.

Mantzavinos D. Removal of cinnamic acid derivatives from aqueous effluents by Fenton and Fenton-like processes as an alternative to direct biological treatment. Water Air Soil Pollution Focus. 2003a;3:211–21.

Mantzavinos D. Removal of benzoic acid derivatives from aqueous effluents by the catalytic decomposition of hydrogen peroxide. Process Safe Environ. 2003b;81:99–106.

Mantzavinos D, Kalogerakis N. Treatment of olive mill effluents part I: organic matter degradation by chemical and biological processes-an overview. Environ Int. 2005;31:289–95.

Mari I, Ehaliotis C, Kotsou M, Balis C, Georgakakis D. Respiration profiles in monitoring the composting of by-products from the olive oil agro-industry. Bioresour Technol. 2003; 87(3):331–6.

Mari I, Ehaliotis C, Kotsou M, Chatzipavlidis I, Georgakakis D. Use of sulfur to control pH in composts derived from olive processing by-products. Compost Sci Util. 2005;13(4):281–7.

Marques I-P, Teixeira A, Rodrigues L, Martins Dias S, Novais JM. Anaerobic treatment of olive mill wastewater with digested piggery effluent. Water Res. 1998;70(5):1056–61.

Marsilio V, Di Giovacchino L, Lombardo N, Briccoli-Bati C. First observations on the disposal effects of olive mills vegetation waters on cultivated soil. Acta Horticulturae. 1990; 286:493–8.

Martínez-Nieto L, Garrido-Hoyos S-E, Camacho Rubio F, García-Pareja M-P, Ramos-Cormenzana A. Biological purification of waste products from olive extraction. Bioresour Technol. 1993;43(3):215–9.

Martinez-Toledo M-V, Gonzalez-Lopez J, Rodelas B, Pozo C, Salmeron V. Production of poly-β-hydroxybutyrate by *Azotobacter chroococcum* H23 in chemically defined medium and alpechin medium. J Appl Bacteriol. 1995;78(4):413–8.

Martín-Martín A, Borja-Padilla R, Chica A. Kinetic study of an anaerobic fluidized bed system used for the purification of fermented olive mill wastewater. J Chem Technol Biotechnol. 1993;56(2):155–62.

Mascolo A, Cucurachi A, Di Giovacchino L, Ranalli A. Disposal of waste water from olive oil production, by means of activated sludge plants for treatment of urban sewage, Annali dell'Istituto Sperimentale per la Elaiotecnica, 9-1981-198. 1990.

Massadeh M-I, Modallal N. Ethanol production from olive mill wastewater (OMW) pretreated with *Pleurotus sajor-caju*. Energ Fuel. 2008;22(1):150–4.

Mebirouk M, Sbai L, Lopez M, Gonzalez J. Olive oil mill wastewaters pollution abatement by physical treatments and biodegradation with *phanerochaetae chrysosporium*. Environ Technol. 2006;27(12):1351–6.

Mercade M-E, Manresa M-A. The use of agroinustrial by-products for biosurfactant production. J Am Oil Chem Soc. 1994;71(1):61–4.

Mercade M-E, Manresa M-A, Robert M, Espuny M-J, de Andres C, Guinea J. Olive oil mill effluent. New substrate for biosurfactant production. Bioresour Technol. 1993;43(1):1–6.

Michailof C, Stavropoulos G-G, Panayiotou C. Enhanced adsorption of phenolic compounds, commonly encountered in olive mill wastewaters, on olive husk derived activated carbons. Bioresour Technol. 2008;99:6400–8.

Michelakis N, Klapaki G, Kasapakis G. Olive vegetation water management by evaporation basins. Acta Horticulturae. 1999;474:757–60.

Mignone NA. Biological inhibition/toxicity control in municipal anaerobic digestion facilities. 2005; http://www.awpca.net/Biological Inhibition.pdf.

Minh D-P, Gallezot P, Besson M. Degradation of olive oil mill effluents by catalytic wet air oxidation: 1. Reactivity of p-coumaric acid over Pt and Ru supported catalysts. Appl Catal B-Environ. 2006;63(1–2):68–75.

Minh D-P, Aubert G, Gallezot P, Besson M. Degradation of olive oil mill effluents by catalytic wet air oxidation: 2-oxidation of p-hydroxyphenylacetic and p-hydroxybenzoic acids over Pt and Ru supported catalysts. Appl Catal B-Environ. 2007a;73(3):236–46.

Minh D-P, Aubert G, Gallezot P, Besson M. Treatment of olive oil mill wastewater by catalytic wet air oxidation. 3. Stability of supported ruthenium catalysts during oxidation of model pollutant p-hydroxybenzoic acid in batch and continuous reactors. Appl Catal B-Environ. 2007b; 75(1–2):71–7.

Minh D-P, Gallezot P, Azabou S, Sayadi S, Besson M. Catalytic wet air oxidation of olive oil mill effluents 4. Treatment and detoxification of real effluents. Appl Catal B-Environ. 2008;84:749–57.

Miranda M-A, Galindo F, Amat A-M, Arques A. Pyrylium salt-photosensitized degradation of phenolic contaminants derived from cinnamic acid with solar light: correlation of the observed reactivities with fluorescence quenching. Appl Catal B-Environ. 2000;28:127–33.

Miranda M-A, Galindo F, Amat A-M, Arques A. Pyrylium salt-photosensitised degradation of phenolic contaminants present in olive oil wastewaters with solar light: part II benzoic acid derivatives. Appl Catal B-Environ. 2001;30:437–44.

Miranda M-A, Marín M-L, Amat A-M, Arques A, Seguí S. Pyrylium saltphotosensitized degradation of phenolic contaminants present in olive oil wastewater with solar light: part III tyrosol and p-hydroxyphenylacetic acid. Appl Catal B-Environ. 2002;35:167–74.

Mitrakas M, Papageorgiou G, Docoslis A, Sakellaropoulos G. Evaluation of various pretreatment methods for olive oil mill wastewater. Eur Water Pollut Contr. 1996;6(6):10–6.

Mohorcic M, Teodorovic S, Golob V, Friedrich J. Fungal and enzymatic decolourisation of artificial textile dye baths. Chemosphere. 2006;63(10):1709–17.

Molina Alcaide E, Nefzaoui A. Recycling of olive oil by-products: possibilities of utilization in animal nutrition. Int Biodeter Biodegr. 1996;38(3–4):227–35.

Monteagudo J-M, Carmona M, Duran A. Photo-Fenton-assisted ozonation of p-coumaric acid in aqueous solution. Chemosphere. 2005;60(8):1103–10.

Moreno E, Quevedo-Sarmiento J. Ramos-Cormenzana A. Antibacterial activity of washwater from olive mills. In: Cheremisinoff P-N, editor. Encyclopaedie of environmental control technology, 3 (Chapter 26), Gulf Publications, Houston, TX, USA; 1990. p.731–6.

Moreno-Castilla C, Carrasco-Marín F, López-Ramón M-V, Alvarez-Merino M-A. Chemical and physical activation of olive-mill waste water to produce activated carbons. Carbon. 2001; 39(9):1415–20.

Morisot A. Utilisation des margines per épandage. L'Olivier. 1979;19(1):8–13.

Morisot A, Tournier J-P. Répercussions agronimiques de l' épandage d'effluents et déchets de moulins à huile d' olive. Agronomie. 1986;6(3):235–41.

Negro Alvarez M-J, Solano M-L. Laboratory composting assays of the solid residue resulting from the flocculation of oil mill wastewater with different lignocellulosic residues. Compost Sci Util Sample Doc. 1996;4(4):62–71.

Niaounakis M, Halvadakis CP. Olive processing waste management: literature review and patent survey, Waste management series, vol. 5. 2nd ed. UK: Elsevier Ltd; 2006.

Nogales R, Thompson R, Calmet A, Benítez E, Gómez M, Elvira C. Feasibility of microcomposting residues from olive oil production obtained using two-stage centrifugation. J Environ Sci Health A Tox Hazard Subst Environ Eng. 1998;A33(7):1491–506.

Nogales R, Melgar R, Guerrero A, Lozada G, Benítez E, Thompson R, et al. Growth and reproduction of *Eisenia andrei* in dry olive cake mixed with other organic wastes. Pedobiologia. 1999;43(6):744–52.

Novotny C, Dolezalovna L, Novak M. The production of lipase by some *Candida* and *Yarrowia* yeasts. J Basic Microbiol. 1988;28:221–7.

Ntaikou I, Kourmentza C, Koutrouli E, Stamatelatou K, Zampraka A, Kornaros M, et al. Exploitation of olive oil mill wastewater for combined bio-fuels and biopolymers production. Bioresour Technol. 2008. doi:10.1016/j.biortech.2008.12.001.

Ntougias S, Papadopoulou K-K, Zervakis G-I, Kavroulakis N, Ehaliotis C. Suppression of soil-borne pathogens of tomato by composts derived from agro-industrial wastes abundant in Mediterranean regions. Biol Fert Soils. 2008;44(8):1081–90.

Obied H-K, Allen M-S, Bedgood D-R, Prenzler P-D, Robards K, Stockmann R. Bioactivity and analysis of biophenols recovered from olive mill waste. J Agric Food Chem. 2005;53(4):823–37.

Oliveira de J-S. Effluents from olive oil extracting plants subsidies for the study of the obtention of sealable products with simultaneous elimination of BOD. 12th International Congress of Agro-Food Industries, Athens, Greece; 1974.

Olivieri G, Marzocchella A, Salatino P, Giardina P, Cennamo G, Sannia G. Olive mill wastewater remediation by means of *Pleurotus ostreatus*. Biochem Eng J. 2006;31(3):180–7.

Olori L, de Fulvio S, Morgia P. Acque di vegetazione. Le problematiche ambientalie gli aspetti igienico-sanitari. Inquinamento. 1990;1:40–6.

Ota Y, Gomi K, Sato S, Sugiura T, Minoda Y. Purification and some properties of cell-bound lipase from *Saccharomycopsis lipolytica*. Agr Biol Chem. 1990;46:2885–93.

Papadelli M, Roussis A, Papadopoulou K, Venieraki A, Chatzipavlidis I, Katinakis P, Balis C. Biochemical and molecular characterization of an *Azotobacter vinelandii* strain with respect to its ability to grow and fix nitrogen in olive mill wastewater. Proceedings of Olive Oil Processes and By Products Recycling; 1995 Sep 10–13; Granada, Spain. Int Biodeter Biodegr. 1996;38(3–4):179–81.

Papadimitriou E-K, Chatzipavlidis I, Balis C. Application of composting to olive mill wastewater treatment. Environ Technol. 1997;18(1):101–7.

Papafotiou M, Phsyhalou M, Kargas G, Chatzipavlidis I, Chronopoulos J. Olive-mill wastes compost as growing medium component for the production of poinsettia. Scientia Horticulturae. 2004;102(2):167–75.

Papaioannou D. A method of processing waste gases from the drying of olive presscake. Biol Waste. 1988;24(2):137–45.

Paraskeva P, Diamadopoulos E. Technologies for olive mill wastewater (OMW) treatment: a review. J Chem Technol Biotechnol. 2006;81:1475–85.

Paraskeva C-A, Papadakis V-G, Kanellopoulou D-G, Koutsoukos P-G, Angelopoulos K-C. Membrane filtration of olive mill wastewater and exploitation of its fractions. Water Environ Res. 2007a;79(4):421–9.

Paraskeva C-A, Papadakis V-G, Tsarouchi E, Kanellopoulou D-G, Koutsoukos P-G. Membrane processing for olive mill wastewater fractionation. Desalination. 2007b;213:218–29.

Paredes C. Compostaje del alpechín. Una solucion agricola para la reduccion de su impacto ambiental CEBAS- CSIC. 1998.

Paredes M-J, Moreno E, Ramos-Cormezana A, Martinez J. Characteristics of soil after pollution with wastewaters from olive oil extraction plants. Chemosphere. 1987;16(7):1557–64.

Paredes C, Cegarra J, Sánchez-Monedero M-A, Galli E. Composting of fresh and pond-stored olive-mill wastewater by the Rutgers system. In: Bertoldi M, de Sequi P, Lemmes B, Papi T, editors. The science of composting. 2nd ed. Glasgow: Blackie; 1996. p. 1266–70.

Paredes C, Ceggara J, Roing A, Sánchez-Monedero M-A, Bernal M-P. Characterization of olive mill wastewater (alpechin) and its sludge for agricultural purposes. Bioresour Technol. 1999;67:111–5.

Passeri A, Schmidt M, Haffner T, Wray V, Lang S, Wagner F. Marine biosurfactants. IV. Production, characterization and biosynthesis of an anionic glucose lipid from the marine bacterial strain MM1. Appl Microbiol Biotechnol. 1992;37(3):281–6.

Pérez J, de la Rubia T, Hamman O-B, Martínez J. *Phanerochaete flavido-alba* laccase induction and modification of manganese peroxidase isoenzyme pattern in decolorized olive oil mill wastewaters. Appl Environ Microbiol. 1998;64(7):2726–9.

Petarca L, Vitolo S, Bresci B. Pyrolysis of concentrated olive mill vegetation waters. In: Kaltschmitt M, Bridgwater AV, editors. Proceedings of international conference on biomass gasification and pyrolysis. UK: CPL Press; 1997. p.374–81.

Petruccioli M, Servili M, Montedoro G, Federici F. Development of a recycle procedure for the utilization of vegetation waters in the olive oil extraction process. Biotechnol Lett. 1988;10(1):55–60.

Piperidou C-I, Chaidou C-I, Stalikas C-D, Soulti K, Pilidis G, Balis C. Bioremediation of olive oil mill waste water; chemical alterations induced by *Azotobacter vinelandii*. J Agric Food Chem. 2000;48(5):1941–8.

Polcaro A-M, Mascia M, Palmas S, Vacca A. Electrochemical oxidation of p-hydroxybenzoic, and protocatechuic acids at a dimensional stable anode (DSA) in the presence of NaCl. Ann Chim. 2002;92(10):1015–23.

Pompei C, Codovilli F. Risultati preliminari sul trattamento di depurazione delle acque di vegetazione delle olive per osmosi inversa. Scienza e Tecnologica degli Alimenti. 1974;4(IV6):363–4.

Potoglou D, Kouzeli-Katsiri A, Haralambopoulos D. Solar distillation of olive mill wastewater. Renew Energ. 2003;29(4):569–79.

Pozo C, Martinez-Toledo M-V, Rodelas B, Gonzalez-Lopez J. Effects of culture conditions on the production of polyhydroxyalkanoates by *Azotobacter chroococcum* H23 in media containing a high concentration of alpechín (wastewater from olive oil mills) as primary carbon source. J Biotechnol. 2002;97(2):125–31.

Ragazzi E, Veronese G. Indagini sui componenti fenolici degli oli di oliva. Rivista Italiana Delle Sostanze Grasse. 1982;58(LVIII):443–52.

Ranalli A. Microbial treatment of oil mill waste waters. Grasas Acetes. 1992;43:16–21.

Reis M-T-A, Freitas O-M-F, de Ferreira L-M, Carvalho J-M-R. Extraction of 2-(4-hydroxyphenyl) ethanol from aqueous solution by emulsion liquid membranes. J Membrane Sci. 2006; 269:161–70.

Reverberi M, Di Mario F, Tomati U. β-Glucan synthase induction in mushrooms grown on olive mill wastewaters. Appl Microbiol Biotechnol. 2004;66(2):217–25.

Rice-Evans A-C, Miller J-N, Paganda G. Antioxidant properties of phenolic compounds. Trends Plant Sci. 1997;2(4):152–9.

Rinaldi M, Rana G, Introna M. Olive-mill wastewater spreading in southern Italy: effects on a durum wheat crop. Field Crops Research. 2003;84:319–26.

Rivas F-J, Beltrán F-J, Gimeno O, Frades J. Treatment of olive oil mill wastewater by Fenton's reagent. J Agric Food Chem. 2001a;49:1873–80.

Rivas F-J, Gimeno O, Portela J-R, de la Ossa E-M, Beltrán F-J. Supercritical water oxidation of olive oil mill wastewater. Ind Eng Chem Res. 2001b;40(16):3670–4.

Rivas F-J, Frades J, Alonso M-A, Montoya C, Monteagudo J-M. Fenton's oxidation of food processing wastewater components. Kinetic modeling of protocatechuic acid degradation. J Agric Food Chem. 2005;53(26):10097–104.

Rodis PS, Karathanos VT, Mantzavinou A. Partitioning of olive oil antioxidants between oil and water phases. J Agric Food Chem. 2002;50(3):596–601.

Rodríguez Pérez S, García Oduardo N, Bermúdez Savón R-C, Fernández Boizán M, Augur C. Decolourisation of mushroom farm wastewater by *Pleurotus ostreatus*. Biodegradation. 2008;19(4):519–26.

Roig A, Cayuela M-L, Sánchez-Monedero M-A. The use of elemental sulphur as organic alternative to control pH during composting of olive mill wastes. Chemosphere. 2004;57(9): 1099–105.

Rosa M-F, Vieira A-M. Perspectivas e limitac̣ões no tratamento e utilizac̣ão das águas residuais de lagares de azeite: situac̣ão portuguesa, Boletim de Biotechnologia, (52), Dec 1995.

Rovatti M, Bisi M, Ferraiolo G. High added value products from difficult wastes. Resour Conserv Recy. 1992;7(4):271–83.

Rozzi A, Malpei F. Treatment and disposal of olive mill effluents. Proceedings of Olive Oil Processes and By-products Recycling, 1995 Sep 10–13; Granada, Spain. Int Biodeter Biodegr. 1996;38(3–4):135–44.

Rozzi A, Limoni N, Menegatti S, Boari G, Liberti L, Passino R. Influence of sodium and calcium alkalinity on UASB treatment of olive mill effluents, part 1: preliminary results. Process Biochem. 1988;23(3):86–90.

Rozzi A, Di Pinto A-C, Limoni N, Tomei M-C. Start-up and operation of anaerobic digesters with automatic bicarbonate control. Bioresour Technol. 1994;48(3):215–9.

Rubia T, de la González-López J, Martínez M-V, Ramos-Cormenzana A. Flavonoids and the biological activity of *Azotobacter vinelandii*. Soil Biol Biochem. 1987;19:223–4.

Rubia T, de la Lucas M, Martínez J. Controversial role of fungal laccases in decreasing the antibacterial effect of olive mill waste-waters. Bioresour Technol. 2008;99(5):1018–25.

Russo C. A new membrane process for the selective fractionation and total recovery of polyphenols, water and organic substances from vegetation waters (VW). J Membrane Sci. 2007;288(1–2):239–46.

Sabbah I, Marsook T, Basheer S. The effect of pretreatment on anaerobic activity of olive mill wastewater using batch and continuous systems. Process Biochem. 2004;39(12):1947–51.

Sainz H, Benítez E, Melgar R, Alvarez R, Gómez M, Nogales R. Biotransformación y valorización agrícola de subproductos del olivar -orujos secos y extractados- mediante vermicompostaje. 2000;7(2):103–11.

Sánchez-Villasclaras S, Martínez Sancho M-E, Espejo Caballero M-T, Delgado Pérez A. Production of microalgae from olive mill wastewater. Proceedings of Olive Oil Processes and By Products Recycling; 1995 Sep 10–13; Granada, Spain. Int Biodeter Biodegr. 1996;38(3–4):245–47.

Sanjust E, Pompei R, Rescigno A, Rinaldi A, Ballero M. Olive milling wastewaters as a medium for growth of four *Pleurotus* species. Appl Biochem Biotechnol. 1991;31(3):223–35.

Santori F, Cicalini A-R. Process of olive-mill waste water phytodepuration and relative plant. EP1216963 A. 26 Jun 2002.

Sarika R, Kalogerakis N, Mantzavinos D. Treatment of olive mill effluents Part II. Complete removal of solids by direct flocculation with poly-electrolytes. Environ Int. 2005;31:297–304.

Saroj D-P, Kumar A, Bose P, Tare V, Dhopavkar Y. Mineralization of some natural refractory organic compounds by degradation and ozonation. Water Res. 2005;39:1921–33.

Sayadi S, Ellouz R. Decolourization of olive mill wastewater by the white rot fungus *Phanerochaete chrysosporium*: involvement of the lignin-degrading system. Appl Microbiol Biotechnol. 1992;37:813–7.

Sayadi S, Ellouz R. Roles of lignin peroxidase and manganese peroxidase from *Phanerochaete chrysosporium* in the decolorization of olive mill wastewaters. Appl Environ Microbiol. 1995;61(3):1098–103.

Sciancalepore V, Pizzuto P, Stefano G, de Piacquadio P, Sciancalepore R. Compostaggio della sansa di olive dei nuovi impianti a due fasi. Rifiuti solidi. 1994;8(VIII):6.

Sciancalepore V, Stefano G, de Piacquadio P, Sciancalepore R. Compostaggio in ambiente protetto del residuo della lavorazione delle olive con impianti ad estrazione bifasica. Ingegneria Ambientale. 1995;24(XXIV):11–2.

Sciancalepore V, Colangelo M, Sorlini C, Ranalli G. Composting of effluent from a new two-phase centrifuge olive mill. Microbial characterization of the compost. Toxicol Environ Chem. 1996;55(1–4):145–58.

Scioli C, Vallaro L. The use of Yarrowia lipolytica to reducepollution in olive mill wastewaters. Water Res. 1997;31:2520–4.

Sellami F, Hachicha S, Chtourou M, Medhioub K, Ammar E. Maturity assessment of composted olive mill wastes using UV spectra and humification parameters. Bioresour Technol. 2008a; 99(15):6900–7.

Sellami F, Jarboui R, Hachicha S, Medhioub K, Ammar E. Co-composting of oil exhausted olive-cake, poultry manure and industrial residues of agro-food activity for soil amendment. Bioresour Technol. 2008b;99:1177–88.

Setti L, Maly S, Iacondini A, Spinozzi G, Pifferi P-G. Biological treatment of olive milling waste waters by *Pleurotus ostreatus*. Annali di Chimica (Rome). 1998;88(3–4):201–10.

Shabou R, Zairi M, Kallel A, Aydi A, Dhia H-B. Assessing the effect of an olive mill wastewater evaporation pond in Sousse, Tunisia, Environmental Geology, Article in Press; 2008. p. 1–8.

Shabtai Y, Daya-Mishne N. Production, purification and properties of a lipase from a bacterium *(Pseudomonas aeruginosa* YS-7) capable of growing in water restricted environments. Appl Environ Microbiol. 1992;58:174–80.

Shammas N-K. Olive extraction waste treatment in Lebanon. Effluent Water Treat J. 1984;24(10):388–9, 391–392.

Sierra J, Marti E, Montserrat G, Cruanas R, Garau M-A. Characterisation and evolution of a soil affected by olive oil mill wastewater disposal. Sci Total Environ. 2001;279:207–14.

Siggelakis C. In: Elaiodentro: ellinikes kai ksenes poikilies. 6th edn. Elaiourgiki. 1982. p.34.

Siskos D. Wastewater treatment plant for olive oil processing effluents comprising a rotating biological contactor with the addition of linear or circular motion. WO9935097 A. 15 Jul 1999.

Skerratt G, Ammar E. The application of reedbed treatment technology to the treatment of effluents from olive oil mills. Final report. Country/project number Tunisia 066599003ZH010; 1999.

Smeltzer M, Hart M, Iandolo J. Quantative spectrophotometric assay of a novel thermostable lipase. Appl Environ Microbiol. 1992;58:2815–9.

Stamatelatou K, Kopsahelis A, Blika P-S, Paraskeva C, Lyberatos G. Anaerobic digestion of olive mill wastewater in a periodic anaerobic baffled reactor (PABR) followed by further effluent

purification via membrane separation technologies. J Chem Technol Biotechnol. 2009;84: 909–17.
Sutherland I-W. Microbial biopolymers from agricultural products: production and potential. Int Biodeter Biodegr. 1996;38:249–61.
Tan Y-Y-J, Hashim M-A, Ramachandran K-B. Biomass acclimatization to sequentially varying substrates in an upflow anaerobic sludge blanket (UASB) bioreactor. Water Qual Res J Can. 2006;41(4):437–48.
Tomati U, Galli E, Fiorelli F, Pasetti L. Fertilizers from composting of olive-mill wastewaters. Int Biodeter Biodegr. 1996;38(3–4):155–62.
Tomati U, Belardinelli M, Galli E, Iori V, Capitani D, Mannina L, et al. NMR characterization of the polysaccharidic fraction from *Lentinula edodes* grown on olive mill waste waters. Carbohydr Res. 2004;339(6):1129–34.
Tsonis S-P. Start up mode of seasonally operating units for the anaerobic digestion of olive oil mill wastewater. In: Bhaskar N, editor. Proceedings of international conference on environmental pollution ICEP-1; Lisbon, Portugal, Apr 1991, 2, p.733–740. Interscience Enterprises; 1991.
Tsonis SP. Olive oil mill wastewater as carbon source in post anoxic denitrification. Water Sci Technol. 1997;36(2–3):53–60.
Tsonis SP. Tolerable amounts of olive oil mill wastewater in sewerage treatment plants. Fourth International Symposium on Waste Management Problems in Agro-Industries, Istanbul-Turkey, 23–25 Sep 1998.vol 2 p. 189–96.
Tsonis S-P, Grigoropoulos S-G. Anaerobic treatability of olive oil mill wastewater. Water Sci Technol. 1993;28(2):35–44.
Tsonis S-P, Tsola V-P, Grigoropoulos S-G. Systematic characterization and chemical treatment of olive oil mill wastewater. Procceedings of the 4th International Congress on Environmental Pollution and its Impact on Life in the Mediterranean Region, Kavala, Greece, 6–11 September 1987 Sep 6–11; Kavala, Greece. Toxicol Environ Chem. 1989;20(1):437–57.
Turano E, Curcio S, Paola M-G, de Calabró V, Iorio G. An integrated centrifugation-ultrafiltration system in the treatment of olive mill wastewater. J Membrane Sci. 2002;109:519–31.
Tziotzios G, Michailakis S, Vayenas D-V. Aerobic biological treatment of olive mill wastewater by olive pulp bacteria. Int Biodeter Biodegr. 2007;60:209–14.
Ubay G, Ozturk I. Anaerobic treatment of olive mill effluents. In: Proceedings of the 2nd IAWQ International Conference on Pretreatment of Industrial Wastewaters; 1996 Oct 16–18; Athens, Greece. Water Sci Technol. 1997;36(2–3):287–94.
Uyanik S. Process performance and bacterial population dynamics in conventional and modified anaerobic baffled reactors treating industrial wastewaters. Ph.D. thesis, University of Newcastle; 2001.
Velioğlu S-G, Curi K, Çamillar S-R. Laboratory experiments on the physical treatment of olive oil wastewater. Int J Dev Technol. 1987;5(1):49–57.
Vigo F, Giordani M, Capannelli G. Ultrafitrazione di acque di vegetazione da frantoi di olive. Rivista Italiana Delle Sostanze Grasse. 1981;58(LVIII):70–3.
Vigo F, Avalle L, Paz M. Smaltimento delle acque di vegetazione provenienti da frantoi di olive. Studio del processo di ossidazione electrochimica. Rivista Italiana Delle Sostanze Grasse. 1983a;60(LX):125–31.
Vigo F, de Paz M, Avalle L. Ultrafitrazione di acque di vegetazione da frantoi di olive. Esperineza gestionale in impianto semi-pilota. Rivista Italiana Delle Sostanze Grasse. 1983b;60(LX):267–72.
Vinciguerra V, D'Annibale A, Delle Monache G, Sermanni G-G. Degradation and biotransformation of phenolic compounds of olive waters by the white rot basidiomycete *Lentinus edodes*. Med Fac Landbouw (Univ Gent). 1993;58(4):1811–4.
Vinciguerra V, D'Annibale A, Delle Monache G, Sermanni G-G. Correlated effects during the bioconversion of waste olive water by *Lentinus edodes*. Bioresour Technol. 1995;51:221–6.
Visioli F, Galli C. Biological properties of olive oil phytochemicals. Crit Rev Food Sci Nutr. 2002;42(3):209–21.

Vitolo S, Petarca L, Bresci B. Treatment of olive oil industry wastes. Bioresour Technol. 1999;67(2):129–37.

Vlyssides A-G, Loizidou M, Zorpas A. Characteristics of solid residues from olive oil processing as bulking material for co-composting with industrial wastewaters. J Environ Sci Heal A. 1999;34(3):737–48.

Wang T-S-C, Yang T-K, Chuang T-T. Soil phenolic acids as plant growth inhibitors. Soil Sci. 1967;103:239–46.

Widsten P, Kandelbauer A. Laccase applications in the forest products industry: a review. Enzyme Microb Technol. 2008;42(4):293–307.

WWF. EU policies for olive farming. Unsustainable on all counts, WWF and BirdLife Joint Report; 2001. p. 16.

Yacob S, Shirai Y, Hassan M-A, Wakisaka M, Subash S. Start-up operation of semi-commercial closed anaerobic digester for palm oil mill effluent treatment. Process Biochem. 2005;41:962–4.

Yesilada O, Fiskin K. Degradation of olive mill waste by *Coriolus versicolor*, Turkish. J Biol. 1996;20(1):73–9.

Yesilada O, Fiskin K, Yesilada E. The use of white rot fungus Funallia trogii (Malatya) for the decolorization and phenol removal from olive mill wastewater. Environ Technol. 1995;16:95–100.

Yesilada O, Sik S, Sam M. Biodegradation of olive oil mill wastewater by *Coriolus versicolor* and *Funalia trogii*: effects of agitation, initial COD concentration, inoculum size and immobilization. World J Microb Biot. 1998;14(1):37–42.

Zervakis G, Balis C. Bioremediation of olive oil mill wastes through the production of fungal biomass. In: Royse D-J, editor. Proceedings of the 2nd International conference on mushroom biology and mushroom products. University Park: Pennsylvania State University, College of Agricultural Sciences; 1996. p. 311–23.

Zervakis G, Yiatras P, Balis C. Edible mushrooms from olive oil mill wastes. Proceedings of Olive Oil Processes and By-Products Recycling; 1995 Sep 10–13; Granada, Spain. Int Biodeter Biodegr. 1996;38(3–4):237–43.

Zhang X, Liu J-H, Lu T-S. Industrial application of liquid membrane separation for phenolic wastewater treatment. Water Treat. 1987;2:127–35.

Zouari N. Decolorization of olive oil mill effluent by physical and chemical treatment prior to anaerobic digestion. J Chem Technol Biotechnol. 1998;73(3):297–303.

Part II
Safety and Quality Considerations

Chapter 5
Safety Considerations of Nutraceuticals and Functional Foods

Semih Otles and Ozlem Cagindi

5.1 Introduction

"Let food be the medicine and medicine be the food" was embraced 2,500 years ago by Hippocrates, the father of medicine, and therefore the conception that foods might provide therapeutic benefits is not new. Recent knowledge, however, supports the hypothesis that, besides fulfilling nutrition needs, diet may modulate various functions in the body and may exhibit detrimental or beneficial roles in some diseases (Hasler 2002).

Functional foods are one of the fastest-growing segments of the food industry. Current interest by consumers in taking a proactive role in maintaining their health has produced a profusion of products on the market that contain dietary ingredients intended to produce a positive effect on the health of the consumer. These ingredients are included in food products marketed in the form of both dietary supplements and "functional foods" or "nutraceuticals".

The term functional food is used to describe nutrients that exhibit an effect on physiologic processes besides their established nutritional function. Functional foods exist at the interface between food and drugs, and therefore offer great potential for health improvement and prevention of diseases when ingested as part of a balanced diet (Hilliam 2000).

The discovery, development and marketing of these nutraceutical products is currently one of the fastest growing segments of the food industry (Hardy 2000).

S. Otles (✉)
Department of Food Engineering, Ege University
Bornova, Turkey
e-mail: semih.otles@ege.edu.tr

O. Cagindi
Department of Food Engineering, Celal Bayar University, Manisa, Turkey
e-mail: ozlem.cagindi@bayar.edu.tr

This trend is driven by the current consumer perception that "natural is good" and also, to some extent, by other factors such as the increasing cost of many pharmaceuticals (Zink 1997). Due to the rapid expansion in this area an imbalance exists between the increasing number of claims and products on the one hand, and the development of policies to regulate their application and safety on the other (Hasler et al. 2001).

Functional foods and nutraceutical ingredients are defined legally as natural substances that may be used individually, in combination, or even added to food or beverage for a particular technologic purpose or health benefits and must have an adequate safety profile that demonstrates safety for consumption by humans. Risk of toxicity or adverse effects of medical drugs led to the search for safer nutraceutical and functional food based approaches for health management. This resulted in a world-wide nutraceutical revolution. The option of health and disease management by natural means has been embraced by a significant portion of the world population. Rapid growth in research on nutraceuticals and functional foods is an integral and vital component of the revolution. Proof of efficacy and safety are two key sets of information required for the successful use of nutraceuticals and functional foods in the management of human health and well-being. The inventory of modern and sophisticated technologies available to derive such information is rapidly expanding in this age of technology driven health care. It is appropriate to revisit nutraceuticals and functional food research in light of these emerging technologies. This chapter will focus on the safety perspectives of nutraceutical and functional foods world.

5.2 What Are Functional Foods?

Japan is the birthplace of the term 'functional food' (Kubomara 1998). Japan has been at the forefront of the development of functional foods since the early 1980s when systematic and large-scale research programs were launched and funded by the Japanese government on systematic analysis and development of food functions, analysis of physiological regulation of function by food and analysis of functional foods and molecular design. As a result of a long decision making process, to establish a category of foods for potential enhancing benefits as part of a national effort to reduce the escalating cost of health care, the concept of foods for specified health use (FOSHU) was established in 1991.These foods, which are intended to be used to improve people's health and for which specific health effects or claims are allowed to be displayed, are included as one of the categories of foods described in the Nutrition Improvement Law as foods for special dietary use. According to the Japanese Ministry of Health and Welfare, FOSHU are:

- Foods that are expected to have a specific health effect due to relevant constituents, or foods from which allergens have been removed, and
- Foods where the effect of such an addition or removal has been evaluated scientifically and permission to make claims regarding the specific beneficial effects on health expected from their consumption has been granted.

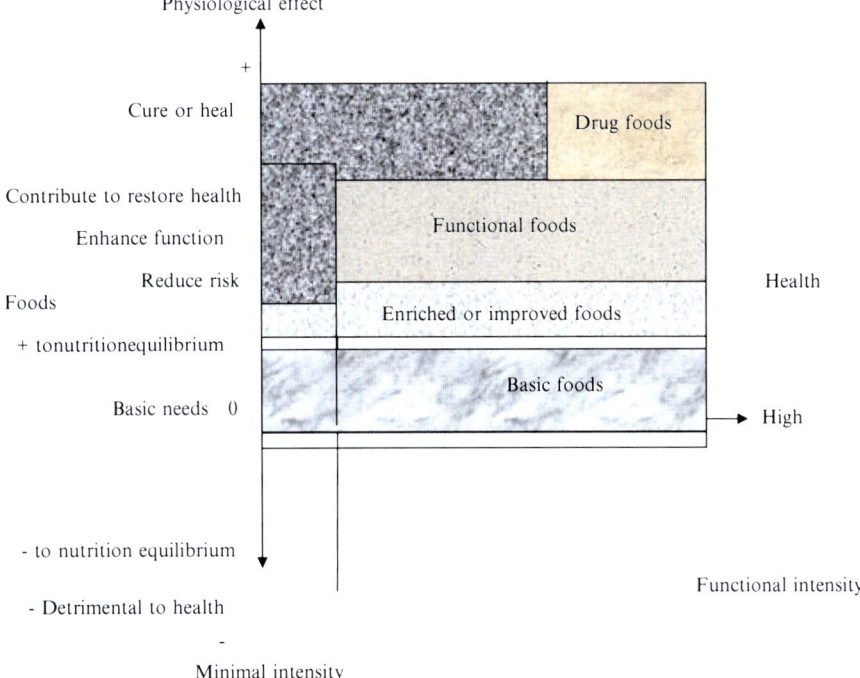

Fig. 5.1 The frontier of the functional food universe

Foods identified as FOSHU are required to provide evidence that the final food product, but not isolated individual component or components, is likely to exert a health or physiological effect when consumed as part of an ordinary diet. Finally, FOSHU products should be in the form of ordinary foods (i.e. not pills or capsules).

In the meantime, but mainly in the 1990s, a variety of terms, more or less related to the Japanese FOSHU, have appeared worldwide. In addition to functional foods, these include more terms such as 'nutraceuticals', 'designer foods', 'pharmafoods', 'medifoods', 'vitafoods', etc., but also the more traditional 'dietary supplements' and 'fortified foods'.

"Functional foods," "nutraceuticals," "pharmaconutrients," and "dietary integrators" are all terms used incorrectly and indiscriminately for nutrients or nutrient-enriched foods that can prevent or treat diseases.

In an effort to distinguish between functional or medical foods and drugs, the term "nutraceutical" was coined in 1989 by the Foundation for Innovation in Medicine (FIM 1992) to cover "any substance that may be considered a food or part of a food, and provides medical or health benefits, including the prevention and treatment of disease." The frontier of the functional food universe is given in Fig. 5.1 (Doyon and Labrecque 2008). Nutraceuticals are clearly not drugs, which are pharmacologically active substances that will potentiate, antagonize, or otherwise modify any physiological or metabolic function. In 1996 the Foundation for

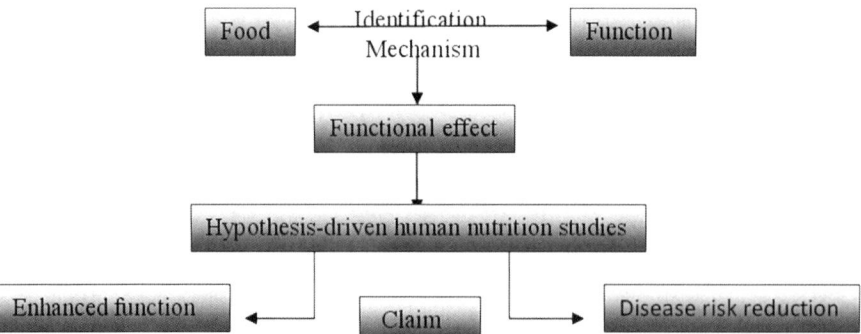

Fig. 5.2 The strategy for functional food development (Roberfroid 2000)

Innovation in Medicine (FIM) coined the term "nutraceutical" as a food or parts of foods that provide medical-health benefits including the prevention and/or treatment of disease. Such products may range from isolated nutrients, dietary supplements and diets to genetically engineered designer foods, functional foods, herbal products and processed foods such as cereals, soups and beverages (De Felice 1996).

The scope of nutraceuticals is significantly different from functional foods for several reasons. These include:

- Whereas the prevention and treatment of disease (i.e., medical claims) are relevant to nutraceuticals, only reduction of disease, not the prevention and treatment of disease, is involved with functional foods.
- Whereas nutraceuticals include dietary supplement as well as other type of foods, functional foods should be in the form of ordinary food.

Functional foods have been developed in virtually all food categories. The strategy for functional food development is given in Fig. 5.2. (Roberfroid 2000). According to alternative classification systems some functional products :

1. "Add well to your life", e.g., improve the regular stomach and colon functions (pre- and probiotics) or "improve children's life" by supporting their learning capability and behavior. It is difficult, however to find good biomarkers for cognitive, behavioral and psychological functions.
2. Functional food is designed for reducing an existing health risk problem such as high cholesterol or high blood pressure.
3. Consists of those products, which "makes your life easier" (e.g., lactose-free, gluten-free products) (Mäkinen-Aakula 2006).

Consumers' interest for functional foods Major reasons for increased interest in functional foods are increased health care costs, recent legislation, and scientific discoveries (Milner 2000). Growing inclination towards functional foods has been noted in the past decade and several factors responsible for such consumer's awareness are delineated underneath:

Health-conscious baby-boomers and the "self-care" movement

- Staggering healthcare costs associated with chronic diseases of aging
- Advances in technology, particularly nutritional genomics
- Changes in food regulations
- Market opportunity
- Development of new scientific findings linking foods and/or food components to optimal health (Hasler 1996)
- New food processing, retailing, and distribution Technologies
- Changing consumer demands and social attitudes
- Scientific evidence of health benefits of certain ingredients and
- Search for new opportunities to add value to existing products and to increase profits (Stanton et al. 2001).

5.3 Safety of Nutraceuticals and Functional Foods

Food safety is receiving more attention than ever before by governments and policy makers, health professionals, the food industry, the biomedical community, and last but not least, the public (Crutchfield and Roberts 2000; Crutchfield and Weimer, 2000; Kaferstein and Abdussalam 1999; Woteki et al. 2001). For most consumers in the United States (Food Marketing Institute 2000) and Europe (Food Marketing Institute 1995), safety has become one of the most important attributes of food. Their concern over food quality has intensified in recent years, and prompted heated debate about the integrity and safety of produce. Consumer concern, fuelled by several food scares, the political arena, international trade, and the farming industry has influenced food purchasing patterns (Buzby 2001). One such aspect has been the expansion of demand for organically grown food.

Functional foods have been expected to play an important role in modern nutrition, which is expected to promote health and reduce the risk of chronic diseases. Functional foods must fulfill all standards of food safety assessment. However, for this type of food, the concept of benefits versus risk of long-term intake has to be investigated further, developed and validated. The safety of intake of low or high amounts of nutrients and high amounts of non-nutrients related to long-term consumption of functional food, as well as interactions between food components and biological processes have to be characterized. Protocols for pre-marketing nutrition studies on functional food and post-marketing monitoring are needed. Taking into account the safety aspects, i.e. the possibility of increased contamination, adverse metabolic effects, excessive stimulation of immunological systems and potential gene transfer. The manufacturer of probiotics is obliged to present clinical documentation showing the absence of these (previously mentioned) effects. In particular, probiotic bacteria must not significantly increase the risk of transfer of antibiotic resistance and cannot possess hemolytic activity (Nowicka and Naruszewicz 2004).

The requirements of functional food safety indicated by FAO/WHO impose on the manufacturers the obligation to conduct placebo-controlled clinical studies and to evaluate their results in four phases:

1. Safety
2. Efficiency
3. Effectiveness
4. Surveillance.

The safety of functional foods still requires thorough investigation. The novelty of raw materials and processes employed, as well as the potential use in the diet, has yet to be examined. Safety is guaranteed by the application of various regulations control any novelty in food components and the respective manufacturing processes. The test schemes selected to evaluate safety will depend on the type of food novelty and whether or not it can be considered to be functional,, the evaluation of its safety will be dictated by the specific legislation that controls the particular material (EC 1997). To assess a functional food's safety, a dietitian, nutritionist, food science specialist or other health professionals must be able to answer several specific questions (Hasler et al. 2001):

- What is (are) the functional ingredient(s)?
- How much of the ingredient is present per serving?
- What is a typical serving?
- How often is the functional food eaten?
- Has the manufacturer conducted safety studies on the ingredient(s) (in animals and humans)? Is published peer reviewed research on the ingredient(s) available?
- Will the manufacturer provide you safety information or background on published research?
- Does the functional ingredient interact with prescription drugs?

A functional food's efficacy must be able to answer several specific questions:

- What is the scientific evidence that the functional food is effective?
- Are there measurable health benefits from the functional ingredient?
- Have well-designed and controlled human clinical intervention trials been conducted?
- Are studies published in peer-reviewed journals?
- If it has been found effective, does a single serving provide enough of the functional ingredient to make a difference?

Clinical data supporting benefits of functional foods are even less abundant than for herbs. There are, however, 12 health claims that have been approved by the FDA as a result of a body of scientific evidence showing significant agreement to back up the claim. In order to use the health claim, the product must provide a specified minimum amount of the functional ingredient. Other functional ingredients are present in amounts proven effective, but several servings a day must be consumed to achieve an effective dose. While research may strongly suggest a benefit for a specific functional food or ingredient, it's important to distinguish between research using a whole food and that utilizing an isolated food ingredient (Hasler et al. 2001).

5 Safety Considerations of Nutraceuticals and Functional Foods

The social, food, nutrition and biomedical sciences allow for increasingly rational product development to fit consumers' health needs. The development of new food products brings with it elements of the unknown and therefore risk. However, when it comes to foodstuffs and beverages, it is expected that the risk is negligible. Monitoring and surveillance is a requisite of any community that is exposed to products launched for purported physiological or health reasons to optimize the risk–benefit relationships of functional foods with definable health properties, the following approach has been recommended by a Joint WHO/FAO Working Group on Novel Foods in Nutrition Health and Development: Benefits, Risks and Communication, and published in the Metung Report (Clugston et al. 2002).

1. Consider the health outcome in question
2. Select a plant food or foods, which confer these characteristics, preferably with an established food cultural base
3. Formulate a food for trial
4. Carry out a risk evaluation
5. Conduct a food trial using biomarkers and/or health outcomes
6. Develop an appropriate monitoring and surveillance strategy
7. Seek regulatory approach as novel food for safety
8. Formulate a food-based educational and informational framework, with or without health claims.

The General Accounting Office (GAO 2000) has made the following recommendations regarding the safety of functional foods:

- Develop and promulgate regulations or other guidance for industry on the evidence needed to document the safety of new dietary ingredients in dietary supplements;
- Develop and promulgate regulations or other guidance for industry on the safety related information required on labels for dietary supplements and functional foods; and
- Develop an enhanced system to record and analyze reports of health problems associated with functional foods and dietary supplements.

Functional food ingredients may impart health-promoting effects other than general nutrition. Compared to dietary supplement ingredients, functional foods are subject to unique regulatory requirements and require different safety criteria. Under the Dietary Supplement Health and Education Act (DSHEA) of 1994, newly marketed ingredients after that date must demonstrate safety to a "reasonable expectation of no harm [emphasis added]" as part of a new dietary ingredient notification to the Food and Drug Administration (FDA) (US CFR 190.6 2008). Although a functional food ingredient may have been marketed previously as a dietary supplement ingredient, its safety data set has been augmented and made available to regulators and external expert panels to demonstrate compliance with the standard of "reasonable certainty of no harm [emphasis added]" (US CFR 170.3(i); Griffiths 2007; Griffiths and Teske 2007). A functional food ingredient differs from many Food Chemicals Codex (FCC) ingredients, because FCC originally contained chemicals added to foods to achieve a desired technical function, e.g., as processing aids (USP 2008).

Firstly, the optimal levels of the majority of the biologically active components currently under investigation have yet to be determined. Furthermore, a number of animal studies show that some of the same phytochemicals (e.g., allyl isothiocyanate), highlighted for their cancer-preventing properties, have been shown to be carcinogenic at high concentrations. The doctrine that "All substances are poisons ,, the right dose differentiates a poison from a remedy" is even closer to reality now, given the proclivity for dietary supplements (Arvanitoyannis and Houwelingen-Koukaliaroglou 2005).

The benefits and risks to individuals and populations must be weighed carefully when considering the widespread use of physiologically-active functional foods. For example, the risks of recommending the increased intake of compounds (e.g., isoflavones) that may modulate estrogens' metabolism Have to be characterized. Soy phytoestrogens may represent a "double-edged sword" because of reports that genistein may actually promote certain types of tumors in animals (Rao et al. 1997). The knowledge of the toxicity of functional food components is crucial to decrease the *risk: benefit* ratio. Other substances for which safety concerns have been raised are comfrey, chaparral, lobelia, germander, aristolochia, ephedra (ma huang), L-tryptophan, germanium, magnolia-stephania, and stimulant laxative ingredients, such as those found in dieter's teas. The herb comfrey, for example, contains certain alkaloids that can cause serious liver damage. Consumers should not take any product containing comfrey either orally or as a suppository and should not apply comfrey products to wounded skin (www.ftc.gov/bcp/conline/pubs/health/frdheal.htm). Even some vitamins and minerals, when consumed in excessive quantities, can cause problems. For example, high intakes of vitamin an over a long period can reduce bone mineral density, cause birth defects, and lead to liver damage, according to the National Academy of Sciences (Arvanitoyannis and Houwelingen-Koukaliaroglou 2005).

The claimed benefits attributed to a product should be justified by reference to appropriate studies. Consuming white bread to which fish oils are added is not the same as eating oily fish, and health benefits associated with oily fish consumption should not be mistaken with other products. Likewise, giving the impression that omega-3 fatty acid-fortified white bread can adequately substitute for oily fish consumption, or that fiber-enriched soft drinks can substitute a diet rich in naturally-occurring dietary fiber, is rather confusing and undermines attempts to encourage the consumption of oily fish, fruits, vegetables, and wholegrain foods. Indeed, if the product is of poor nutritional quality, then supplementing it with a functional ingredient (e.g., fiber in soft drinks) only serves to reinforce poor dietary habits. Besides, the addition or promotion of high contents of vitamin A in foods (e.g., as one of the so-called 'ACE' vitamins) without mentioning the danger of abnormalities in pregnant females consuming high amounts of this nutrient is extremely hazardous. Care is also required to avoid excessive intakes of other components (e.g., iron, vitamin D, and folic acid), as a continuously growing number of products get enriched (Arvanitoyannis and Houwelingen-Koukaliaroglou 2005).

Health-conscious consumers are increasingly seeking functional foods in an effort to control and improve their own health and well-being. The field of functional foods, however, is in its infancy. Claims about health benefits of functional foods must be based on sound scientific criteria (Clydesdale 1997). A number of factors, such as the complexity of food substances, effects on food, compensatory metabolic changes that may occur with dietary changes, and lack of surrogate markers of disease development, complicate considerably the establishment of a strong scientific foundation. Further research is required to substantiate the potential health benefits of those foods for which the diet health relationships are not sufficiently scientifically validated. Research into functional foods will not advance public health, unless the benefits of the foods are effectively and clearly communicated to the consumer. Furthermore, foods that have novel health claims that are supported by solid scientific substantiation have the potential to become increasingly important components of a healthy lifestyle. Such foods can be extremely beneficial to the public and the food industry. When health claims are clearly supported by sound scientific evidence, consumers' confidence grows.

Therefore, it is vital to conduct clinical studies on people, not only on animals, in order to guarantee the safety of products and substantiate the functionality of active components (Arvanitoyannis and Houwelingen-Koukaliaroglou 2005).

5.3.1 The Evaluation of Safety

The demonstration of safety for a functional ingredient is guided by several fundamental concepts (Kruger and Mann 2003):

1. Functional ingredients are biologically active and may therefore produce a range of outcomes in the body, at various levels of intake, from suboptimal physiologic action to therapeutic effect to frank toxicity. Understanding the mechanisms for pharmacologic activity and for toxicological potential is important to predict the consequences of exposure at different dose levels.
2. Functional ingredients are a diverse class of compounds and may be represented by single component ingredients, or complex herbal extracts or products derived from novel sources or processes; the compositional analysis for each of these types of products is a critical determinant of the approach to the determination of safety for the ingredient. Unique safety issues are associated with each of these types of products and must be dealt with on a case-by-case basis.
3. The intended use and potential exposure to a functional ingredient must be compared to its determined safe level of ingestion; depending upon the compound, historical exposure and/or scientific studies (animal toxicology, absorption, distribution, metabolism and excretion (ADME), clinical trials) may be used to determine that safe level; the margin of safety between the intended level of ingestion and a potentially toxic level may be very small.
4. Similar to drugs, there is a potential for contraindicated drug and food interactions; these should be determined, if possible, to identify potential safety concerns.

The assurance of safety for functional ingredients is complicated by a potential for adverse effects derived not only from toxicity that can be produced by any impurities present, but also from physiologic activity. In addition, a functional ingredient can be the "active principal" in a complex and uncharacterized mixture. The impurities and/or interactions among chemical constituents of mixtures must also be evaluated for their potential to produce toxicity.

Animal toxicology studies for functional ingredients should be used to determine the toxicological endpoints of concern as necessary (repeat dose toxicity with target organ evaluation; mutagenicity; carcinogenicity; developmental toxicity; reproductive toxicity; immunotoxicity; neurotoxicity, etc.), as well as input appropriate endpoints into the design of clinical trials. The primary evidence of safety, however, must be derived from human clinical trials due to limitations in the animal models themselves with respect to factors such as pharmacokinetic and pharmacologic handling of the drug. The limitation in extrapolating from animal models to humans must be considered when evaluating the safety of functional ingredients and thus, clinical substantiation of safety is critical. It is critical to determine on a case-by-case basis specific elements of the animal study design, such as species, dose level, duration, and route of administration and control groups in order to evaluate whether findings from the animal study can be used to extrapolate to humans (Kruger and Mann 2003).

A strategy for safety determination of a functional ingredient should rely on an understanding of both the exposure and structural analysis of the constituents. The extent of toxicology testing necessary for supplements is defined by both the characterization of components (both active and impurities) as well as the pre-existing data base of safety that can be used to predict, with reasonable certainty, the potential for adverse health effects resulting from ingestion at the dose levels recommended. In compositional terms, the type of product may be divided into the following categories:

1. Conventional, synthetic or extracted single component ingredients;
2. Plant extracts or complex mixtures containing the ingredient; and
3. Products derived from novel sources or processes, e.g., products of fermentation or biotechnology (Kruger and Mann 2003).

Animal models can be used to ascertain the target organs and effects that are produced as a result of toxicity. The following criteria must be met to derive a safe level of exposure without additional toxicology testing:

- Active components and related substances are well characterized and there is an adequate understanding of the lack of potential for toxicity at the human dose levels recommended based upon existing data from the literature.
- Impurities are well characterized and there is an adequate understanding of the lack of potential for toxicity based upon existing data from the literature.
- Manufacturing process is standardized and reproducible (Kruger and Mann 2003).

The primary objective of long-term repeat-dose toxicity studies in animals is to identify the target organs and/or systems for toxicity and the threshold doses for producing toxic effects. Acute (single-dose) studies may not be adequate to support the conclusion that multiple administrations are non-toxic because they were not designed to monitor the usual parameters of toxicity (e.g., clinical pathology and histopathology) or take into consideration the effect of more frequent dosing (Kruger and Mann 2003).

In the development of a compound that is a single molecular entity; pharmacokinetic studies can demonstrate the systemic exposure and can facilitate dose – response comparison between animals and humans.

Reproductive toxicology studies, such as those on fertility/reproductive performance, teratology, and prenatal/ perinatal development in animals, provide information on the potential a botanical drug product has to produce toxicity during the different stages of reproductive and developmental processes.

Carcinogenicity studies may or may not be needed, depending on the duration of intended exposure or any specific cause for concern.

Studies that screen for modes and sites of action of the ingredient are very useful in evaluating pharmacologic activity on organs and/or systems. These studies can provide explanation, where appropriate and necessary, of toxic findings in animal studies.

For functional ingredients, the primary evidence of safety must be derived from human clinical trials due to limitations in the animal models themselves with respect to factors such as pharmacokinetic and pharmacologic handling of the substances. The limitation in extrapolating from animal models to humans must be considered when evaluating the safety of functional ingredients and thus, clinical substantiation of safety is critical. It is important, however, to evaluate the quality and not just the quantity of clinical evidence available to support safety of the functional ingredient (Kruger and Mann 2003).

5.4 Regulation of Nutraceuticals and Functional Foods

Functional foods need to be safe according to all criteria defined in current food regulations. But in many cases, new concepts and new procedures will need to be developed and validated to assess functional food risks. In Europe, some, but certainly not all, functional foods will be classified as 'novel foods' and consequently will require the decision tree assessment regarding safety that is described in the EU Novel Food Regulation (European Commission Novel Food Directive, 97/258/CEE).

It is often quite ambiguous whether a new product should be labeled as food, supplement, or drug. These categories are regulated differently; for new foods and drugs, an FDA approval is always required, whereas, dietary supplements are more lightly regulated. Functional foods exist at the interface between foods and drugs. There is no provision in the existing food regulations for foods intended to be consumed to prevent disease. As the current situation stands, dietary supplements will be marketed under the Dietary Supplement Health and Education Act, and foods will be marketed under the Nutrition Labeling and Education Act. Applicable legislation, regulations, and guidelines is given in Table 5.1, and Compliance, enforcement, and penalties is given in Table 5.2.

Table 5.1 Applicable legislation, regulations, and guidelines (www.scisoc.org/aacc/funcfood)

Jurisdiction	Legislation	Regulations	Guidelines
European Union	Adopted Legislation agreed at EU level may not be implemented within member states (countries), published in "L" series of EU Official Journal. Proposal for a directive on claims has been under review for several years but appears stalled.	Regulation exists (optional implementation by member states) in the form of directives.	NA
United States	Nutrition Labelling and education Act (NLEA), Separate law exists for dietary supplements – Dietary Supplement Health Education Act.	7-page NLEA act accessible through Internet Some allowable label claims under DSHEA remain to be defined.	Three publications exist, in addition to Federal Register. Further advice is available from the FDA Office of Food Labelling and the Office of Special Nutritionals.
Japan	Sale of FOSHU ("functional foods") is governed by the Nutrition Improvement Act which also regulates conduct of a National Nutrition Survey, established a Nutrition Consultation Office and appointment/certification of Nutrition Instructors. Act provides for approval and labelling of Special Nutritive Foods, Enriched Foods and Foods for Special Dietary Uses.	FOSHU (foods for specified health use) exists under regulatory umbrella of "Special Nutritive Foods," enabling manufacturers to establish basis for functional (health benefit) claims.	Japan Health Food and Nutrition Food Association is authorized by Japan's Ministry of Health and Welfare to provide guidance to the food industry in assembling data for formal submission for licensing of new FOSHU to the MHW.
Canada	Primary statute is Food and Drugs Act and Regulations, defining "food" and "drug," prohibitions governing sale of foods.	Extensive regulations to the Food and Drug Act can be amended without reference to Parliament as necessary to prevent deception or to injury to the health of the consumer. However, matters pertaining to health promotion and disease prevention are considered to be beyond the reach of the F and D Act, originally passed for, among other things, food safety and food integrity.	Guide for Food Manufacturers and Advertisers, Guidelines for Nutritional Labelling, Guidelines for Safety Assessment of Novel Foods and others. Requirements also exist for pre-market review of food additives, infant formulae and irradiated foods.

5 Safety Considerations of Nutraceuticals and Functional Foods

Table 5.2 Compliance, enforcement, and penalties (www.scisoc.org/aacc/funcfood)

Jurisdiction	Jurisprudence	Level of enforcement	Penalties
European Union	A body of jurisprudence in many areas is evolving as a result of decisions made by the European Court of Justice.	Adopted legislation is in the form of Commission or Council directives or regulations which have been agreed upon at the Community level and published in the "L" Series of the Official Journal. Such legislation may or may not be implemented in all of the 15 member states. Challenges respecting implementation are ultimately resolved by the European Court of Justice.	Imprisonment and fines
United States	Pertinent court rulings and discussions of constitutional issues concerning the NLEA are contained in a number of Federal Register regulations.	High level of voluntary compliance. Enforcement measures are not clearly declared by FDA, which relies primarily upon field inspections and consumer/industry complaints to identify non-compliant products.	Imprisonment and fines
Japan	NA	Industry self-regulation as government has delegated the determination of standards, licensing of certain products and review of labelling, nutritional and health claims to industry bodies.	NA
Canada	Limited. Most prosecutions have involved the alleged adulteration of food products with prohibited substances and the manufacture of foods under unsanitary conditions. Compositional and identity standards for food products in the Food and Drug Regulations have been enforceable only when the standardized product crosses an interprovincial or International border.	Encouragement of voluntary compliance with regulation and guidelines. Compliance is monitored by periodic field inspection and related product analysis. Enforcement is normally achieved through "voluntary" measures encouraged by federal authorities. Enforcement can be achieved through product recalls and seizures and in rare cases, prosecution under the Food and Drugs Act.	Liability on summary conviction or on conviction upon indictment, for fines and/or imprisonment.

European legislation however, does not consider functional foods as specific food categories, but rather a concept (Coppens et al. 2006; Stanton et al. 2005). Therefore, the rules to be applied are numerous and depend on the nature of the foodstuff. The General Food Law Regulation is applicable to all foods. In addition, legislation on dietetic food genetically modified organism (GMO), food supplements or on novel foods may also be applicable to

Functional foods depend on the nature of the product and on their use. In the EU, rather than regulating the product group per se, legislative efforts currently being developed are directed towards restricting the use of health claims on packages and in marketing (EC 2006; Niva 2007). According to the EU regulation on nutrition and health claims made on foods (EC No. 1924/2006), a list of authorised claims has to be published for all member states, and nutrient profiles also has to be established for foods containing health claims. Health claims can be "function claims" and "reduction of disease risk claims".

To complement the food cGMPs, FDA recently promulgated cGMPs for dietary supplement ingredients (FDA 2008). Unfortunately, at present FDA's cGMPs for foods, functional foods, and dietary supplements specify neither robust analytical methods nor reference standards, both of which are the foundation of public standards in USP compendia. The lack of standards is particularly problematic for ingredients that are derived from herbs and botanicals. Further, food and dietary supplement cGMPs use the term specifications without ever really defining what is meant. In 1938, when FFDCA was enacted, the US did not have a food ingredient compendium.

Therefore, early twentieth century law addressed food adulteration generically, e.g., with statements that food must be "wholesome" and "not ordinarily injurious to health" and would be deemed adulterated if it contained an "added poisonous or deleterious substance, which may render it injurious to health" (FFDCA 402(a)(1) 2008).

In contrast to cGMPs, USP–NF and FCC include detailed monographs that cover the identity and purity of dietary supplements ingredients and food ingredients, respectively. For example,

USP–NF and FCC monographs for botanically based ingredients include requirements governing a broad range of contaminants, including those that arise from microbes, heavy metals, organic volatile impurities, and pesticides. In addition, the compendia rely on General Chapters that specify in detail the types of analytical technologies that should be used to asses product quality, e.g., Identification of Articles of Botanical Origin and a series of chapters devoted to microbial purity in nutritional and dietary supplements, including functional foods (USP–NF 2008). These and other tests and chapters in USP–NF and FCC establish public standards, including specifications (advanced analytical tests, procedures, and acceptance criteria). Acceptance criteria are usually objective numerical decisions that may refer to a comparison to a validated reference material of known purity and known performance by the analytical procedure.

5.5 Conclusion

Current interest by consumers in taking a proactive role in maintaining their health has produced a profusion of products on the market that contain dietary ingredients intended to produce a positive effect on the health of the consumer. These ingredients are included in food products marketed in the form of both dietary supplements and "functional foods" or "nutraceuticals".

Functional foods and nutraceutical ingredients and those legally defined as natural substances that may be used individually, in combination, or even added to food or beverage for a particular technologic purpose or health benefits, must have an adequate safety profile demonstrating the safety for consumption by humans. Risk of toxicity or adverse effects of medical drugs lead us to the consideration of safer nutraceutical and functional food based approaches for health management.

References

Arvanitoyannis I, Houwelingen-Koukaliaroglou M. Functional foods: a survey of health claims, pros and cons, and current legislation. Crit Rev Food Sci Nutr. 2005;45:385–404.

Buzby JC. Effects of food-safety perceptions on food demand and global trade. In: Regmi A, editor. Changing structure of global food consumption and trade. Washington: United States Department of Agriculture (USDA), Economic Research Service; 2001. p. 55–66.

Clugston G, Lupien JR, Savige GS, Winarno FG, Wahlqvist M, Okada A, Editors. 2002. Novel foods in nutrition health and development: benefits, risks and communication. Metung, Australia, 11–14 November 2001, Asia Pacific J Clin Nutr. 2002;Suppl 11:S97–S229.

Clydesdale FM. A proposal for the establishment of scientific criteria for health claims for health claims for functional foods. Nutr Rev. 1997;55:413–22.

Coppens P, Da Silva MF, Pettman S. European regulations on nutraceuticals, dietary supplements and functional foods: a framework based on safety. Toxicology. 2006;221:59–74.

Crutchfield SR, Roberts T. Food safety efforts accelerate in the 1990's. Food Rev. 2000;23:44–9.

Crutchfield SR, Weimer J. Nutrition policy in the 1990s. Food Review, 2000;23:38–43.

De Felice SL. The need for a research-intensive nutraceutical industry: what can congress do? (the claims research connection). In: Shaw S, editor. Functional food, nutraceutical or pharmaceutical? London: IBC; 1996. p. 15–26.

Doyon M, Labrecque J. Functional foods: a conceptual definition. Br Food J. 2008;110(11):1133–49.

EC. Regulation (EC) No. 1924/2006 of the European parliament and of the council of 20 December 2006 on nutrition and health claims made on foods. Off J Eur Union. 2006;12:3–18.

EC. Regulation (EC) No. 258/97 of the European parliament and of the council of 27 January concerning novels foods and novel foods ingredients, Off J Eur Commun. 1997;43:1–7.

European Commission Novel Food Directive, 97/258/CEE.

FDA. Current good manufacturing practice in manufacturing, packaging, labeling, or holding operations for dietary supplements. http://frwebgate3.access.gpo.gov/cgi-bin/waisgate.cgi? (2008) WAISdocID=36094430629+101+0+&WAISaction=retrieve.essed 29 Jul 2008.

US Federal Food, Drug, and Cosmetic Act (FFDCA), Section 402(a)(1). 2008. http://www.fda.gov/opacom/laws/fdcact/fdcact4.htm Accessed 29 Jul 2008.

FIM. The foundation for innovative medicine. The nutraceuticals initiative: a proposal for economic and regulatory reform. Food Tech. 1992;6:77.

Food Marketing Institute. Trends in Europe – consumer attitudes & the supermarket. Washington: Food Marketing Institute (FMI); 1995.

Food Marketing Institute. Trends in the United States – consumer attitudes & the supermarket. Washington: Food Marketing Institute (FMI); 2000.

GAO. Report to congressional committees: food safety: improvements needed in overseeing the safety of dietary supplements and functional foods. Washington: US General Accounting Office; 2000.

Griffiths JC. How safe is safe? Dietary supplements and the reasonable expectation of certainty. Food law drug institute update. 2007;2:22–24.

Griffiths JC, Teske S. Are functional foods GRAS? Asia Food J. 2007;10 (November):28–30.

Hardy G. Nutraceuticals and functional foods: introduction and meaning. Nutrition. 2000;16:688–9.

Hasler CM. Functional foods: the western perspective. Nutr Rev. 1996;54:6–10.

Hasler CM. Functional foods: benefits, concerns and challenges – a position paper from the American council on science and health. J Nutr. 2002;132:3772–81.

Hasler C, Moag-Stahlberg A, Webb D, Hudnall M. How to evaluate the safety, efficacy, and quality of functional foods and their ingredients. J Am Diet Ass. 2001;101:733–6.

Hilliam N. Functional foods: how big is the market? World of Food Ingredients. 2000;12:50–3.

Kaferstein F, Abdussalam M. Food safety in the 21st century. Bull World Health Org. 1999;77:347–51.

Kruger CL, Mann SW. Safety evaluation of functional ingredients. Food Chem Toxicol. 2003;41:793–805.

Kubomara K. Japan redefines functional foods. Prepared Foods. 1998;167:129–32.

Mäkinen-Aakula M. Trends in functional foods dairy market. In: Proceedings of the third functional food net meeting. 2006.

Milner JA. Functional foods: the US perspective. Am J Clin Nutr. 2000;71:1654–9.

Niva M. All foods affect health: understandings of functional foods and healthy eating among health-oriented Finns. Appetite. 2007;48:384–93.

Nowicka G, Naruszewicz M. Assessing health claims for functional foods. In: Arnoldi A, editor. Functional foods, cardiovascular disease and diabetes. England: Woodhouse Publishing in Food Science and Technology, CRS Press; 2004. p. 10–7.

Rao CV, Wang CX, Simi B. Enhancement of experimental colon cancer by genistein. Cancer Res. 1997;57:3717–22.

Roberfroid MB. Defining functional foods. In: Gibson GR, Williams CM, editors. Functional foods: concept to product. Cambridge: Woodhead Publishing Limited; 2000. p. 9–28. ISBN 978-0849308512.

Stanton C, Gardiner G, Meehan H, Collins K, Fitzgerald GP, Lynch B, et al. Market potential for probiotic. Am J Clin Nutr. 2001;73:476–83.

Stanton C, Ross RP, Fitzgerald GF, Van Sinderen D. Fermented functional foods based on probiotics and their biogenic metabolites. Curr Opin Biotechnol. 2005;16:198–203.

US CFR; Section 170.3(i). 2008. http://edocket.access.gpo.gov/cfr_2002/aprqtr/21cfr170.3.htm Accessed 29 Jul 2008.

US CFR; Section 190.6. 2008. http://edocket.access.gpo.gov/cfr_2008/aprqtr/21cfr190.6.htm Accessed 29 Jul 2008.

USP. Food chemicals codex. Rockville: United States Pharmacopeial Convention; 2008.

USP. USP 31–NF 26. Rockville: The United States Pharmacopeial Convention; 2008.

Woteki CE, Facinoli SL, Schor D. Keep food safe to eat: healthful food must be safe as well as nutritious. J Nutr. 2001;131:502–9.

Zink DL. The impact of consumer demands and trends on food processing. Emerg Infect Dis. 1997;3:467–9.

Chapter 6
Consumer Behavior: Determinants and Trends in Novel Food Choice

Mona Elena Popa and Alexandra Popa

6.1 Introduction

In recent years, several food technologies emerged giving rise to new products on the market. These so-called novel foods are based on sound scientific research, but their success on the market depends on how they are perceived by the actors in the food supply chain, and especially by the consumers. The aim of this chapter is to review the determinants and trends in consumer food choice regarding novel food technologies.

A novel food is defined as a food or food ingredient that does not have a significant history of consumption within the EU before 15 May 1997 and falls under the novel food regulation [Regulation (EC) No. 258/97; European Commission 1997]. In this chapter, the novel foods considered are products improved using the following technologies:

1. Biotechnology (or genetic modification, GM) – range of tools which improve domesticated plants, animals, or microbes to enhance their traits with regard to ease or efficiency of production or their end-use qualities and characteristics
2. Voluntary food fortification – addition of food substances that provide health benefits beyond their nutritional value, transforming these products into so-called functional foods
3. Ultrahigh pressure (UHP) – physical process that inactivates all vegetative spore forms of microorganisms if the pressure applied is high enough
4. Pulsed electric field (PEF) – nonthermal food preservation technology that is based on the use of electric fields to eradicate food borne pathogens and to control spoilage microorganisms in food

M.E. Popa (✉) • A. Popa
University of Agronomic Sciences and Veterinary Medicine Bucharest,
Bucharest, Romania
e-mail: monapopa@agral.usamv.ro

5. Food irradiation – the process by which ionizing radiation at controlled levels is used to treat food
6. Ohmic treatment – an advanced thermal processing method wherein the food material, which serves as an electrical resistor, is heated by passing electricity through it
7. Nanotechnology – technology that works at the atomic or molecular level of food packages and even food products

The above-mentioned technologies were driven by consumer needs for convenience, freshness, safety, and more natural products with fewer additives and preservatives, and minimally processed (Hendrickx et al. 2002). Nonetheless, they have faced different levels of consumer acceptance over time and across nations. Thus, food policies dealing with novel foods should be dynamic and adapted to the domestic reality.

6.2 Consumer Behavior and Food Choice

In this section, some notions regarding consumer sciences will be presented. The importance of consumer behavior in food choice is highlighted with respect to new product buying and the main determinants of novel food choice.

6.2.1 New Product Buying

Generally, consumers consider that traditional foods (that have often been eaten for thousands of years) are safe (World Health Organization 2002). When new foods are developed, they are received with reticence by the general public. People are often suspicious of these innovations (Evans et al. 2006), especially radical ones such as the case of genetically modified foods. The adoption process takes time and depends on the social system of a particular country or region. Nevertheless, McCarthy (1977) cited in Evans et al. (2006) summarized the characteristics of five adopter categories: innovators, early adopters, early majority, late majority, and laggards. This is an important base for segmenting markets, for better marketing strategies, and for improved policy-making processes by targeting each category.

Rogers (1959, 1983) cited in Evans et al. (2006) proposed several criteria useful for determining the likely acceptance of new products: compatibility (the degree to which a product is consistent with consumers' current values, cognitions, and behaviors), relative advantage (the degree to which an item has a sustainable, competitive, differential advantage over other products), trialability (the degree to which a product can be tried on a limited basis or divided into small quantities for an inexpensive trial), observability (the degree to which a product or its effects can be sensed by other consumers), speed (how fast the benefits of the product are experienced by the consumer), simplicity (the degree to which a product is easy for a consumer to understand and use), perceived risk (if the perceived risk for the new product is low, it is more likely to be adopted), product symbolism (what the product means to the consumer), and aesthetic and hedonistic innovations (appealing to sensory and pleasure-seeking needs). In this respect, Novoselova et al. (2007) recommended "future studies on food technology adoption…to develop integrated chain models including all chain participants as well as consumers; and for these models to explicitly account for uncertainty and the trade-offs between concerns and benefits."

Moreover, technological modification of food and food production can evoke a negative response among consumers, especially in the absence of good communication on risk assessment efforts and cost – benefit evaluations (World Health Organization 2002). The acceptance of a technology depends on the consumer's perception of benefits and risks. These include the impact of the technology on taste, convenience, and nutritional value, the perceived safety of the process or technology, the magnitude of the risk the technology reduces, and the effect of the technology on the environment. Public acceptance is influenced by perceived credibility of data, rigor of regulatory policy, impartial action of regulators, and demonstrated responsibility of industry (Bruhn 2007). Social amplification processes may generate public concern about hazards that are judged as low risks by experts. Such social amplification processes can be observed in the domain of genetically modified foods in various European countries. There are presently no indications that such an amplification process must be expected in the domain of nanotechnology food (Siegrist 2008). An important issue relates to the observation that risk estimates associated with novel foods may differ depending on whether more emphasis is placed by the individual on the results of technical risk assessment or on an individual's perceptions of risk associated with different hazards. Consumer perceptions of benefits associated with novel foods also differ. Perceptions of what constitutes both risk and benefit appear to be important determinants of consumer acceptability of particular products (Van Puttena et al. 2006).

Successful consumer-driven food product development resembles the task of hitting a moving target. The goal to produce exactly the food that is positively perceived by the consumer is a very complex process. Not only characteristics of the individual and the food product are important, but so are variables related to the context in which the food is to be consumed, such as time and place, and environmental variables such as the family background and societal features. Consumer-driven food product development is a must for food companies to stay in the market. In this process the companies have to take account of societal changes, such as mass-individualization, globalization, and broadening of the quality concept, while also keeping track of possible new technologies that offer opportunities for their product range. Nowadays, many new product introductions still fail. Close cooperation between marketers and food technologists is imperative to be successful in the application of these methods, but also in consumer-driven food product development in general (Linnemann et al. 2006).

Among the different factors that influence the acceptance or rejection of foods, the attitudes, beliefs, and opinions of their potential consumers are relevant and can in some cases be decisive. The influence of these factors on food choice and purchase is especially important in the acceptance of some types of foods (organic or ecological, genetically modified or functional) that are presented to the consumer as a possible alternative to conventional food. The main qualitative and quantitative methods applicable to the investigation of the consumers' opinions and attitudes toward food are in-depth interviews, focus groups, and questionnaire-based surveys (Barrios and Costell 2004).

New technologies that successfully address technical problems may face barriers in marketing and consumer acceptance. Products will have the greatest likelihood of success when developers listen to consumer needs, respond to consumer concerns, and offer tangible benefits. Researchers have demonstrated that statements about the benefits associated with a particular food or food processing technique will reduce concerns toward the food or technology and improve both its acceptance and the likelihood of its consumption. Factual information about new technologies, clear statements about the safety and benefits, and exposure to the produce can be expected to improve expected liking and increase the likelihood of consumer acceptance (Bruhn 2007).

Consumer acceptance is crucial to the development of successful food products. As a consequence, consumers' attitudes toward new food technologies should be taken into account at an early stage of product development. Consumers' risk perception may differ from experts' assessments. The views of both laypeople and experts must be taken into account concurrently in order to be successful in the market. Informing consumers about the food technology used for a given product may even affect its sensory appeal. Laypeople may not possess the knowledge necessary for assessing the risks and benefits associated with novel foods. Trust in the industry is, therefore, an important factor affecting the acceptance of novel foods. Cultural and social norms influence trust in industry and attitudes toward food technologies. Environmental attitudes and the Food Neophobia Scale are only weakly related to the perception of new food technologies. The most important variable seems to be personal importance of naturalness. Consumers who show a strong preference for organic food and for whom a natural food production is important negatively evaluate new food technologies. The public must be informed about the benefits of new technologies. It is important that this information stems from a trusted source. An approach such as labeling or public participation may not be very efficient and may not have the intended effects (Siegrist 2008).

6.2.2 Determinants of Novel Food Choice

Depending on the region of the world, people often have different attitudes to food. In addition to nutritional value, food often has societal and historical connotations, and in some instances may have religious importance (World Health Organization 2002). Figure 6.1 summarizes some general determinants of novel food choice.

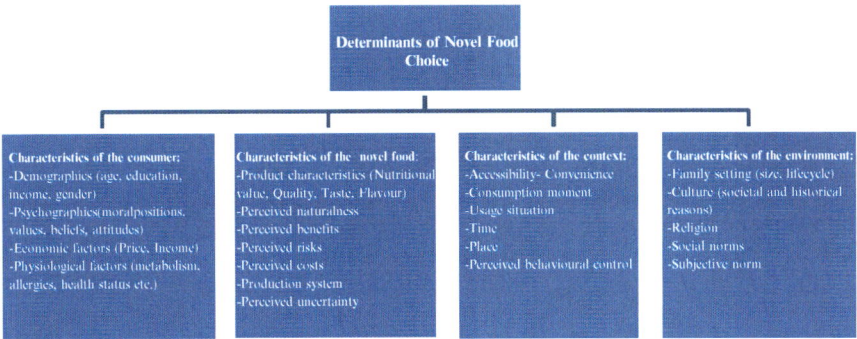

Fig. 6.1 Overview of the main determinants of novel food choice

A vital insight is the set of linkages between people's attitudes, their claimed prospective behavior, and their actual behavior. Many studies invite respondents to indicate what they think they would do in a particular situation, i.e., to claim a prospective behavior. The evidence suggests that, in general, there are likely to be strong links between people's attitudes and the choices they say they will make, but there is a much weaker link between either of these and what they will actually do when confronted with a choice in the "real world." A key reason for this is that attitudes (toward new technologies or about ethics/morals in general) do not exist or operate in isolation. In reality, and in particular at the point of making food choices, individuals are concerned with, for example, price, convenience, accessibility, taste, and quality. These contextual factors are either internal or external. Internal factors consist of values, beliefs, attitudes, moral positions, and so forth. These may be "irrational," but they are real. External factors comprise economic factors, family setting, social norms, and so forth (Lyndhurst 2009).

Some recent food-technology-based innovations have been adopted easily by consumers, whereas others have essentially been rejected (Cardello 2003). Prominent examples of the latter are genetically modified foods in Europe and food irradiation (Henson 1995 cited in Ronteltap et al. 2007).

Consumers' fears of novel food technologies are documented. The ability to identify population segments that have greater or lesser neophobia, thus enabling identification of early adopters of innovative products, would be useful. The Food Neophobia Scale is a useful tool for assessing reactions to ethnic foods (and sensation seeking) but is less suitable for assessing receptivity to foods produced by novel technologies. Therefore, there is a need of a new psychometric tool that identifies food technology neophobia: the Food Technology Neophobia Scale has the basic requirement of a valid measure of "attitudes to new food technologies." This scale was constructed for application to food product testing of novel foods to identify segments of consumers willing to accept foods produced using novel technologies (Cox and Evans 2008).

Ronteltap et al. (2007) developed a conceptual framework for consumer acceptance of technology-based food innovations: the framework distinguishes "distal" and "proximal" determinants of acceptance. Distal factors (characteristics of the innovation, the consumer, and the social system) influence consumers' intention to accept an innovation through proximal factors (perceived cost – benefit considerations, perceptions of risk and uncertainty, social norm, and perceived behavioral control). In this framework, consumers' actual adoption of innovation in the food area is ultimately determined by their intention to use it. At the proximal level, adoption (intention) is determined by (1) perceived costs and benefits, (2) perceived risk and uncertainty, (3) subjective norm, and (4) perceived behavioral control. However, these perceptions are affected by a set of more distal determinants, namely, (1) features of the innovation, (2) consumer characteristics, and (3) characteristics of the social system of which the consumer is part. Whereas innovation and consumer characteristics have a direct relationship with perceptions of the proximal determinants, the characteristics of the social system affect the framework more generically. Communication is an important means linking innovation features to consumer perceptions. It is appreciated that this communication may need to be adjusted depending on the characteristics of the consumer (segment) at which the innovation is targeted.

Consumers appear to be cautious about accepting novel technologies applied to foods because of perceived risks and lack of benefits. Generally, addressing "information deficit" did not overcome aversion to novel technologies applied to food concepts (Cox et al. 2007).

It is worth bearing in mind that consumers have been making choices about "novel food technologies" for many years. Tinned foods and frozen foods, for example, were once novel and took many years to achieve saturation acceptance by the general public. To this day, myths persist about these "technologies," and even refrigeration is treated with suspicion in some communities. One challenging feature of "opinion" is its relationship to "knowledge." The evidence is ambiguous. In some cases increased knowledge leads to more positive views, in some cases more negative views; in some cases it leads to firmer views, in some cases to more malleable views. Knowledge can be both direct (through exposure to and/or experience of a new technology) and indirect (through receiving information). The latter is clearly mediated by the degree of trust in the relevant information channel; the former is in part a function of the setting in which direct experience occurs. Levels of awareness differ between different technologies, between different groups in society, and between countries.

6.3 Novel Food Technologies and Emerging Trends

This section comprises a presentation of reviewed literature on novel foods and consumer acceptance at the global level and a brief overview of some food policies for novel foods around the world.

6.3.1 Brief Overview of Novel Technologies

In the area of food and nutrition, various technological applications, which have resulted in novel foods, have emerged in the past few decades. Many of these were related to prolonging the shelf life of foods and enhancing their safety, e.g., pasteurization, and novel food packaging. A food preservation technology that has failed to reach widespread adoption is food irradiation. Despite the fact that the scientific community recognized food irradiation as a safe and effective process, significant consumer resistance has inhibited the application of the technology (Henson 1995 cited in Ronteltap et al. 2007).

Since the earliest days of agricultural biotechnology development, scientists have envisioned harnessing the power of genetic engineering to enhance nutritional and other properties of foods for consumer benefit. The first generation of agricultural biotechnology products to be commercialized, however, were more geared toward so-called input traits, GMs that make insect, virus, and weed control easier or more efficient. These first products were rapidly adopted by US farmers, and now account for most of the soybeans, cotton, and corn grown in the USA. Agricultural biotechnology innovations aimed directly at consumers, sometimes collectively referred to as output traits, have been longer in development. As the technology advances and we learn more about the genes and biochemical pathways that control those attributes that could offer more direct consumer benefits, the genetically engineered food with more direct consumer benefits moves closer to reality (PEW Initiative on Food and Biotechnology 2007; World Health Organization 2005).

The technology of nutrigenomics focuses on developing an improved understanding of the relationship between human nutrition and human genetics in order to maintain and promote health and prevent the occurrence of disease. The innovative feature of nutrigenomics in nutritional sciences is that it is based on knowledge of the individual's genetic constitution. It could provide opportunities for the development of food products or dietary advice tailored to the nutritional needs of specific groups of consumers, or even individuals. Some authors have optimistically argued that these so-called personalized foods will shift the food market from a technology push into a consumer pull system, where innovation development is determined by the demand side of the market. In that situation, the consumer's preference for optimal health is a major driver for food choice, and as a consequence for how foods are produced. Despite its promise, nutrigenomics may also raise consumer concerns regarding the impact of human genetics on the integrity of nature, privacy, and control of sensitive information (Ronteltap et al. 2007).

One category of potential products aimed at consumers is those products with added health benefits, also known as "functional foods." The term "functional food" means different things to different people, but generally refers to foods that provide health benefits beyond basic nutrition (PEW Initiative on Food and Biotechnology 2007). In recent years, many of the new food technologies and food innovations have been targeted at the promotion of good health. Fortification and restoration of foods were used as methods to add nutrients to food products that did not contain them originally or had lost them in the manufacturing process. Functional foods constitute a range of novel foods designed to deliver benefits beyond nutritional value to the person consuming them, and the so-called nutraceuticals provide medical or health benefits by maintaining and modifying bodily functions (Ronteltap et al. 2007).

It was mainly the advances in understanding the relationship between nutrition and health that resulted in the development of the concept of functional foods, which means a practical and new approach to achieve optimal health status by promoting the state of well-being and possibly reducing the risk of disease (Siro et al. 2008). The European Commission's Concerted Action on Functional Food Science in Europe, coordinated by International Life Science Institute Europe, defined functional food as follows: "a food product can only be considered functional if together with the basic nutritional impact it has beneficial effects on one or more functions of the human organism thus either improving the general and physical conditions or/and decreasing the risk of the evolution of diseases. The amount of intake and form of the functional food should be as it is normally expected for dietary purposes. Therefore, it could not be in the form of pill or capsule just as normal food form." In contrast to this latter statement, since 2001 products categorized as foods for specified health use in Japan can also take the form of capsules and tablets, although a great majority of products are still in more conventional forms.

As an enabling technology, nanotechnology has vast potential to revolutionize agriculture and food systems. Two key applications of nanotechnology are particularly relevant: sensing technology designed to alert growers and others in the agricultural community to soil, moisture, nutrient, pathogen, and related conditions pertinent to growing crops, and nano-enabled technologies intended to enhance performance by improving the delivery of nutrients and agricultural chemicals, thereby diminishing the volume of these materials introduced into the environment (OECD 2009b).

6.3.2 Consumer Acceptance of Novel Foods

As shown in the previous section, consumer acceptance is critical for the market success of novel foods.

The Asian Food Information Centre conducted a consumer research study in five countries (China, India, Japan, Philippines, and South Korea) to provide insights into how consumers in Asia perceive the use of biotechnology to produce foods and how likely consumers are to accept the various benefits biotechnology-derived foods may bring. It appeared that confidence in the safety of the food supply is on average neutral to mildly positive. More accurate food labels seem to be critical in order to increase the consumer confidence level in food safety. Food poisoning, country of origin, pesticide residues, and food additives are the top food safety concerns in all countries. Consumers in the five countries expressed little to no concern about the safety of food biotechnology. An improved understanding of the direct consumer benefits of biotechnology foods raises consumer acceptance. Consumers in the food-producing countries strongly believe that food biotechnology will bring benefits in the next few years, whereas consumers in the food-importing countries are rather unsure about the future potential of biotechnology food. The survey underscores the fact that learning about the benefits of biotechnology foods has a significant impact on consumers' perception and acceptance. Science-based information promoting an understanding of the consumer benefits of biotechnology foods should be communicated to the public using easy to understand language (Asian Food Information Centre 2008).

In a 2009 study requested by the Food Standards Agency (UK), Lyndhurst (2009) highlighted that UK public attitudes are unsupportive of the novel food technologies reviewed (GM, cloning, nanotechnology, food irradiation, food fortification, synthetic biology, and novel food processes). In general, levels of understanding about the technologies are low. Key factors in determining both fixed and malleable attitudes are perceptions of benefit, perceptions of risk/threat, and moral/ethical position. Moral and ethical concerns loom larger for some technologies than others (notably cloning and GM). The use of the word "perceptions" is key: public levels of trust (in government, the media, scientists) are generally low and powerfully influence the nature of beliefs about and attitudes toward novel food technologies. Of all the factors influencing public attitudes, those linked to health may be the most significant. Where new technologies are thought or believed to potentially contribute to better health, they are likely to be viewed more favorably; and, conversely, where there is a perception of threat to health, they are perceived more negatively (Lyndhurst 2009).

Verbeke et al. (2009) reported on consumers' reactions toward calcium-enriched fruit juice, omega-3-enriched spread, and fiber-enriched cereals, each with a nutrition claim, health claim, and reduction of disease risk claim. Positive attitudes toward functional foods and familiarity with the concrete functional product category boosted the claim type and product ratings, whereas perceived control over the individual consumer's own health and perceiving functional foods as a marketing scam decreased the appeal of all products.

Flavonoids from fruit and vegetables are currently widely studied as components that have the potential to provide multiple health benefits. Lampila et al. (2009) examined consumer perceptions of flavonoids. In general, the term "flavonoid" was unfamiliar. After consumers had received information about the possible health benefits, positive attitudes toward flavonoids were expressed. Relevant issues for the acceptance of flavonoids were the natural occurrence and the health benefits associated with common diseases. However, the need to enhance flavonoid content was questioned since fruit and vegetables were perceived to be already healthy with the natural flavonoid content; additionally, consumers had perceptions of risk and uncertainty associated with breeding and processing methods. Familiar processing methods were said to be most acceptable for enhancing flavonoid content. Consumer knowledge of the health effects of flavonoids is limited, and thus there is a need to inform consumers about them. The challenge in informing consumers about the benefits of flavonoids is to maintain the natural image of fruit-based products.

Cardello (2003) has provided evidence that consumer concerns toward the technologies used to process and preserve foods may be an important variable influencing the expected liking/disliking for the product. Cardello (2003) has also provided evidence that factual information about these technologies, clear statements about the safety and benefit of these technologies, and exposure to visual product characteristics can all improve expected liking and increase the chances of consumer acceptance upon initial trial of the product. Future research on novel food processing and preservation technologies, including genetic engineering, irradiation, and other emerging nonthermal technologies, should move forward from simple assessments of consumer concerns about the technologies and begin to focus more directly on questions and issues related to the consumer's expected liking/disliking for the products of these technologies. In this way, we can begin to develop both a better understanding of the variables that directly influence the acceptance of these products and more effective marketing and informational strategies to improve their likely success in tomorrow's marketplace.

The success of new food processing technologies is highly dependent on consumers' acceptance. Nielsen et al. (2009) studied European consumers' perceptions of two new processing technologies and food products produced by means of these novel technologies. To accomplish this, a qualitative study on consumer attitudes toward high-pressure processing (HPP) and PEF processing of food was conducted. The results show that consumers perceived the main advantages of HPP and PEF products to be the products' naturalness, improved taste, and high nutritional value, whereas the main disadvantage was the lack of information about the PEF and HPP products. The results of the participants' evaluation of the PEF and HPP processes showed that environmental friendliness and the more natural products were seen as the main advantages, but there were concerns about body and health, the higher price of the products, and the lack of information about the technologies, and there was a general skepticism. The study also shows that north European participants were a bit more skeptical toward PEF and HPP products than the east European participants.

According to Cardello et al. (2007), perceived risks associated with the technologies were the most important factors influencing interest in use. Among the emerging technologies assessed, irradiation and GM resulted in the greatest negative effect on likely use, whereas HPP produced the most positive effect. The term "cold preservation" had positive associations for all groups, but "minimally processed" had negative associations. Implications of the data for the marketing of foods processed by innovative and emerging technologies were discussed.

The food industry is currently interested in a variety of novel production and processing technologies that may result in economic and improved-quality products. However, consumer attitudes toward and conceptions of these new technologies can greatly influence their success in the marketplace. The results of this study show that "perceived risks" of the technologies are the most important determinant of interest in their use by consumers. This finding and other findings uncovered in this study suggest that industry must be vigilant in its knowledge of consumer attitudes toward these processes in order to avoid unexpected failure of these products upon market introduction.

New food technologies enable innovations in the food sector. Not all technologies, however, are equally accepted by consumers. According to Siegrist (2008), there are hardly any discussions about HPP of food. In contrast, other food technologies, such as gene technology, are not well accepted in Europe. For yet other food technologies, public acceptance is an open question. An example of the latter is nanotechnology, which has the potential to generate radically new food products and is increasingly being employed in the areas of food production and packaging. How the public will react to nanotechnology food products is unclear.

According to a Eurobarometer survey in 2006 (Memo/07/130 The European Commission 2007), one in two Europeans believes that biotechnology will improve their quality of life. The survey showed widespread support for medical and industrial applications of biotechnology, although there was a more negative position on genetically modified foods (58% against, 42% in favor). The same survey showed that the EU is more trusted than national governments on issues of regulating biotechnology. This supports previous findings from other economic surveys that the way of asking people may strongly affect their responses. Hence, further research is needed to compare consumer preferences collected via stated-preference and experimental methods with data derived from the food market or alternatively, as long as GM-labeled food products are nonexistent, with data derived from field experiments

Europeans are the most skeptical group with regard to genetically modified food. The analysis of preference dispersion indicates that it is particularly the large fraction of the population opposing genetically modified food which distinguishes European consumers from American consumers. Consumers in Asia seem to be somewhat polarized in their views, as they are more likely to be pro-GM and less likely to be indifferent between GM and non-GM. The results show furthermore that the differences across regions arise from diverging preferences over time. Whereas the aversion to genetically modified food is steeply increasing in Europe, it is only gently increasing in America and is even decreasing in the rest of the world (Dannenberg 2009).

It is clear that consumers want freedom of choice when buying foods and some of them say "yes" to GM when offered that freedom. Overall, people seem not to be able to recognize genetically modified food in spite of the labeling requirements. But this does not appear to be a problem as people are in general not careful to avoid these products, a conclusion supported by the scant attention paid to labels. However, people do react differently toward GM-free-labeled products, suggesting that those products are chosen with greater thought on the part of consumers who want them. In practice, shoppers frequently behaved differently from the way they say they would do. One third of the respondents were wrong in their perceptions about their purchases of genetically modified foods, whereas another third did not know what they had bought (European Commission: Framework 6 2008).

Furthermore, consumers are not homogenous, and what is perceived as a desirable benefit by one consumer may be regarded as irrelevant by another. In the case of novel foods, such products should be marketed specifically to consumers who want to buy them, under conditions of fully informed consumer choice (Van Puttena et al. 2006).

The issue of public acceptance of genetically modified foods, and indeed the emerging biosciences more generally, is considered in the context of risk perceptions and attitudes, public trust in regulatory institutions, scientists, and industry, and the need to develop communication strategies that explicitly include public concerns rather than exclude them (Frewer et al. 2004). Increased public participation has been promoted as a way of increasing trust in institutional practices associated with the biosciences, although questions still arise as to how to best utilize the outputs of such exercises in policy development. This issue will become more of a priority as decision-making systems become more transparent and open to public scrutiny. It is recommended that new methods are developed in order to integrate public values more efficaciously into risk analysis processes, specifically with respect to the biosciences and to technology implementation in general. Jensen and Sandoe (2002) cited in Frewer et al. (2004) observed that the decline in public confidence in food safety matters continues, despite the creation of new food safety institutions such as the European Food Safety Authority. In part, they argue, this is because communications about food safety issues (including GM) that are based on scientific risk assessments do not reassure the public. If public confidence is to be regained, it is important to explicitly incorporate public concerns into the risk analysis process, perhaps through developing new and influential methods of public engagement and consultation. Once public concerns (and the values on which they are based) are understood, they can be more effectively introduced into risk assessment and risk management practices.

In terms of policy recommendations, it is becoming very apparent that institutional transparency, coupled with the integration of public concerns into policy development and implementation, will facilitate the introduction of emerging technologies and their applications into society. For this strategy to be successful, it is important to understand what is driving public concern, and to integrate this into policy development rather than dismiss it as irrational as has sometimes happened in the past (Frewer et al. 2004).

Delgado-Gutierrez and Bruhn (2008) evaluated the attitudes of food and health professionals to three new food processing technologies (UHP, PEF, and ohmic heating) that have been developed to respond to consumer demands such as superior taste, longer shelf life, higher nutritional content, health benefits, and environment-friendly processing. Educational brochures were developed based on the current literature and consumer information needs. An Web survey was conducted to determine information needs of food and health professionals and to assess their satisfaction with the data provided in the educational brochures. Health professionals have a positive attitude toward UHP, PEF, and ohmic heating. No negative attitudes or discomfort was reported. More people indicated they were very comfortable with UHP than with any other technology, and as a consequence this technology received the highest scores of likelihood to be included in health professionals' recommendations.

Consumers are interested in receiving information about new developments in food and nutrition; however, consumers may be confused because of the multiple sources and quality and content of information (Delgado-Gutierrez and Bruhn 2008). Media reports frequently do not provide enough information or provide conflicting information, leading to consumer food and nutrition misinformation. One way to avoid consumer confusion is through educational programs. To influence consumer attitudes and behavior, people need to trust the information source. Food and health professionals are considered legitimate sources of information by consumers. Therefore, educational activities should be led by professionals in food and health science because other sources may not provide sufficient or accurate information. These activities could contribute to more accurate understanding and, under appropriate circumstances, increased acceptance of new technological approaches. The acceptance of a novel technology is highly dependent on the information provided to the consumer. Acceptance may be increased if consumers perceive direct benefits from use of the technology. Food attributes, such as good taste, microbiologically safe, absence of pesticides and additives, and extended shelf life, form part of consumer demands and expectations.

6.3.3 Food Policies for Novel Foods

The biological sciences are adding value to a host of products and services, producing what some have labeled the "bioeconomy." From a broad economic perspective, the bioeconomy refers to the set of economic activities relating to the invention, development, production, and use of biological products and processes. If it continues on course, the bioeconomy could make major socioeconomic contributions in all countries. These benefits are expected to improve health outcomes, boost the productivity of agriculture and industrial processes, and enhance environmental sustainability. The bioeconomy's success is not, however, guaranteed: harnessing its potential will require coordinated policy action by governments to reap the benefits of the biotechnology revolution (OECD 2009a).

When new foods are developed by natural methods, some of the existing characteristics of foods can be altered, in either a positive or a negative way. National food authorities may be called upon to examine traditional foods, but this is not always the case. Indeed, new plants developed through traditional breeding techniques may not be evaluated rigorously using risk assessment techniques. With genetically modified foods most national authorities consider that specific assessments are necessary. Specific systems have been set up for the rigorous evaluation of genetically modified organisms and genetically modified foods relative to both human health and the environment. Similar evaluations are generally not performed for traditional foods. Hence, there is a significant difference in the evaluation process prior to marketing for these two groups of food (World Health Organization 2002).

In the case of the first genetically modified foods introduced onto the European market, the products were of no apparent direct benefit to consumers (not cheaper, no increased shelf life, no better taste). The potential for genetically modified seeds to result in bigger yields per cultivated area should lead to lower prices. However, public attention has focused on the risk side of the risk – benefit equation. Consumer confidence in the safety of food supplies in Europe has decreased significantly as a result of a number of food scares that took place in the second half of the 1990s that are unrelated to genetically modified foods. This has also had an impact on discussions about the acceptability of genetically modified foods. Consumers have questioned the validity of risk assessments, with regard to both consumer health and environmental risks, focusing in particular on long-term effects. Consumer concerns have triggered a discussion on the desirability of labeling genetically modified foods, allowing an informed choice. The public concerns about genetically modified food and genetically modified organisms in general have had a significant impact on the marketing of genetically modified products in the EU. In fact, they have resulted in the so-called moratorium on approval of genetically modified products to be placed on the market. Marketing of genetically modified food and genetically modified organisms in general is the subject of extensive legislation. European Community legislation has been in place since the early 1990s.

The way governments have regulated genetically modified foods differs. In some countries genetically modified foods are not yet regulated. Countries which have legislation in place focus primarily on assessment of risks to consumer health. Countries which have provisions for genetically modified foods usually also regulate genetically modified organisms in general, taking into account health and environmental risks, as well as control- and trade-related issues (such as potential testing and labeling regimes). In view of the dynamics of the debate on genetically modified foods, legislation is likely to continue to evolve (World Health Organization 2002).

Codex Alimentarius is developing principles for the human health risk analysis of genetically modified foods. The premise of these principles dictates a premarket assessment, performed on a case-by-case basis and including an evaluation of both direct effects (from the inserted gene) and unintended effects (that may arise as a consequence of insertion of the new gene).

FAO member countries, especially developing ones, look to FAO to provide sound and unbiased advice on the safety of genetically modified food, and AGNS, in collaboration with international bodies such as Codex Alimentarius, has been involved in a wide range of biotechnology-related issues, including science-based safety evaluation and risk assessment systems to objectively determine the benefits and risks of genetically modified food, recommendations for the labeling of foods obtained through biotechnology, assessment of nutritional aspects of food derived from modern biotechnology, and detection of protein and/or DNA in genetically modified food (FAO – AGNS 2009).

Labeling in the EU is mandatory for products derived from modern biotechnology or products containing genetically modified organisms. Legislation also addresses the problem of accidental contamination of conventional food by genetically modified material. It introduces a 1% minimum threshold for DNA or protein resulting from GM, below which labeling is not required. In 2001, the European Commission adopted two new legislative proposals on genetically modified organisms concerning traceability, reinforcing current labeling rules and streamlining the authorization procedure for genetically modified organisms in food and feed and for their deliberate release into the environment.

The release of genetically modified organisms into the environment and the marketing of genetically modified foods have resulted in a public debate in many parts of the world. This debate is likely to continue, probably in the broader context of other uses of biotechnology (e.g., in human medicine) and their consequences for human societies. For instance, the humanitarian crisis in southern Africa has drawn attention to the use of genetically modified food as food aid in emergency situations. A number of governments in the region raised concerns relating to environmental and food safety fears. Although workable solutions have been found for distribution of milled grain in some countries, other countries have restricted the use of genetically modified food as food aid and obtained commodities which do not contain genetically modified organisms (World Health Organization 2002).

Most of the countries have a system for the safety assessment and regulation of novel foods and feeds, including products of modern biotechnology. Commercialization of a number of biotechnology products, particularly new crop varieties, has moved into international trade. In this context, international harmonization of regulatory assessment of novel foods and feeds will ensure the protection of human and animal health (OECD 2009a).

Nanotechnology applications in the agricultural sector raise many interesting and vexing policy issues. Although the potential benefits of nanotechnology to agriculture and the environment are well recognized, regulators will be challenged to ensure that sufficient measures are in place to balance the potential for adverse environmental impacts resulting from the release or potential release and/or degradation of nanomaterials into the environment with the many benefits nanotechnology offers. Global regulatory and governance systems differ considerably throughout the world with respect to managing agriculture and food safety. Care will need to be taken to identify and consider the unique geographical and societal differences that will materially affect the ability of various governance systems to balance these risks and benefits responsibly and comprehensively. Effective policies will also need to recognize limitations on the available data points that will inform regulators and policy-makers of all relevant aspects of the environmental and life cycle implications of nanotechnology and nanomaterials applications in the agricultural sector, and to ensure appropriate measures are in place to ensure sustainable agricultural practices that incorporate nanotechnology (OECD 2009b).

The Advisory Committee on Novel Foods and Processes is a non-statutory, independent body of scientific experts that advises the Food Standards Agency (UK) on any matters relating to novel foods (including genetically modified foods) and novel processes (including food irradiation). The committee carries out safety assessments of any novel food or process submitted for approval under Novel Food Regulation (EC) 258/97 (ACNFP 2007). Until April 2004, the scope of this regulation included all foods produced using genetically modified organisms. However, genetically modified foods are now subject to approval under a separate regulation. Approval of genetically modified foods now involves centralized risk assessments, which are the responsibility of the European Food Safety Authority.

Although there is little controversy about many aspects of biotechnology and its application, genetically modified organisms have become the target of a very intensive and, at times, emotionally charged debate. FAO (2000) recognizes that genetic engineering has the potential to help increase production and productivity in agriculture, forestry, and fisheries. It could lead to higher yields on marginal lands in countries that today cannot grow enough food to feed their people. However, FAO is also aware of the concern about the potential risks posed by certain aspects of biotechnology. FAO supports a science-based evaluation system that would objectively determine the benefits and risks of each individual genetically modified organism. This calls for a cautious case-by-case approach to address legitimate concerns for the biosafety of each product or process prior to release of the product.

Functional foods are found virtually in all food categories; however, products are not homogeneously scattered over all segments of the growing market. The development and commerce of these products is rather complex, expensive, and risky, as special requirements must be met. Besides potential technological obstacles, legislative aspects as well as consumer demands need to be taken into consideration when developing functional food. In particular, consumer acceptance has been recognized as a key factor to successfully negotiate market opportunities (Siro et al. 2008). European legislation, however, does not consider functional foods as specific food categories, but considers them rather as a concept. Therefore, the rules to be applied are numerous and depend on the nature of the foodstuff. The General Food Law Regulation is applicable to all foods. In addition, legislation on dietetic food, genetically modified organisms, food supplements, or novel foods may also be applicable to functional foods depending on the nature of the product and on its use. In the EU, rather than regulation of the product group per se, legislative efforts currently being developed are directed toward restricting the use of health claims on packages and in marketing. According to the EU regulation on nutrition and health claims made for foods [Regulation (EC) No. 1924/2006], a list of authorized claims has to be published for all member states, and nutrient profiles also have to be established for foods for which health claims are made. Health claims can be "function claims" and "reduction of disease risk claims."

How a product is categorized has a substantial implication for how it is marketed, the safety and labeling standards it must meet, and what requirements it must fulfill for regulatory review and clearance or approval. If a functional food product is marketed for a therapeutic purpose (e.g., to treat a disease), it will be subject to regulation as a "drug." If a product is subject to regulation as a "food," it may be further classified as a conventional food, dietary supplement, food for special dietary use (including infant formula), or medical food, again depending upon its intended use and other factors. The application of biotechnology to foods expressly to improve nutritional and health characteristics holds great potential. The use of modern biotechnology to enhance human and other animal food will likely not change these regulatory paradigms, but may challenge the boundaries of some of the regulatory classifications (PEW Initiative on Food and Biotechnology 2007).

A list of EU-approved health claims was expected to be in place by 2010. Until that time, claims would remain subject to general food labeling legislation that prohibits claims that are untrue or otherwise misleading to the consumer. Claims that state or imply that a food can prevent, treat, or cure a disease will continue to be prohibited (ACNFP 2007).

6.4 Conclusions

The purpose of this chapter was to explore consumer attitudes to novel food products. It focused on reviewing literature on consumer acceptance of novel food technologies and the determinants in novel food choice. Moreover it presented an overview at the policy level.

The literature review findings revealed that public and private institutions have to develop clear definitions, standards, policies, and strategies for novel foods. In this way, consumers' food choice will become easier, the levels of acceptability will increase, and the information asymmetry will decrease; hence, the demand for novel foods will increase. If consumers receive mixed messages from policy-makers and stakeholders, their level of involvement and motivation will decrease.

The full benefits of agricultural biotechnology (Hoban 2004), but also of all novel food technologies, will only be realized if consumers and food manufacturers consider it safe and beneficial. Although a few internationally comparable public opinion surveys have been conducted on this issue, the available evidence suggests that public attitudes differ sharply both between and within countries and are evolving over time. Consumer attitudes toward novel foods differ across and within countries. Comparisons of results from different studies must be made with caution, because of the sensitivity of such studies to the particular circumstances surrounding their design and administration. In general, consumers in the USA and Asia are more favorably disposed toward biotechnology than Europeans, although significant minorities in most countries express reservations. Trends in the USA and Europe have fluctuated over time, with European views growing generally more negative throughout the 1990s before turning slightly more positive in the most recent survey. How these trends will evolve is a matter for continuing research.

Dannenberg (2009) concluded that genetically modified food products in Europe may have a chance only as a niche product, at least for the time being, whereas they may rapidly spread out in other regions of the world. The uncertainty and skepticism of European consumers, on the one hand, and the upcoming cultivation and trading of genetically modified crops, on the other hand, will be a challenge for European politicians in the near future. The question of why European consumers persist in their distrust of this new technology remains to be answered by future research. It may be, for example, that they respond differently to the work of environmental and consumer protection organizations or to food scandals, e.g., the BSE crisis in Europe or the melamine crisis in China. Although the analysis provides interesting information about consumer preferences in North America, Europe, and Asia, the metastudy includes only a few observations for Australia/Oceania and Africa and no observations for South America. Further research is needed to analyze consumer preferences in all these regions and their development over time.

6.4.1 Emerging Trends

Several factors will drive the emerging bioeconomy by creating opportunities for investment. In addition to the use of biotechnology to meet the challenge of environmentally sustainable production, a major driver is increasing population and per capita income, particularly in developing countries. The latter trends, combined with rapid increases in educational achievement in China and India, indicate not only that the bioeconomy will be global, but that the main markets for biotechnology in primary production (agriculture, forestry, and fishing) and industry could be in developing countries (OECD 2009a). Obtaining the full benefits of the bioeconomy will require purposive goal-oriented policy. This will require leadership, primarily by governments, but also by leading firms, to establish goals for the application of biotechnology to primary production, industry, and health; to put in place the structural conditions required to achieve success, such as obtaining regional and international agreements, and to develop mechanisms to ensure that policy can flexibly adapt to new opportunities.

The rapidly developing field of nanotechnology is likely to become yet another source of exposure to a range of materials incorporating nanotechnology such as paints, cosmetics, electronics, clothing, food, medicines, and diagnostics. Depending on their planned application, these engineered nanomaterials may be inhaled, ingested, absorbed through the skin, or intentionally injected in a medical procedure. In addition, although there are many proposed potential beneficial outcomes of nanotechnology on the environment (remediation, self-cleaning surfaces), there is currently a paucity of data to know for sure if these materials could have undesirable effects on the environment. Nanotechnology is widely perceived as one of the key technologies of the twenty-first century, and accordingly there have been huge advances in and increased funding of global technological research on nanomaterials. An ever-increasing range of potential nanotechnology applications and higher production rates will undoubtedly result in increased human and environmental exposure; hence, there is growing consensus among scientists, regulators, industry, and the public that we need to know more about the possible harmful or adverse effects of nanoparticles on human health and the environment.

Nanotechnology has tremendous potential to contribute to humans flourishing in socially just and environmentally sustainable ways. However, nanotechnology is unlikely to realize its full potential unless its associated social and ethical issues are adequately attended (Sandler 2009).

For all the other novel food technologies, e.g., food irradiation, UHP, PEF, and ohmic treatment, there is a need for more marketing knowledge. Nonetheless, there is no doubt that functional foods are one of the most promising and dynamically developing segments of the food industry (Siro et al. 2008).

6.4.2 Recommendations for Future Research

It is clear that there is a great deal more that needs to be found out about public attitudes (in particular, what lies behind "top line" attitudes, and how the various drivers of behavior interact); and it is clear, too, that engagement with the public about novel food technologies will require very careful thought and extreme tact (Lyndhurst 2009).

On the basis of the review and conclusions, the recommendations focus on future research needs. There are limited data on how attitudes toward any of the technologies looked at have changed over time. There is growing understanding that populations are not homogenous – people have different concerns and preferences and set the information they receive about new technologies in different personal contexts. With this in mind, there would be benefit in differentiating populations on the basis of values and further examining the drivers of difference.

Although there has been some work which examines the effects of giving people information about the risks and benefits of particular technologies, to date little research has been on how people weigh these up. Further examination of how people reach their viewpoints would further the understanding of how to communicate any information about novel technologies and provide insight into how people might react to, for example, health scares or particular publicity. Moreover, the difference between food shoppers and food consumers should be taken into account in future studies.

Studies have repeatedly found that a large proportion of the population is undecided about novel food technologies and yet very little research has been done into the source of this indifference. Examining the source of indifference in further detail would seem a priority.

A gap in the research appeared to be the public's understanding of the role of different agents within the novel food technology arena. In particular, the public's understanding of the role of regulators seemed under researched, and it is recommended that research in this area could support the development of any public engagement work.

Finally, the relationship between, on the one hand, rational, scientific, factual, evidence-based perceptions, and understanding and, on the other hand, emotional, irrational, ethical, and values-based perceptions is highlighted. In the context of public attitudes toward novel food technologies, both perspectives are real and valid – there is no right or wrong per se. Any organization seeking to pursue public engagement around novel food technologies will need to respect this reality and adopt a neutral stance.

References

ACNFP. Annual report. http://www.food.gov.uk/multimedia/pdfs/acnfpanrep07 (2007).
Asian Food Information Centre. Food biotechnology: consumer perceptions of food biotechnology in Asia. http://www.whybiotech.com/resources/tps/AsiaConsumerPerceptions.pdf (2008).
Barrios EX, Costell E. Review: use of methods of research into consumers' opinions and attitudes in food research. Food Sci Technol Int. 2004;10(6):359–71.
Bruhn CM. Enhancing consumer acceptance of new processing technologies. Innov Food Sci Emerg Technol. 2007;8:555–8.
Cardello AV. Consumer concerns and expectations about novel food processing technologies: effects on product liking. Appetite. 2003;40:217–33.
Cardello AV, Schutz HG, Lesher LL. Consumer perceptions of foods processed by innovative and emerging technologies: a conjoint analytic study. Innov Food Sci Emerg Technol. 2007;8:73–83.
Cox DN, Evans G. Construction and validation of a psychometric scale to measure consumers' fears of novel food technologies: the food technology neophobia scale. Food Qual Prefer. 2008;19:704–10.
Cox DN, Evans G, Lcase IIJ. The influence of information and beliefs about technology on the acceptance of novel food technologies: a conjoint study of farmed prawn concepts. Food Qual Prefer. 2007;18:813–23.

Dannenberg A. The dispersion and development of consumer preferences for genetically modified food – a meta-analysis. Ecol Econ. 2009;68:2182–92.
Delgado-Gutierrez C, Bruhn CM. Health professionals' attitudes and educational needs regarding new food processing technologies. J Food Sci Educ. 2008;7:78–83.
European Commission. http://ec.europa.eu/food/food/biotechnology/novelfood/ (1997).
European Commission: Framework 6, Project No. 518435. Do European consumers buy GM foods? http://www.whybiotech.com/resources/tps/DoConsumersBuyGMFoods.pdf (2008).
Evans M, Jamal A, Foxall G. Consumer behaviour. Chichester: Wiley; 2006.
FAO. FAO statement on biotechnology. http://www.fao.org (2000).
FAO–AGNS. Biotechnology (GM food). http://www.fao.org/ag/agn/agns/biotechnology_en.asp (2009).
Frewer L, Lassen J, Kettlitz B, Scholderer J, Beekman V, Berdal KG. Societal aspects of genetically modified foods. Food Chem Toxicol. 2004;42:1181–93.
Hendrickx MEG, Knorr DW. Ultra high pressure treatments of foods. Springer; 2002.
Hoban TJ. Public attitudes towards agricultural biotechnology, FAO. http://www.fao.org/es/esa (2004).
Lampila P, van Lieshout M, Gremmen B, Lähteenmäki L. Consumer attitudes towards enhanced flavonoid content in fruit. Food Res Int. 2009;42:122–9.
Linnemann AR, Benner M, Verkerk R, van Boekel MAJS. Consumer-driven food product development. Trends Food Sci Technol. 2006;17:184–90.
Lyndhurst B. An evidence review of public attitudes to emerging food technologies, Food Standards Agency. http://www.food.gov.uk/multimedia/pdfs/emergingfoodteches.pdf (2009).
Memo/07/130, Brussels, 11 Apr 2007. Life sciences and biotechnology – a key sector for Europe's competitiveness and sustainability. http://europa.eu/rapid/pressReleasesAction.do? reference=MEMO/07/130&format=HTML&aged=0&language=EN&guiLanguage=fr. (2007).
Nielsen HB, Sonne AM, Grunert KG, Banati D, Pollak-Toth A, Lakner Z, et al. Consumer perception of the use of high-pressure processing and pulsed electric field technologies in food production. Appetite. 2009;52:115–26.
Novoselova TA, Meuwissen MPM, Huirne RBM. Adoption of GM technology in livestock production chains: an integrating framework. Trends Food Sci Technol. 2007;18:175–88.
OECD. The bioeconomy to 2030: designing a policy agenda. http://www.oecd.org/futures/bioeconomy/2030 (2009a).
OECD. Conference on potential environmental benefits of nanotechnology: fostering safe innovation-led growth. http://www.oecd.org/dataoecd/4/45/43289415.pdf (2009b).
PEW Initiative on Food and Biotechnology. Application of biotechnology for functional foods. http://www.pewtrusts.org/uploadedFiles/wwwpewtrustsorg/Reports/Food_and_Biotechnology/PIFB_Functional_Foods.pdf (2007).
Ronteltap A, van Trijp JCM, Renes RJ, Frewer LJ. Consumer acceptance of technology-based food innovations: lessons for the future of nutrigenomics. Appetite. 2007;49:1–17.
Sandler R. Nanotechnology: the social and ethical issues. Washington, DC: Pew Charitable Trusts; 2009.
Siegrist M. Factors influencing public acceptance of innovative food technologies and products. Trends Food Sci Technol. 2008;19:603–8.
Siro I, Kapolna E, Kapolna B, Lugasi A. Functional food. Product development, marketing and consumer acceptance – a review. Appetite. 2008;51:456–67.
Van Puttena MC, Frewer LJ, Gilissen LJWJ, Gremmen B, Peijnenburg AACM, Wichers HJ. Novel foods and food allergies: a review of the issues. Trends Food Sci Technol. 2006;17:289–99.
Verbeke W, Scholderer J, Lähteenmäki L. Consumer appeal of nutrition and health claims in three existing product concepts. Appetite. 2009;52:684–92.
World Health Organization. Modern food biotechnology, human health and development: an evidence-based study. Geneva: WHO; 2005.
World Health Organization. 20 questions on genetically modified (GM) foods. http://www.who.int/foodsafety/publications/biotech/en/20questions_en.pdf (2002). Accessed 8 July 2009.

Part III
Novel Process Technologies with a Green/Environmental Slant

Chapter 7
Recent Advances in the Microencapsulation of Oils High in Polyunsaturated Fatty Acids

S. Drusch, M. Regier, and M. Bruhn

7.1 Microencapsulation Techniques

Encapsulation can be defined as the inclusion of small solid particles, liquid droplets, or gases in a coating material. Encapsulation techniques are used in a variety of food and nonfood applications. In the food sector, protection of an active ingredient, masking of undesirable properties, and the controlled release of substances are the major aims of encapsulation techniques. New microencapsulation techniques continuously emerge and the growing market for functional foods is the major driving force behind innovation in this field (Frost & Sullivan 2005). Generally, encapsulation techniques can be divided into three classes: chemical techniques such as molecular inclusion and interfacial polymerization; physical techniques such as spray-drying, freeze-drying, air-suspension coating, and extrusion; and physico-chemical techniques such as coacervation and liposome entrapment (Kunz et al. 2003). The principal technologies used for encapsulation of lipophilic food ingredients are spray-drying, coacervation, and extrusion. The structure of microcapsules prepared by these techniques is referred to as matrix type with the core material homogenously dispersed in the wall material. A scanning electron micrograph of a typical matrix-type microcapsule is shown in Fig. 7.1.

Micro- and nanoencapsulation techniques for the protection and controlled release of functional food ingredients and aroma compounds have been extensively

---

S. Drusch (✉)
Beuth University of Applied Sciences, Berlin, Germany
e-mail: drusch@beuth-hochschule.de

M. Regier
University of Applied Sciences, Trier, Germany

M. Bruhn
University of Kiel, Kiel, Germany

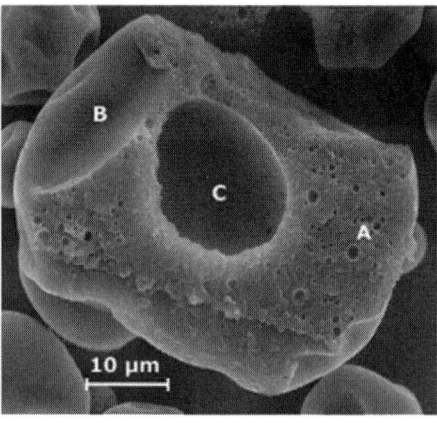

Fig. 7.1 Scanning electron micrograph of fish oil microencapsulated in a matrix of *n*-octenylsuccinate-derivatized starch and glucose syrup by spray-drying. *A* oil droplets, *B* wall material, *C* air inclusion

reviewed (Desai and Park 2005; Garg et al. 2006; Gharsallaoui et al. 2007; Gouin 2004; Lopez-Rubio et al. 2006; Madene et al. 2006; Ré 1998; Schrooyen et al. 2001; Ubbink and Krüger 2006; Vega and Roos 2006; Zeller et al. 1999). This chapter focuses on the most recent developments in microencapsulation of oils rich in long-chain polyunsaturated fatty acids.

7.1.1 Microencapsulation by Spray-Drying

7.1.1.1 The Selection of a Suitable Core Material

Selection of a core material with superior sensory properties and its stability or stabilization, respectively, is particularly important when encapsulating food ingredients by spray-drying. In most cases, the wall material is water-soluble and therefore the core material is released from the capsule during food preparation or upon chewing in the oral cavity. A possible off-flavor developing during deterioration of the core material will consequently impair the sensory perception and product acceptability. The selection of the core material is of utmost importance in the case of encapsulation of oils for nutritional purposes. All of these oils, such as borage oil, evening primrose oil, sea buckthorn oil, and oils from marine sources, contain a high amount of polyunsaturated fatty acids, which are highly susceptible to oxidation. Their stability during the process of encapsulation and storage can be improved by efficient stabilization.

A frequently cited ternary combination of antioxidants and synergists for the stabilization of oils rich in polyunsaturated fatty acids is the mixture of ascorbyl palmitate, lecithin, and tocopherols (Hamilton et al. 1998; Kulås and Ackman 2001). It has also recently been shown that efficient stabilization might also be achieved by choosing a combination of tocopherols rich in γ-tocopherol or δ-tocopherol and low in α-tocopherol, by including tocopherol-sparing synergists such as ascorbyl palmitate and carnosic acid from rosemary extract and metal-chelating agents (Drusch et al. 2008a).

In the process of microencapsulation, autoxidation occurs in the first stages of the microencapsulation process itself during emulsification and spray-drying. Serfert et al. (2008) investigated the efficacy of antioxidants during the microencapsulation process, and trace metal chelation by, e.g., Citrem or lecithin in combination with ascorbyl palmitate proved to be of particular importance in the emulsion, but not during storage of the microencapsulated oil. As described for the bulk oil, in the microencapsulated oil, the addition of rosemary extract rich in carnosic acid to ternary blends of tocopherols, ascorbyl palmitate, and lecithin or Citrem significantly retarded autoxidation (Serfert et al. 2008).

The oxidative stability of an emulsified oil can also be controlled by emulsifier type, concentration, and location (Donnelly et al. 1998; Fomuso et al. 2002; Hu et al. 2003). If emulsion droplets are surrounded by cationic emulsifiers, prooxidative metal cations are repelled, which leads to stabilization of the core material (Mancuso et al. 1999). A strengthening of the emulsion interface and the introduction of a positive charge by using lecithin and chitosan for emulsification of fish oil has recently been described. Both emulsions (Klinkesorn et al. 2005a) and microcapsules prepared by spray-drying of these emulsions (Klinkesorn et al. 2005b) showed a significant lower rate of autoxidation during storage. Specific literature on stabilization of the different core materials is available; some reviews have also recently been published (Moure et al. 2001; Naguib et al. 2003; Nenadis et al. 2003; Yanishlieva and Marinova 2001).

7.1.1.2 The Selection of a Suitable Combination of Wall Materials

An overwhelming amount of literature is available on the microencapsulation properties of different wall materials for lipophilic food ingredients. Generally, the wall materials can be classified into gums (gum arabic, locust bean gum, agar), lipids (wax, palm fat), proteins (gelatine, milk proteins, soy protein), polysaccharides (starch, xanthan, pullulan, guar gum, alginate), monosaccharides, disaccharides, and oligosaccharides (hydrolyzed starch, lactose), and cellulose and its derivatives (carboxymethylcellulose, methylcellulose).

The different characteristics of the individual wall materials need to be evaluated prior to selection of an individual compound or a mixture for a specific application. Physicochemical features of the wall material such as molecular weight, viscosity in solution, film-forming properties, and chemical features determining interactions in multicomponent matrices determine the stability of the core material and the structural integrity of the capsule. Important features of the wall material have been summarized by Desai and Park (2005) and among these are:

- Good rheological properties at high concentration and easy workability during encapsulation
- The ability to disperse or emulsify the active material and stabilize the emulsion produced
- Nonreactivity with the material to be encapsulated both during processing and on prolonged storage

- The ability to seal and hold the active material within its structure during processing or storage
- The ability to completely release the solvent or other materials used during the process of encapsulation under drying or other desolventization conditions
- The ability to provide maximum protection to the active material against environmental conditions (Desai and Park 2005)

Major emphasis must be put on the stability of the capsule structure during storage. Rapid dehydration of the emulsion during spray-drying generally leads to an amorphous solid microstructure of the wall material. In the amorphous glassy state, the mobility of molecules as well as permeability, a key step in diffusion-controlled reactions such as the Maillard reaction and lipid oxidation, is reduced. Nevertheless, oxygen permeation through glassy carbohydrate matrices still occurs and is the rate-limiting step in terms of autoxidation (Andersen et al. 2000; Orlien et al. 2000). Drusch et al. (2009) recently reported a difference in the oxidative stability of fish oil encapsulated in carbohydrate-based matrices with different molecular weight profiles. Using positron annihilation lifetime spectroscopy, they detected differences in the size of free volume elements in the amorphous matrices and hypothesized that these differences are linked to a change in oxygen diffusivity.

In contrast, matrix molecular mobility is not considered to be an important factor determining the stability of the core material as long as no collapse of the matrix occurs (Grattard et al. 2002; Selim et al. 2000). Crystallization of the capsule matrix leads to an increase in molecular mobility, with subsequent coalescence of the core material and its release from the matrix. The core material migrates to the surface of the microcapsule and leads to a change in surface composition. A change in surface composition, i.e., an increase in the amount of surface fat, after crystallization of the carbohydrate in the capsule wall has been described, e.g., by Elofsson and Millqvist-Fureby (2003). The loss of protection at temperatures above the specific glass transition temperature caused by crystallization of the capsule matrix has been described for encapsulated β-carotene (Elizalde et al. 2002), sea buckthorn oil (Partanen et al. 2002, 2005), and fish oil (Drusch et al. 2006). Crystallization of carbohydrates from the amorphous matrix may be retarded by combining sugars and polymers or sugars and divalent cations (Buera et al. 2005). An excellent review on the concept of glass transition temperature is that of LeMeste et al. (2002).

A positive effect of interactions between the individual wall constituents used for encapsulation has been reported by Rusli et al. (2006) and Augustin et al. (2006). In both studies, a heat-treated protein-carbohydrate mixture was used as an encapsulating agent. For both caseinate (Augustin et al. 2006) and whey or soy protein (Rusli et al. 2006) an increase in oxidative stability compared with encapsulation using the wall material without heating was reported. The authors explained the increase in stability by the development of antioxidative Maillard reaction products during heating of the wall material. However, Drusch et al. (2009) recently suggested that other factors apart from the antioxidative activity of the glycated proteins must significantly contribute to this effect. Finally, the wall material for encapsulation is strongly associated with the choice of homogenization and drying conditions, which is discussed in the following sections.

7.1.1.3 Emulsion Preparation

It is generally known that according to Stokes law oil the droplet size of the emulsion, the density of the different phases, and viscosity affect emulsion stability. The composition of the oil–water interface, interfacial thickness and strength, localization of antioxidants in the interface, metal-chelating ability, and oxygen- and radical-scavenging ability play a key role in stabilizing the emulsified material, and these factors also play an important role in the stability of the microencapsulated core material (Rusli et al. 2006).

Results on the influence of oil droplet size on the oxidative stability of emulsified oils have been contradictory. Lethuaut et al. (2002) reported that oxygen consumption and formation of conjugated dienes was increased in emulsions with large interfacial area. A similar observation has been reported for whey-protein-stabilized sunflower oil emulsions with average oil droplet sizes ranging from 1.9 to 0.5 μm. Although there was no clear correlation, the number of conjugated dienes tended to increase with decreasing oil droplet size (Kiokias et al. 2007). No effect of oil droplet size on the oxidative stability was detected in structured lipid-based oil-in-water emulsions with a droplet size ranging from 0.3 to 2.7 μm (Osborn and Akoh 2004). Finally, a higher stability for a low oil droplet size was reported by Nakaya et al. (2005). The hydroperoxide content in emulsions with a mean droplet size of 0.8 μm was significantly lower than that in emulsions with a mean droplet size of 12.8 μm for up to 120 h of oxidation time. A reduced mobility of the triacylglycerols due to a "wedge effect" associated with hydrophobic acyl residues of emulsifiers was proposed as a possible mechanism to explain differences in oxidative stability. Let et al. (2007) recently suggested that the composition of the interface is more important than the total surface area itself.

Only little information on the influence of oil droplet size and interfacial composition on the oxidative stability of encapsulated lipophilic food ingredients is available. In the case of proteins, structural changes at high homogenization pressure may lead to conformational changes and a compositional change of the oil–water interface. Integrity of the interface is maintained during drying and results in a more efficient encapsulation (Rusli et al. 2006). For encapsulation of flavors such as orange oil, a small droplet size may be favorable. Soottitaniwat et al. (2003) reported a lower retention of orange oil during encapsulation and a higher amount of surface oil for increased oil droplet size in the parent emulsion. The authors speculated that shearing and rupture of large oil droplets during atomization may be responsible for these effects. Concerning the influence of oil droplet size on the stability of the encapsulated lipophilic component during storage, available data indicate that a small oil droplet size is favorable. Minemoto et al. (2002) encapsulated linoleic acid into a matrix of maltodextrin and Tween 85. Linoleic acid oxidized faster in capsules with an oil droplet size of 1.6 μm compared with 1 μm. However, the encapsulation efficiency of 60% was very low. In encapsulated methyl linoleate, stability increased when microcapsules were prepared from emulsions with small oil droplet size (Nakazawa et al. 2008). A lower surface coverage with oil as reported for microcapsules prepared from nanoemulsions as reported by Jafari et al. (2008) as well as a modified interfacial composition might contribute to this effect.

7.1.1.4 Suitable Process Conditions to Optimize Core Material Retention and Physicochemical Characteristics

Process conditions include selection of the dryer itself, the type of atomization, and the relative humidity and temperature of the drying air. The selection of suitable process conditions greatly depends on the parent emulsion composition and its characteristics. The latter is the reason for contradictory results on the influence of individual process parameters on microencapsulation efficiency and core material stability, making it difficult to derive general recommendations for unit operations for microencapsulation by spray-drying.

Generally, at elevated inlet air temperature, ballooning may occur (Finney et al. 2002). Walton and Mumford (1999) described the effect of ballooning for skin-forming materials such as milk powder as follows: "A skin covered the whole droplet surface virtually instantaneously; this was rapidly followed by internal bubble nucleation. The bubbles expanded to violently distort, and eventually rupture, the skin surface causing the particle to collapse, shrivel and then re-inflate. This cycle was repeated three or four times until most of the internal moisture had evaporated. As the skin dried out and hardened, vaporization of residual moisture inflated the particle permanently to form a hollow particle with, in the majority of cases, a relatively smooth surface structure." High total solids in the parent emulsion may facilitate ballooning. High total solids leads to fixation of the particle structure early during the drying process (Finney et al. 2002). Thermal expansion of the air and water inside the droplet erases dents due to ballooning (Danviriyakul et al. 2002). Walton and Mumford (1999) emphasized that a high surface tension of the droplet but low viscosity is required to return to the spherical droplet to maintain particle integrity. In addition, it has been reported that small droplet formation associated with a large surface area hastens the drying process and influences microencapsulation efficiency (Rusli et al. 2006).

Apart from the most critical parameter, inlet air temperature, air inclusion is also influenced by air inclusion in the feed emulsion and air inclusion during atomization, steam formation during drying, and the rate of drying. The latter is determined by the difference between inlet and outlet air temperature (Finney et al. 2002). A strong influence of processing conditions on the physicochemical properties of fish oil encapsulated with n-octenylsuccinate-derivatized starch has been reported by Drusch and Schwarz (2006), including the phenomenon of ballooning (Fig. 7.2). Ballooning was associated with an increase in vacuole volume and a decrease in wall thickness of the microcapsules. Furthermore, high inlet and outlet air temperature led to ballooning and autoxidation of the encapsulated and nonencapsulated core material during the drying process

Small and dense particles are expected to have good oxygen barrier properties and thus possibly protect the core material. However, Finney et al. (2002) observed that large less dense particles are not necessarily associated with a decrease in stability. Orange oil encapsulated in a modified starch showed longer shelf life after drying at 220°C/100°C compared with the stability of encapsulated orange oil dried

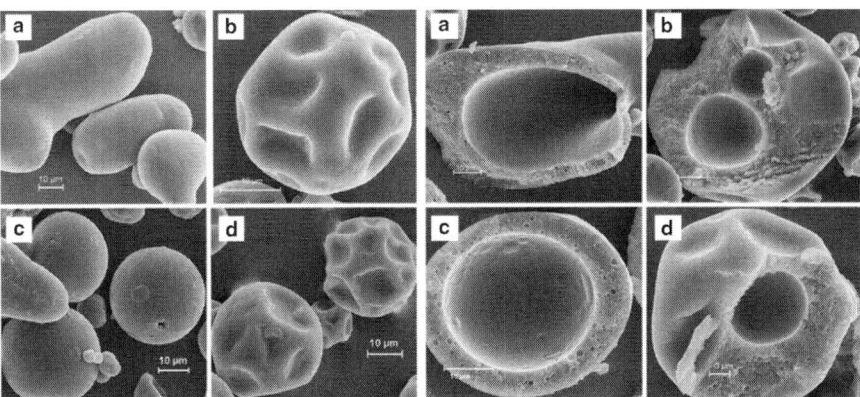

Fig. 7.2 Scanning electron micrographs of microencapsulated fish oil prepared with different types of modified starch (*A, B* medium viscosity; *C, D* low viscosity) and dried at 210°C/90°C (*A, C*) or 160°C/60°C (*B, D*) (Adapted from Drusch and Schwarz 2006)

at 170°C/80°C. A similar observation has been reported by Buffo and Reineccius (2000). Finney et al. (2002) stated that higher retention of flavors at elevated outlet temperature is only observed for highly volatile compounds. The reason is a lower relative humidity and thus a more rapid drying. Slow rotational speed when using a rotating disk for atomization led to a larger droplet size and thus particle size during encapsulation of linolenic acid into a matrix of gum arabic. Oxidation of linoleic acid proceeded more slowly in large particles (Fang et al. 2005).

Removal of water during rapid drying can result in conformational changes of high molecular weight emulsifying agents such as proteins. Subsequently, destabilization of the emulsion and an increase in oil droplet size may occur (Rusli et al. 2006). Low molecular weight carbohydrates such as glucose syrup with a dextrose equivalent of 36 may compensate for this effect by stabilizing the protein during dehydration (Danviriyakul et al. 2002). When maltodextrin with a dextrose equivalent of 10 was used, this stabilizing effect was missing and an increase in oil droplet size after drying was observed. Danviriyakul et al. (2002) also reported an interrelation between oil droplet size and microencapsulation efficiency. The amount of surface fat in the microcapsules increased sevenfold when the oil droplet size increased from 0.5 to 1.2 µm. The same authors observed an increase in the amount of surface fat from 2% to 25% when the carbohydrate source was changed from a glucose syrup with a dextrose equivalent of 36 to a maltodextrin with a dextrose equivalent of 10. Drying of the droplets is associated with an increase in viscosity due to a phase transition from a liquid to rubbery and finally a glassy state. Since the glass transition temperature of carbohydrates with a low dextrose equivalent is higher than the glass transition temperature of carbohydrates with a high dextrose equivalent, phase transition occurs earlier in the drying process, thus retarding the formation of impermeable microregions entrapping the core material (Danviriyakul et al. 2002).

7.1.1.5 The Role of Nonencapsulated Core Material

In a study by Rusli et al. (2006), microcapsules with a low proportion of nonencapsulated oil had lower stability than microcapsules with a high proportion of nonencapsulated oil. The authors hypothesized that a reduced oil droplet size in microcapsules with high microencapsulation efficiency was the reason for the decreased stability. A small oil droplet size is related to a high surface area and for a given amount of wall material subsequently to a thinner film of encapsulating material (Rusli et al. 2006). Nonencapsulated core material is frequently not correlated with the shelf life of the microencapsulated product as was also shown for orange oil (Finney et al. 2002), methyl linoleate, (Minemoto et al. 1997), and fish oil (Drusch and Berg 2008) and led some authors to the misleading conclusion that nonencapsulated core material is not important in terms of product stability or shelf life. Confocal laser scanning microscopy, scanning electron microscopy, and different extraction procedures revealed that the extractable oil in microencapsulated fish oil is mainly located on the surface and in oil droplets close to the surface, but is also located in inner parts of the microcapsule (Drusch and Berg 2008).

Irrespective of its amount, in certain applications such as encapsulation of oils rich in polyunsaturated fatty acids nonencapsulated core material may strongly impair product acceptability. Autoxidation of polyunsaturated fatty acids leads to the development of volatile secondary oxidation products. These volatiles may have a very low odor threshold and consequently negatively affect sensory properties of the microcapsules (Jacobsen 1999; Lee et al. 2003; Venkateshwarlu et al. 2004). Furthermore, the nonencapsulated fat is directly associated with flowability and wettability of the microcapsules (Vega and Roos 2006).

For these reasons, strategies to eliminate the nonencapsulated core material are required. One possibility is to mask nonencapsulated core material by applying a secondary coating of the microcapsules. Secondary coating within a fluidized bed of starch granules inside the spray-dryer for encapsulated fish oil has been successfully applied and has been patented (Schaffner 2004). Typical coating materials applied in the food industry include gums, cellulose derivatives, modified starches proteins, and lipids and waxes. Secondary coating with improved stability has been described, e.g., for flavor compounds encapsulated into a matrix of gum arabic, maltodextrin, and modified starch coated with gellan gum (Cho and Park 2002) or butter oil encapsulated with sucrose using hydrogenated stearins (Onwulata et al. 1998). In both cases moisture uptake and thus stability was improved. However, when choosing a coating material for secondary coating of polyunsaturated fatty acids, one also needs to consider oxygen barrier properties of the coating material. Generally, lipids provide good protection against moisture, but not against oxygen. Gas permeability can be reduced by using proteins as coating material (McHugh and Krochta 1994). An alternative strategy is the removal of nonencapsulated oil from the microcapsule. Valentinotti et al. (2006) recently patented a technology to remove nonencapsulated oil by washing microcapsules with limonene.

In conclusion, spray-drying is still a valuable, straightforward, and inexpensive technology for encapsulation of lipophilic food ingredients. Concise analysis of the

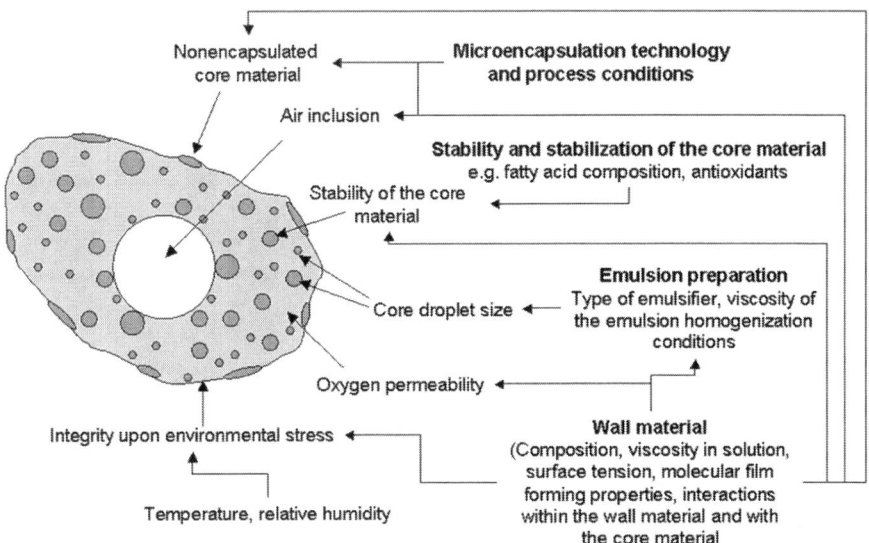

Fig. 7.3 Factors determining the stability of lipophilic functional ingredients and interacting product and process characteristics

available literature showed that success in microencapsulation by spray-drying depends on a variety of variables, which are summarized in Fig. 7.3. Process and product variation strongly influences microcapsule properties, and process modification and additional processing steps may be required for successful development of new applications.

7.1.2 Microencapsulation by Extrusion

Surprisingly few scientific studies deal with the microencapsulation of polyunsaturated fatty acids by extrusion and also even more generally with the encapsulation of lipophilic functional food ingredients by extrusion. Microencapsulation by extrusion faces the same critical factors and challenges as described for microencapsulation by spray-drying. An increased stability of the core material due to a decrease in the free volume is claimed by companies in the aroma industry, but reliable scientific data have not been presented. It is well accepted that oxygen permeation in glassy matrices occurs. Orlien et al. (2006) described that glassy matrix yields only partial protection against lipid oxidation as it allows permeation of oxygen and other small molecules. At low temperature, lipid oxidation was rate-determining, whereas at higher temperature, oxygen permeation through the matrix with a lower energy of activation became rate-determining.

Gray et al. (2008) recently encapsulated a mixture of free fatty acids rich in linoleic acid into a starch-based matrix using a corotating, intermeshing, twin-screw extruder. The authors described an increase in fatty acid oxidation during storage, when small cracks were visible in the glassy matrix. Furthermore, the authors emphasized the significance of nonencapsulated core material at the surface of the extrudate for the overall oxidative status. The latter is another similarity to encapsulated oils prepared by spray-drying. To overcome sensory problems associated with the oxidation of nonencapsulated core material, Valentinotti et al. (2006) developed a patented encapsulation process in which the extrudate is washed with a limonene-rich extract from citrus fruits. Van Lengerich (2002) described extrusion of a precoated core material into a starch-based matrix or extrusion with subsequent coating of the extrudate.

7.1.3 Microencapsulation by Gelation and Phase Separation of Proteins and Biopolymers

The microencapsulation techniques covered in this section are based on gelation and phase-separation phenomena of single biopolymers and/or proteins as well as mixtures thereof. The resulting microcapsules are sometimes referred to as hydrogel particles (McClements et al. 2007). This generalization is scientifically not precise, since phase-separation phenomena such as simple or complex coacervation do not result in formation of hydrogels. Hydrogels are generally defined as hydrophilic polymeric networks capable of immobilizing large amounts of a liquid phase, usually water.

Microcapsules based on a single biopolymer can be produced by phase separation (simple coacervation) or gelation. Gelled emulsion particles are mainly used to modulate sensory perception and to deliver aroma compounds released in the oral cavity upon chewing (Malone et al. 2003). Ionotropic gelation or postgelation cross-linking of the biopolymers or proteins alters diffusibility and digestibility and thus facilitates the possibility for microencapsulation of oils rich in polyunsaturated fatty acids. Bustos et al. (2003) encapsulated krill oil via ionotropic gelation of a Tween 20- and chitosan-based emulsion.

An alternative method for microcapsule formation is the gelation of a biopolymer in an oil–water–oil emulsion. The inner oil phase contains the core material and the double emulsion is prepared in two subsequent homogenization steps. The biopolymer is gelled and facultatively cross-linked within this secondary emulsion and the outer oil phase is subsequently removed with an organic solvent. Using this method, Cho et al. (2003) encapsulated fish oil in capsules based on soy protein isolate. Capsules were either cross-linked with transglutaminase ot heat-gelled, or treated with a combination of both. Owing to the comparably long reaction time, loss of core material was rather high when transglutaminase cross-linking was applied. In contrast, the highest stability of the core material was achieved with transglutaminase cross-linking, and also a sustained release was only achieved when enzyme treatment was applied.

The dominating technology in this field is complex coacervation. Complex coacervation is a phase-separation process based on the simultaneous desolvation of oppositely charged polyelectrolytes induced by modification of the medium (Ducel et al. 2004). The technique can be used to deposit or to build up the complex coacervates on core material particles with subsequent hardening and/or drying. As reviewed by de Kruif et al. (2004), a number of oppositely charged proteins and hydrocolloids have successfully been used for the formation of complex coacervates. Lamprecht et al. (2001) encapsulated eicosapentaenoic acid ethyl ester using gelatine as an emulsifier and subsequent coacervate formation with gum arabic. Ethanol hardening of the complex coacervate, cross-linking with dehydroascorbic acid, and spray-drying of the complex coacervate were compared, and ethanol hardening provided the highest storage stability for the encapsulated core material. Plant phenolics represent an alternative class of cross-linkers for food applications (Strauss and Gibson 2004). Weinbreck et al. (2004) highlighted that the size of the complex coacervate and the mode of cross-linking affect the release of the encapsulated core material in vivo. An interesting technology combing complex coacervation and spray-drying to prepare dry microcapsules with an extremely low amount of nonencapsulated core material has been patented (Yan et al. 2004). The core material is emulsified with an emulsifying polymer such as gelatine. Coacervate formation is achieved by adding a specific amount of polyphosphate, adjustment of pH, and modification of temperature. Stepwise addition of polyphosphate leads to the formation of multiple coacervate shells. An alternative method to increase the protection against atmospheric oxygen is to encapsulate the coacervate in a glassy matrix by extrusion (Bouquerand et al. 2007).

7.1.4 Concluding Remarks on Microencapsulation Techniques

Ubbink and Krüger (2006) emphasized that new applications in the food sector are particularly challenging, since in the food sector product safety, appearance, storage conditions, ease of preparation, freshness, and sensory properties of the product are not to be compromised by incorporation of the active ingredient. The different requirements concerning the protective environment resulting from the type of core material, its controlled release, and the properties of the specific food matrix necessitate individual concepts for different encapsulants and applications. A top-down approach with a precise definition of the goals of the microencapsulation process is therefore required prior to development of an application.

Significant improvement of existing microencapsulation technologies can be achieved by focusing on stabilization and improvement of the composition and nanostructural features of existing capsules. In this context, a more detailed investigation of the influence of the emulsion interface and its active modification as well as the use of "intelligent" wall material systems actively participating in the stabilization of the core material are promising fields of future research.

7.2 Energetic Aspects of Microencapsulation

The principal operations in nearly all the technologies presented are the preparation of an oil-in-water emulsion and in a later stage of the process a drying step, most frequently spray-drying. In terms of environmental aspects, energy efficiency is one of the key issues and is discussed in this section.

7.2.1 Emulsification

During emulsification, droplets of an insoluble phase (A) are dispersed within another phase (B), the so-called continuous phase. Disregarding the "energy that is gained" by enhancing the entropy during the mixing of the two phases (which only plays an important role in nanoemulsions with droplet sizes much below 100 nm and strong surfactants), the minimum energy that has to be afforded for the droplet generation is that for building the new interface S_I between the dispersed and the continuous phase. This work W can be expressed as

$$W = \sigma \cdot S_I, \tag{7.1}$$

where σ is the interfacial tension between the two phases, which is mainly dependent on their composition (e.g., can be effectively reduced by surfactants or emulsifiers). By a simple calculation, the typical range of values of that work can be estimated. Using Eq. 7.1 and calculating the newly formed interface S_I from the number n of droplets built and with $x_{1,2}$ as their Sauter diameter (one of several mean diameters possible; Löffler and Raasch 1992), one obtains for the work

$$W = \sigma \cdot S_I = \sigma \cdot n \cdot \pi x_{1,2}^2. \tag{7.2}$$

The number of droplets is expressed by the ratio of the volume V_A (of the complete phase A) and the volume v of one droplet with Sauter diameter $v = \dfrac{\pi}{6} x_{1,2}^3$:

$$n = \frac{6 \cdot V_A}{\pi \cdot x_{1,2}^3}. \tag{7.3}$$

This can be inserted into Eq. 7.2, which yields

$$W = \sigma \cdot \frac{6 \cdot V_A}{\pi \cdot x_{1,2}^3} \cdot \pi \cdot x_{1,2}^2 = \frac{6 \cdot V_A \cdot \sigma}{x_{1,2}}. \tag{7.4}$$

Using $\phi = \dfrac{V_A}{V}$, the volume fraction of the dispersed phase, the work or energy density E_V that is needed to form the new interface is

$$E_V = \frac{W}{V} = \frac{6 \cdot \phi \sigma}{x_{1,2}}. \tag{7.5}$$

As an example, for the generation of oil droplets of diameter $x_{1,2} = 1$ μm within pure water where the interfacial tension is approximately 0.005 N/m with a dispersed phase fraction of $\phi = 50\%$, the necessary energy density according to Eq. 7.5 is

$$E_V = \frac{6 \times 0.5 \times 0.005 \text{N/m}}{10^{-6} \text{m}} = 150 \times 10^3 \frac{\text{N}}{\text{m}^2} = 1.5 \, bar.$$

The energy density has the same dimension as pressure, so this value can be compared with a typical value for the pressure drop during high-pressure emulsification of 150 bar, for example. Thus, the energy efficiency η of such a system is very low at 1%. In reality, this efficiency is even lower, when the reduced interfacial tension in real emulsions (due to emulsifiers) is taken into account. It has been shown recently (Kempa et al. 2006) that emulsifiers do not facilitate the interface generation (as was believed previously) but are mainly useful for stabilizing the newly built droplets against coalescence.

Similar low efficiencies as in high-pressure homogenization exist for other emulsification techniques where large droplets are disrupted by shear and elongational forces as in rotor-stator and ultrasonic emulsification machines. These low efficiencies point out that most of the energy in these devices is not used for interface generation but is used for turbulence and thus for heat production. In most of the cases up to now this heat is only emitted and is not used for energy recovery.

A different principle of droplet dispersion is used in membrane emulsification and was recently used in microchannel emulsification. In both techniques the phase to be dispersed is pressed by a rather low pressure through pores or channels into the continuous phase. On a theoretical basis, much higher efficiencies should be achieved. In traditional membrane emulsification (Schröder and Schubert 1999) droplets are generated from fluid "jets" (emitted from the pores) by shear forces from the continuous phase, passing over the membrane surface. To achieve small droplet sizes, the continuous phase flow has be high, so high dispersed phase volume fractions can only be achieved by large membranes and/or recirculation. Both methods lead to higher energy consumption (pressure drops) for pumping the continuous phase or the emulsion and thus reduced efficiencies (compared with the theoretical efficiencies). Nevertheless, the minimum droplet size is limited by approximately the pore size of the membrane. That means that in order to increase the droplet size, the continuous flow can be reduced, whereas to achieve smaller droplet sizes it may be necessary to have another membrane with smaller pores. Similar effects also occur for premix emulsification, where a pre-emulsion is completely pressed through a membrane, decreasing the droplet size.

In microchannel emulsification (Sugiura et al. 2002) the droplets are generated at the end of specially formed microchannels by fluid dynamic instabilities, so high shear forces of the continuous phase and thus high energy consumption are not

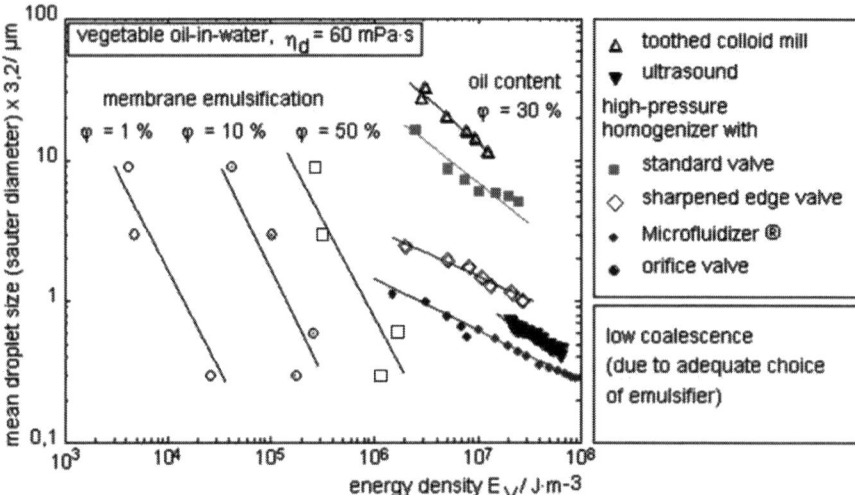

Fig. 7.4 Comparison of various homogenization processes by means of energy density (From Schubert et al. 2006)

Table 7.1 Emulsification processes and their effect on droplet size (process function) (Adapted from Schuchmann and Schuchmann 2005)

Process	Flow regime in dispersing zone	Dependency of the droplet size on energy density
Rotor-stator machines: colloid mill, toothed ring dispersion machine	Laminar shear turbulent	$x_{1,2} \sim E_V^{-1}$ $x_{1,2} \sim E_V^{-0.25...-0.4}$
High-pressure homogenizers (depending on geometry)	Turbulent laminar elongation cavitation	$x_{1,2} \sim E_V^{-0.25...-0.4}$ $x_{1,2} \sim E_V^{-1}$
Ultrasound	Cavitation, turbulent	$x_{1,2} \sim E_V^{-0.6}$
Membrane	Laminar droplet formation	$x_{1,2} \sim E_V^{-1}$

needed. Nevertheless, problems in scaling up the microscale setups and thus the possible emulsion volumes still hinder its possible wide industrial use.

For the choice of the most economic emulsification technique the energy density concept has been developed: It was shown that in continuous emulsification (with not too long residence times within the dispersing zone) the desired Sauter diameter $x_{1,2}$ of an emulsion can be expressed as a function of the energy density E_V. In Fig. 7.4 and Table 7.1 this is depicted for some typical emulsification machines using the example of an oil-in-water emulsion. The diameter calculated from Eq. 7.5 is also shown, giving the minimum diameter possible at the corresponding energy density $x_{1,2} = \dfrac{6 \cdot \phi \cdot \sigma}{E_V}$ but depending on the interfacial tension and the phase fraction, which can approximately be achieved by membrane emulsification.

The comparison makes clear that apart from membrane emulsification (and the microchannel emulsification; not presented in this section) the efficiency becomes

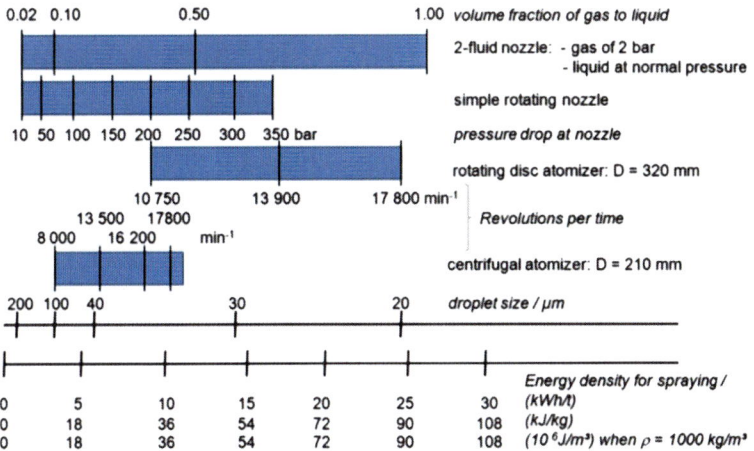

Fig. 7.5 Energy densities for spraying (Adapted from Schuchmann and Schuchmann 2005)

even lower when the droplet size is reduced. Thus, it is very important to choose the right technique.

7.2.2 Spray-Drying

The process of spray-drying can be divided into three unit operations that are normally coupled in the process: the spraying of the liquid (solution or dispersion), the drying, which means the evaporation of the liquid within a noncondensing gas, and the separation of the dried solid particles from that gas. Depending on the particle size, the total energy used is divided into different fractions in these unit operations: the smaller the particles, the more energy has to be invested in spray generation and solid–gas separation, whereas the energy for evaporation is more or less independent of the droplet size.

For the generation of droplets/sprays, as in the case of emulsification, the area of the interface (here between liquid and gas) is largely increased, this being the reason for short drying times. This droplet formation is generally done in rotating disc (so-called) atomizers with different diameters and rotation speeds or in nozzles where only the liquid or the liquid and a gas are pressed through. Also, ultrasound dispersion is being investigated. Typical specific energies E_m (based on mass, $E_m T = \dfrac{E_\lambda}{\rho}$, where ρ is the mass density) are given in Fig. 7.5, showing that as in the case of emulsification a lot of energy is dissipated into heat and not used for surface generation. But in this case the heat is helpful for increasing the temperature of the liquid, thus aiding the evaporation. Besides, the energy densities for spraying are often by at least an order of magnitude smaller than the energy density used for

Table 7.2 Selected food trends and related market opportunities (Adapted from Steenkamp 1997)

Food trend	Opportunities
Growing health concerns	Fresh products, salads
	Low-calorie and light products
	Functional foods
Growing convenience orientation	Frozen food
	Ready-to-eat foods
	Food away from home
Hedonism	Taste
	Lifestyle- and image-enhancing products
Polarization: price consciousness versus enjoyment and fun	Store brands, generic brands
	Gourmet, exotic, ethnic, and luxury food
Growing environmental and ethical concerns	Regional products
	Free-range meat
	Organic food
	Fair-trade-certified products
	Reduced CO_2 emission

heating the liquid and evaporation (e.g., $h_{V,water} = 2{,}260$ kJ/kg), so they do not play such an important role as in emulsification.

Whereas the heat that is stored in the dry product cannot be simply reused, the energy stored in the gas or vapor can be recovered by condensation and preheating the liquid to be sprayed, simultaneously drying the gas. Nevertheless it is often much more efficient to increase the solid concentration in the liquid dispersion in advance of drying by nonthermal methods as, for example, in membrane processes.

7.3 Consumer Trends and Market Perspectives

The development and marketing of new food products is an important although risky strategy of modern food companies in saturated markets. A main contributor to the success of new food products is knowledge about the consumer. There has been an increasing focus on so-called consumer-driven product development and one often talks about the need to listen to the consumers' voice (Ruder et al. 2007).

In this regard, Table 7.2 gives an overview of selected food trends and consumers' needs and desires concerning food demand, such as growing health concerns, growing convenience orientation, hedonism, polarization, and growing environmental and ethical concerns. Additionally, this table provides some examples of related marketing opportunities. Health-conscious consumers are expected to buy more fresh products, low-calorie products, and/or functional food than consumers in other consumer segments.

According to Euromonitor International (2006), health consciousness is one of the dominant drivers of consumer markets in the twenty-first century. Consumers' growing health concerns are caused by a lack of physical exercise, a life full of stress, time pressure, and most importantly the rapid aging of the Western population.

Normally, older people have more health problems than young people. Consequently, they are particularly interested in products that can contribute to or restore their health (Steenkamp 1997). In addition, empirical results indicate that people who maintain a healthy lifestyle tend to be female and better educated (Divine and Lepisto 2005).

In the context of healthy nutrition, functional foods play a specific role. They are intended not only to satisfy hunger and provide humans with necessary nutrients, but also to prevent nutrition-related diseases and increase physical and mental well-being of consumers (Menrad 2003). The benefits for consumers are wide-ranging, and are associated with different advantages regarding human health and quality of life (Frewer et al. 2003). The motivation behind consumers' functional food choice differs. They may use functional food in the context of a healthy lifestyle or as a means to compensate for an unhealthy one (De Jong et al. 2003). Furthermore, functional foods are a rather convenient way to follow a healthy lifestyle. According to Euromonitor International (2006), convenience is another key reason why consumers choose functional foods. This example shows that the food trends mentioned above and shown in Table 7.2 are partially interrelated (Steenkamp 1997). However, health and convenience are considered as the two biggest trends in today's food industry (Divine and Lepisto 2005). Functional food has actually been reported as the top trend (Verbeke 2005).

Between 2004 and 2005, the functional food market grew by 13.4% globally (Euromonitor International 2006). Food industry companies have rather high expectations for food products that meet consumers' demand for a healthy lifestyle (Menrad 2003). Therefore, to expand the markets for healthy products and functional foods, food companies may focus on ways of making their products more convenient for time-pressured consumers (Divine and Lepisto 2005). Additionally, to be successful in the market the products should not only bring positive health effects, they should also have sensory characteristics that are attractive to the target group (Ruder et al. 2007).

According to industry consultants Frost & Sullivan (2005), the future for omega-3 polyunsaturated fatty acids – in comparison with all functional food ingredients currently available – looks most promising. In 2007, heart-related and circulatory conditions, including general heart health, blood pressure, and stroke, were the most important health concerns of Consumers in the US (International Food Information Council 2007). There is scientific evidence that the consumption of long-chain omega-3 polyunsaturated fatty acids reduces the risk of coronary artery disease. In this regard, many countries have issued recommendations for increasing their consumption. The American Health Association advocates, for example, the consumption of fish high in omega-3 fatty acids (Melanson et al. 2005). The UK dietary guideline for cardiovascular disease acknowledges their importance in reducing heart disease risks (Ruxton et al. 2004). In many European countries, increasing sales of products with polyunsaturated fatty acids were paralleled by more favorable ratios of polyunsaturated to saturated fat in national diets.

Omega-3 fatty acids have also shown great performance in the nutritional supplement market. The European nutritional supplement market is presumed to be the most developed. One reason might be that consumers' acceptance is highest in regions that have traditionally used marine oils such as northern Europe and Scandinavia. Fortified foods and beverages are also starting to gain acceptance (Frost & Sullivan 2004). That means omega-3 fatty acids are expected to become more and more an important nutrient in a wide range of products: breads, milk, juices, tortillas, chocolate, yoghurt drinks, spreads, peanut butter, eggs, meat, etc. It is estimated that there were 723 omega-3-bearing product launches in Europe in 2007 compared with 291 in 2005. In North America there were 315 launches in 2005 and 541 in 2007 (Starling 2008b).

New food technologies such as microencapsulation improve the palatability as well as the stability of omega-3 polyunsaturated fatty acids (Whelan and Rust 2006). The most important factor of market success is the ability to develop and market a high-quality food product. Analyses of previous successes and failures of functional foods suggest that specifically a lack of competencies in the optimization of sensory properties of the product, a lack of negotiation power in securing retail distribution for the products, and a lack experience with attractive pricing and packaging of foods are responsible for failures (Bech-Larsen and Scholderer 2007; Menrad 2003). Furthermore, as mentioned above, scientific evidence and perceived consumers' need from a public health perspective are not a guarantee for market success (Weststrate et al. 2002).

Presently, very few omega-3 functional foods and beverages have moved beyond niche sales level. The omega-3 market is not as strong as expected. Globally, the number of mainstream products is low. In Spain, a milk (Puleva) has been on the market for about 10 years and in Australia an omega-3 bread (Tip-Top) has gained over 10% of the sliced white bread market (Starling 2008). Additionally, infant nutrition has become a successful sector for omega-3 functional foods (Starling 2008). Empirical studies indicate that Consumers in the US are increasingly aware of specific health benefits associated with various functional foods. But consumers tend to think of certain foods or food categories that contain healthy components rather than the functional ingredients themselves. That means consumers may identify fish, fish oil, certain nuts, and flax as being good for their hearts and other conditions, but they are not able to articulate that omega-3 polyunsaturated fatty acids are the healthful ingredient that they all have in common (International Food Information Council 2007). Consequently, food manufacturers and marketers have to educate consumers. They have to make them understand and accept omega-3 fatty acids as "good fats" that provide a wide range of benefits that are unmatched by any other food ingredients (Frost & Sullivan 2004) and that not all omega-3 fatty acids are the same. α-Linolenic acid is not the same thing as eicosapentaenoic acid and docosahexaenoic acid (Whelan and Rust 2006).

To sum up, according to Weststrate et al. (2002), the following factors are important prerequisites for success of a functional food:

- Consumer need and awareness
- Consumer acceptance of a food solution

- Powerful communication of the health benefits to the consumer
- Uncompromised taste
- Optimal convenience
- Adequate retail or out-of-home availability
- Proven safety and efficacy
- Acceptable price level
- Assurance and support from different sources, including scientific opinion leader
- A clear regulatory framework for making claims

Finally, manufacturers have to take into account that the application of modern food technology may constitute an unaccepted risk from the perspective of consumers. Consumers may perceive functional foods as less natural than conventional products. Accordingly, functional foods are avoided by those who value naturalness in food choices (Frewer et al. 2003). However, commercial development of new and innovative sources of omega-3 polyunsaturated fatty acids (algal oils, including oils from organic sources, genetically modified plants) provides a cheaper and safer supply compared with traditional sources such as fish. This may result in a rapid development of nontraditional foods that are enriched or fortified (Whelan and Rust 2006). To increase market sales it is important that consumers perceive a match in functional foods between the carrier food and the functional ingredient with which it is enriched (Bech-Larsen and Scholderer 2007). For future marketing strategies, an approach that integrates insights into consumer needs and desires (market pull) and structured scientific research (science push) will give the greatest chance of real innovations (Weststrate et al. 2002).

References

Andersen AB, Risbo J, Andersen ML, Skibsted LH. Oxygen permeation through oil-encapsulating glassy food matrix studied by ESR line broadening using a nitroxyl spin probe. Food Chem. 2000;70:499–508.

Augustin MA, Sanguansri L, Bode O. Maillard reaction products as encapsulants for fish oil powders. J Food Sci. 2006;71:E25–32.

Bech-Larsen T, Scholderer J. Functional foods in Europe: consumer research, market experiences and regulatory aspects. Trends Food Sci Technol. 2007;18:231–4.

Bouquerand P-E, Dardelle G, Gouin S, Schleifenbaum B, Trophardy G. Encapsulated active ingredients, methods of preparation and their use. 2007. WO2007/026307.

Buera P, Schebor C, Elizalde B. Effects of carbohydrate crystallization on stability of dehydrated foods and ingredient formulations. J Food Eng. 2005;67:157–65.

Buffo R, Reineccius GA. Optimization of gum acacia/modified starch/maltodextrin blends for the spray drying of flavors. Perfum Flavor. 2000;25:45–54.

Bustos R, Romo L, Yánez K, Díaz G, Romo C. Oxidative stability of carotenoid pigments and polyunsaturated fatty acids in microparticulate diets containing krill oil for nutrition of marine fish larvae. J Food Eng. 2003;56:289–93.

Cho YH, Park J. Characteristics of double-encapsulated flavor powder prepared by secondary fat coating process. J Food Sci. 2002;67:968–72.

Cho Y-H, Shim HK, Park J. Encapsulation of fish oil by an enzymatic gelation process using transglutaminase cross-linked proteins. J Food Sci. 2003;68:2717–23.

Danviriyakul S, McClements DJ, Decker E, Nawar WW, Chinachoti P. Physical stability of spray-dried milk fat emulsion as affected by emulsifiers and processing conditions. J Food Sci. 2002;67:2183–9.

De Jong N, Ocké MC, Branderhorst HAC, Friele R. Demographic and lifestyle characteristics of functional food consumers and dietary supplement users. Br J Nutr. 2003;89:273–81.

de Kruif CG, Weinbreck F, de Vries R. Complex coacervation of proteins and anionic polysaccharides. Curr Opin Colloid Interface Sci. 2004;9:340–9.

Desai KGH, Park HJ. Recent developments in microencapsulation of food ingredients. Dry Technol. 2005;23:1361–94.

Divine RL, Lepisto L. Analysis of a healthy lifestyle consumer. J Consum Mark. 2005;22:275–83.

Donnelly JL, Decker E, McClements DJ. Iron-catalyzed oxidation of menhaden oil as affected by emulsifiers. J Food Sci. 1998;63:997–1000.

Drusch S, Berg S. Extractable oil in microcapsules prepared by spray-drying: localisation, determination and impact on oxidative stability. Food Chem. 2008;109:17–24.

Drusch S, Schwarz K. Microencapsulation properties of two different types of n-octenylsuccinate-derivatised starch. Eur Food Res Technol. 2006;222:155–64.

Drusch S, Serfert Y, Van Den Heuvel A, Schwarz K. Physicochemical characterization and oxidative stability of fish oil encapsulated in an amorphous matrix containing trehalose. Food Res Int. 2006;39:807–15.

Drusch S, Groß N, Schwarz K. Efficient stabilisation of bulk fish oil rich in long chain polyunsaturated fatty acids. Eur J Lipid Sci Technol. 2008a;110:351–9.

Drusch S, Berg S, Scampicchio M, Serfert Y, Somoza V, Mannino S, Schwarz K. Role of glycated caseinate in stabilisation of microencapsulated lipophilic functional ingredients. Food Hydrocolloid. 2008b. doi:10.1016/j.foodhyd.2008.07.004.

Drusch S, Rätzke K, Serfert Y, Steckel H, Scampicchio M, Voigt I, Schwarz K, Mannino S. Differences in free volume elements of the carrier matrix affect the stability of microencapsulated lipophilic food ingredients. Food Biophys. 2009;4:42–48.

Ducel V, Richard J, Saulnier P, Popineau Y, Boury F. Evidence and characterization of complex coacervates containing plant proteins: application to the microencapsulation of oil droplets. Colloids Surf A Physicochem Eng Aspects. 2004;232:239–47.

Elizalde B, Herrera ML, Buera MP. Retention of b-carotene encapsulated in a trehalose-based matrix as affected by water content and sugar crystallization. J Food Sci. 2002;67:3039–45.

Elofsson U, Millqvist-Fureby A. Stability of spray-dried protein-stabilized emulsions – effects of different carbohydrate additives. Spec Publ R Soc Chem. 2003;284:265–74.

Euromonitor International. Functional products meet the demands of today's consumer. 2006. http://www.euromonitor.com/Articles.aspx?folder=Functional_products_meet_the_demands_of_todays_consumer&print=true. Accessed 13 Oct 2008.

Fang X, Shima M, Adachi S. Effects of drying conditions on the oxidation of linoleic acid encapsulated with gum arabic by spray drying. Food Sci Technol Res. 2005;11:380–4.

Finney J, Buffo R, Reineccius GA. Effects of type of atomization and processing temperature on the physical properties and stability of spray-dried flavours. J Food Sci. 2002;67:1108–14.

Fomuso LB, Corredig M, Akoh CC. Effect of emulsifier on oxidation properties of fish oil-based structured lipid emulsions. J Agric Food Chem. 2002;50:2957–61.

Frewer L, Scholderer J, Lambert N. Consumer acceptance of functional foods: issues for the future. Br Food J. 2003;105:714–31.

Frost & Sullivan. European omega-3 and omega-6 PUFA ingredients market. Research review. 2004. http://www.frost.com/prod/servlet/report-brochure.pag?id=B329-01-00-00-00. Accessed 14 Oct 2008.

Frost & Sullivan, editor. Opportunities in the microencapsulated food ingredient market. 2005. http://www.frost.com/prod/servlet/report-brochure.pag?id=B716-01-00-00-00. Accessed 26 May 2006.

Garg ML, Wood LG, Singh H, Moughan PJ. Means of delivering recommended levels of long chain n-3 polyunsaturated fatty acids in human diets. J Food Sci. 2006;71:R66–71.

Gharsallaoui A, Roudaut G, Chambin O, Voilley A, Saurel R. Applications of spray-drying in microencapsulation of food ingredients. Food Res Int. 2007;40:1107–21.

Gouin S. Microencapsulation: industrial appraisal of existing technologies and trends. Trends Food Sci Technol. 2004;15:330–47.

Grattard N, Salaün F, Champion D, Roudat G, Le Meste M. Influence of physical state and molecular mobility of freeze-dried maltodextrin matrices on the oxidation rate of encapsulated lipids. J Food Sci. 2002;67:3002–10.

Gray DA, Bowen SE, Farhat I, Hill SE. Lipid oxidation in glassy and rubbery-state starch extrudates. Food Chem. 2008;106:227–34.

Hamilton RJ, Kalu C, McNeill GP, Padley FB, Pierce JH. Effects of tocopherols, ascorbyl palmitate, and lecithin on autoxidation of fish oil. J Am Oil Chem Soc. 1998;75:813–22.

Hu M, McClements DJ, Decker EA. Lipid oxidation in corn oil-in-water emulsions stabilized by casein, whey protein isolate, and soy protein isolate. J Agric Food Chem. 2003;51:1696–700.

International Food Information Council. Consumer attitudes toward functional foods/foods for health, executive summary. 2007. http://www.ific.org/research/foodandhealthsurvey.cfm. Accessed 10 Oct 2008.

Jacobsen C. Sensory impact of lipid oxidation in complex food systems. Fett/Lipid. 1999;101:484–92.

Jafari SM, Assadpoor E, Bhandari B, He Y. Nano-particle encapsulation of fish oil by spray-drying. Food Res Int. 2008;41(2):172–83.

Kempa L, Schuchmann HP, Schubert H. Tropfenzerkleinerung und Tropfenkoaleszenz beim mechanischen Emulgieren mit Hochdruckhomogenisatoren. Chem Ing Tech. 2006;78:765–8.

Kiokias S, Dimakou C, Oreopoulou V. Effect of heat treatment and droplet size on the oxidative stability of whey protein emulsions. Food Chem. 2007;105:94–100.

Klinkesorn U, Sophanodora P, Chinachoti P, McClements DJ, Decker E. Increasing the oxidative stability of liquid and dried tuna oil -in-water emulsions with electrostatic layer-by-layer deposition technology. J Agric Food Chem. 2005a;53:4561–6.

Klinkesorn U, Sophanodora P, Chinachoti P, McClements DJ, Decker E. Stability of spray-dried tuna oil emulsions encapsulated with two-layered interfacial membranes. J Agric Food Chem. 2005b;53:8365–71.

Kulås E, Ackman RG. Protection of a-tocopherol in nonpurified and purified fish oil. J Am Oil Chem Soc. 2001;78:197–203.

Kunz B, Krückeberg S, Weißbrodt J. Chancen und Grenzen der Mikroverkapselung in der modernen Lebensmittelverarbeitung. Chem Ing Tech. 2003;75:1733–40.

Lamprecht A, Schäfer UF, Lehr CM. Influences of process parameters on preparation of microparticle used as a carrier system for w-3 unsaturated fatty acid ethyl esters used in supplementary nutrition. J Microencapsul. 2001;18:347–57.

Le Meste M, Champion D, Roudat G, Blond D, Simatos D. Glass transition and food technology: a critical appraisal. J Food Sci. 2002;67:2444–58.

Lee H, Kizito SA, Weese SJ, Craig-Schmidt MC, Lee Y, Wei C-I, et al. Analysis of headspace volatile and oxidized volatile compounds in DHA-enriched fish oil on accelerated oxidative storage. J Food Sci. 2003;68(7):2169–77.

Let MB, Jacobsen C, Soerensen A-DM, Meyer AS. Homogenization conditions affect the oxidative stability of fish oil enriched milk emulsions: lipid oxidation. J Agric Food Chem. 2007;55:1773–80.

Lethuaut L, Métro F, Genot C. Effect of droplet size on lipid oxidation rates of oil-in-water emulsions stabilized by protein. J Am Oil Chem Soc. 2002;79:425–30.

Löffler F, Raasch J. Grundlagen der Mechanischen Verfahrenstechnik. Wiesbaden: Vieweg Verlag; 1992.

Lopez-Rubio A, Gavara R, Lagaron JM. Bioactive packaging: turning foods into healthier foods through biomaterials. Trends Food Sci Technol. 2006;17:567–75.

Madene A, Jacquot M, Scher J, Desobry S. Flavour encapsulation and controlled release – a review. Int J Food Sci Technol. 2006;41:1–21.

Malone ME, Appelqvist IAM, Norton IT. Oral behavior of food hydrocolloids and emulsions. Part 2. Taste and aroma release. Food Hydrocolloid. 2003;17:775–84.

Mancuso JR, McClements DJ, Decker E. The effects of surfactant type, pH, and chelators on the oxidation of salmon oil-in-water emulsions. J Agric Food Chem. 1999;47:4112–6.

McClements DJ, Decker E, Weiss J. Emulsion-based delivery systems for lipophilic bioactive compounds. J Food Sci. 2007;72:R109–24.

McHugh TH, Krochta JM. Milk-protein based edible films and coatings. Food Technol. 1994;48:97–103.

Melanson SF, Lewandrowski EL, Flood JG, Lewandrowski KB. Measurement of organochlorines in commercial over-the-counter fish oil preparations – implications for dietary and therapeutic recommendations for omega-3 fatty acids and a review of literature. Arch Pathol Lab Med. 2005;129(1):74–7.

Menrad K. Market and marketing of functional food in Europe. J Food Eng. 2003;56:181–8.

Minemoto Y, Adachi S, Matsuno R. Comparison of oxidation of methyl linoleate encapsulated with gum arabic by hot-air-drying and freeze-drying. J Agric Food Chem. 1997;45:4530–4.

Minemoto Y, Hakamata K, Adachi S, Matsuno R. Oxidation of linoleic acid encapsulated with gum arabic or maltodextrin by spray-drying. J Microencapsul. 2002;19:181–9.

Moure A, Cruz JM, Franco D, Domínguez JM, Sineiro J, Domínguez H, et al. Natural antioxidants from residual sources. Food Chem. 2001;72:145–71.

Naguib YMA, Hari SP, Passwater Jr R, Huang D. Antioxidant activities of natural vitamin E formulations. J Nutr Sci Vitaminol. 2003;49:217–20.

Nakaya K, Ushio H, Matsuwaka S, Shimizu M, Ohshima T. Effect of oil droplet size on the oxidative stability of oil-in-water emulsions. Lipids. 2005;40:501–7.

Nakazawa R, Shima M, Adachi S. Effect of oil-droplet size on the oxidation of microencapsulated methyl linoleate. J Oleo Sci. 2008;57:225–32.

Nenadis N, Zafiropoulou I, Tsimidou M. Commonly used food antioxidants: a comparative study in dispersed systems. Food Chem. 2003;82:403–7.

Onwulata CI, Konstance RP, Holsinger VH. Properties of single- and double-encapsulated butteroil powder. J Food Sci. 1998;63:100–3.

Orlien V, Andersen AB, Sinkko T, Skibsted LH. Hydroperoxide formation in rapeseed oil encapsulated in a glassy food model as influenced by hydrophilic and lipophilic radicals. Food Chem. 2000;68:191–9.

Orlien V, Risbo J, Rantanen H, Skibsted LH. Temperature-dependence of rate of oxidation of rapeseed oil encapsulated in a glassy food matrix. Food Chem. 2006;94:37–46.

Osborn HT, Akoh CC. Effect of emulsifier type, droplet size, and oil concentration on lipid oxidation in structured lipid-based oil-in-water emulsions. Food Chem. 2004;84:451–6.

Partanen R, Yoshii H, Kallio H, Yang B, Forssell P. Encapsulation of sea buckthorn kernel oil in modified starches. J Am Oil Chem Soc. 2002;79:219–23.

Partanen R, Hakala M, Sjövall O, Kallio H, Forssell P. Effect of relative humidity on the oxidative stability of microencapsulated sea buckthorn seed oil. J Food Sci. 2005;70:37–43.

Ré MI. Microencapsulation by spray-drying. Dry Technol. 1998;16:1195–236.

Ruder J, Åström A, Hall G, Bruhn M. Healthy snacks for adolescents – a case on consumer-driven product-development. Proceedings of the Nordic Consumer Policy Research Conference. 2007. http://www.consumer2007.info/wp-content/uploads/children2-%20Ruder.pdf. Accessed 14 Oct 2008.

Rusli JK, Sanguansri L, Augustin MA. Stabilization of oils by microencapsulation with heated protein-glucose syrup mixtures. J Am Oil Chem Soc. 2006;83:965–72.

Ruxton CHS, Reed SC, Simpson MJA, Millington KJ. The health benefits of omega-3 polyunsaturated acids: a review of the evidence. J Hum Nutr Diet. 2004;17:449–59.

Schaffner D. Process for the manufacture of powderous preparations of fat-soluble substances. 2004. WO2004/062382.

Schröder V, Schubert H. Influence of emulsifier and pore size on membrane emulsification. In: Food emulsions and foams. Cambridge: Royal Society of Chemistry; 1999.

Schrooyen PMM, van der Meer R, de Kruif CG. Microencapsulation: its application in nutrition. Proc Nutr Soc. 2001;60:475–9.

Schubert H, Engel R, Kempa L. Principles of structured food emulsions: novel formulations and trends. IUFoST 13th World Congress of Food Science and Technology, Nantes. 2006. doi: 10.1051/IUFoST:20061343.

Schuchmann HP, Schuchmann H. Lebensmittelverfahrenstechnik. Weinheim: Wiley-VCH; 2005.

Selim K, Tsimidou M, Biliaderis CG. Kinetic studies of degradation of saffron carotenoids encapsulated in amorphous polymer matrices. Food Chem. 2000;71:199–206.

Serfert Y, Drusch S, Schwarz K. Chemical stabilisation of microencapsulated oils rich in long chain polyunsaturated fatty acids. Food Chem. 2008. Accepted for publication.

Soottitantawat A, Yoshii H, Furuta T, Ohkawara M, Linko P. Microencapsulation by spray drying: influence of emulsion size on the retention of volatile compounds. J Food Sci. 2003;68:2256–62.

Starling S. Markets: who is buying omega-3 and in what form? 2008. http://www.nutraingredients.com/Consumer-Trends/Markets-Who-is-buying-omega-3-and-in-what-form. Accessed 10 Oct 2008.

Starling S. Omega-3 reality check. 2008. http://www.nutrition.org.uk/upload/10%20Key%20Facts(4).pdf. Accessed 10 Oct 2008.

Steenkamp J-BEM. Dynamics in consumer behavior with respect to agricultural and food products. In: Wierenga B, Van Tiburg A, Grunert K, Steenkamp J-BEM, Michel W, editors. Agricultural marketing and consumer behavior in a changing world. Boston: Kluwer; 1997. p. 143–88.

Strauss G, Gibson SM. Plant phenolics as cross-linkers of gelatin gels and gelatin-based coacervates for use as food ingredients. Food Hydrocolloid. 2004;18:81–9.

Sugiura S, Nakajima M, Seki M. Effect of channel structure on microchannel emulsification. Langmuir. 2002;18:5708–12.

Ubbink J, Krüger J. Physical approaches for the delivery of active ingredients in foods. Trends Food Sci Technol. 2006;17:244–54.

Valentinotti S, Armanet L, Porret J. Encapsulated polyunsaturated fatty acids. 2006. US2006/0134180.

van Lengerich BH. Embedding and encapsulation of sensitive components into a matrix to obtain discrete controlled release particles. 2002. US2002/0044968.

Vega C, Roos YH. Spray-dried dairy and dairy-like emulsions – compositional considerations. J Dairy Sci. 2006;89:383–401.

Venkateshwarlu G, Let MB, Meyer AS, Jacobsen C. Chemical and olfactometric characterization of volatile flavor compounds in a fish oil enriched milk emulsion. J Agric Food Chem. 2004;52:311–7.

Verbeke W. Consumer acceptance of functional foods: socio-demographic, cognitive and attitudinal determinants. Food Qual Prefer. 2005;16:45–57.

Walton DE, Mumford CJ. The morphology of spray-dried particles. The effect of process variables upon the morphology of spray-dried particles. Trans IChemE. 1999;77A:442–60.

Weinbreck F, Minor M, de Kruif CG. Microencapsulation of oils using whey protein/gum arabic coacervates. J Microencapsul. 2004;21:667–79.

Weststrate JA, Van Popperl G, Verschuren PM. Functional foods, trends and future. Br J Nutr. 2002;88:233–5.

Whelan J, Rust C. Innovative dietary sources of n-3 fatty acids. Annu Rev Nutr. 2006;26:75–103.

Yan N, Jin Y, Moulton S, Perrie T-C. Microcapsules having multiple shells and method for the preparation thereof. 2004. WO2004/041251.

Yanishlieva NV, Marinova EM. Stabilisation of edible oils with natural antioxidants. Eur J Lipid Sci Technol. 2001;103:752–67.

Zeller BL, Saleeb FZ, Ludescher RD. Trends in development of porous carbohydrate food ingredients for use in flavor encapsulation. Trends Food Sci Technol. 1999;9:389–94.

Chapter 8
Biocontrol of Foodborne Bacteria*

Lynn McIntyre, J. Andrew Hudson, Craig Billington, and Helen Withers

8.1 Introduction

Consumers dislike the use of chemical preservatives in their food, and with some there is an associated public health risk. This has increased the pressure for these chemicals to be removed from food and for the identification and adoption of novel and more "natural" means of preservation. Although there are a variety of approaches to using natural preservatives, the most often adopted approach to date has been to use biocontrol.

The New Zealand Parliamentary Commissioner for the Environment defines biocontrol as "Using biological means (such as parasites, viruses or predators) to control a pest" (http://www.pce.govt.nz/reports/pce_reports_glossary.shtml). For the purposes of this review, the pest is primarily the pathogen of concern in a particular food, but biocontrol can also be applied to spoilage organisms and some examples of this will be given.

Potential candidates for biocontrol in foods include bacteriophages, bacteriocins (peptides secreted by one species that inhibit another), siderophores (molecules that sequester iron), quorum sensing (QS; where the presence of an autoinducer of one species may influence the growth of another), competitive organisms, and various microbe- and plant-derived antimicrobials. Of these, siderophores might be considered as a candidate biocontrol mechanism for lower-iron foods such as seafood (Gram and Melchiorsen 1996) and fruits and vegetables (Henry et al. 1991) but they are not discussed further here.

---

*An earlier version of this chapter was published in Food New Zealand, 7(5):25–32.

L. McIntyre (✉)
Harper Adams University College, Edgmond, Newport, Shropshire, TF10 8NB, UK
e-mail: lmcintyre@harper-adams.ac.uk

J.A. Hudson • C. Billington
Institute of Environmental Science and Research Ltd,
PO Box 29 181, Christchurch 8540, New Zealand

H. Withers
AgResearch MIRINZ, Ruakura Research Centre, Private Bag 3123, Hamilton, New Zealand

What follows is a review of current and potential food-related applications of selected biocontrol approaches.

8.2 Viral Predators: Bacteriophages

Bacteriophages (phages) are viruses that exclusively infect and kill bacteria. Their structure is simple, comprising nucleic acid contained within a protein capsule that may include a lipid component (Fig. 8.1).

Lytic phages, which are the phage group most likely to be of use as biocontrol agents, kill their hosts soon after infection. This "lytic cycle" is shown in Fig. 8.2. When a phage and its host meet (step 1, Fig. 8.2), there needs to be a specific interaction between the host and the phage for infection to occur, i.e., any given phage will only infect a specific group of hosts. During infection, nucleic acid is passed from the phage into the host through the bacterial cell wall (step 2). The nucleic acid is expressed to produce proteins which take over the host cell's functions, forcing it to replicate phage nucleic acid (step 3) and then to produce new phage particles (step 4). When, typically, around 100 new phages have been made, they are released

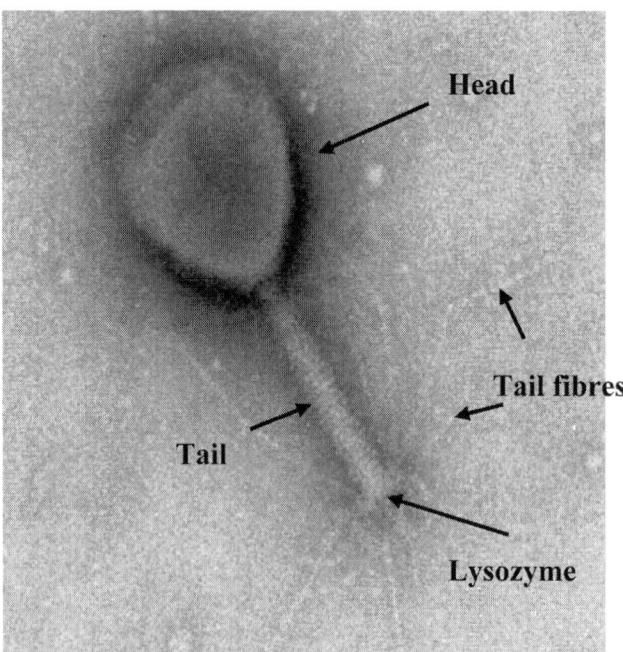

Fig. 8.1 Electron micrograph of phage T4. The head contains the DNA, the tail fibers interact with the bacterium, and three lysozyme molecules at the base plate degrade the cell wall. In phage T4 the tail is contractile and acts to push the tail core into the cell wall. (The image was taken by the Electron Microscope Unit of the University of Otago, Dunedin)

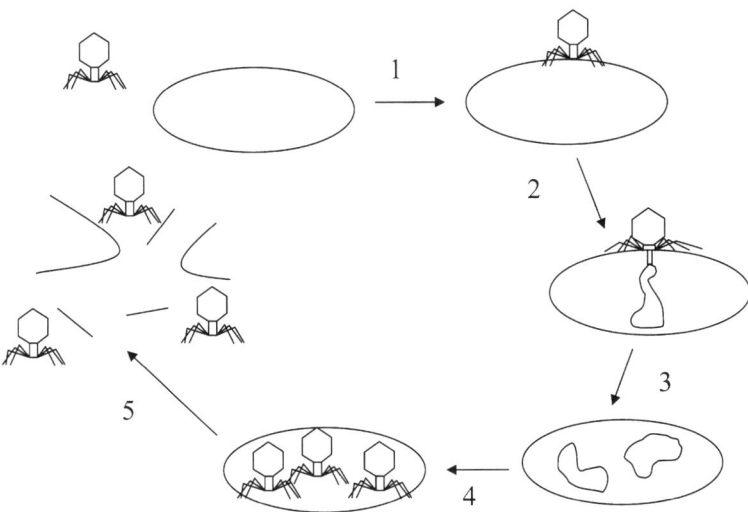

Fig. 8.2 The lytic cycle of bacteriophage infection (not to scale). 1 adsorption, 2 entry of nucleic acid into the cell, 3 nucleic acid replication, 4 phage particle replication, 5 host cell lysis

from the host cell by fatal disruption of the cell wall (lysis, step 5), and each progeny phage particle can then infect another cell.

The potential for using phages to kill bacteria has been recognized for almost 100 years, with early pioneering work being conducted at around the time of the First World War. Phage therapy for treating human disease was not routinely adopted in Western countries, and the discovery of antibiotics saw an end to the early era of phage therapeutics, apart from in the Soviet Union and eastern Europe. In more recent years, however, with the threat of antibiotic-resistant bacteria emerging as a serious problem, the potential human clinical use of phages has received renewed attention. Flowing from this has been interest in their use in a variety of other applications, one of which is as a control of pathogens in foods (Greer 2005). Indeed, commercial production of phages for use in foods has begun. For example the antilisterial phage Listex™ P100 (Carlton et al. 2005) is being produced in the Netherlands (http://www.ebifoodsafety.com) and has now received FDA approval. EcoShield™ (formerly ECP-100) and ListShield™ (formerly LMP-102) are commercial products for the control of *Esherichia coli* O157 and *Listeria monocytogenes* respectively (http://www.intralytix.com/Intral_FAQ.htm). ListShield™ has FDA approval.

Several food studies have now been published (Hudson et al. 2005), and a summary of some of these studies is shown in Table 8.1. In many of these food studies, inactivation was only achieved by using a high phage concentration. In addition, inactivation has been reported to occur at temperatures beneath the growth minimum of the host. These facts suggest that the inactivation reported is not caused through the lytic cycle in situ (as shown in Fig. 8.2), as that would require the host cells to be growing. It is possible that the reduction occurs through lysis from without, whereby

Table 8.1 Examples of investigations into the biocontrol of pathogens on foods

Host bacterium	Food	Effect of biocontrol	References
Campylobacter	Raw and cooked meat	Reduced host numbers; dependent on conditions	Bigwood et al. (2008)
	Chicken skin	Reduced host numbers	Atterbury et al. (2003)
Cronobacter sakazakii	Reconstituted infant milk powder	Reduced host numbers/reduced growth	Kim et al. (2007)
Escherichia coli O157:H7	Beef	Reduced host numbers/eradicated host	O'Flynn et al. (2004)
	Cantaloupes	Pathogen numbers reduced at 4°C but not at 20°C	Sharma et al. (2009)
	Lettuce	Pathogen numbers reduced at 4°C	
	Tomato, spinach, broccoli	Reduction in pathogen numbers maintained over 168 h	Abuladze et al. (2008)
	Minced beef	Reduction in pathogen numbers after 24 h	
Listeria monocytogenes	Smear-ripened cheese	Growth reduction or eradication. Concentration dependent	Carlton et al. (2005)
	Smear-ripened cheese	Growth reduction or eradication. Concentration dependent	Schellekens et al. (2007)
	Beef	No effect	Dykes and Moorhead (2002)
	Cut melon	Growth reduction or eradication. Concentration dependent	Leverentz et al. (2004)
	Cut melon	Growth reduction	Leverentz et al. (2003)
	Cut apple	No effect	
	Ham	Growth retarded when phage used alone. Final concentration reduced with protective culture at 10°C and growth prevented at 4°C	Holck and Berg (2009)
	Various liquid foods	Reduction sustained over incubation period	Guenther et al. (2009)
	Various solid foods	Initial reduction then regrowth	
	Salmon fillet	Reduction sustained over incubation period	Soni and Nannapaneni (2010)
	Channel catfish	Reduction sustained over incubation period	Soni et al. (2010)

(continued)

Table 8.1 (continued)

Host bacterium	Food	Effect of biocontrol	References
Salmonella	Mustard and broccoli seeds	Reduced growth	Pao et al. (2004)
	Alfalfa seed sprouts	Initial but temporary inhibition	Kocharunchitt et al. (2009)
	Mung and alfalfa bean sprouts	Reduced growth. More effective in co-culture with antagonistic bacterium	Ye et al. (2010)
	Frankfurters	Reduced growth	Whichard et al. (2003)
	Cheddar cheese	Growth reduction during manufacture	Modi et al. (2001)
	Raw and cooked meat	Reduced host numbers	Bigwood et al. (2008)
	Chicken skin	Reduced numbers/eradicated host	Goode et al. 2003
	Chicken and turkey carcasses	Reduced carcass prevalence	Higgins et al. (2005)
	Cut melon	Growth reduction	Leverentz et al. (2001)
	Cut apple	No effect	
	Tomatoes	No effect	Ye et al. (2009)
Staphylococcus aureus	UHT milk, acid, and rennet curd	Reduced host numbers	García et al. (2007)
	Raw milk	No reduction in concentration	O'Flaherty et al. (2005)
	Heat-treated milk	Host cells eliminated	
	Pasteurized milk	Control over a short time period	García et al. (2009)
	Raw milk	Some inhibition of growth	

large numbers of attached phages kill the cell (Delbrück 1940). An example of data which could be explained by this is the inactivation of *Salmonella* on melon which was reported to occur at 5°C and at a phage to host ratio of 200 (Leverentz et al. 2001). However, other explanations are also possible since some consideration needs to be given to the probability of phage and host coming into contact at low host cell concentrations (Kasman et al. 2002). In liquids, data suggest that high phage concentrations are needed to ensure that most cells become infected when they are present at low concentrations, which is likely to be the case for foods (Bigwood et al. 2009). On surfaces, a similar situation may apply (Hagens and Offerhaus 2008). As this occurs under conditions where the host is not growing, it is possible that the cell becomes infected and later dies on the resumption of growth.

The host specificity of the phage is important, as within each species of pathogen there are numerous subtypes, all of which need to be controlled. An effective phage treatment should therefore have a "Goldilocks" host range; not too narrow and not too broad. An example of a phage with an almost perfect host range is Felix O1, which lyses 96–99.5% of *Salmonella* serovars (Lindberg 1967). In contrast, *Campylobacter* phage Cj6 is a phage with a poor host range, i.e., it infects only the host on which it was isolated (Bigwood et al. 2008). Work with spoilage organisms has shown that if the host range is too narrow, biocontrol does not work (Greer and Dilts 1990). One possibility that would overcome this limitation is to use combinations of phages, or cocktails, and this has proven successful for controlling *E. coli* O157:H7 on beef (O'Flynn et al. 2004).

The direct application of phages to foods is only one biocontrol approach that might be adopted. Many pathogens contaminate meat either directly or indirectly from the animal's feces during processing. The use of phages to destroy pathogens within the intestinal tracts of food animals could therefore be a way of reducing the number of pathogens on meat. The inactivation of both *Salmonella* and *Campylobacter* in chickens has been demonstrated (Atterbury et al. 2007; Loc Carrillo et al. 2005).

It may also be possible to use phages to decontaminate the hides of meat animals prior to slaughter (Callaway et al. 2006). Although there are no published data, our experience in the laboratory and with housed animals supports this notion. However, trials in commercial conditions need to be undertaken to assess the efficacy of the approach.

A small number of articles have reported the use of phages in tandem with another biocontrol agent to control bacteria on food. For example, the use of *Lactobacillus sakei* as a protective culture in combination with phage P100 was shown to be superior to use of the phage alone for the control of *L. monocytogenes* on ham (Holck and Berg 2009). On growing tomato plants, a combination of *Enterobacter asburiae* and phages reduced colonization of the resulting fruits by *Salmonella*, although most of the effect was attributed to the bacterial antagonist (Ye et al. 2009). No controlling effect was demonstrated on postharvest fruits, although the data were not expressed in units of surface area or weight of fruit, making their interpretation difficult with respect to assessing whether the phage inoculum was sufficiently high. The same problem is encountered with another attempt to control *L. monocytogenes* on beef cubes with nisin used in combination with phages (Dykes and Moorhead 2002). Nisin showed a synergistic activity with phages for control of *L. monocytogenes* on melon (Leverentz et al. 2003) but not on apple, where there was no phage-related reduction. In pasteurized milk a synergy was also observed between nisin and phages for the control of *Staphylococcus aureus* (Martínez et al. 2008), although the occurrence of nisin-resistant cells was noted. Although this approach has produced mixed results to date, it is intuitively attractive as control is exerted by two mechanisms, thereby decreasing the possibility of resistant mutants growing.

Another approach is the application of purified phage-derived antimicrobials. These include enzymes (endolysins) produced by lytic phages which degrade the bacterial cell wall from the inside out, allowing phages to burst out of the cell (step 5 in Fig. 8.2). These enzymes are generally very host specific (Fischetti 2005) and can also work from the outside, at least for Gram-positive pathogens. Purified endolysins have been shown to be effective for the control of *S. aureus* in pasteurized milk at 37°C (Obeso et al. 2008). Gram-negative bacteria, on the other hand, have an outer membrane which hinders access of the enzyme to the cell wall. However, this does not seem to be insurmountable, with peptidoglycan degradation by a *Pseudomonas* endolysin having been demonstrated (Paradis-Bleau et al. 2007).

Public perception may pose an obstacle to the use of phages in foods. Would consumers respond favorably to knowing that viruses had been added to their food? Maybe they would. It was indicated that some people see phages as "green" and environmentally friendly (Fox 2005). Phages may also be seen as a natural alternative to chemical preservatives. Early phage therapy pioneers demonstrated safety by ingesting preparations themselves, and the safe intake of phage T4 by volunteers has been reported (Bruttin and Brüssow 2005). Whatever the perceptions concerning safety (discussed in a review by Atterbury 2009), it is a fact that phages are a normal component of foods (Hudson et al. 2009b; Tsuei et al. 2007) and so are being ingested by everyone every day.

Another possible objection to the use of phages is the possibility that they will mediate horizontal transfer of genes encoding, for example, virulence traits or antibiotic resistance. A possible way to mitigate this is to use phages which contain either no nucleic acid or severely damaged nucleic acid. In work recently published (Hudson et al. 2010) UV-treated T4 phages were shown to kill *E. coli* in broth culture and on the surface of food. UV-treated phages may inactivate cells in three ways: (1) lysis from without; (2) by a single phage infecting a cell and causing it to stop growing; and (3) by "multiplicity reactivation," where multiple phages inactivating the same cell contribute sufficient undamaged DNA that viable phages can be produced by the cell, so killing it.

Given the possibilities presented above, it is apparent that phages and phage-derived technologies present a toolbox of techniques employing a spectrum from intact viable phages to enzymes derived from them, any of which could be applied along the farm-to-fork continuum. The trick will be to identify which tool is best suited to which segment of the value chain.

8.3 Bacterial Predators: *Bdellovibrio*

A number of predatory bacteria are capable of attacking and killing bacteria in a manner similar to that described previously for phages. The best characterized of these is the Gram-negative *Bdellovibrio bacteriovorus*, so called because of its parasitic leech-like behavior (Latin *bdello*) and curved (*vibrio*) shape. First isolated from soil by Stolp and Starr in 1962, this ubiquitous organism has since been found in a variety of environments, including freshwater, saltwater, brackish water, sewage, and mammalian intestines.

Current knowledge of the predatory replication cycle of *B. bacteriovorus* has been reviewed by Lambert et al. (2006), who described a recent renaissance in research activity. This is attributed, in part, to the first description of the entire genome sequence of *B. bacteriovorus* strain HD100 by researchers at the Max Plank Institute for Developmental Biology (Rendulic et al. 2004).

The replication cycle commences with the "attack phase," during which *Bdellovibrio* cells are small, monoflagellated, and extremely fast moving. Following a collision with a Gram-negative prey cell, the attacking cell attaches to and invades the prey cell using "twitching-type" pili to pull it through holes in the cell wall generated using a blend of hydrolytic enzymes. Internalization of *Bdellovibrio* in the prey cell's periplasm results in loss of the flagellum and a morphology change to produce a nonmotile elongated "growth phase" form in a joint predator–prey structure known as a bdelloplast. Replication then occurs in the bdelloplast, followed by prey lysis and escape back into the environment.

Given the reported absence of horizontal gene transfer during predation and an inability to invade mammalian cells, *Bdellovibrio* has been suggested as a "living antibiotic," or at the very least a useful source of novel antimicrobial substances (Rendulic et al. 2004; Lambert et al. 2006). With its varied environmental niches and a relatively broad host range, it has particular appeal in the battle against Gram-negative pathogens such as *Salmonella* and *E. coli* O157 in both food and water.

Unsurprisingly, given the previous lack of interest in this organism, there are relatively few applied research publications to date. Fratamico and Cooke (1996) reported the ability of *Bdellovibrio* isolates to control *E. coli* O157:H7 and *Salmonella* spp. on stainless steel surfaces, suggesting the potential for application to food processing equipment and food. The capacity of *B. bacteriovorus* strain 109 J to attack *E. coli* biofilms was reported some time later by Kadouri and O'Toole (2005). A 7 $\log_{10}$ reduction in planktonic *E. coli* numbers was found after 24 h of exposure, whereas biofilm-associated pathogen numbers declined by 4 $\log_{10}$ units. Although cells in this biofilm were clearly able to survive predation to a far greater extent, this work does suggest that *Bdellovibrio* could go some way towards reducing the issues associated with biofilms in the food industry and in other situations. Further research of this type is clearly necessary to determine how whole *Bdellovibrio* cells or derived antimicrobials can be successfully applied to specific situations.

8.4 Competitive Enhancement Techniques, Protective Cultures, and Antimicrobial Metabolites

Competitive enhancement techniques, protective cultures, and antimicrobial metabolites offer a number of options in terms of a farm-to-fork approach to food safety. Competitive enhancement techniques utilize antagonistic bacteria that can be administered before harvest to food animals to either inhibit or reduce intestinal colonization and carriage of human pathogens. After harvest, protective cultures and metabolites can be added directly to foods during formulation and processing to create an additional hurdle either through manipulating the composition of the food's microflora to the detriment of pathogens or by introducing inhibitory substances active against them, or both. Antagonistic bacteria have also been evaluated as microbial cleaners to control pathogens such as *Listeria* in processing environments (Zhao et al. 2006) and as a treatment for mastitis. Given the potential breadth of application, significant research efforts have been devoted to the characterization and commercialization of both bacteria and metabolites for a variety of purposes.

In all cases, regardless of the application being considered, defining these microorganisms is important to ensure that cultures do not contain potentially pathogenic subpopulations or virulence and antimicrobial resistance genes that could be transferred to either the animal or the human host microflora (Wagner 2006). As major components of the microflora of both healthy mature animals and foods, lactic acid bacteria (LAB) have been most extensively investigated as competitive enhancement and protective culture candidates. They are generally nonpathogenic and produce antimicrobial compounds, including acids, hydrogen peroxide, diacetyl, and bacteriocins, that give them an additional edge against related organisms that might otherwise predominate in a particular environmental niche. However, other nonpathogenic microorganisms such as fungi and yeasts can also be suitable (Vandeplas et al. 2010).

8.4.1 Competitive Enhancement Techniques

Competitive enhancement techniques (as defined by Callaway et al. 2008) include competitive exclusion cultures that are typically native to the animal species in question, and probiotics (also referred to as direct-fed microbials) such as lactobacilli and bifidobacteria that are not necessarily isolated from animals. These microorganisms may act to control or inhibit other microbes via mechanisms including (1) nutrient competition, (2) production of inhibitory compounds, (3) immunostimulation, and (4) competition for binding sites. The development of competitive enhancement techniques has been prompted in part by the increased use of antibiotics and the consequent development of resistance amongst microbes (Anderson et al. 2006; Diez-Gonzalez and Schamberger 2006), and various aspects have recently been reviewed by Callaway et al (2008), Gaggìa et al. (2010), and Vandeplas et al. (2010).

The most common uses of competitive exclusion and probiotics/direct-fed microbials include the prevention of colonization of *Salmonella* in poultry using microbes such as lactobacilli, enterococci, pediococci, and *Saccharomyces* (Al-Zenki et al. 2009; Vandeplas et al. 2010). Improved animal productivity both alone (Stephens et al. 2010) and in concert with *E. coli* O157:H7 reduction (Younts-Dahl et al. 2005; Stephens et al. 2007) has also be observed in cattle. The application of competitive exclusion cultures works particularly well in chicks with naïve essentially sterile intestinal tracts, whereas probiotics are suggested to be most useful in poultry when beneficial intestinal bacteria are in decline as a consequence of stress (Vandeplas et al. 2010). More recent research has focused on the growing problem of *Campylobacter* infection of poultry (Zhang et al. 2007; Santini et al. 2010). Prebiotic nutrients may also be included in treatments where they either promote the growth of probiotics in the intestinal tract or, in the case of mannanoligosaccharides, bind Gram-negative pathogens such as *E. coli* and promote their excretion from the intestinal tract (Newman 1994). The use of non-LAB organisms includes generic *E. coli* which can inhibit *E. coli* O157 through the production of colicins (Schamberger and Diez-Gonzalez 2004). Colicins have also been described that inhibit other Shiga-toxin-producing *E. coli* serotypes such as O26 and O111 (Murinda et al. 1996).

For the development of successful commercial products, the choice of strain is an important factor (Stephens et al. 2007) along with its survival characteristics, particularly under acidic conditions, and the method of administration (Anderson et al. 2006). Available commercial products include Bovamine and Lactoedge for cattle, and Aviguard, Avifree, Broilact, MSC, Preempt (or CF-3), and Fastrack for poultry (Callaway et al. 2008; Schneitz 2005). The Canadian biotech company CanBiocin currently markets an LAB-based product called Procin®, developed to prevent *E. coli* scours in pigs.

8.4.2 Protective Cultures and Antimicrobial Metabolites

Current consumer awareness regarding the application of beneficial bacteria or their metabolites to foods is probably limited to the use of LAB as probiotics in yoghurts and dairy drinks to improve human gut health. However, LAB protective cultures or their associated metabolites can be added directly to foods to enhance food safety and have been extensively reviewed (Goktepe 2006; Hudson et al. 2009a). The dose required for effective antagonism in the food in question and the potential effects on sensory properties (Smith et al. 2005) are key considerations.

Several species of bifidobacteria have been applied to fish, seafood, poultry, and dairy foods to extend shelf life and protect against pathogens. Reported shelf life extensions ranged from 3 to 14 days depending on the food and the mode of application, and can be enhanced by the combined use of acid salts such as sodium acetate and potassium sorbate (Goktepe 2006). Similarly, lactobacilli and pediococci have been exploited in a variety of food applications, including ready-to-eat meats (Amézquita and Brashears 2002; Bredholt et al. 1999; 2001), ground beef (Smith et al. 2005), vegetables (Scolari and Vescovo 2004), and dairy products (Millette et al. 2006). Their antimicrobial effect has been attributed to the production of organic acids and a variety of antimicrobial compounds.

The elucidation of antimicrobial compounds produced by bacteria has been an area of intense research, particularly the production of bacteriocins, antimicrobial peptides which produce pores in the cytoplasmic membrane, ultimately inhibiting energy production and biosynthesis activities. These offer another means of control, either through the direct use of cultures known to produce certain compounds or via the addition of fermentates containing antimicrobials.

The best known bacteriocin, nisin, is produced by *Lactococcus lactis*. It has an unusually broad host spectrum against both Gram-positive spoilage and pathogenic bacteria including *L. monocytogenes*, *Bacillus*, and *Clostridium*, making it a particularly attractive food additive for processed food products. Although nisin has been approved for use in an estimated 50 countries, it does demonstrate some limitations in relation to food type and pH (Deegan et al. 2006). Hence, research activities have continued unabated to isolate and characterize bacteriocins produced by other bacteria.

Despite substantial research efforts, the availability of commercialized bacterial cultures and metabolite-based antimicrobial products is not as extensive as might be anticipated. Often this is simply related to a failure to reproduce preliminary in vitro successes in more complex food systems (Hudson et al. 2009a). Danisco has been successful in this endeavor and the company produces and markets a range of protective cultures and defined and undefined fermentates containing antimicrobial metabolites for food applications. Cultures and undefined fermentates are seen as more-consumer-friendly approaches not requiring extensive labeling (Deegan et al. 2006). Defined (purified) fermentates, on the other hand, need to be labeled as additives.

The Danisco HOLDBAC™ range of protective cultures can be used in dairy and meat products (including fermented versions) to control the growth of fungal and bacterial undesirables, including *L. monocytogenes* (Delves-Broughton et al. 2007). CanBiocin markets Micocin®, a US Department of Agriculture-approved LAB product with activity against *Listeria* for use in processed meat products. Examples of defined antimicrobials include powdered nisin, marketed as Nisaplin®, and Natamax®, a broad spectrum antifungal compound produced by the fermentation of *Streptomyces natalensis*. Another defined bacteriocin, ALTA™ 2431, produced by *Pediococcus acidilactici*, is commercially manufactured by Kerry Bioscience (Deegan et al. 2006).

Undefined fermentates are marketed as Microgard® fermented milk and sugar powders which can be labeled as "cultured milk" or "cultured dextrose" (Delves-Broughton et al. 2007). These products, typically produced by fermentations using *Propionibacterium freudenreichii* or *L. lactis*, are multifactorial in their activity, and can be used in a range of foods, including dairy products, salad dressings, ready-to-eat meals, and baked goods.

Although the use of antagonistic bacteria and/or bacteriocins is not a total food safety solution, they can be combined with other preservation systems, e.g. modified atmospheres (Mataragas et al. 2003), bioactive packaging and high pressure processing (Deegan et al. 2006), and bacteriophages (Reynolds 2007). Research to date suggests that they are most useful as one hurdle in combination with other preservation options, and their future success relies on a greater understanding of the mechanisms of action occurring in food systems.

8.5 Antimicrobials of Plant Origin

Although the potential for antimicrobial effects from medicinal plant extracts has been appreciated for centuries, it is only relatively recently that plant-derived food ingredients have been explored for these properties. The most frequently used form of plant extract tested for antimicrobial activity in foods is the essential oils. These are aromatic oily liquids, composed of up to 60 individual components that are usually isolated by steam distillation (Burt 2004). Essential oils have been reported to be active in vitro against a wide range of foodborne pathogenic organisms, including *E. coli* O157:H7, *Salmonella*, *Shigella*, *Campylobacter*, *L. monocytogenes*, *S. aureus*, *Bacillus cereus*, and *Vibrio* species. They are generally more effective against Gram-positive than Gram-negative microorganisms, and can be active against fungi.

The most frequently described group of inhibitory compounds are phenolic-based, with aromatic and isothiocyanate-based components also having been isolated. Synergism between individual components from the same essential oil is possible, as single fractions are sometimes less efficacious than whole extracts. Some of the compounds are the same as those involved in plant host defense against microorganisms (Holley and Patel 2005).

The mechanisms of action of the essential oils are not fully understood. It is known that phenolic components appear to integrate into the cytoplasmic membrane of the microorganism, causing increased proton and potassium permeability (Perez-Conesa et al. 2006). It has been suggested that a lower sensitivity of Gram-negative bacteria to essential oils may be due to their relatively impermeable outer membrane (Fisher and Phillips 2006).

The simplest and most reported method for delivery of essential oil antimicrobial activity is incorporation directly into the food, with up to 8 log unit kills reported (Fisher et al. 2007). This is an active area of research, with many foods being studied, such as meat (Govaris et al. 2010), fruit salads (Belletti et al. 2008), vegetable salads (Gutierrez et al. 2009), and fruit juices (Duan and Zhao 2009). However, the fat composition of foods has been shown to influence the effectiveness of essential oils. For example, in low-fat cheese, bay, clove, cinnamon, and thyme essential oils at 1% reduced *L. monocytogenes* to less than 1 $\log_{10}$ cfu ml^{-1} at 4°C, but only clove oil was effective in full-fat cheese. In further experiments, *Salmonella* Enteritidis incubated at 10°C was less affected by fat content, with thyme oil being the only ineffective essential oil in full-fat cheese (Smith-Palmer et al. 2001). It is thought that the hydrophobic nature of essential oils means that they tend to partition in fat and oil phases and decrease the effective concentration in the water phase. Micellar encapsulation of essential oils may deliver a higher dose to the bacterial cell membrane in these types of food environments (Gaysinsky et al. 2005), and this encapsulation strategy may also have potential for the decontamination of surface-adherent biofilms containing pathogens (Perez-Conesa et al. 2006).

Addition of essential oils to foods can also be used in combination with modified-atmosphere packaging (MAP). The quality and safety of MAP-packed table grapes was significantly improved by addition of eugenol and thymol (Valero et al. 2006). Chitosan and thyme have been used in combination with MAP to extend the shelf life of ready-to-cook chicken kebabs (Giatrakou et al. 2010). The combination of MAP and oregano essential oil has proven effective in improving the quality and shelf life of both rainbow trout fillets (Mexis et al. 2009) and bologna sausages (Viuda-Martos et al. 2010).

Direct addition of essential oils to food can result in reduced pathogen populations, but may also alter the food's sensory characteristics. Incorporating essential oils into active packaging could be an alternative way of delivering the antimicrobial action to the food without negatively impacting on organoleptic properties. Whey protein films containing 1–4% (w/v) essential oils from oregano, rosemary, and garlic were effective in inhibiting *E. coli* O157:H7, *S. aureus*, *Salmonella* Enteritidis, and *L. monocytogenes* on laboratory media (Seydim and Sarikus 2006). The incorporation of tea tree essential oil into chitosan films was found to slow the growth of *L. monocytogenes* on media at 10°C (Sánchez-González et al. 2010). A trial of alginate-based edible films containing essential oils of oregano, cinnamon, and savory reduced *Salmonella* Typhimurium and *E. coli* O157:H7 growth on beef muscle slices by 1–2 $\log_{10}$ cfu cm^{-2} over 5 days at 4°C (Oussalah et al. 2006). The incorporation of essential oils and their active fractions was also found to significantly reduce tensile strength and elasticity of alginate–apple puree edible films, but did not affect vapor or oxygen permeability (Rojas-Grau et al. 2007).

Another potential intervention point for the use of essential oils is in the feed of food animals. Several polysaccharide-emulsified essential oils were evaluated for introduction into pig diets for the preslaughter control of *Salmonella* Typhimurium DT104 within the gut (Si et al. 2006a, b). Essential oils were stable at low pH and active at minimum bactericidal concentrations similar to those of dietary antibiotics used in swine rearing in some countries (100–300 µg ml^{-1}). Carvacrol and thymol were found to have inhibitory activity similar to that of cinnamaldehyde on laboratory media; however, after they had been mixed with pig food, only cinnamaldehyde retained some activity against *Salmonella*. Essential oils may also have a role in controlling pathogens during poultry production (reviewed in Brenes and Roura 2010).

Other plant materials with potential antimicrobial use in foods include extracts from berries, tea, and wood smoke. Grape skins, juice, and seed extracts have been found to be highly inhibitory to *L. monocytogenes* (Rhodes et al. 2006). Wine has also been found to rapidly inactivate *Campylobacter jejuni* (Carneiro et al. 2008). Both Gram-negative and Gram-positive pathogens were inhibited by Nordic berry extracts (Puupponen-Pimia et al. 2005), with cloudberry and raspberry the best inhibitors. Whole tea infusions and flavanoid extracts were inhibitory when applied to *B. cereus* cultures (Friedman et al. 2006). Although there are some concerns about the carcinogenic effects of wood smoking on food, liquid wood smoke that has had suspect compounds removed appears to retain antimicrobial activity and has been tested successfully for inhibition of *Listeria* on salmon and trout, and molds on cheese (Holley and Patel 2005).

Plant extracts are promising antimicrobial agents, with activities often rivaling those of synthetic chemicals. However, a significant barrier for the widespread adoption of essential oils or their components is that at the concentrations required for effective control, the sensory quality of foods may be altered. The use of active packaging, multihurdle strategies with lowered doses, or combination with existing treatments offers promise to negate these concerns.

8.6 Biocontrol Through Quorum Sensing

Bacteria live in complex environments that are constantly changing. They do not live in isolation but often live within mixed communities that contain more than one species of bacteria as well as other microorganisms such as fungi and yeasts. To survive, bacteria must be able to sense and interpret their environment, making adaptive responses that promote either proliferation or survival. Those that cannot adapt or

Fig. 8.3 Bacterial quorum sensing regulates target gene expression, resulting in altered cell activities

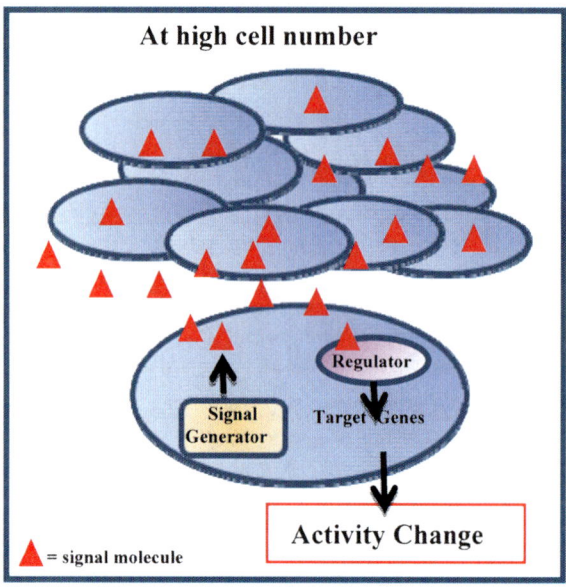

respond to change do not survive. Those that do are now well adapted to their new environment. This is especially true for bacteria that appear in processed foods.

Cell signaling, or quorum sensing (QS), within bacterial communities is a well-established phenomenon that orchestrates coordinated behavior by bacteria through the detection of small signal molecules known as autoinducers (Williams et al. 2007). QS allows for cell-density-dependent regulation of many different bacterial activities, including surface attachment, biofilm formation, expression of virulence factors, and secondary metabolite production (Fig. 8.3; Fuqua et al. 2001; Williams et al. 2007). Many of these QS-regulated microbial activities are involved in food spoilage and are essential for the survival of pathogens within the food matrix (Ammor et al. 2008).

A range of cell signaling molecules have been identified that allow bacteria to communicate with each other as well as with other species of bacteria, often manipulating or controlling their activities (Qazi et al. 2006; Hogan et al. 2004). In addition, molecules produced by a host plant or animal can be detected by bacteria, influencing the microbial composition of the resident microflora as well as affecting their day-to-day activities (Dudler and Eberl 2006). Many of the bacterial species found in food are capable of producing and/or detecting signaling molecules (Liu et al. 2006; Medina-Martinez et al. 2006; Nychas et al. 2009). A wide range of QS molecules have been detected in a variety of food products (Ammor et al. 2008; Medina-Martinez et al. 2006; Liu et al. 2006).

Biocontrol strategies that exploit bacterial QS provide an opportunity to (1) down-regulate microbial spoilage activity and thereby increase shelf life and (2) alter microbial activity such that survival of the targeted microorganism is unlikely. The main advantage with the first approach is that it does not leave open an opportunity for other undesirable microorganisms to colonize the niche, but downregulates the expression of enzymes such as proteases and lipases, thus limiting damage to the food. In the case of foodborne pathogens, the second approach may be the most desirable outcome. Furthermore, using the bacterium's own QS system against itself minimizes the possibility that the bacteria will adapt and become resistant to the QS inhibitor used. Interventions targeting bacterial QS in food are currently largely unexplored.

A number of different classes of molecules that block QS have been sought, including enzymes that degrade the signal (e.g. lactonases), signal analogs, and signal antagonists as well as the use of naturally derived signal molecules to stimulate an inappropriate response leading to cell death (Dong et al. 2002; Castang et al. 2004; Müh et al. 2006; Ren et al. 2001; Qazi et al. 2006). However, many of these inhibitory compounds thus far identified are chemically reactive and therefore likely to be unstable in complex food matrices. Furthermore, their use in food is questionable owing to potential toxicity for humans. However, a number of quorum quenching molecules have been identified which are derived from plants commonly used in the food industry today. Extracts of garlic have been shown to block QS by *Pseudomonas aeruginosa,* in a cystic fibrosis mouse model, limiting the production of biofilm and thereby aiding clearance of the bacteria (Bjarnsholt et al. 2005). Similarly, vanilla extracts have been shown to interfere with QS of *Chromobacterium violaceum,* suggesting that the consumption of vanilla-containing foods may be beneficial (Choo et al. 2006). Novak and Fratamico (2004) demonstrated that the presence of ascorbic acid, an AI-2 analog, resulted in the reduction of toxin production by *Clostridium perfringens* as well as a decrease in spore production when added to ground meat extract. Another food additive that exhibits QS-inhibiting abilities is cinnamaldehyde (Niu et al. 2006). Low concentrations of cinnamaldehyde, similar to those used in food, were shown to block detection of two different bacterial cell signaling molecules by competing with the signal molecule itself or signal precursors. Although these three QS inhibitors are already used as food additives, their antibacterial activity potential within different food substrates needs to be explored.

Medina-Martinez et al. (2007) showed that *Yersinia enterocolitica* produced QS signals when grown on meat and fish extracts but not when grown on certain vegetable extracts, suggesting that QS systems are active but are substrate-dependent and most probably organism-dependent. Furthermore, not all QS inhibitors are able to block QS systems and prevent food spoilage. Rasch et al. (2007) showed that some QS inhibitors were able to decrease proteolytic activity of *Pectobacterium,* but their activity did not prevent spoilage of bean sprouts by these bacteria. Furthermore, garlic extract did not prevent spoilage of bean sprouts by *Pectobacterium,* but rather appeared to induce it. It is clear that spoilage is multifactorial, with other regulatory pathways involved. It may be that QS inhibition will be the "magic bullet" in some situations, whereas in others it may not be successful. What is clear is that for each organism the regulatory role played by QS needs to be understood and the significance in specific food matrices evaluated; it is only then that QS inhibition can be effective in the fight against bacteria in our food.

8.7 Conclusions

Consumer expectations of pathogen-free food with an acceptable shelf life without the use of synthetic chemicals present major challenges for food manufacturers given the currently available technologies. The biocontrol approaches reviewed here offer some particular advantages by being more targeted than conventional treatments, both leaving the good bugs alone and reducing wastage, and are perceived as more "natural" and "green." Although no one biocontrol approach currently offers a complete solution, collectively they create a toolbox of options employing various mechanisms of attack which could be applied individually or in combination. Indeed, as our knowledge of the relationship between exposure and disease expands through quantitative risk assessment activities, it is becoming more apparent that "king hit" reductions are not necessarily the best approach to reducing human disease incidence. The use of biocontrol strategies in food, therefore, has great potential to achieve more realistic outcomes, although much work is still to be done to bring these from the laboratory bench to market.

References

Abuladze T, Li M, Menetrez MY, Dean T, Senecal A, Sulakvelidze A. Bacteriophages reduce experimental contamination of hard surfaces, tomato, spinach, broccoli and ground beef by *Escherichia coli* O157:H7. Appl Environ Microbiol. 2008;74:6230–8.

Al-Zenki SF, Al-Nasser AY, Al-Saffar AE, Abdullah FK, Al-Bahouh ME, Al-Haddad AS, et al. Effects of using a chicken-origin competitive exclusion culture and probiotic cultures on reducing *Salmonella* in broilers. J Appl Poult Res. 2009;18:23–9.

Amézquita AA, Brashears MM. Competitive inhibition of *Listeria monocytogenes* in ready-to-eat meats by lactic acid bacteria. J Food Prot. 2002;65:316–25.

Ammor MS, Michaelidis C, Nychas G-JE. Insights into the role of quorum sensing in food spoilage. J Food Prot. 2008;7:1510–25.

Anderson RC, Genovese KJ, Harvey RB, Callaway TR, Nisbet DJ. Preharvest food safety applications of competitive exclusion cultures and probiotics. In: Goktepe I, Juneja VK, Ahmedna M, editors. Probiotics in food safety and human health. Boca Raton: Taylor & Francis; 2006. p. 273–84.

Atterbury R. Bacteriophage biocontrol in animals and meat products. Microb Biotechnol. 2009;2:601–12.

Atterbury RJ, Connerton PL, Dodd CER, Rees CED, Connerton LF. Application of host-specific bacteriophages to the surface of chicken skin leads to a reduction in recovery of *Campylobacter jejuni*. Appl Environ Microbiol. 2003;69:6302–6.

Atterbury RJ, Van Bergen MAP, Oritz F, Lovell MA, Harris JA, De Boer A, et al. Bacteriophage therapy to reduce *Salmonella* colonization of broiler chickens. Appl Environ Microbiol. 2007;73:4543–9.

Belletti N, Lanciotti R, Patrignani F, Gardini F. Antimicrobial efficacy of citron essential oil on spoilage and pathogenic microorganisms in fruit-based salads. J Food Sci. 2008;73:M331–8.

Bigwood T, Hudson JA, Billington C, Carey-Smith GV, Heinemann JA. Phage inactivation of foodborne pathogens on cooked and raw meat. Food Microbiol. 2008;25:400–6.

Bigwood T, Hudson JA, Billington C. Influence of host and phage concentration on the inactivation of foodborne pathogenic bacteria by two bacteriophages. FEMS Microbiol Lett. 2009;291:59–64.

Bjarnsholt T, Jensen PO, Rasmussen TB, Christophersen L, Calum H, Hentzer M, et al. Garlic blocks quorum sensing and promotes rapid clearing of pulmonary *Pseudomonas aeruginosa* infections. Microbiology. 2005;151:3873–80.

Bredholt S, Nesbakken T, Holck A. Protective cultures inhibit growth of *Listeria monocytogenes* and *Escherichia coli* O157:H7 in cooked, sliced, vacuum- and gas- packaged meat. Int J Food Microbiol. 1999;53:43–52.

Bredholt S, Nesbakken T, Holck A. Industrial application of an antilisterial strain of *Lactobacillus sakei* as a protective culture and its effect on the sensory acceptability of cooked, sliced, vacuum-packaged meats. Int J Food Microbiol. 2001;66:191–6.

Brenes A, Roura E. Essential oils in poultry nutrition: main effects and modes of action. Anim Feed Sci Technol. 2010;158(1–2):1–14.

Bruttin A, Brüssow H. Human volunteers receiving *Escherichia coli* T4 orally: a safety test of phage therapy. Antimicrob Agents Chemother. 2005;49:2874–8.

Burt S. Essential oils: their antibacterial properties and potential applications in foods – a review. Int J Food Microbiol. 2004;94(3):223–53.

Callaway TR, Edrington TS, Anderson RC, Harvey RB, Genovese KJ, Kennedy CN, et al. Probiotics, prebiotics and competitive exclusion for prophylaxis against bacterial disease. Anim Health Res Rev. 2008;9:217–25.

Callaway TR, Edrington TS, Braddan AD, Kutter EM, Anderson RC Nisbet DJ. "Use of bacteriophage to reduce *E. coli* O157:H7 in the intestinal tract and the hides of cattle". Presented at VTEC 2006, Melbourne, Australia 2006.

Carlton RM, Noordman WH, Biswas B, de Meester ED, Loessner MJ. Bacteriophage P100 for control of *Listeria monocytogenes* in foods: genome sequence, bioinformatic analysis, oral toxicity study, and application. Regul Toxicol Pharmacol. 2005;43:301–12.

Carneiro A, Couto JA, Mena C, Queiroz J, Hogg T. Activity of wine against *Campylobacter jejuni*. Food Control. 2008;19:800–5.

Castang S, Chantegrel B, Deshayes C, Dolmazon R, Gouet P, Haser R, et al. *N*-Sulfonyl homoserine lactones as antagonists of bacterial quorum sensing. Bioorg Med Chem Lett. 2004;14:5145–9.

Choo JH, Rukayadi Y, Hwang JK. Inhibition of bacterial quorum sensing by vanilla extract. Lett Appl Microbiol. 2006;42:637–41.

Deegan LH, Cotter PD, Hill C, Ross P. Bacteriocins: biological tools for bio-preservation and shelf-life extension. Int Dairy J. 2006;16:1058–71.

Delbrück M. The growth of bacteriophage and lysis of the host. J Gen Physiol. 1940;23:643–60.

Delves-Broughton J, Weber G, Elsser D. "Natural preservation of foods using bacterial metabolites and live addition of bacteria". Food Science Central. (2007) http://www.foodsciencecentral.com/fsc/ixid14740.

Diez-Gonzalez F, Schamberger GP. Use of probiotics in preharvest food safety applications. In: Goktepe I, Juneja VK, Ahmedna M, editors. Probiotics in food safety and human health. Boca Raton: Taylor & Francis; 2006. p. 251–72.

Dong YH, Gusti AR, Zhang Q, Xu JL, Zhang LH. Identification of quorum-quenching *N*-acyl homoserine lactonases from *Bacillus* species. Appl Environ Microbiol. 2002;68:1754–9.

Duan J, Zhao Y. Antimicrobial efficiency of essential oil and freeze-thaw treatments against *Escherichia coli* O157:H7 and *Salmonella enterica* Ser. Enteritidis in strawberry juice. J Food Sci. 2009;74(3):M131–7.

Dudler R, Eberl L. Interactions between bacteria and eukaryotes via small molecules. Curr Opin Biotechnol. 2006;17:268–73.

Dykes GA, Moorhead SM. Combined antimicrobial effect of nisin and listeriophage against *Listeria monocytogenes* in broth but not in buffer or on raw beef. Int J Food Microbiol. 2002;73:71–81.

Fischetti VA. Bacteriophage lytic enzymes: novel anti-infectives. Trends Microbiol. 2005;13:491–6.

Fisher K, Phillips CA. The effect of lemon, orange and bergamot essential oils and their components on the survival of *Campylobacter jejuni*, *Escherichia coli* O157, *Listeria monocytogenes*, *Bacillus cereus* and *Staphylococcus aureus in vitro* and in food systems. J Appl Microbiol. 2006;101:1232–40.

Fisher K, Rowe C, Phillips CA. The survival of three strains of *Arcobacter butzleri* in the presence of lemon, orange and bergamot essential oils and their components *in vitro* and on food. Lett Appl Microbiol. 2007;44:495–9.

Fox JL. Therapy with phage: mirage or potential barrage of products? ASM News. 2005;71:453–5.

Fratamico PM, Cooke PH. Isolation of Bdellovibrios that prey on *Escherichia coli* O157:H7 and *Salmonella* species and application for removal of prey from stainless steel surfaces. J Food Saf. 1996;16:161–73.

Friedman M, Henika PR, Levin CE, Mandrell RE, Kozukue N. Antimicrobial activities of tea catechins and theaflavins and tea extracts against *Bacillus cereus*. J Food Prot. 2006;69:354–61.

Fuqua C, Parsek MR, Greenberg EP. Regulation of gene expression by cell-to-cell communication: acyl-homoserine lactone quorum sensing. Annu Rev Genet. 2001;35:439–68.

Gaggìa F, Mattarelli PA, Biavati B. Probiotics and prebiotics in animal feeding for safe food production. Int J Food Microbiol. 2010. doi:10.1016/j.ijfoodmicro.2010.02.031.

García P, Madera C, Martínez B, Rodríguez A. Biocontrol of *Staphylococcus aureus* in curd manufacturing processes using bacteriophages. Int Dairy J. 2007;17:1232–9.

García P, Madera C, Martínez B, Rodríguez A, Suárez JE. Prevalence of bacteriophages infecting *Staphylococcus aureus* in dairy samples and their potential as biocontrol agents. J Dairy Sci. 2009;92:3019–26.

Gaysinsky S, Davidson PM, Bruce BD, Weiss J. Growth inhibition of *Escherichia coli* O157:H7 and *Listeria monocytogenes* by carvacrol and eugenol encapsulated in surfactant micelles. J Food Prot. 2005;68:2559–66.

Giatrakou V, Ntzimani A, Savvaidis IN. Combined chitosan-thyme treatments with modified atmosphere packaging on a ready-to-cook poultry product. J Food Prot. 2010;73(4):663–9.

Goktepe I. Probiotics as biopreservatives for enhancing food safety. In: Goktepe I, Juneja VK, Ahmedna M, editors. Probiotics in food safety and human health. Boca Raton: Taylor & Francis; 2006. p. 285–307.

Goode D, Allen VM, Barrow PA. Reduction of experimental *Salmonella* and *Campylobacter* contamination of chicken skin by application of lytic bacteriophages. Appl Environ Microbiol. 2003;69:5032–6.

Govaris A, Solomakos N, Pexara A, Chatzopoulou PS. The antimicrobial effect of oregano essential oil, nisin and their combination against *Salmonella* Enteritidis in minced sheep meat during refrigerated storage. Int J Food Microbiol. 2010;137(2–3):175–80.

Gram L, Melchiorsen J. Interaction between fish spoilage bacteria *Pseudomonas* sp. and *Shewanella putrefaciens* in fish extracts and on fish tissue. J Appl Bacteriol. 1996;80:589–95.

Greer GG. Bacteriophage control of foodborne bacteria. J Food Prot. 2005;68:1102–11.

Greer GG, Dilts BD. Inability of a bacteriophage pool to control beef spoilage. Int J Food Microbiol. 1990;10:331–42.

Guenther S, Huwyler D, Richard S, Loessner MJ. Virulent bacteriophage for efficient biocontrol of *Listeria monocytogenes* in ready-to-eat foods. Appl Environ Microbiol. 2009;75:93–100.

Gutierrez J, Barry-Ryan C, Bourke P. Antimicrobial activity of plant essential oils using food model media: efficacy, synergistic potential and interactions with food components. Food Microbiol. 2009;26:142–50.

Hagens S, Offerhaus ML. Bacteriophages-new weapons for food safety. Food Technol. 2008;62(4):46–54.

Henry MB, Lynch JM, Fermor TR. Role of siderophores in the biocontrol of *Pseudomonas tolaasii* by fluorescent pseudomonad antagonists. J Appl Bacteriol. 1991;70:104–8.

Higgins JP, Higgins SE, Guenther KL, Huff W, Donoghue AM, Donoghue DJ, et al. Use of a specific bacteriophage treatment to reduce *Salmonella* in poultry. Poult Sci. 2005;84:1141–5.

Hogan DA, Vik A, Kolter R. A *Pseudomonas aeruginosa* quorum sensing molecule influences *Candida albicans* morphology. Mol Microbiol. 2004;54:1212–23.

Holck A, Berg J. Inhibition of *Listeria monocytogenes* in cooked ham by virulent bacteriophages and protective cultures. Appl Environ Microbiol. 2009;75:6944–6.

Holley RA, Patel D. Improvement in shelf-life and safety of perishable foods by plant essential oils and smoke antimicrobials. Food Microbiol. 2005;22:273–92.

Hudson JA, Billington C, Carey-Smith G, Greening G. Bacteriophages as biocontrol agents in food. J Food Prot. 2005;68:426–37.

Hudson JA, Billington C, McIntyre L. Biological control of human pathogens on produce. In: Fan X, Niemira BA, Doona CJ, Feeherry FE, Gravani RB, editors. Microbial safety of fresh produce. Ames: IFT Press/Wiley-Blackwell; 2009a. p. 205–24.

Hudson JA, McIntyre L, Billington C. Bacteriophages and the control of bacteria in food. In: Adams T, editor. Contemporary trends in bacteriophage research. New York: Nova Scientific; 2009b. p. 11–45.

Hudson JA, Bigwood T, Premaratne A, Billington C, Horn B, McIntyre L. Potential to use UV-inactivated bacteriophages to control foodborne pathogens. Foodborne Pathog Dis. 2010;7:687–93.

Kadouri D, O'Toole GA. Susceptibility of biofilms to *Bdellovibrio bacteriovorus* attack. Appl Environ Microbiol. 2005;71:4044–51.

Kasman LM, Kasman A, Westwater C, Dolan J, Schmidt MG, Norris JS. Overcoming the phage replication threshold: a mathematical model with implications for phage therapy. J Virol. 2002;76:5557–64.

Kim K-P, Klumpp J, Loessner MJ. *Enterobacter sakazakii* bacteriophages can prevent bacterial growth in reconstituted infant formula. Int J Food Microbiol. 2007;115:195–203.

Kocharunchitt C, Ross T, McNeill DL. Use of bacteriophages as biocontrol agents to control *Salmonella* associated with seed sprouts. Int J Food Microbiol. 2009;128:453–9.

Lambert C, Morehouse KA, Chang C-Y, Sockett RE. *Bdellovibrio*: growth and development during the predatory cycle. Curr Opin Microbiol. 2006;9:639–44.

Leverentz B, Conway WS, Alavidze Z, Janisiewicz WJ, Fuchs Y, Camp MJ, et al. Examination of bacteriophage as a biocontrol method for *Salmonella* on fresh-cut fruit: a model study. J Food Prot. 2001;64:1116–21.

Leverentz B, Conway WS, Camp MJ, Janisiewicz WJ, Abuladze T, Yang M, et al. Biocontrol of *Listeria monocytogenes* on fresh-cut produce by treatment with lytic bacteriophages and a bacteriocin. Appl Environ Microbiol. 2003;69:4519–26.

Leverentz B, Conway WS, Janisiewicz WJ, Camp MJ. Optimizing concentration and timing of a phage spray application to reduce *Listeria monocytogenes* on honeydew melon tissue. J Food Prot. 2004;67:1682–6.

Lindberg AA. Studies of a receptor for Felix O-1 phage in *Salmonella minnesota*. J Gen Microbiol. 1967;48:225–33.

Liu M, Gray JM, Griffiths MW. Occurrence of proteolytic activity and N-acyl-homoserine lactone signals in the spoilage of aerobically chill-stored proteinaceous raw foods. J Food Prot. 2006;69:2729–37.

Loc Carrillo CM, Atterbury RJ, El-Shibiny A, Connerton PL, Dillon E, Scott A, et al. Bacteriophage therapy to reduce *Campylobacter jejuni* colonization of broiler chickens. Appl Environ Microbiol. 2005;71:6554–63.

Martínez B, Obeso JM, Rodríguez A, García P. Nisin-bacteriophage coresistance in *Staphylococcus aureus*. Int J Food Microbiol. 2008;122:253–8.

Mataragas M, Drosinos EH, Metaxopoulos J. Antagonistic activity of lactic acid bacteria against *Listeria monocytogenes* in sliced cooked cured pork shoulder stored under vacuum or modified atmosphere at 4±2°C. Food Microbiol. 2003;20:259–65.

Medina-Martinez MS, Uyttendaele M, Meireman S, Debevere J. Screening of N-acyl-L-homoserine lactone production by bacteria isolated from fresh foods. Commun Agric Appl Biol Sci. 2006;71:209–12.

Medina-Martinez MS, Uyttendaele M, Meireman S, Debevere J. Relevance of N-acyl-L-homoserine lactone production by *Yersinia enterocolitica* in fresh foods. J Appl Microbiol. 2007;102:1150–8.

Mexis SF, Chouliara E, Kontominas MG. Combined effect of an oxygen absorber and oregano essential oil on shelf life extension of rainbow trout fillets stored at 4°C. Food Microbiol. 2009;26:598–605.

Millette M, Luquet FM, Lacroix M. *In vitro* growth control of selected pathogens by *Lactobacillus acidophilus*- and *Lactobacillus casei*-fermented milk. Lett Appl Microbiol. 2006;44:314–9.

Modi R, Hirvi Y, Hill A, Griffiths MW. Effect of phage on survival of *Salmonella* Enteritidis during manufacture and storage of cheddar cheese made from raw and pasteurised milk. J Food Prot. 2001;64:927–33.

Müh U, Hare BJ, Duerkop BA, Schuster M, Hanzelka BL, Heim R, et al. A structurally unrelated mimic of a *Pseudomonas aeruginosa* acyl-homoserine lactone quorum-sensing signal. Proc Natl Acad Sci USA. 2006;103:16948–52.

Murinda SE, Roberts RF, Wilson RA. Evaluation of colicins for inhibitory activity against diarrheagenic *Escherichia coli* strains including serotype O157. Appl Environ Microbiol. 1996;62:3196–202.

Newman K. Mannan-oligosaccharides: natural polymers with significant impact on the gastrointestinal microflora and the immune system. In: Lyons TP, Jacques KA, editors. Proceedings of Alltech's 10th anniversary symposium: biotechnology in the feed industry. Nottingham: Nottingham University Press; 1994. p. 167–74.

Niu C, Afre S, Gilbert ES. Subinhibitory concentrations of cinnamaldehyde interfere with quorum sensing. Lett Appl Microbiol. 2006;43:489–94.

Novak JS, Fratamico PM. Evaluation of ascorbic acid as a quorum-sensing analogue to control growth, sporulation and entertoxin production in *Clostridium perfringens*. J Food Sci. 2004;69:72–8.

Nychas G-JE, Dourou D, Skandamis P, Koutsoumanis K, Baranyi J, Sofos J. Effect of microbial cell-free meat extract on the growth of spoilage bacteria. J Appl Microbiol. 2009;107:1819–29.

O'Flynn G, Ross RP, Fitzgerald GF, Coffey A. Evaluation of a cocktail of three bacteriophages for biocontrol of *Escherichia coli* O157:H7. Appl Environ Microbiol. 2004;70:3417–24.

Obeso JM, Martínez B, Rodríguez A, García BE. Lytic activity of the recombinant staphylococcal bacteriophage ΦH5 endolysin active against *Staphylococcus aureus* in milk. Int J Food Microbiol. 2008;128:212–8.

O'Flaherty S, Coffey A, Meaney WJ, Fitzgerald GF, Ross RP. Inhibition of bacteriophage K proliferation on *Staphylococcus aureus* in raw bovine milk. Lett Appl Microbiol. 2005;41:274–9.

Oussalah M, Caillet S, Salmieri S, Saucier L, Lacroix M. Antimicrobial effects of alginate-based film containing essential oils for the preservation of whole beef muscle. J Food Prot. 2006;69:2364–9.

Pao S, Randolph SP, Westbrook EW, Shen H. Use of bacteriophages to control *Salmonella* in experimentally contaminated sprout seeds. J Food Sci. 2004;69:M127–30.

Paradis-Bleau C, Cloutier I, Lemieux L, Sanschagrin F, Laroche J, Auger M, et al. Peptidoglycan lytic activity of the *Pseudomonas aeruginosa* phage ΦKZ gp144 lytic transglycosylase. FEMS Microbiol Lett. 2007;266:201–9.

Perez-Conesa D, McLandsborough L, Weiss J. Inhibition and inactivation of *Listeria monocytogenes* and *Escherichia coli* O157:H7 colony biofilms by micellar-encapsulated eugenol and carvacrol. J Food Prot. 2006;69:2947–54.

Puupponen-Pimia R, Nohynek L, Hartmann-Schmidlin S, Kahkonen M, Heinonen M, Maatta-Riihinen K, et al. Berry phenolics selectively inhibit the growth of intestinal pathogens. J Appl Microbiol. 2005;98:991–1000.

Qazi S, Middleton B, Muharram SH, Cockayne A, Hill P, O'Shea P, et al. *N*-Acyl homoserine lactones antagonize virulence gene expression and quorum sensing in *Staphylococcus aureus*. Infect Immun. 2006;74:910–9.

Rasch M, Rasmussen TB, Andersen JB, Persson T, Nielsen J, Givskov M, et al. Well-known quorum sensing inhibitors do not affect bacterial quorum sensing-regulated bean sprout spoilage. J Appl Microbiol. 2007;102:826–37.

Ren D, Sims JJ, Wood TK. Inhibition of biofilm formation and swarming of *Escherichia coli* by (5Z)-4-bromo-5-(bromomethylene)-3-butyl-2(5H)-furanone. Environ Microbiol. 2001;3:731–6.

Rendulic S, Jagtap P, Rosinus A, et al. A predator unmasked: life cycle of *Bdellovibrio bacteriovorus* from a genomic perspective. Science. 2004;303:689–92.

Reynolds G. Bacteria effective against fresh cut fruit contamination. 2007. http://www.foodnavigator-usa.com/news/printNewsBis.asp?id=75142.

Rhodes PL, Mitchell JW, Wilson MW, Melton LD. Antilisterial activity of grape juice and grape extracts derived from *Vitis vinifera* variety Ribier. Int J Food Microbiol. 2006;107:281–6.

Rojas-Grau MA, Avena-Bustillos RJ, Olsen C, Friedman M, Henika PR, Martin-Belloso O, et al. Effects of plant essential oils and oil compounds on mechanical, barrier and antimicrobial properties of alginate-apple puree edible films. J Food Eng. 2007;81:634–41.

Sánchez-González L, González-Martínez C, Chiralt A, Cháfer M. Physical and antimicrobial properties of chitosan-tea tree essential oil composite films. J Food Eng. 2010;98:443–52.

Santini C, Baffoni L, Gaggìa F, Granata M, Gasbarri R, Di Gioia D, et al. Characterisation of probiotic strains: an application as feed additives in poultry against *Campylobacter jejuni*. Int J Food Microbiol. 2010. doi:10.1016/ijfoodmicro.2010.03.039.

Schamberger P, Diez-Gonzalez F. Characterization of colicinogenic *Escherichia coli* strains inhibitory to enterohemorrhagic *Escherichia coli*. J Food Prot. 2004;67:486–92.

Schellekens M, Wouters MJ, Hagens S, Hugenholtz J. Bacteriophage P100 application to control *Listeria monocytogenes* on smeared cheese. Milchwissenschaft. 2007;62:284–7.

Schneitz C. Competitive exclusion in poultry – 30 years of research. Food Control. 2005;16:657–67.

Scolari G, Vescovo M. Microbial antagonism of *Lactobacillus casei* added to fresh vegetables. Ital J Food Sci. 2004;16:465–75.

Seydim AC, Sarikus G. Antimicrobial activity of whey protein based edible films incorporated with oregano, rosemary and garlic essential oils. Food Res Int. 2006;39:639–44.

Sharma M, Patel JR, Conway WS, Ferguson S, Sulakvelidze A. Effectiveness of bacteriophages in reducing *Escherichia coli* on fresh-cut cantaloupes and lettuce. J Food Prot. 2009;72:1481–5.

Si W, Gong J, Chanas C, Cui S, Yu H, Caballero C, et al. *In vitro* assessment of antimicrobial activity of carvacrol, thymol and cinnamaldehyde towards *Salmonella* serotype Typhimurium DT104: effects of pig diets and emulsification in hydrocolloids. J Appl Microbiol. 2006a;101:1282–91.

Si W, Gong J, Tsao R, Zhou T, Yu H, Poppe C, et al. Antimicrobial activity of essential oils and structurally related synthetic food additives towards selected pathogenic and beneficial gut bacteria. J Appl Microbiol. 2006b;100:296–305.

Smith L, Mann JE, Harris K, Miller MF, Brashears MM. Reduction of *Escherichia coli* O157:H7 and *Salmonella* in ground beef using lactic acid bacteria and the impact on sensory properties. J Food Prot. 2005;68:1587–92.

Smith-Palmer A, Stewart J, Fyfe L. The potential application of plant essential oils as natural food preservatives in soft cheese. Food Microbiol. 2001;18:463–70.

Soni KA, Nannapaneni R. Bacteriophage significantly reduces *Listeria monocytogenes* on raw salmon fillet tissue. J Food Prot. 2010;73:32–8.

Soni KA, Nannapaneni R, Hagens S. Reduction of *Listeria monocytogenes* on the surface of fresh channel catfish fillets by bacteriophage Listex P100. Foodborne Pathog Dis. 2010;7:427–34.

Stephens TP, Loneragan GH, Chichester LM, Brashears MM. Prevalence and enumeration of *Escherichia coli* O157 in steers receiving various strains of *Lactobacillus*-based direct-fed microbials. J Food Prot. 2007;70:1252–5.

Stephens TP, Stanford K, Rode LM, Booker CW, Vogstad AR, Schunicht OC, et al. Effect of a direct-fed microbial on animal performance, carcass characteristic and the shedding of *Escherichia coli* O157 by feedlot cattle. Anim Feed Sci Technol. 2010;158:65–72.

Tsuei A-C, Carey-Smith GV, Hudson JA, Billington C, Heinemann JA. Prevalence and numbers of coliphages and *Campylobacter jejuni* bacteriophages in New Zealand foods. Int J Food Microbiol. 2007;116:121–5.

Valero D, Valverde JM, Martinez-Romero D, Guillen F, Castillo S, Serrano M. The combination of modified atmosphere packaging with eugenol or thymol to maintain quality, safety and functional properties of table grapes. Postharvest Biol Technol. 2006;41:317–27.

Vandeplas S, Dubois Dauphin R, Beckers Y, Thonart PA, Théwis A. *Salmonella* in chicken: current and developing strategies to reduce contamination at farm level. J Food Prot. 2010;73:774–85.

Viuda-Martos M, Ruiz-Navajas Y, Fernández-López J, Pérez-Álvarez JA. Effect of orange dietary fibre, oregano essential oil and packaging conditions on shelf-life of bologna sausages. Food Control. 2010;21:436–43.

Wagner RD. Efficacy and food safety considerations of poultry exclusion products. Mol Nutr Food Res. 2006;50:1061–71.

Whichard JM, Sriranganathan N, Pierson FW. Suppression of *Salmonella* growth by wild-type and large-plaque variants of bacteriophage Felix O1 in liquid culture and on chicken frankfurters. J Food Prot. 2003;66:220–5.

Williams P, Winzer K, Chan WC, Cámara M. Look who's talking: communication quorum sensing in the bacterial world. Philos Trans R Soc B. 2007;362:1119–34.

Ye J, Kostrzynska M, Dunfield K, Warriner K. Evaluation of a biocontrol preparation consisting of *Enterobacter asburiae* JX1 and a lytic bacteriophage cocktail to suppress the growth of *Salmonella* Javina associated with tomatoes. J Food Prot. 2009;72:2284–92.

Ye J, Kostrzynska M, Dunfield K, Warriner K. Control of *Salmonella* on sprouting mung bean sprouts and alfalfa seeds by using a biocontrol preparation based on antagonistic bacteria and lytic phages. J Food Prot. 2010;73:9–17.

Younts-Dahl SM, Osborn GD, Galyean ML, Rivera JD, Loneragan GH, Brashears MM. Reduction of *Escherichia coli* O157 in finishing beef cattle by various doses of *Lactobacillus acidophilus* in direct-fed microbials. J Food Prot. 2005;68:6–10.

Zhang G, Ma L, Doyle MP. Potential competitive exclusion bacteria from poultry inhibitory to *Campylobacter jejuni* and *Salmonella*. J Food Prot. 2007;70:867–73.

Zhao T, Podtburg TC, Zhao P, Schmidt BE, Baker DA, Cords BA, et al. Control of *Listeria* spp. by competitive-exclusion bacteria in floor drains of a poultry processing plant. Appl Environ Microbiol. 2006;72:3314–20.

Chapter 9
Plant Extracts as Natural Antifungals: Alternative Strategies for Mold Control in Foods

Virginia Fernández Pinto, Andrea Patriarca, and Graciela Pose

9.1 Introduction

Fungi can contaminate foods in the field, at harvest, and in different stages of storage and processing. Common deterioration signs are rotting and the development of off-odors and off-flavors. The most important aspect of spoilage caused by fungi is the formation of toxic secondary metabolites called mycotoxins, which may have harmful effects on human and animal health.

Fungal spoilage is an increasing economic problem in the food industry. Chemical antifungals are becoming less attractive as food preservatives, owing to the development of resistance and to the stricter legal regulations concerning the permitted concentrations. Finally, consumers now tend to demand products preserved through natural processes.

Food preservation is a continuous battle waged against microorganisms that spoil food or make it unsafe. The food industry has sought to replace food preservation techniques such as heat treatments, salting, acidification, drying, and chemical preservation with new, less invasive techniques (high-pressure, pulsed electric field, pulsed light, oscillating magnetic fields, ultrasound, and UV treatments) that satisfy increased consumer demand for natural, nutritious, and tasty food products. In spite of intensive research efforts and investments, very few of these novel techniques have been used by the food industry.

The banning of chemical additives by EU and other regulatory bodies has driven the food industry and food research institutions towards investigation of numerous natural antimicrobial compounds that exist in plants and which have evolved as host

V.F. Pinto (✉) • A. Patriarca
Universidad de Buenos Aires, Buenos Aires, Argentina
e-mail: virginia@qo.fcen.uba.ar

G. Pose
Universidad Nacional de Quilmes, Buenos Aires, Argentina

defense mechanisms. This antimicrobial activity has been demonstrated mainly in vitro. Two aspects that are crucial for the use of natural compounds are often overlooked: namely, the induced changes in organoleptic and textural properties of the food caused by addition of these materials and their interaction with food ingredients and the influence of this interaction on antimicrobial efficacy. In many cases, the concentrations of the antimicrobial compounds in herbs and spices are too low to be effective without adverse effects on the sensory characteristics of the food products (Devlieghere et al. 2004).

Fungicides are the primary means of controlling postharvest diseases. About 23 million kilograms of fungicides are used annually to treat fruit and vegetables. It is generally accepted that production and marketing of these products would not be possible without such use of fungicides. However, as harvested fruit and vegetables are commonly treated with fungicides to retard postharvest diseases, there is a greater likelihood of direct human exposure to them. In addition, synthetic fungicides can leave significant residues in treated commodities. Development of resistance to commonly used fungicides within populations of postharvest pathogens has also become a significant problem. For example, many synthetic fungicides are used to control blue mold rot of citrus fruit. However, acquired resistance by *Penicillium italicum* and *Penicillium digitatum* to fungicides used on citrus fruit has become a matter of concern in recent years. The adverse effects of the use and overuse of synthetic fungicides means that alternative strategies need to be developed to reduce losses due to postharvest decay that are perceived as safe by the public and pose negligible risk to human health and to the environment. Bioload reduction achieved through sanitation, the use of nonselective fungicides (sodium carbonate, sodium bicarbonate, active chlorine, and sorbic acid), and physical treatments such as heat treatment, low-temperature storage, hot water treatments, and radiation can significantly lower the disease pressure on harvested commodities. Harvesting and handling techniques that minimize injury to the commodity, along with storage conditions that are optimum for maintaining host resistance, will also aid in suppressing disease developing after harvest. However, none of these treatments are significantly effective, and they many cause damage to the commodities. Thus, replacement of synthetic fungicides by natural products, particularly of plant origin, which are nontoxic and specific in their action is gaining considerable attention.

Biological compounds are comparatively biodegradable and most of them leave virtually no residues in nature. Several plants have 1,000 years of history and have proven nontoxicity, at least at levels that are commonly ingested. This safety feature is very important in formulations of such products because it has an impact on the cost of development and registration of new pesticides. The research and development cost of botanical fungicides from discovery to marketing is much less than that of chemical fungicides. This expense is in large part due to the concern over possible high toxicities of chemical fungicides that necessitate long-term toxicological testing on experimental animals. Biological products, because of their target specificity, typically require only short-term toxicology studies. Although the exploitation of natural products to protect the postharvest decay of perishable products is just beginning, these products have the potential to be safe fungicides soon replacing the synthetic ones. In view of the merits of botanicals as postharvest fungicides, the products that are efficacious during in vitro tests should be properly tested for their practical potency using in vivo trials, organoleptic tests, and the safety limit profile (Tripanthi and Dubey 2004).

9.2 Plant Extracts

The essential oils and extracts from different plants are extensively used in the perfume, beverage, and food industries. These materials exhibit antimicrobial activities against a variety of bacteria and fungi. The potential antimicrobial properties of plants have been related to their ability to synthesize, through secondary metabolism, several chemical compounds of relatively complex structures with antimicrobial activity. These include phenolics, alkaloids, flavonoids, isoflavonoids, tannins, coumarins, glycosides, terpenes, phenyl propanes, and organic acids. These compounds are often toxic to insects, pathogens, and other organisms. The increased production of phenolics, often identified as condensed tannins, is a plant response to the feeding of chewing insects and has been extensively studied and well documented. Additionally, certain plant natural products are highly effective against fungicide-resistant microorganisms. For example, natural compounds such as cinnamaldehyde and salicylaldehyde are effective against four strains of *Fusarium sambucinum* resistant to the synthetic fungicide thiabendazole (Ahn et al. 2005).

Plant products of recognized antimicrobial spectrum can appear in food conservation systems as the main antimicrobial compounds or as adjuvants to improve the action of other antimicrobial compounds. In general, the inhibitory action of natural products on mold cells involves cytoplasm granulation, cytoplasmic membrane rupture, and inactivation and/or inhibition of synthesis of intercellular and extracellular enzymes. These actions can occur as isolated events or in a continuous way and culminate with the inhibition of mycelium germination. The occurrence of phytochemicals that do not show antimold activity could suggest that in some situations certain phytochemicals have an antimicrobial action only when they are acting in a synergistic way with other constituents of plants extracts. Effective phytocompounds are expected to be far more advantageous than synthetic pesticides, as they are easily decomposable, are not environmental pollutants, and posses no residual or phytotoxic properties (Boyraz and Ozcan 2006).

9.2.1 Crude Extracts

The fungi that cause biodeterioration can easily develop resistance traits against a single active component. However in instances where the plant extracts contain different antimicrobial ingredients possessing fungitoxic potency due to the synergistic effects of different components, this phenomenon is less likely to occur.

Plant extracts are potentially useful additives for food preservation as they are likely to prolong shelf life and improve the quality of stored food products (Shukla et al. 2008).

The traditional preparation of medicines from herbs often involves a broth or tea. However, much of the antifungal research conducted to date has assessed ethanol or methanol extracts, whereas only a few studies have utilized aqueous extracts, a closer approximation to traditional medicine. This is due to alcohol being a general solvent that tends to provide better extraction of compounds with a variety of polarities. Thus, aqueous extracts may not contain some of the less polar compounds (Webster et al. 2008).

The activity of crude extracts from plants against diverse microorganisms has been investigated by Tegegne et al. (2008), who developed a screening program which involved testing crude extracts of more than 3,000 South African plants for antifungal activity. *Agapanthus africanus* (L.) Hoffm., an evergreen plant indigenous to South Africa, was identified as one of the most potent. Crude methanolic extracts of all different plant parts markedly inhibited the mycelial growth of all test fungi, in vitro, at a concentration of 1 g/l.

Lee et al. (2007) examined the effect of different herb extracts alone or in combinations on some fungal strains, such as *Aspergillus niger*, *Botrytis cinerea*, *Fusarium moniliforme*, *Glomerella cingulata*, and *Phyllosticta caricae*. The extracts from *Cinnamomum cassia*, *Coptidis rhizoma*, and *Curcuma longa* exhibited a significant inhibitory effect on fungal growth, and special attention was given to these extracts because of their pleasant flavor. It was also found that the various combinations of herb extracts have a higher inhibitory effect towards the fungi tested than that of individual extract. These results indicate that the combinations of extracts of *C. cassia*, *C. longa*, and *C. rhizoma* have a synergistic effect on the fungi tested. Therefore, the combined herb extract is preferential for incorporation into various food products where a naturally antimicrobial additive is desired.

9.2.2 Medicinal Plants

Since antiquity, medicinal plants have been well-known natural sources of treatment of various diseases. In Asia, Latin America, Africa, and India about 20,000 plant species are known to be used for medicinal purposes. The extensive use of natural plants as primary remedies, owing to their pharmacological properties, is quite common in various parts of the world. Plants such as *Withania somniferum*, known as *ashwagandha* in ayurvedic traditional medicine, *Aegle marmelos* (bael tree), *Piper nigrum* (black pepper), and *Panax notoginseng* (ginseng) are known to have antifungal activity owing to lactones, terpenoids, alkaloids, and saponins, respectively. Most of the studies on extracts or essential oils of medicinal plants were performed on phytopathogenic fungi. The bael tree (*A. marmelos*) is indigenous to India, and various parts of the plant are used in traditional Indian medicine for the treatment of many diseases. The bael tree essential oil exhibited a considerable degree of inhibition of spore germination of various phytopathogenic fungi, including

Fusarium and *Alternaria* (Rana et al. 1997). Extracts of another plant used in Indian medicine, *Pseudarthria viscida* (L.), also effectively inhibit *Fusarium*, *Alternaria*, and *Aspergillus niger* (Deepa et al. 2004). Many species of *Hypericum* (*Hypericaceae* or *Guttiferae*) are used in folk medicine in various regions of the world, including Turkey. In recent years, *Hypericum* species have attracted much attention because of their antidepressant, antianxiety, antiviral, antimicrobial, and wound healing activities. Acetone and methanol extracts of *Hypericum linarioides* showed inhibitory effects on the growth of species of *Fusarium* and *Alternaria* (Cakir et al. 2005).

Quiroga et al. (2001) investigated the antifungal activity of the ethanolic extracts of ten Argentinean plants used in native medicine, and found that alcoholic extracts of *Zuccagnia punctata* and *Larrea divaricata* have considerable in vitro activity against all filamentous fungi tested, which included members of the *Aspergillus*, *Penicillium*, and *Fusarium* genera. The leaf pulp of *Aloe vera*, commonly known as aloe vera gel, has been used since early times for a host of curative purposes in humans and animals. The results of Saks and Barkai-Golan (1995) indicate that the gel extract from leaves of *Aloe vera* possess an inhibitory effect on spore germination and mycelial growth of four common postharvest fungi: *P. digitatum*, *Alternaria alternata*, *B. cinerea*, and *Penicillium expansum*.

Adenocalymma alliaceum Miers. (family Bignoniaceae), commonly known as "garlic creeper," is native to the Amazon rain forests of South America. The leaves and flowers are widely consumed by Brazilians as a substitute for garlic. The plant has a number of traditional medicinal properties: research has documented its analgesic, antiarthritic, anti-inflammatory, antipyretic, antirheumatic, antitussive, depurative, laxative, and anthelmintic (vermifugal) properties. The most sensitive fungal strains against *Adenocalymma* aqueous extract were *Mucor* sp., *Dreschlera* sp., *Fusarium roseum*, and *Humicola* sp. The extract prevented spore germination of *Aspergillus flavus* completely from a concentration of 15 mg/ml upwards (Shukla et al. 2008).

9.2.3 Spices and Herbs

Spices and herbs were originally added to foods to change or improve taste, but can also be used to enhance shelf life because of the antimicrobial properties they possess. The antimicrobial activity of spices and herbs and their essential oils is well recognized. Essential oils of thyme, cinnamon, bay, oregano, rosemary, sage, vanillin, and clove all possess antimicrobial activity, and several studies have reported results on their preservative action.

Some studies claim that the phenolic compounds present in spices and herbs might play a major role in their antimicrobial effects. Shan et al. (2007) reported a highly positive relationship between antibacterial activity and total phenolic content in a large number of spice and herb extracts. They also demonstrated that many plant extracts containing high levels of phenolics also possess strong antibacterial activity. They could be a potential source of inhibitory substances against some foodborne pathogens as well as antioxidant agents.

9.2.4 Essential Oils and Oleoresins

Essential oils are aromatic oily liquids obtained from plant material (flowers, buds, seeds, leaves, twigs, bark, herbs, wood, fruits, and roots). They are obtained principally using advanced distillation technology from spices and herbs. Modern distillation techniques ensure that the most valuable fractions are preserved in the distillate. These components are highly odorous, volatile compounds of plant material that possess characteristic aroma and flavor. An estimated 3,000 essential oils have been identified, of which about 300 are commercially important, destined chiefly for the flavors and fragrances market. The greatest use of essential oils is in food (as flavorings), perfumes (fragrances and aftershaves), and pharmaceuticals (for their functional properties). Individual components of essential oils are also used as food flavorings, either extracted from plant material or synthetically manufactured. It has long been recognized that some essential oils have antimicrobial properties, but the relatively recent enhancement of interest in "green" consumerism has led to a renewal of scientific interest in these substances. Besides antibacterial properties, essential oils or their components have antiviral, antimycotic, antitoxigenic, antiparasitic, and insecticidal properties. These characteristics may be related to the function of these compounds in plants (Burt 2004).

Plant essential oils are usually mixtures of several components. The oils with high levels of eugenol (allspice, clove bud and leaf, bay, and cinnamon leaf), cinnamamic aldehyde (cinnamon bark, cassia oil), and citral are usually strong antimicrobials. Activity of sage and rosemary is due to borneol and other phenolics in the terpene fraction. The volatile terpenes carvacrol, p-cymene, and thymol are probably responsible for the antimicrobial activity of oregano, thyme, and savory. In sage the terpene thejone and in rosemary a group of terpenes (borneol, camphor, 1,8-cineole, α-pinene, camphone, verbenonone, and bornyl acetate) are responsible for the activity (Holley and Patel 2005).

The composition of essential oils from different parts of the same plant can also differ widely. For example, oils obtained from the seeds of coriander (*Coriandrum sativum* L.) have quite a different composition from oils of cilantro (coriander), which is obtained from the immature leaves of the same plant.

Although essential oils perform well in in vitro antimicrobial assays, it has generally been found that a higher concentration of essential oils is needed to have the same effect in foods. The greater availability of nutrients in foods compared with laboratory media may enable microorganisms to repair damaged cells faster. Not only are the intrinsic properties of the food (fat/protein/water content, antioxidants, preservatives, pH, salt and other additives) relevant in this respect, the extrinsic determinants (temperature, packaging in vacuum/gas/air, characteristics of microorganisms) can also influence sensitivity (Burt 2004).

Oleoresins are the true essence of spices in their most concentrated form and the most preferred and convenient substitute for raw spices in the processed food industry. Oleoresins are spice or herb extracts produced by solvent extraction containing volatile and nonvolatile components. The isolated nonvolatile components consist of several chemical groups such as carotenoids, steroids, alkaloids, anthocyanins, and glycosides. This fraction can be essential for taste, mouthfeel, texture, and antioxidant properties of the foods. Oleoresins differ from essential oils as they account for all the flavoring ingredients of a particular spice. They are free of microorganisms and may be standardized to a desired degree of flavor strength. When edible films are enriched with essential oils, the drying temperatures usually employed to form an edible coating are high enough to volatilize a high percentage of the aromatic components. The advantage of substituting essential oils for their corresponding food-grade oleoresins could lie in the introduction of other nonvolatile components, positively affecting food quality.

9.2.5 Antifungal Potential of Plant Extracts Against Toxigenic Fungi

Essential oils should find a practical application in the inhibition of mycotoxin production in stored grains. The presence and growth of fungi in food may cause spoilage and result in a reduction in quality and quantity. The use of natural antimicrobial compounds is important not only for the preservation of food but also in the control of human and plant diseases of microbial origin.

Aflatoxin-producing molds are widely distributed in nature and are frequently the cause of contamination in human food resources. Aflatoxins are secondary metabolites produced by toxigenic strains of *A. flavus* and *Aspergillus parasiticus*. These fungi grow rapidly on a variety of natural substrates, and consumption of contaminated food can pose serious health hazards to human and animals. *Apergillus* species are known to produce aflatoxins in a variety of food and feedstuffs. Chemicals are currently used to limit the growth of hazardous fungi such as *A. flavus* and *A. parasiticus* in stored foods. In recent years there has been considerable interest in the preservation of grains through the use of essential oils that are known to retard growth and mycotoxin production. Many investigators used essential oils such as cinnamon, peppermint, basil, and thyme to protect maize kernels against *A. flavus* infection, without affecting germination and corn growth (Soliman and Badeaa 2002). The findings indicate that essential oils should have a practical application in the inhibition of mycotoxin production in food products and could be added to grain in storage to protect it from fungal infection.

A. parasiticus, as food contaminant, has received little attention as far as essential oils are concerned. Allameh et al. (2001) investigated the effect of neem leaf extract on the aflatoxin production by *A. parasiticus*. The neem (*Azadirachta indica* Juss) is an ornamental tree of the arid areas of Asia and East Africa. The plant extract at a concentration of 50% v/v (1:1 with the culture medium) caused more than 90% inhibition in aflatoxin production. Essential oils of *Thymus eriocalyx* and *Thymus X-porlock* were inhibitory to fungal development and aflatoxin production by *A. parasiticus* at low concentrations (250 ppm) (Rasooli and Owlia 2005). Both fungal growth and aflatoxin biosynthesis by *A. parasiticus* were suppressed by essential oils from *Rosmarinus officinalis* and *Trachyspermum copticum* L. The inhibitory effect of the oils increased in proportion to their concentrations. Aflatoxin production was significantly inhibited at lower than fungistatic concentration of both oils (Rasooli et al. 2008). Essential oils could be safely used as preservative materials on some kinds of foods, not only to prevent fungal growth but also to reduce the risk of mycotoxins. These oils could be used as a substitute for chemical fungicides as they are natural and are nontoxic to humans.

9.3 Hurdle Technology

Hurdle technology, which involves the use of a combination of preservation approaches, has generally been successful in controlling pathogens and maintaining food quality during storage, but food safety issues remain. Food preservation is performed during food processing in an attempt to maintain raw material quality to provide safe products that have a low spoilage potential. This is mainly achieved through purposely designed processing that varies from one product to another. Hence, in preservation processes the general aim is to employ combination processes where, for example, a mild heat stress and a low concentration of preservatives are combined to fulfill the objectives. Combination of preservation treatments permits the required level of protection to be reached, while at the same time retaining the organoleptic qualities of the products. The natural preservatives' potential in combination with other treatments may lead to the development of novel mild preservation regimes (Brul and Coote 1999).

9.3.1 Physical and Chemical Methods Used To Reduce the Microbial Load of Fresh Products

With respect to microbiological safety and extension of the lifespan of products that are consumed and commercialized fresh, such as fruits and vegetables, numerous physical and chemical methods that reduce the microbial load can be applied. Each method has advantages and disadvantages depending upon the type of product, mitigation protocol, and other variables.

The traditional method to extend the useful life of these products is storage at refrigeration temperatures. Nevertheless, these temperatures alone cannot be relied on to prevent the development of molds causing spoilage, since many molds are capable of developing at low temperatures.

Other methods are the physical removal of microorganisms by washing and brushing and the application of solutions of hypochlorite, chlorine dioxide (ClO_2), alkaline compounds, organic acids, peracetic acid, and hydrogen peroxide, among others. In the first case, the products are washed superficially by oscillating brushes for the physical removal of soil and microorganisms. This is frequently performed with a detergent, followed by rinsing with potable water. Nevertheless, the brushing process also removes a portion of the natural waxy cuticle on the surface of the product that acts as a barrier to microorganisms. In the second case, hypochlorite has a long history of use. The material corrodes the equipment and chlorine residues might have adverse effects on health. The major advantages that chlorine dioxide has over hypochlorous acid (HOCl) or hypochlorites are its lower levels of reactivity with organic matter, superior antimicrobial activity in media at neutral pH, and lower levels of chlorinated residues. Nevertheless, the lesser stability of chlorine dioxide might be a problem. Another commonly used method of food preservation is acidification. The organic acids and their esters are widespread in nature; they are found in fruits, examples being citric acid in the citrus fruits, benzoic acid in blueberries, and sorbic acid in the fruit of the ash tree (*Fraxinus* species). Several organic acids are found in spices. Lactic acid is also found in animal tissues. Today, certain organic acids are used to help the preservation of diverse products. The organic acids are highly effective and inexpensive means used for the prevention of growth of yeasts and fungi that cause food spoilage. Peracetic acid is a strong oxidizing agent that is much less corrosive than chlorine compounds. In addition, peracetic acid has a greater spectrum of action and is effective in the presence of organic matter. Peracetic acid is used at low concentrations and is environmentally friendly as it rapidly breaks down to water, oxygen, and acetic acid. Peracetic acid does not stain and has long-term stability when stored as a concentrate.

Hydrogen peroxide owes its antimicrobial activity to its properties as an oxidizing agent and to its capacity to generate other cytotoxic oxidizing species such as hydroxyl radicals. Their bactericidal, sporicidal, and inhibitory activities, coupled with rapid breakdown to nontoxic products, make it an attractive alternative for application in numerous commodities and in the food industry (FDA/CFSAN 2001).

9.4 A Case Study

The case study concerns the application of mild preservation treatments combined with plant extracts to control *Alternaria* diseases in tomato.

9.4.1 Hurdle Technology and Its Application to Fruits and Vegetables: Combined Methods

A variety of mitigation regimens and sanitizers are available to reduce microbial populations on different types of produce. Washing and sanitizing efficiencies depend on several factors, such as the characteristics of the surface of the produce, water quality, cleaner/sanitizer used, contact time, and presence and type of scrubbing action. On the basis of reported data, it is likely that different sanitation mitigation strategies are needed for different produce items. The concept of using multiple intervention methods is analogous to hurdle technology, where two or more preservation technologies are used to prevent growth of microorganisms in or on foods. To adequately address safety issues associated with fresh produce, it is necessary to enhance the quantity and quality of research on mitigation strategies. A few requisites include determination of additive, antagonistic, or synergistic effects of sanitation treatments when used in combination. The hurdle concept can be applied to the sanitization of fruits and vegetables. The principal form of marketing of these products is as fresh produce. Thus, these products are perishable and highly susceptible to fungal diseases. The spores spread, remain on the surface, and germinate when mature, even in refrigeration conditions. A great quantity of fungicides have been used to extend the useful life of these products; nevertheless, there is a trend to opt for organic produce. With such a trend, the essential oils and the oleoresins may prove to be a viable alternative. In combination with other nonharmful and organically acceptable methods, such as previous washing with solutions of organic acids or hydrogen peroxide, they might turn out to be an excellent alternative to extend the shelf life of fruits and vegetables.

9.4.2 Tomato and Tomato Products

Tomatoes are one of the most used fruits in the world with one of the highest processing volumes in the world. The production is destined for fresh consumption and for industry, where the tomatoes are converted into tomato puree, pulp, juice, sauces, etc. Around 60% of the industrialized tomato fraction is used in the production of tomato paste, which is reprocessed by the adding of water, spices, and other ingredients, and then is packed and sold as "tomato puree." Because of their thin skin, tomatoes are very susceptible to fungal decay, and *Alternaria* is the most common fungus found on moldy tomatoes. *A. alternata* is the causative agent of black mold of ripe tomato fruits, a disease that frequently causes substantial loss of tomatoes, especially those used for canning.

The genus *Alternaria* contains numerous species that are both saprophytic and pathogenic with regard to many plants, including several plants used for food. During the colonization of the host, the pathogen can produce toxic metabolites in the infected plants that are hazardous to human health. *Alternaria* species are known to produce at least 70 secondary metabolites. Some species, in particular *A. alternata*, can produce mycotoxins in infected plants. High levels of toxins were found in fruits and vegetables such as apples, mandarins, peppers, olives, and tomatoes. *Alternaria* toxins have also been found in several other agricultural commodities, including wheat, sunflower seeds, oilseed rape, sorghum, and pecans.

Considering that *Alternaria* spp. are the most frequent fungal species invading tomatoes and their toxigenic capacity, the presence of this fungus must be prevented in tomatoes and tomato products (Terminiello et al. 2006).

9.4.3 In Vitro Antifungal Activity of Organic Acids, Sanitizers, and Oleoresins

The effect of organic acids and sanitizers on the load reduction of *Alternaria* spores has been investigated. The efficiency of these sanitizers and organic acids was challenged with a suspension of *Alternaria* strains (10^4 spores/ml) isolated from tomatoes. The spore suspension was mixed with 0.05%, 0.5%, and 1% solutions of acetic acid, citric acid, salicylic acid, sodium benzoate, sodium lactate, and potassium sorbate. After 30 s, the solutions were neutralized and the number of surviving spores was determined on potato dextrose agar. One logarithmic reduction of the number of *Alternaria* spores was achieved with 0.05% sodium benzoate, 0.05% salicylic acid, 0.5% acetic acid, 0.5% potassium sorbate, and 1% sodium lactate solutions. An inhibitory effect of 100% was observed with 0.5% of salicylic acid.

Peracetic acid and hydrogen peroxide were assayed with the same method. The concentrations employed were 0.05%, 0.10%, and 0.3% for peracetic acid and 1.5% and 3.0% for hydrogen peroxide. At a low exposure time (30 s) no inhibition was observed. One logarithmic reduction was observed with 0.05% peracetic acid and 1.5% hydrogen peroxide after 15 min. An inhibitory effect of 100% was observed only with 0.3% peracetic acid and 3% hydrogen peroxide.

The inhibitory effect of mace, oregano, paprika, and pepper oleoresins on the mycelial growth of *Alternaria* strains was also evaluated. The inoculum of *Alternaria* spores was placed on potato dextrose agar plates and concentrations of 0.1, 3.0, 7.0, 10.0, 14.0, and 25.0 mg/ml of each oleoresin were added. The plates were incubated for 7 days at 25°C. The inhibitory effect was measured on the basis of the percent radial growth compared with control colonies. Oregano and pepper oleoresins showed the highest inhibitory effect. At concentrations of 7 mg/ml for oregano and 14 mg/ml for pepper, 100% inhibition was observed. Inhibition was also observed at the lowest concentrations of both oleoresins.

9.4.4 Application of Disinfection Processes to Tomato Fruits

A first disinfection was performed by washing the fruits with potable water and then dipping them in a solution of 10:1 water–sodium hypochlorite. The tomatoes were then inoculated by puncture with a solution of 10^4 *Alternaria* spores per milliliter and were then dried for 1 h in a laminar flow cabinet. After they had been dried, the fruits were dipped into the selected solutions for 2 min and kept in an incubator at 12°C and 90% relative humidity for 10 days (Fig. 9.1).

The solutions with the best performance in the in vitro assays were applied either alone or in combination on tomato fruits.

The treatments assayed were:

1. Acetic acid 0.5%
2. 0.1% acetic acid plus 0.05% salicylic acid

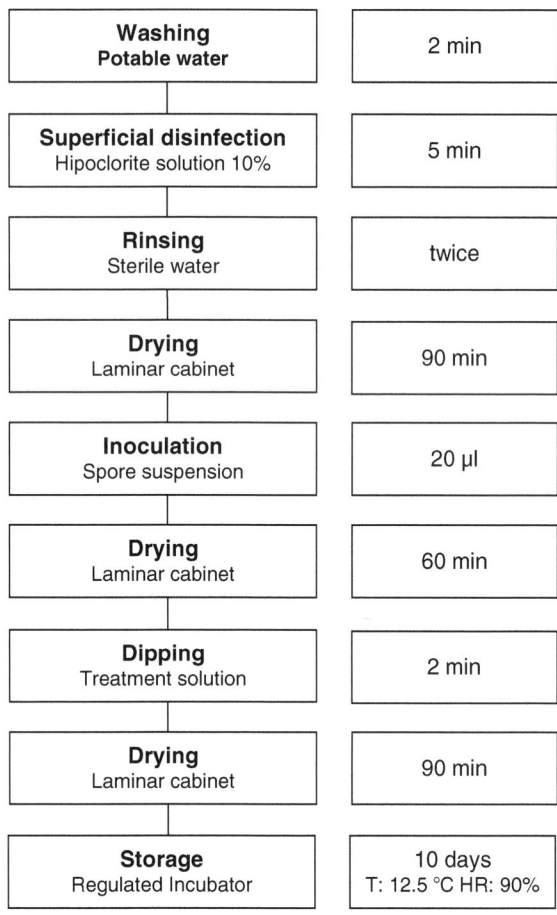

Fig. 9.1 Method of application of selected treatments on tomatoes. *RH* relative humidity

3. 0.3% peracetic acid
4. 1.5% hydrogen peroxide
5. 3.0% hydrogen peroxide
6. 7 mg/ml oregano oleoresin
7. 0.5% acetic acid plus 7 mg/ml oregano oleoresin

The best results were obtained by a first treatment with 0.5% acetic acid plus 7 mg/ml oregano oleoresin. This combined treatment did not permit fungal growth, did not affect the organoleptic properties of the fruit, closed the fruit injuries, and improved the visual quality of the tomatoes by making them glossier and healthier (Fig. 9.2).

9 Plant Extracts as Natural Antifungals...

Fig. 9.2 (**a**) Control inoculated tomatoes; (**b**) inoculated tomatoes treated with 0.5% acetic acid plus oregano oleoresin at 7 mg/ml

9.5 Conclusion

The current world trend for trying to achieve a healthier lifestyle has led to increased demand for convenient fresh foods free from additives with high nutritional value. Fresh-cut unprocessed fruit and vegetable offer great advantages to consumers who perceive them as good because of their high quality and less wastage coupled with reasonable price. To achieve quality fresh-cut produce and the best possible safety levels it is necessary to implement improved strategies through introduction of or combination of sustainable techniques for sanitation. Plant extracts in combination with other safe treatments are a good alternative for the prevention of fungal infection of fresh products and could be applied before harvest, during processing, and during storage or shipping.

References

Ahn Y, Lee H, Oh H, Kim H, Lee Y. Antifungal activity and mode of action of Galla rhois-derived phenolics against phytopathogenic fungi. Pestic Biochem Physiol. 2005;81:105–12.

Allameh A, Razzaghi Abyaneh M, Shams M, Rezaee MB, Jaimand K. Effects of neem leaf extract on production of aflatoxins and activities of fatty acid synthetase, isocitrate dehydrogenase and glutathione S-transferase in *Aspergillus parasiticus*. Mycopathologia. 2001;154:79–84.

Boyraz N, Ozcan M. Inhibition of phytopathogenic fungi by essential oil, hydrosol, ground material and extract of summer savory (*Satureja hortensis* L.) growing wild in Turkey. Int J Food Microbiol. 2006;107(3):238–42.

Brul S, Coote P. Preservatives in food. Mode of action and microbial resistance mechanisms. Int J Food Microbiol. 1999;50:1–17.

Burt S. Essential oils: their antibacterial properties and potential applications in foods-a review. Int J Food Microbiol. 2004;94:223–53.

Cakir A, Kordali S, Kilic H, Kaya E. Antifungal properties of essential oil and crude extracts of *Hypericum linarioides* Bosse. Biochem Syst Ecol. 2005;33:245–56.

Deepa MA, Narmatha Bai V, Basker S. Antifungal properties of *Pseudarthria viscida*. Fitoterapia. 2004;75:581–4.

Devlieghere F, Vermeiren L, Debevere J. New preservation technologies: possibilities and limitations. Int Dairy J. 2004;14:273–85.

FDA/CFSAN . Analysis and evaluation of preventive control measures for the control and reduction/elimination of microbial hazards on fresh and fresh-cut produce. http://www.fda.gov/Food/ScienceResearch/ResearchAreas/SafePracticesforFoodProcesses/ucm090977.htm (2001).

Holley RA, Patel D. Improvement in shelf-life and safety of perishable foods by plant essential oils and smoke antimicrobials. Food Microbiol. 2005;22:273–92.

Lee S, Chang K, Su M, Huang Y, Jang H. Effects of some Chinese medicinal plant extracts on five different fungi. Food Control. 2007;18:1547–54.

Quiroga EM, Sampietro AR, Vattuone MA. Screening antifungal activities of selected medicinal plants. J Ethnopharmacol. 2001;74:89–96.

Rana BK, Singh UP, Taneja V. Antifungal activity and kinetics of inhibition by essential oil isolated from leaves of *Aegle marmelos*. J Ethnopharmacol. 1997;57(1):29–34.

Rasooli I, Owlia P. Chemoprevention by thyme oils of *Aspergillus parasiticus* growth and aflatoxin production. Phytochemistry. 2005;66:2851–6.

Rasooli I, Hadi Fakoor M, Yadegarinia D, Gachkar L, Allameh A, Bagher Rezaei M. Antimycotoxigenic characteristics of Rosmarinus officinalis and Trachyspermum copticum L. essential oils. Int J Food Microbiol. 2008;122:135–9.

Saks Y, Barkai-Golan R. Aloe vera gel activity against plant pathogenic fungi. Postharvest Biol Technol. 1995;6:159–65.

Shan B, Cai Y, Brooks J, Corke H. The in vitro antibacterial activity of dietary spice and medicinal herb extracts. Int J Food Microbiol. 2007;117:112–9.

Shukla R, Kumar A, Prasad CS, Srivastava B, Dubey NK. Antimycotic and antiaflatoxigenic potency of *Adenocalymma alliaceum* Miers. on fungi causing biodeterioration of food commodities and raw herbal drugs. Int Biodeterior Biodegrad. 2008;62:348–51.

Soliman KM, Badeaa RI. Effect of oil extracted from some medicinal plants on different mycotoxigenic fungi. Food Chem Toxicol. 2002;40:1669–75.

Tegegne G, Pretorius JC, Swart WJ. Antifungal properties of *Agapanthus africanus* L. extracts against plant pathogens. Crop Prot. 2008;27:1052–60.

Terminiello L, Patriarca A, Pose G, Fernández Pinto V. Occurrence of alternariol, alternariol monomethyl ether and tenuazonic acid in Argentinean tomato puree. Mycotoxin Res. 2006;22(4):236–40.

Tripanthi P, Dubey N. Exploitation of natural products as an alternative strategy to control postharvest fungal rotting of fruit and vegetables. Postharvest Biol Technol. 2004;32:235–45.

Webster D, Taschereau P, Belland RJ, Sand C, Rennie RP. Antifungal activity of medicinal plant extracts; preliminary screening studies. J Ethnopharmacol. 2008;115:140–6.

Chapter 10
Reduction of Mycotoxin Contamination by Segregation with Sieves Prior to Maize Milling

Ana M. Pacin and Silvia L. Resnik

10.1 Introduction

Mycotoxins are toxic, structurally diverse, secondary metabolites produced by a wide range of molds which infect food, feed, and agricultural commodities. Mycotoxin contamination is one of the major problems detected in maize commercialization. The levels of contamination change from year to year and with the season and are also influenced by geographical variations.

Although a large number of mycotoxins are known to be produced by toxigenic fungi, especially under suitable culture conditions, relatively few occur naturally in maize that are considered to have health and/or economic impact.

These toxic compounds could be harmful to different "target organs," such as liver, hematological tissues, and kidney, through biochemical mechanisms that interfere with protein synthesis and/or carbohydrate metabolism (Jolly et al. 2007; Meky et al. 2003; Pestka and Smolinski 2005; Shephard 2008; Wagacha and Muthomi 2008).

Food safety is one of the principal requisites of the world food market; hence, producers and governments must ensure the quality and safety of foodstuffs. The control of mycotoxins in cereal products is one such requisite. The Food and Agriculture Organization (FAO) estimates that at least 25% of world cereal production is contaminated with mycotoxins (JECFA 2001).

The present limits and their enforcement in international trade have created greater awareness of the problem of mycotoxin contamination in maize and the maize processing industry. Consequently, international attention has become more

---

A.M. Pacin (✉)
CIC, Fundación de Investigaciones Científicas Teresa Benedicta de la Cruz,
Buenos Aires, Argentina
e-mail: fundacion@ictbdelacruz.org.ar

S.L. Resnik
CIC, Universidad de Buenos Aires, Buenos Aires, Argentina

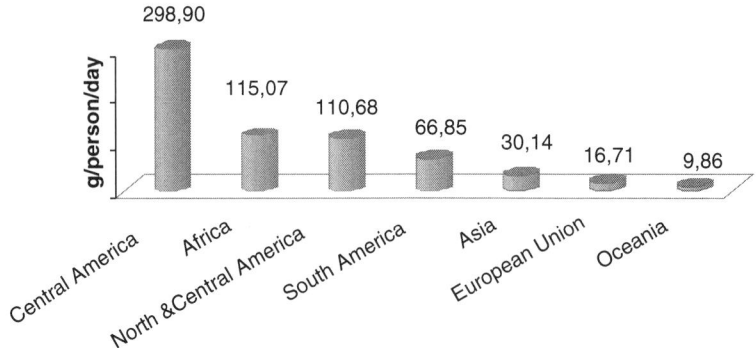

Fig. 10.1 Maize consumption prepared from Food Balance Sheets (2002)

Table 10.1 Amounts of proteins, carbohydrates, and fats per 100 g of maize (Taken from Pacin 2006)

Proteins	Amount (g)	Carbohydrates	Amount (g)	Fats	Amount (g)
Maize	9.6	Flour maize	92.0	Maize	3.5
Flour maize	9.0	Flour maize	76.0	Flour maize	3.4
Flour maize	8.7	Flour maize	74.5	Flour maize	1.0
		Maize	73.9		

focused on the occurrence of mycotoxins and the establishment of measures for contamination control.

Maize has been an important part of the human food supply dating back thousands of years to the great Mesoamerican civilizations of the Aztecs, Mayans, and Incas. By the time Europeans arrived in the Americas in the late fifteenth century, maize was well established as an agricultural crop throughout South America, Central America, North America, and the Caribbean islands

Maize is used for animal feed and human food and to different extents, which are country-dependent. As an example, Mexico has maize as a staple food, but maize consumption is as low as 13 g per person per day in Chile. Figure 10.1 shows maize consumption in different geographical regions obtained from the Food Balance Sheet (2002) published by FAO.

From the nutritional point of view, in Table 10.1, the wide variability of proteins, carbohydrates, and fats in maize flour are shown. Maize composition depends on various factors, such as soil composition, climatic conditions, management, and harvest region. Different hybrids have been developed around the world, and a range of compositions are known as shown in Table 10.2.

The strategies for reducing mycotoxin concentrations in maize are currently under development. Some approaches are directed toward resistance to infection or infection reduction in the grain, whereas others are aimed at detoxification of contaminated maize. To reduce contamination by natural mycotoxins, strategies for grain handling have been developed or are under development (Abramson et al. 2005; Ariño et al. 2009; Bandyopadhyay et al. 2007; Biernasiak et al. 2006;

Table 10.2 Range of chemical compositions of different types of maize, based on "Maize from human nutrition" (Food and Agriculture Organization of the United Nations (FAO) 1992)

Component	Content (% wet basis)
Moisture	9.5–12.3
Ash	1.2–2.9
Protein	5.2–13.7
Crude fiber	0.8–2.9
Carbohydrates	66.0–75.9

Blandino et al. 2009; Bullerman and Bianchini 2007; Castells et al. 2005; Cenkowski et al. 2007; Dorner 2009; Fandohan et al. 2006; Hazel and Patel 2004; Ioos et al. 2005; JECFA 2001; Kabak 2009; Lauren and Smith 2001; Meister and Springer 2004; Miller 2008; Neira et al. 1997; Pacin et al. 2009; Park and Kim 2006; Pirgozliev et al. 2003; Scudamore and Banks 2003; Samar et al. 2007; Trigo-Stockli 2002; Vigier et al. 2001; Wagacha and Muthomi 2008).

Several issues have to be taken into account when discussing grain mycotoxin contamination levels. The issues include sampling plan design, occurrence of mycotoxins, and analytical method for evaluating handling procedures.

This focus of this chapter is on sieving process, with the purpose of analyzing the potential reduction of mycotoxin contamination in the cleanup step before maize storage.

10.2 Brief Review of Concepts Related to Mycotoxin Contamination

Cereals can be contaminated in the field and after harvest by fungi that may produce mycotoxins, and crop handling practices do not adequately control fungal contamination. There is some fungus contamination associated with maize, as shown in Table 10.3, that is known to be toxigenic (Baird et al. 2008; Broggi et al. 2007; Folcher et al. 2009; González et al. 1987, 1988, 1995, 2001; Moltó et al. 1997; Pacin et al. 2001; Resnik et al. 2003a, b; Torres et al. 1998; Solovey et al. 1999).

The most frequently isolated toxigenic fungi (Logrieco et al. 2002; Pacin et al. 2002, 2003, Pacin & Resnik 2007; Resnik et al. 1995, 1996a) are *Fusarium verticillioides*, *Aspergillus flavus*, and *Fusarium graminearum*, and if the climatic conditions are favorable, these molds produce fumonisins, aflatoxins, deoxynivalenol, and zearalenone. The chemical structures are shown in Fig. 10.2. The total aflatoxin contamination level is the sum of the contamination levels of aflatoxin B_1, aflatoxin B_2, aflatoxin G_1, and aflatoxin G_2.

The groups of fumonisins that occur in nature belong to the fumonisin B series. The most important ones are fumonisins B_1, B_2, and B_3. In general, total fumonisins is considered to be the sum of fumonisins B_1, B_2, and B_3.

On the other hand, aflatoxins B_1 and B_2 are more frequent in temperate weather, whereas in tropical countries aflatoxins G_1 and G_2, can also be found at high levels.

Table 10.3 Toxins able to be produced by the maize isolated fungal species in Argentina

Fungi	More frequent toxins	Other toxins
Alternaria alternata	Tenuazonic acid, alternariol, alternariol monomethyl ether	Altenuen, altertoxins I and II
Aspergillus flavus	Ciclopiazonic acid, aflatoxins B_1, B_2, G_1, and G_2	
Aspergillus niger	Ochratoxin A, malformins	
Fusarium equiseti	ZEA, T2, HT2, NEO, DON, 15-acetylDON, NIV	FX, equisetin, enniatin, culmorin, clamidosporol, butenolide, zearalenone 4-sulfato, fusarocromanona,
Fusarium graminearum	DON, 3-acetylDON, 15-acetylDON, ZEA	NIV, T2, HT2, DAS, NEO, butenolide, culmorin, sambucinol, calonectrin, fusarins
Fusarium heterosporum	T2	
Fusarium oxysporum	MON	ZEA, wortmannin, NIV, FX, sambucynin
Fusarium semitectum	ZEA, DAS, NIV, T2, NEO	FX, equisetina, beauvericin
Fusarium sporotrichioides	ZEA, T2, HT2, T-2 triol, T-2 tetraol	
Fusarium subglutinans	MON	Fumonisin B_1
Fusarium verticillioides	Fumonisins B_1, B_2, B_3, and B_4	Fusarins A, C, D, E, and F, fusaric acid
Fusarium proliferatum	Fumonisins B_1, B_2, and B_3, MON	Fusaric acid
Penicillium citrinum	Citrinin	
Penicillium funiculosum	Patulin	

ZEA zearalenone, *T2* T2 toxin, *HT2* HT2 toxin, *NEO* neosolaniol, *DON* deoxynivalenol, *NIV* nivalenol, *MON* moniliformin, *DAS* diacetoxyscirpenol, *FX* fusarenon X

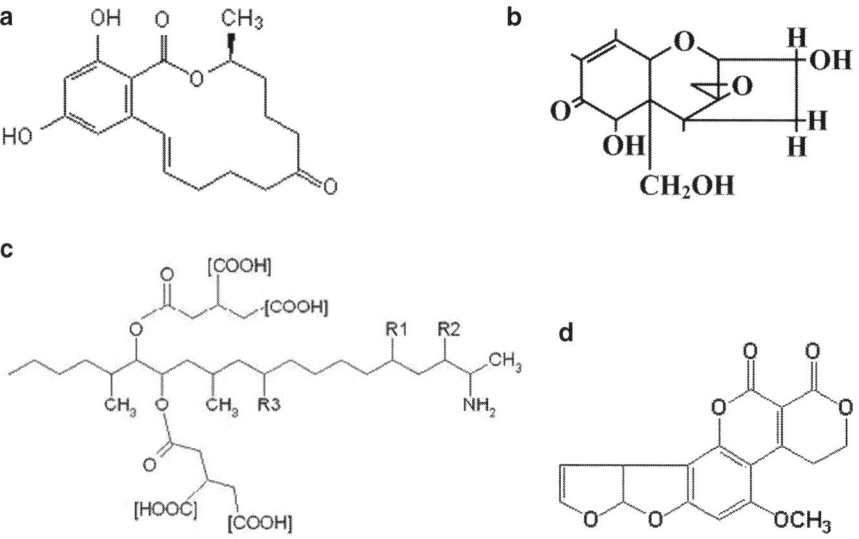

Fig. 10.2 Chemical structure of zearalenone (**a**), deoxynivalenol (**b**), fumonisins (**c**), and aflatoxins (**d**)

Zearalenone and fumonisins are widely distributed, but maize contaminated by deoxynivalenol is found more frequently in the north than in the south of the world.

The analytical method and sampling plans must be taken into account to evaluate grain mycotoxin contamination levels.

The use of reliable and robust analytical methods that are effective for both raw maize and maize screenings is vital for studying the effect of processing on the mycotoxin contamination level. AOAC International and CEN (the European Standardization Committee, the European equivalent of the ISO) have a number of standardized methods of analysis for mycotoxins that have been validated in formal interlaboratory method validation studies, and this number is gradually growing. The latest edition of *Official Methods of Analysis of AOAC International* contains approximately 40 validated methods for mycotoxin determination. The CEN has produced a document that provides specific criteria for various mycotoxin methods. The reader is referred to the following references for information on methods, validation aspects, proficiency testing, and use of certified reference materials: Berthiller et al. (2007), Castegnaro et al. (2006), Krska et al. (2008), Macarthur et al. (2006), Samar and Resnik (2002), Trucksess and Pohland (2002), Trucksess (2006), and Van Egmond and Jonker (2004).

Appropriate sampling plans are essential to ensure that the analytically derived results for a sample are representative of the true concentration of a lot (Blanc 2006; Champeil et al. 2004; Coker 1998; Coker et al. 2000; Johansson et al. 2000a, b, c; Macarthur et al. 2006, Resnik et al. 2003d, Spanjer 2007; Spanjer et al. 2006; Whitaker 2004, 2006; Whitaker et al. 1998, 2007). The mycotoxin concentration of a lot is usually estimated by measuring the mycotoxins in a small portion of a lot. The mycotoxin concentration in the lot is assumed to be the same as in the sample. The only way to determine the true value of contamination is to analyze all the lot, which involves use and destruction of the whole lot, which is not economically feasible. There is therefore always a degree of uncertainty about the true value of the lot. The two important uncertainties are accuracy and precision. Accuracy is defined as the closeness of the measured values and the true contamination level. Precision is associated with variability and is usually associated with the mycotoxin distribution among contaminated particles in the lot. Collecting a representative sample is therefore critical to obtain accurate estimates of mycotoxin concentration. Studies have shown the most important component of test result variability is lack of sample homogeneity. Figure 10.3 shows the distribution of the coefficients of variation for aflatoxin level assessment (adopted from Whitaker 2004).

The total variability associated with a mycotoxin test procedure is equal to the sum of the sampling, sample preparation, and analytical variances associated with each step of the mycotoxin test procedure (Samar et al. 2003). For a maize lot with a contamination level of 2 mg/kg total fumonisins, the coefficients of variation associated with sampling (1.1 kg sample), sample preparation (Romer mill and 25 g analytical portion), and analysis by high-performance liquid chromatography were 16.6, 9.1, and 9.7 %, respectively, and were independent of the fumonisin type (B_1, B_2, B_3, or total). The coefficient of variation associated with the total test procedure (sampling, sample preparation, and analysis) was 21.4% (Whitaker et al. 1998).

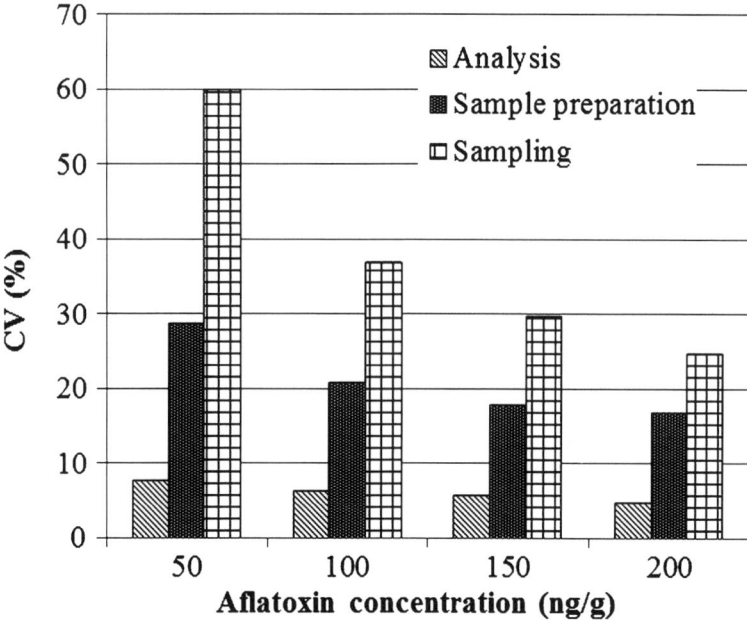

Fig. 10.3 Coefficients of variation (*CV*) for sampling, sample preparation, and analysis when testing shelled maize for aflatoxin. *HPLC* high-performance liquid chromatography. (adopted from Whitaker 2004)

The sampling coefficients of variation depend on the mycotoxin type and type of product studied, but generally show that sampling variance is a function of concentration and decreases with it.

10.3 Brief Review of Concepts Related to Cleaning

Cleaning is the unit operation in which contaminating materials are removed from the grain. The type of contaminants found on raw maize range from stones and metals to microbial products such as fungi and mycotoxins.

The procedures for cleaning are categorized into wet (e.g., soaking, spraying, flotation washing, and ultrasonic cleaning) and dry (separation by air, magnetism, or physical methods) procedures. The selection of a cleaning procedure is determined by the nature of the product to be cleaned and by the types of contaminant to be removed.

Dry cleaning procedures are used for grains that are small, have great mechanical strength, and possess relatively low moisture content. The main groups of equipment used for dry cleaning are classifiers (use a moving stream of air to separate contaminants by differences in their densities, as they are used in most of the new harvesting machines), magnetic separators, and separators based on screening.

Screening, segregation, or sieving is a mechanical process, which implies use of a net made from metal wire with well-defined spacing between the wires (sieve). Only particles with a dimension smaller than the openings can pass through.

Maize sieving can be considered a process with particles changing orientation in a random manner both in time and in space. The physical parameters affecting the cleaning process obtained from the literature are broadly grouped into crop characteristics and machine parameters (Simonyan and Yiljep 2008; Simoyan et al. 2006):

1. Crop factors are crop variety, maturity stage, grain and straw moisture content, bulk density of the grain, and grain diameter.
2. Machine factors are frequency of sieve oscillations, amplitude of oscillation, sieve slope, length and width of the sieve, diameter of the sieve holes, threshing speed, velocity of air, air stream pressure and density, angle of air direction, and terminal velocity of particles.

Physical procedures generate little heat other than that caused by the operation of the machinery for cleaning, so no significant thermal breakdown of mycotoxins is expected at this stage. A better understanding and quantification of the cleaning process would provide a means of predicting cleaning efficiency and reduction of the level of mycotoxins.

10.4 Case Study

Molds and mycotoxins are often concentrated in dust and broken grains or in the outer seed coats of the maize grains, which are more susceptible to fungal infection and toxin contamination (Brekke et al. 1975; Brera et al. 2006; Broggi et al. 2002; Saunders et al. 2001; Scudamore and Banks 2003). So, cracked, broken, and otherwise damaged maize can contain quite high level of mycotoxins, and removing these kernels from normal maize is an important step in controlling residues. Research findings on how the cleaning process affects the retention of mycotoxins in maize have been reported by Fandohan et al. (2006), Jouany (2007), Malone et al. (1998), Saunders et al. (2001), Scudamore and Banks (2003), Sydenham et al. (1994), and Vanara et al. (2009).

In Argentina, the equipment often used for cleaning first removes stones or larger contaminants as bags or cobs (Fig. 10.4). The grain is exposed to air currents to help remove minor contaminating matter. The next stage involves separation by size on sieves to remove maize screenings or broken maize kernels. Compared with other grains, maize has a unique shape and low specific gravity and the maize kernels are the largest cereal seeds.

Figure 10.5 shows a photograph of the equipment evaluated in this case, and in Fig. 10.6, maize kernels can be seen on the sieve.

The sieve dimensions were 1.5-m width and 2.5-m length, with the first sieve of 11 mm at an angle of 7° screening inclination and the next six sieves with a 6° inclination. The maximum output is 105 t/h, and the equipment was operated at an aver-

Fig. 10.4 Sieve size from 10 to 14 mm

Fig. 10.5 Grain cleaning machine: Garelli model NG 1050/T3

age speed of 85 t/h. The sieve oscillation frequency was 3.4 1/s and the amplitude was 56 mm.

The bulk density was 78.40 kg/hl, the maize moisture content was 14.50% on a wet basis, the maize was physiologically mature, and 85% of the grain was retained on an 8-mm sieve.

Information in this case study is related to aflatoxins B_1, B_2, G_1, and G_2, zearalenone, deoxynivalenol, and fumonisins B_1, B_2, and B_3, which were analyzed in maize screening with six sieves of 7 mm. During the cleaning of 1,024 t of maize, 100 samples of unclean maize were taken. Every 2 min, 550 g was collected. In a similar way, 100 samples of screenings (under sieve) and 100 clean maize samples were taken; 944 t of clean maize was obtained.

Fig. 10.6 Maize over the sieve

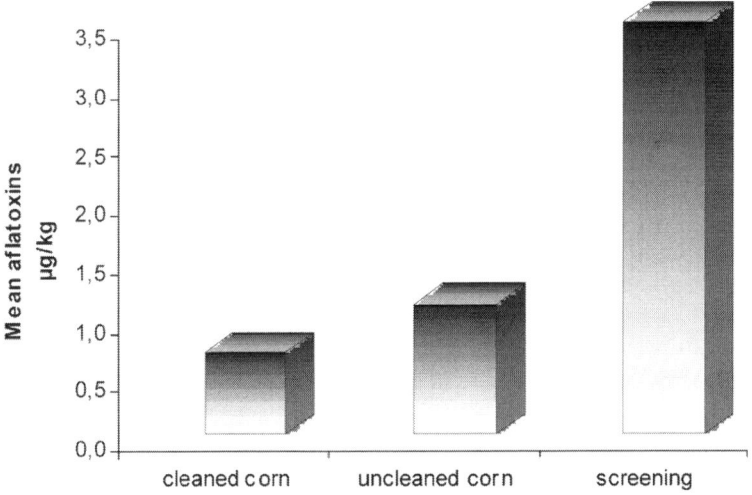

Fig. 10.7 Mean levels of alfatoxins (sieve 7 mm) (Resnik et al. 2003c)

Deoxynivalenol contamination was not detected in all 300 samples. The level of contamination by aflatoxins and zearalenone in the original maize (uncleaned maize) before screening was lower than expected. Nevertheless, some important conclusions could be reached. The mean levels of mycotoxins (Figs. 10.7, 10.8)

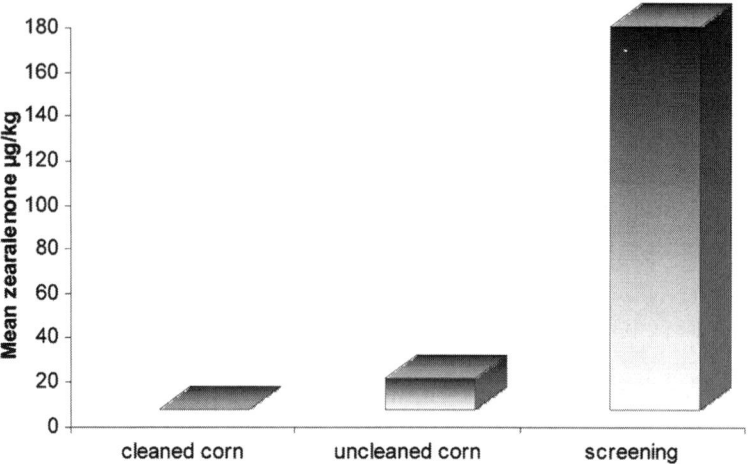

Fig. 10.8 Mean levels of zearalenone (sieve 7 mm)

were lower in clean maize (over sieve) than in uncleaned maize. In both cases in nearly half of the uncleaned maize samples no contamination was detected, but the screenings always had greater mycotoxin levels than uncleaned and cleaned maize.

The levels of fumonisin contamination were quite different from those of the former mycotoxins because all samples were contaminated. But in all cases fumonisin found in grain not retained by the sieve was at greater levels than in clean maize.

The clean maize was less contaminated by fumonisins B_1, B_2, and B_3, but a number of samples still had high levels, rendering them unsafe for consumption.

10.5 Economic Implication

Aflatoxins and other mycotoxins cost maize growers millions of dollars every year. This cleaning process could be useful if it were available on the market. The use of different mesh sizes should be based on the reduction of contamination and fraction yield. The design must take into consideration the need for a compromise between efficiency (yields) and costs.

Cleaning efficiency (purity) was obtained by the procedure detailed by the National Institution of Agricultural Engineering (NIAE 1952):

$$\eta = \frac{G_0}{G_0 + C_{cg}} \times 100, \quad (1)$$

where η is the cleaning efficiency (%), G_0 is the weight (kg) of pure grain at the outlet (cleaned maize), and C_{cg} is the weight (kg) of contaminant (screenings).

In this case study with the 7-mm sieve, 92% cleaning efficiency and 8% screenings were obtained. The concentration of fumonisins was reduced from 7,847.2 µg/kg in the uncleaned maize to 3,638.1 µg/kg in the clean fraction, whereas the levels in the screenings reached 17,754 µg/kg. Through the application of different sieve sizes it is possible to obtain better reduction of mycotoxin contamination in the cleaned maize, but this would imply an enormous quantity of screenings, which would not be possible for foods or feeds. Such retention levels could be impractical and potentially render the process nonviable.

10.6 Safety Issues

Maize utilization can be divided into that which is consumed with minimal preparation and that involving substantial value-added processing. An example of the former is the maize that can be used directly by livestock to produce meat and dairy products. In such a case, contaminated maize screenings could cause harmful effects, so the development of inexpensive processes to diminish such contamination would be beneficial.

European Commission Regulation (EC) No. 1881/2006 sets maximum levels that were implemented in 2006 for certain mycotoxins in foodstuffs. Limits for fumonisins B_1 and B_2 (fumonisin B_1 plus fumonisin B_2) include 2,000 µg/kg in unprocessed maize, 1,000 µg/kg in maize flour, grits, germ, semolina, and refined maize oil, 400 µg/kg in maize-based foods for direct consumption, and 200 µg/kg in processed maize-based foods and baby foods for infants and young children. New regulations on mycotoxin levels in feeds are proposed as a result of research in the field and levels of acceptability are adjusted accordingly.

10.7 Concluding Remarks

Mycotoxins are usually concentrated in the bran and outer layers of grains and are present at reduced levels in the endosperm. The sieving process reduces mycotoxin contamination.

Argentinian maize usually has low contaminations levels of deoxynivalenol, aflatoxins, and zearalenone and over-sieve fractions are almost "clean" (for these levels of aflatoxins and zearalenone), and it could be expected that food and feed produced with this "clean maize" should be free of these mycotoxins. But the situation was not the same in the case of fumonisins, because the heavily contaminated unclean grains, in spite of the reduction of levels in clean maize, were not free of this toxin.

The sieving process reduced the concentration of fumonisins in a step-by-step manner, but could not eliminate all contamination in the final products.

10.8 Further Trends

The fate of mycotoxins during processing has been, in general, a neglected area of study. Most research effort has concentrated on the means of prevention of mycotoxin formation, and this must remain the best defense for protecting the consumer. However, prevention is not always possible, especially for those mycotoxins formed under field conditions, such as aflatoxins, fumonisins, deoxynivalenol, and zearalenone.

Introduction of further legislation for a wider range of mycotoxins in more food commodities means that there is a need to determine how mycotoxins survive processing so that this can be taken into account when setting statutory or guideline limits.

Achieving the correct limit minimizes any unnecessary restriction on the use of valuable food commodities without compromising human health. It is thus expected that there will be a trend toward further study of the fate of those mycotoxins that pose the greatest potential risk for humans. In some instances, it may be possible to introduce modifications to commercial processes that cause a significant reduction of mycotoxin content in the retail product.

References

Abramson D, House J, Nyachoti C. Reduction of deoxynivalenol in barley by treatment with aqueous sodium carbonate and heat. Mycopathologia. 2005;160:297–301.

Ariño A, Herrera M, Juan T, Estopañan G, Carramiñana JJ, Rota C, et al. Influence of agricultural practices on the contamination of maize by fumonisin mycotoxins. J Food Prot. 2009;72: 898–902.

Baird R, Abbas HK, Windham G, Williams P, Baird S, Ma P, et al. Identification of select fumonisin forming Fusarium species using PCR applications of the polyketide synthase gene and its relationship to fumonisin production in vitro. Int J Mol Sci. 2008;9:554–70.

Bandyopadhyay R, Kumar M, Leslie J. Relative severity of aflatoxin contamination of cereal crops in West Africa. Food Addit Contam. 2007;24:1109–14.

Berthiller F, Sulyok M, Krska R, Schuhmacher R. Chromatographic methods for the simultaneous determination of mycotoxins and their conjugates in cereals. Int J Food Microbiol. 2007;119:33–7.

Biernasiak J, Piotrowska M, Libudzisz Z. Detoxification of mycotoxins by probiotic preparation for broiler chickens". Mycotoxin Res. 2006;22:230–5.

Blanc M. Sampling: the weak link in the sanitary quality control system of agricultural products. Mol Nutr Food Res. 2006;50:473–9.

Blandino M, Reyneri A, Colombari G, Pietri A. Comparison of integrated field programmes for the reduction of fumonisin contamination in maize kernels. Field Crop Res. 2009;111:284–9.

Brekke OL, Peplinski AJ, Griffin Jr EL. Cleaning trials for corn containing aflatoxins. Cereal Chem. 1975;52:198–204.

Brera C, Catano C, De Santis B, Debegnach F, De Giacomo M, Pannunzi E, et al. Effect of industrial processing on the distribution of aflatoxins and zearalenone in corn-milling fractions. J Agric Food Chem. 2006;54:5014–9.

Broggi LE, Resnik SL, Pacin AM, Gonzalez HHL, Cano G, Taglieri D. Distribution of fumonisins in dry-milled corn fractions in Argentina. Food Addit Contam. 2002;19:465–9.

Broggi LE, Pacin AM, Gasparovic A, Sacchi C, Rothermel A, Gallay A, et al. Natural occurrence of aflatoxins, deoxynivalenol, fumonisins and zearalenone in maize from Entre Rios Province, Argentina. Mycotoxin Res. 2007;23:59–64.

Bullerman LB, Bianchini A. Stability of mycotoxins during food processing. Int J Food Microbiol. 2007;119:140–6.

Castegnaro M, Tozlovanu M, Wild C, Molinie A, Sylla A, Pfohl-Leszkowicz A. Advantages and drawbacks of immunoaffinity columns in analysis of mycotoxins in food. Mol Nutr Food Res. 2006;50:480–7.

Castells M, Marín S, Sanchis V, Ramos AJ. Fate of mycotoxins in cereals during extrusion cooking: a review. Food Addit Contam. 2005;22:150–7.

Cenkowski S, Pronyk C, Zmidzinska D, Muir WE. Decontamination of food products with superheated steam. J Food Eng. 2007;83:68–75.

Champeil A, Fourbet JF, Dore T. Effects of grain sampling procedures on Fusarium mycotoxin assays in wheat grains. J Agric Food Chem. 2004;52:6049–54.

Coker RD. Design of sampling plans for determination of mycotoxins in foods and feeds. In: Sinha KK, Bhatnagar D, editors. Mycotoxins in agriculture and food safety. New York: Marcel Dekker; 1998. pp. 109–33.

Coker R, Nagler M, Defize P, Derksen G, Buchholz H, Putzka H, et al. The development of sampling plans for the determination of aflatoxin B1 in animal feedingstuffs. J AOAC Int. 2000;83:1252–8.

Dorner JW. Biological control of aflatoxin contamination in corn using a nontoxigenic strain of *Aspergillus flavus*. J Food Prot. 2009;72:801–4.

Fandohan P, Ahouansou R, Houssou P, Hell K, Marasas WFO, Wingfield MJ. Impact of mechanical shelling and dehulling on Fusarium infection and fumonisin contamination in maize. Food Addit Contam. 2006;23:415–21.

Folcher L, Jarry M, Weissenberger A, Gérault F, Eychenne N, Delos M, et al. Comparative activity of agrochemical treatments on mycotoxin levels with regard to corn borers and *Fusarium* mycoflora in maize (Zea mays L.) fields. Crop Prot. 2009;28:302–8.

Food and Agriculture Organization of the United Nations (FAO). Maize in human nutrition. http://www.fao.org/docrep/t0395e/T0395E03.htm (1992).

Food Balance Sheet. Food and Agriculture Organization of the United Nations. http://faostat.fao.org/DesktopDefault.aspx?PageID=346&SelTab=4 (2002).

González HHL, Resnik SL, Vaamonde G. Influence of inoculum size on growth rate and lag phase of fungi isolated from Argentine corn. Int J Food Microbiol. 1987;4:111–7.

González HHL, Resnik SL, Vaamonde G. Influence of temperature on growth rate an lag phase of fungi isolated from Argentine corn. Int J Food Microbiol. 1988;6:179–83.

González HHL, Resnik SL, Boca RT, Marasas WFO. Mycoflora of Argentinian corn harvested in the main production area in 1990. Mycopathologia. 1995;130:29–36.

González HHL, Resnik SL, Pacin AM. Mycoflora of freshly harvested flint corn from northwestern provinces in Argentina. Mycopathologia. 2001;155:207–11.

Hazel CM, Patel S. Influence of processing on trichothecene levels. Toxicol Lett. 2004;153:51–9.

Ioos R, Belhadj A, Menez M, Faure A. The effects of fungicides on Fusarium spp. and *Microdochium nivale* and their associated trichothecene mycotoxins in French naturally-infected cereal grains. Crop Prot. 2005;24:894–902.

JECFA. Safety evaluation of certain mycotoxins in food. 2001. Fifty Sixth Meeting of the Joint FAO/WHO Expert Committee on Food Additives. FAO Food Nutr Pap. 2001;74:419–555.

Johansson AS, Whitaker TB, Hagler Jr WM, Giesbrecht FG, Young JH, Bowman DT. Testing shelled corn for aflatoxin, part I: estimation of variance components. J AOAC Int. 2000a;83:1264–9.

Johansson AS, Whitaker TB, Giesbrecht FG, Hagler Jr WM, Young JH. Testing shelled corn for aflatoxin, part II: modeling the observed distribution of aflatoxin test results. J AOAC Int. 2000b;83:1270–8.

Johansson AS, Whitaker TB, Giesbrecht FG, Hagler Jr WM, Young JH. Testing shelled corn for aflatoxin, part III: evaluating the performance of aflatoxin sampling plans. J AOAC Int. 2000c;83:1279–84.

Jolly PE, Jiang Y, Ellis WO, Awuah RT, Appawu J, Nnedu O, et al. Association between aflatoxin exposure and health characteristics, liver function, hepatitis and malaria infections in Ghanaians. J Nutr Environ Med. 2007;16:242–57.

Jouany JP. Methods for preventing, decontaminating and minimizing the toxicity of mycotoxins in feeds. Anim Feed Sci Technol. 2007;137:342–62.

Kabak B. The fate of mycotoxins during thermal food processing. J Sci Food Agric. 2009;89:549–54.

Krska R, Schubert-Ullrich P, Molinelli A, Sulyok M, MacDonald S, Crews C. Mycotoxin analysis: an update. Food Addit Contam. 2008;25:152–63.

Lauren DR, Smith WA. Stability of the Fusarium mycotoxins nivalenol, deoxynivalenol and zearalenone in ground maize under typical cooking environments. Food Addit Contam. 2001;18:1011–6.

Logrieco A, Mulè G, Moretti A, Bottalico A. Toxigenic Fusarium species and mycotoxins associated with maize ear rot in Europe. Eur J Plant Pathol. 2002;108:597–609.

Macarthur R, Macdonald S, Brereton P, Murray A. Statistical modelling as an aid to the design of retail sampling plans for mycotoxins in food. Food Addit Contam. 2006;23:84–92.

Malone BM, Richard JL, Romer T, Johansson AJ, Whitaker TB 1998. Fumonisin reductionin corn by cleaning during storage discharge. In: O'Brian L, Blakeney AB, Ross AS, Wrigley CW, editors. North Melbourne, Vic. Cereal Chemistry Division, Royal Australian Chemical Institute, Cairns, Australia; 1998. pp. 372–9.

Meister U, Springer M. Mycotoxins in cereals and cereal products - occurrence and changes during processing. J Appl Bot Food Qual. 2004;78:168–73.

Meky FA, Turner PC, Ashcroft AE, Miller JD, Qiao YL, Roth MJ, et al. Development of a urinary biomarker of human exposure to deoxynivalenol. Food Chem Toxicol. 2003;41:265–73.

Miller JD. Mycotoxins in small grains and maize: old problems, new challenges. Food Addit Contam. 2008;25:219–30.

Moltó GA, González HHL, Resnik SL, Pereyra GA. Production of trichothecenes and zearalenone by isolates of Fusarium spp. from Argentinian maize. Food Addit Contam. 1997;14:263–8.

Neira MS, Pacin AM, Martínez EJ, Moltó G, Resnik SL. The effects of bakery processing on natural deoxynivalenol contamination. Int J Food Microbiol. 1997;37:21–5.

NIAE. National Institution of Agricultural Engineers. General report on the NIAE investigation into the technique of testing grain cleaning machinery, part II. Silsoe: NIAE; 1952. pp. 6.

Pacin A. Micotoxinas contaminantes que no deberíamos olvidar. Enfasis Alim. 2006;6:40–8.

Pacin AM, Resnik S Acido ciclopiazónico. In: Soriano del Castillo JM, editor. Micotoxinas en alimentos. Madrid: Díaz de Santos; 2007. pp. 335–52.

Pacin AM, Broggi LE, Resnik SL, González HHL. Mycoflora and mycotoxins natural occurrence in corn from Entre Ríos province, Argentina. Mycotoxin Res. 2001;17:31–8.

Pacin AM, González HHL, Etcheverry M, Resnik SL, Vivas L, Espin S. Fungi associated with commodities from Ecuador. Mycopathologia. 2002;156:87–92.

Pacin AM, Ciancio Bovier E, González HHL, Whitechurch E, Martínez EJ, Resnik SL. Fungal and fumonisins contamination in Argentine maize (Zea mays L.) silo bags. J Agric Food Chem. 2009;57:2778–81.

Pacin AM, Cano G, Resnik SL, Villa D, Taglieri D, Ciancio E. Incidencia de la contaminación por aflatoxinas en maíz argentino, período 1995-2002, IV Congreso Latino americano de Micotoxicología. Seminario Anual Animal. La Habana, Cuba, 24–26 Sep 2003.

Park JW, Kim Y-B. Effect of pressure cooking on aflatoxin B1 in rice. J Agric Food Chem. 2006;54:2431–5.

Pestka JJ, Smolinski AT. Deoxynivalenol: toxicology and potential effects on humans. J Toxicol Environ Health B Crit Rev. 2005;8:39–69.

Pirgozliev SR, Edwards SG, Hare MC, Jenkinson P. Strategies for the control of *Fusarium* head blight in cereals. Eur J Plant Pathol. 2003;109:731–42.

Resnik SL, Costarrica ML, Pacin A. Mycotoxins in Latin America and the Caribbean. Food Control. 1995;6:19–28.

Resnik SL, González HHL, Pacin A, Viora M, Caballero GM, Gros E. Cyclopiazonic acid and aflatoxins production by *Aspergillus flavus* isolated from Argentinian corn. Mycotoxin Res. 1996a;12:61–6.

Resnik S, Neira MS, Pacin A, Martínez E, Apro N, Latreite S. A survey of the natural occurrence of aflatoxins and zearalenone in Argentine field maize: 1983–1994. Food Addit Contam. 1996b;13(1):115–20.

Resnik SL, Taglieri D, Cano G, Pacin AM. Aflatoxinas en las fracciones obtenidas durante la limpieza del maíz. IV Congreso Latinoamericano de Micotoxicología. Seminario Anual Animal. Havana, Cuba. 2003a;9:24–26.

Resnik S, Taglieri D, Ciancio Bouvier E, Cano G, Pacín A. Reducción de micotoxinas en maiz: limpieza. In: Jornadas Bonaerenses de Ciencia y Tecnologia, 17 December 2003, La Plata Provincia de Buenos Aires, Argentina; 2003b.

Resnik, S.L., Villa, D. and Pacin, A. M. Distribución de fumonisinas en el maíz y en las fracciones obtenidas durante la limpieza de maíz. IV Congreso Latinoamericano de Micotoxicología. Seminario Anual Animal. Havana, Cuba. 2003c;9:24–26.

Resnik SL, Villa D, Pacin AM. Muestreo de fumonisinas en maíz: función de distribución. In: Jornadas Bonaerenses de Ciencia y Tecnologia, 17 Dec 2003, La Plata Provincia de Buenos Aires, Argentina; 2003d.

Samar MM, Resnik SL. Analytical methods for trichothecenes surveillance – an overview over the period 1990-2000. Food Sci Technol Int. 2002;8:257–68.

Samar MM, Ferro Fontán C, Resnik SL, Pacin AM, Castillo MD. Distribution of deoxynivalenol in wheat, wheat flour, bran and gluten, and variability associated with the test procedure. J AOAC Int. 2003;86(3):551–6.

Samar M, Resnik SL, González HHL, Pacin AM, Castillo MD. Deoxynivalenol reduction during the frying process of turnover pie covers. Food Control. 2007;18:1295–9.

Saunders DF, Meredith FI, Voss KA. Control of fumonisin: effects of processing. Environ Health Perspect. 2001;109:333–6.

Scudamore KA, Banks JB. The fate of mycotoxins during cereal processing. In: Barug D, van Egmond H, Lopez-Garcia R, van Osenbruggen T, Visconti A, editors. Meeting the mycotoxin menace. Wageningen: Wageningen Academic; 2003. pp. 165–81.

Shephard GS. Impact of mycotoxins on human health in developing countries. Food Addit Contam. 2008;25:146–51.

Simonyan KJ, Yiljep YD. Investigating grain separation and cleaning efficiency distribution of a conventional stationary Rasp-bar sorghum thresher. Agric Eng Int: CIGR eJ. 2008. pp. 1–13.

Simonyan KJ, Yiljep YD, Mudiare OJ. Modeling the grain cleaning process of a stationary sorghum thresher. Agric Eng Int: CIGR eJ. 2006. pp. 1–17.

Solovey MM, Somoza C, Cano G, Pacin A, Resnik S. A survey of fumonisins, deoxynivalenol, zearalenone and aflatoxins in corn-based food products in Argentina. Food Addit Contam. 1999;16:325–9.

Spanjer MC. Sampling for grain quality. Stewart Postharvest Rev. 2007;3(6):1–6.

Spanjer MC, Scholten JM, Kastrup S, Jorissen U, Schatzki TF, Toyofuku N. Sample comminution for mycotoxin analysis: dry milling or slurry mixing? Food Addit Contam. 2006;23:73–83.

Sydenham EW, van der Westhuizen L, Stockenstrom S, Shephard GS, Thiel PG. Fumonisin contaminated maize: physical treatment for the decontamination of bulk shipments. Food Addit Contam. 1994;11:25–32.

Torres A, González HHL, Etcheverry M, Resnik SL, Chulze S. Production of alternariol and alternariol mono-methyl ether by isolates of Alternaria spp. from Argentinian maize. Food Addit Contam. 1998;15(1):56–60.

Trigo-Stockli DM. Effect of processing on deoxynivalenol and other trichothecenes. Adv Exp Med Biol. 2002;504:181–8.

Trucksess MW. Mycotoxins. J AOAC Int. 2006;89:270–84.

Trucksess M, Pohland A. Methods and method evaluation for mycotoxins. Mol Biotechnol. 2002;22:287–92.

Van Egmond HP, Jonker MA. Current regulations governing mycotoxins limits in food. In: Magan N, Olsen M, editors. Mycotoxins in food: detection and control. 1st ed. Cambridge: Woodhead Publishing/CRC Press LLC; 2004. pp. 88–110.

Vanara F, Reyneri A, Blandino M. Fate of fumonisin B1 in the processing of whole maize kernels during dry-milling. Food Control. 2009;20:235–8.

Vigier B, Reid LM, Dwyer LM, Stewart DW, Sinha RC, Arnason JT, et al. Maize resistance to gibberella ear rot: symptoms, deoxynivalenol, and yield. Can J Plant Pathol. 2001;23:99–105.

Wagacha JM, Muthomi JW. Mycotoxin problem in Africa: current status, implications to food safety and health and possible management strategies. Int J Food Microbiol. 2008;124:1–12.

Whitaker, T.B. Sampling for mycotoxins. In Magan, N. and. Olsen (eds.), M. Mycotoxins in Food: Detection and Control First edition, Woodhead Publishing Ltd and CRC Press LLC, Cambridge, England. 2004. pp. 69–87.

Whitaker TB. Sampling foods for mycotoxins. Food Addit Contam. 2006;23:50–61.

Whitaker TB, Trucksess MW, Johansson AS, Giesbrecht FG, Hagler Jr WM, Bowman DT. Variability associated with testing shelled corn for fumonisin. J AOAC Int. 1998;81:1162–8.

Whitaker TB, Doko MB, Maestroni BM, Slate AB, Ogunbanwo BF. Evaluating the performance of sampling plans to detect fumonisin Bi in maize lots marketed in Nigeria. J AOAC Int. 2007;90:1050–9.

Chapter 11
Rational Use of Novel Technologies: A Comparative Analysis of the Performance of Several New Food Preservation Technologies for Microbial Inactivation

Stella M. Alzamora, Jorge Welti-Chanes, Sandra N. Guerrero, and Paula L. Gómez

11.1 Introduction

Consumers are increasingly aware of the health benefits and risks associated with consumption of food. These informed individuals tend to opt for healthy foods that have been subjected to less extreme treatments (less heat and chill damage), with lower levels of salts, fats, acids, and sugars and/or the complete or the partial removal of chemically synthesized additives. There is also an increased demand for convenience foods with fresh attributes and long shelf lives.

To meet these expectations, the food industry and food researchers have developed in the last two decades a range of novel techniques for food preservation, under the concept of minimal processing, that allow better retention of product flavor, texture, color, and nutrient content than comparable conventional treatments (Alzamora and Salvatori 2006). The minimal processing concept includes a series of products and processes that may be grouped in diverse food categories such as minimally processed, with invisible processing, carefully processed, partially processed, and high moisture shelf-stable (Welti–Chanes et al. 1997). These terms are not exactly equivalent but reflect the great diversity of alternatives for high-quality products.

Originally, the manipulation and basic preparation of foods, as well as life permanence in the biological tissue, were the elements that distinguished the minimal processing of foods. Nowadays, the minimal processing concept has evolved, giving a wider approach than those terms used earlier by Rolle and Chism (1987), Shewfelt (1987), Huxsoll and Bolin (1989), Wiley (1994a), and Ohlsson (1994). Manvell (1997)

S.M. Alzamora (✉) • S.N. Guerrero • P.L. Gómez
Universidad de Buenos Aires, Buenos Aires, Argentina

CONICET, Buenos Aires, Argentina
e-mail: alzamora@di.fcen.uba.ar

J. Welti-Chanes
Instituto Tecnológico y de Estudios Superiores de Monterrey, Monterrey, NL, Mexico

situated minimal processing methods within the broader objective of food processing to extend the shelf life of foods. He defined that minimal processing must be a safe balance between desired and adverse effects of preservation processes and, accordingly, a minimal process is "the least possible treatment" to achieve a purpose. Thus, he considered as minimal processes existing traditional techniques that have been improved to obtain safe and high-quality products (e.g., aseptic packaging, ohmic heating, radio-frequency heating, and microfiltration) as well as techniques based on emerging preservation factors (e.g., high pressure, ultrasound, and pulsed electric field). In the same trend, Ohlsson (1996) stated that developments in thermal technologies are considered "minimal" when they minimize quality losses in foods compared with conventional thermal techniques. Lastly, and in partial agreement with these terms, Snyder (2003), taking into account that there were no regulatory definitions for minimally processed foods, gave an operating definition of them as "foods in which the biological, chemical, and physical hazards are at tolerable level."

The expansion of the minimal processing concept has been reflected in new, renewed, and improved products and processes formulated and designed to produce a greater diversity of minimally processed foods. The reviews and/or books of Ahvenainen et al. (1994, 1996, 2002), Singh and Oliveira (1994), Wiley (1994b), Oliveira and Oliveira (1999), Alzamora et al. (2000a), and Ohlsson and Bengtsson (2002) on minimal processing can be used as further reference on the subject.

Many of the modified (for enhanced food quality) traditional preservation procedures and the radically new ones (high hydrostatic pressure, ultrasound, ultraviolet light, ozone, pulsed light, electric pulses, etc.) are effective in inactivating only vegetative cells of bacteria, yeasts, and filamentous fungi. So, emerging preservation procedures have to be included as components or hurdles in combined preservation systems to ensure food safety. To destroy spores, for example, the lethality of a stress factor is strongly affected by the presence/intensity of other lethal/inhibitory factors.

Within the framework of hurdle technology, a huge amount of scientific literature has been published in recent years, indicating the enormous popularity and potential of the concept in the development of emerging combined technologies to aid in producing minimally processed foods.

A general overview of minimal processing techniques is presented in Table 11.1. Three major initiatives are currently being proposed by the research community and the industry to make minimally processed foods with improved quality (Alzamora et al. 1998, 2000b):

1. Optimization of traditional preservation methods to enhance sensorial, nutritional, and microbiological quality of foods, yield, and energy efficiency
2. Development of mild processes by novel combinations of traditional physical and chemical preservation factors, each one applied at low intensity, to obtain products with quality attributes reminiscent of the fresh or native state of a given food but with longer shelf life
3. Development of new techniques to obtain novel foods with fresh-quality attributes by using combinations of emerging preservation factors or combinations of emerging factors with traditional ones, all of them applied at low doses

Table 11.1 Selected minimal processing technologies based on conventional and emerging preservation factors

Enhanced traditional techniques
Aseptic packaging of thermally processed foods (tube, plate heat exchangers, etc.)
Semiaseptic processes (high heat infusion, vacuum steam plus vacuum cooling, etc.)
Ohmic heating
Radio-frequency heating
Microwave heating
Inductive electrical heating
Infrared heating
Microfiltration
Osmotic dehydration
Vacuum dehydration
Microwave and dielectric drying
Cryogenic freezing
Freezing in dynamic dispersion medium
New mild techniques based on novel combinations of traditional preserving factors
Sous vide and cook–chilled processing
Modified/controlled atmosphere packaging
Active packaging techniques
"Ready–to–eat foods"
New techniques based on combinations of emerging preservation factors
Emerging factors
High hydrostatic pressure
Ionizing radiation
High electric field pulses
Ultraviolet light
Light pulses
Natural antimicrobials
Ultrasound
Biopreservation

Microbial cell physiological responses in relation to emerging factors in combination with other constraints are complex and are not fully understood, as in the case of many traditional preservation factors. Predictive modeling in emerging technologies permits quantification of the influence of various hurdles on the microbial behavior, allowing the interaction effects between them – antagonistic, synergistic, or additive – to be precisely discerned. It is not easy to anticipate the type of the interaction resulting from the combination of many factors. Moreover, the characteristics of the interaction can change with storage time or with minimal alterations in the food formulation (Alzamora et al. 2003).

This chapter compares the ability of selected emerging factors to reduce microbial populations. Inactivation/decline kinetic patterns of microorganisms in treatments with conventional/unconventional lethal agents combined with other environmental stress factors are discussed and analyzed on the basis of a rigorous kinetic analysis of survival data.

11.2 Brief Outline of Concepts Related to Microbial Inactivation Kinetics

The introduction and/or optimization of any preservation technology requires scientific data about microbial response. In particular, kinetic parameters and models are essential to develop food preservation processes that ensure safety (McMeekin 2007; Mafart 2005; Velázquez et al. 2005). The kinetic parameters also allow comparison of the ability of different technologies to reduce the microbial population. Accurate model prediction of survival curves would be beneficial to the food industry in selecting the optimum combinations of lethal agents and environmental factors as well as exposure times to obtain desired levels of inactivation while minimizing production costs and maintaining a maximum degree of sensory and nutrient quality. Potential combinations of preservation factors are numerous, but not much quantitative microbiological information is currently available about the combination of alternative processes with traditional constraints. Predictive microbiology provides the tools to compare the impact of different factors on reduction of the microbial population and is an aid to understand biological system behavior, assisting in the development of mechanistic models that will explain microbial activity in terms of properties and events at the cellular and molecular levels.

In thermal and nonthermal inactivation of microorganisms, four commonly types of semilogarithmic survival curves (Fig. 11.1) are usually found: linear curves (first-order inactivation), curves with a shoulder or initial lag period, curves with a tailing region (or biphasic curves) and sigmoid curves (Peleg and Cole 1998; Xiong et al. 1999).

11.2.1 Linear Microbial Inactivation Response

Traditionally, microbial inactivation has been considered a process that follows first-order kinetics (Fig. 11.1a) (Bialka et al. 2008; Avsaroglu et al. 2006). It has been implicitly assumed that all cells or spores have identical resistance to a lethal agent and that each microorganism has the same probability of dying (van Boekel 2002). Following this mechanistic approach, the equation proposed by Chick (1908) has been widely used to calculate sterility in thermal preservation methods:

$$\frac{dN(t)}{dt} = -k(T)N(t) \tag{11.1}$$

where $N(t)$ is the number of the cells at time t and $k(T)$ is a temperature-dependent rate constant.

From this equation, after integration and transformation into a decimal logarithm, the well-known concept of D value or decimal reduction time was derived:

$$\log\left(\frac{N_0}{N(t)}\right) = \frac{t}{D(T)} \tag{11.2}$$

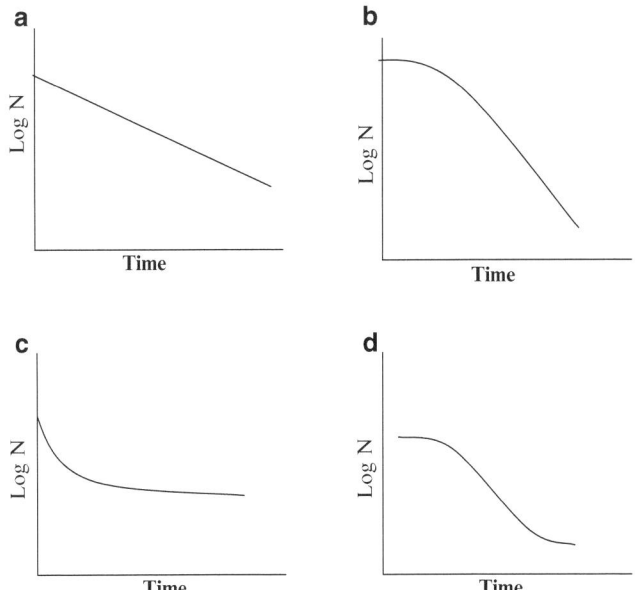

Fig. 11.1 Different inactivation/decline semilogarithmic curves: **a** log-linear; **b** with a shoulder; **c** with a tail; **d** sigmoid

where N_0 is the initial number of the cells, with $D(T) = 2.303/k(T)$.

The temperature dependence of $D(T)$ is expressed using the so-called z value, that is, the increase in temperature that would reduce the decimal reduction time by a factor of 10:

$$\frac{D(T_1)}{D(T_2)} = 10^{\frac{T_2 - T_1}{z}} \qquad (11.3)$$

11.2.2 Nonlinear Microbial Inactivation Response

Although the first-order model has dominated the field of quantitative inactivation microbiology for many years, the decrease of microbial populations does not usually follow first-order kinetics (Peleg 2006). Forcing a straight line through a concave downward semilogarithmic survival curve or a concave upward semilogarithmic survival curve (Fig. 11.1b–d) can result in an overestimate or underestimate of the organism resistance, respectively, and so in overprocessing in the first case or in underprocessing in the second case (Peleg 1999).

Two different approaches, among others, to explain microbial inactivation that have been discussed are the following (Heldman and Newsome 2003):

- Mechanistic: the decay is produced by some molecular or physical mechanisms (i.e., a monomolecular transformation, or an enzyme-catalyzed reaction). This approach is deterministic in nature, all cells behave in the same way, and the death of a single cell is due to a single event.
- Vitalistic: in line with other areas of biological studies, this approach is based on the assumption that the individual microorganisms in a population do not have identical resistances, and that microbial sensitivity to lethal agents is distributed (i.e., biological variability). Shoulders and tails are due to underlying physiological reactions of the cells/spores to lethal conditions.

A variety of alternative models (mechanistic or empirical, deterministic or probabilistic) have been developed to describe curvilinear semilogarithmic inactivation curves, and a number of excellent articles, books, and book chapters are available on the subject (Xiong et al. 1999; McKellar and Lu 2004; McMeekin et al. 1993; Peleg 2006; Sapru et al. 1993; Mafart et al. 2002; Geeraerd et al. 2005).

Two versatile and relatively simple mathematical models that allow quantification of the effects of various lethal agents on microbial inactivation are described.

11.2.2.1 The Weibull Distribution of Resistances Model

It has been shown that the nonlinear models are better than the linear models to describe inactivation kinetics of many microorganisms under thermal or nonthermal processes. The Weibull distribution is a nonlinear function that takes biological variation into account and is used to describe the spectrum of resistances of the population to a lethal agent under different conditions (Peleg and Cole 1998, 2000; van Boekel 2002; Peleg 2006). Survival patterns are explained without assuming the validity of any kinetic model. The Weibull function has been widely used to explain the probabilistic distribution of fracture stresses in brittle materials as well as yield stresses in ductile materials. The main advantages of the model based on the Weibull distribution are its simplicity and its capability of modeling survival curves that are linear and those that contain shoulder or tailing regions. The inactivation mechanism at the molecular level may differ from cell to cell. It is unlikely that all cells in the population behave in the same way and that the death of a single cell is due to a single event. So there will be a distribution of inactivation times and the survival curve should not be treated in kinetic terms but should be considered the cumulative form of the temporal distribution of lethal events. If each individual of a microbial population is inactivated at a specific time (t_{ci}) and if t_{ci} has a continuous distribution, the survival ratio $S(t)$ can be written as

$$S(t) = \int_0^1 f\left[t, t_c(\phi)\right] d\phi \qquad (11.4)$$

where t_c is the time at which the microorganism dies or loses its viability, $S(t)$ is the survival fraction N/N_0 (N_0 is the initial number of the cells and N is the number of the cells at time t), and $f[t, t_c(\varphi)]$ is a function of the exposure time t and the fraction of organisms φ which share any given t_c.

If t_c has a Weibull type distribution

$$\frac{d\phi}{dt_c} = bnt_c^{n-1} \exp(-b't_c^n) \qquad (11.5)$$

$d\phi$ in equation 11.4 from equation 11.5, and solving, the cumulative form of Weibull distribution results:

$$S(t) = \left(\frac{N}{N_0}\right) = exp(-b't^n) \qquad (11.6)$$

Since microbiologists use logarithms of numbers, Eq. 11.6 can be written as follows:

$$log_{10}S(t) = -bt^n \qquad (11.7)$$

where $b = b'/\log_e 10$.

The Weibull model has two parameters: the scale or nonlinear rate parameter b (with dimensions t^n) and the dimensionless shape parameter n, which indicate the overall steepness and the shape of the survival curves respectively.

The studies of Pilavtepe-Çelik et al. (2009), Avsaroglu et al. (2006), San Martin et al. (2007), and Bialka et al. (2008) are examples of the successful application of the Weibull model to describe the microbial inactivation curves.

In Fig. 11.2 the effect of H_2O_2 concentration on semilogarithmic curves of *Escherichia coli* at 25°C and pH 7.2 is shown. Figure 11.2 shows that at a constant time the level of inactivation of *E. coli* increases with increasing H_2O_2 concentration, and that the Weibull model accurately describes inactivation kinetics.

11.2.2.2 The Modified Gompertz Model

Nonlinear survival curves can be modeled using a modified Gompertz equation as follows (Linton et al. 1995):

$$log\left[\frac{N_t}{N_0}\right] = Cexp(-exp(A + Bt)) - Cexp(-exp(A)) \qquad (11.8)$$

where $log(N_t/N_0)$ represents the logarithmic surviving fraction and the three parameter estimates (A, B, C) represent the different regions of the survival curve (Fig. 11.3): the initial shoulder [A (min)]; the maximum death rate [B (min^{-1})]; and

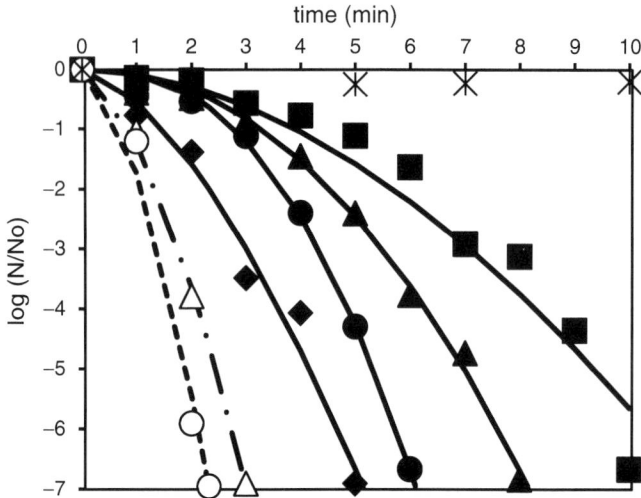

Fig. 11.2 Effect of H_2O_2 concentration on semilogarithmic survival curves of *Escherichia coli* at 25°C and pH 7.2. Experimental values (*points*) and fitted values derived from the Weibull model (*lines*) for the control (x), 0.50% w/v H_2O_2 (■), 0.75% w/v H_2O_2 (▲), 1.00% w/v H_2O_2 (•), 1.50% w/v H_2O_2 (♦), 2.00% w/v H_2O_2 (–.▲.–), and 2.50% w/v H_2O_2 (O) (Adapted from Raffellini et al. 2008

the overall change in the number of survivors (*C*). This equation has been demonstrated to be particularly suitable for sigmoid survival curves (an initial shoulder followed by an exponential phase and a tailing region) and survival curves with a tail or shoulder and has the property that, as time approaches infinity, the lethality approaches – *C* (Juneja and Marks 2003).

Some examples of the successful application of the Gompertz model for predicting the microbial inactivation were obtained by Char et al. (2009) in a study concerned with the effect of the addition of a natural antimicrobial (vanillin) to inhibit the growth of *Listeria monocytogenes* in thermally processed orange juice, by Saucedo-Reyes et al. (2008) for the inactivation of *Listeria innocua* CECT 910 studied in both exponential and stationary growth phases under high hydrostatic pressure (Fig. 11.3), and by Slongo et al. (2009) in a study of the influence of the pressure and holding time on the growth of lactic acid bacteria in vacuum-packaged sliced ham.

11.2.2.3 Comparison of Linear, Weibull, and Gompertz Models for Microbial Inactivation

In a recent study the linear, Weibull, and Gompertz models were compared for the suitability to predict the inactivation of *L. monocytogenes* in ground beef under both isothermal (Fig. 11.4) and dynamic (Fig. 11.5) temperature conditions. Under isothermal

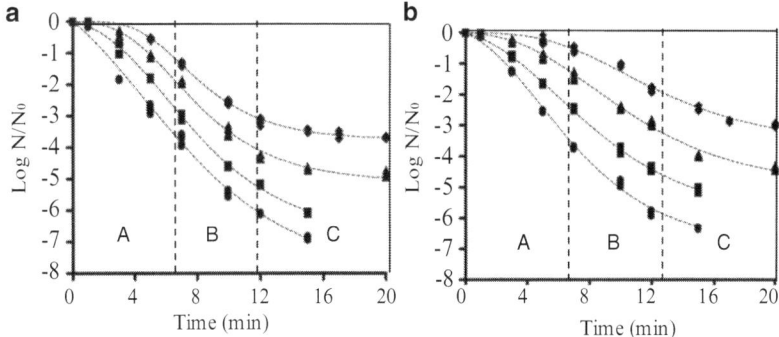

Fig. 11.3 Prediction of the inactivation curves of *Listeria innocua* CECT 910 in the exponential phase (**a**) and the stationary phase (**b**) at 325 MPa (♦), 350 MPa (▲), 375 MPa (■), and 400 MPa (●) with the modified Gompertz model. *A* is the initial shoulder, *B* is the maximum death rate, and *C* is the overall change in the survivor number (Adapted from Saucedo-Reyes et al. 2008)

conditions, a shoulder in the survival curves of *L. monocytogenes* was observed, and therefore the inactivation of the microorganism was better described by the Weibull and the modified Gompertz models (Fig. 11.4). The downward concavity observed was attributed to some biological factors where a thermal process must overcome an initial energy barrier before a lethal effect can be observed. However, under dynamic conditions, only the modified Gompertz model accurately described the survival curves of the microorganism (Fig. 11.5) (Huang 2009). This kind of model comparison is of particular importance in order to define the procedures to evaluate the effective use of new preservation technologies.

11.3 Inactivation Performance of New Preservation Factors

There are a wide range of modern preservation factors that cause physical inactivation of microorganisms at ambient or sublethal temperatures (e.g., high hydrostatic pressure, pulsed electric fields, ultrasound, pulsed light, ultraviolet light, hydrogen peroxide), avoiding the deleterious effects that severe heating has on flavor, texture, and nutritional quality of foods. Table 11.2 lists some new physical food preservation techniques, the proposed mechanisms of action on the microorganisms, and the factors of the technique itself that influence the efficacy of the preservation. Some of these processes are still under development but have commercial potential. Others have completed development and are waiting to be licensed or transferred to industry. Others are applied commercially.

This section will next focus on the mode of action of ultrasound, ultraviolet light, high hydrostatic pressure, high-pressure homogenization, and pulsed electric fields on microorganisms.

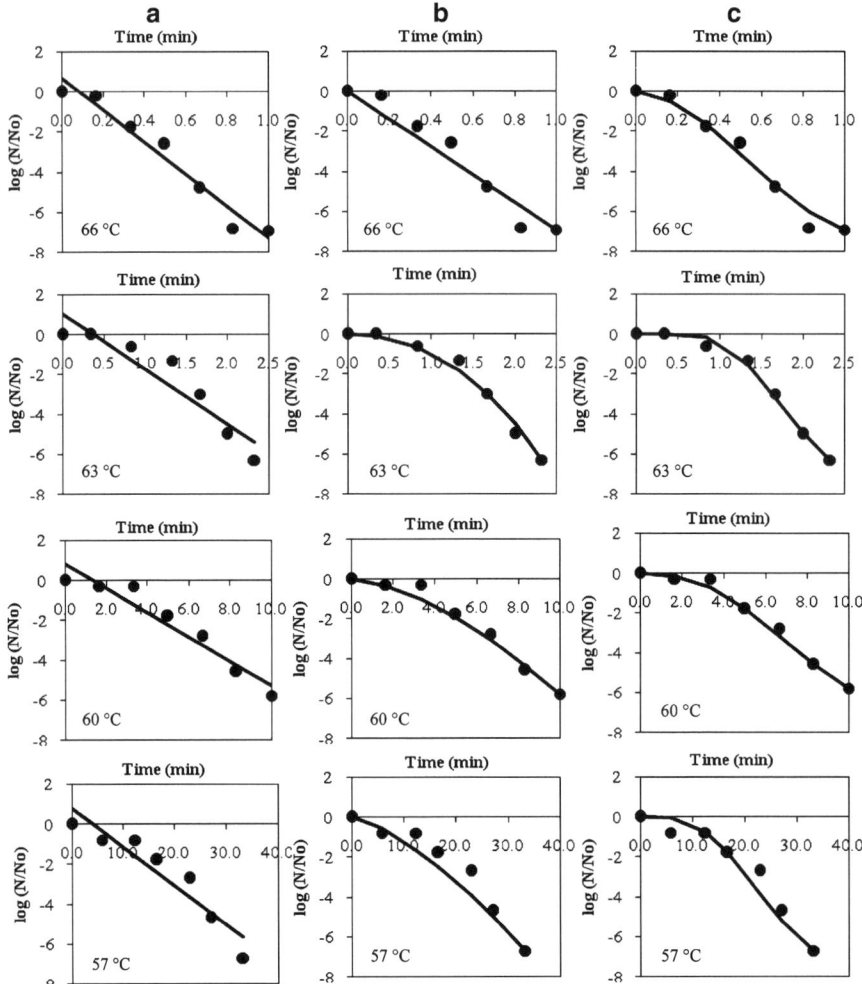

Fig. 11.4 Experimental (●) survival curves of *Listeria monocytogenes* in ground beef at 66°C, 63°C, 60°C, and 57°C, and prediction (–) with the linear model (*column A*), the Weibull model (*column B*), and the modified Gompertz model (*column C*) (Adapted from Huang 2009)

11.3.1 High Hydrostatic Pressure

The response of microorganisms to high hydrostatic pressure treatments can result in inactivation or survival and the cell damage is what determines the effect. The ability of cells to recover after being subjected to high hydrostatic pressure is very important for the efficiency of the process and the evaluation of cell destruction kinetics; if the repair process remains intact, microorganisms can regenerate and grow (Palou et al. 2002; Welti-Chanes et al. 2006). A great variety of mechanisms

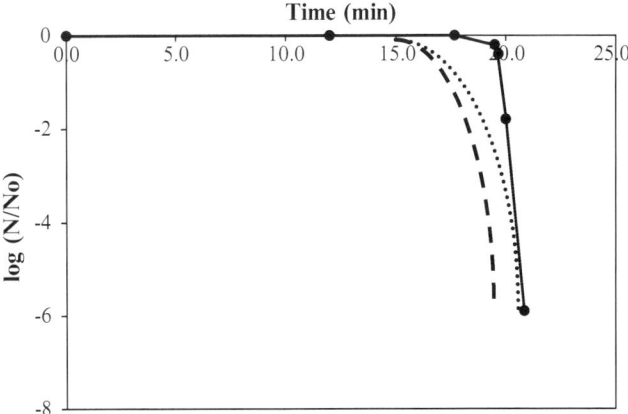

Fig. 11.5 Experimental (●) survival curves of *L. monocytogenes* in ground beef during dynamic heating, and prediction with the linear (...), Weibull (---), and Gompertz (–) models. (Adapted from Huang 2009)

for microbial inactivation by high hydrostatic pressure have been proposed. One of these mechanisms specifies that microorganisms are destroyed owing to irreversible damage in proteins forming cell membranes (Ortega-Rivas 2007; Welti-Chanes et al. 2006, 2005). When the cell membrane is altered, the transport processes related to obtaining nutrients and release of cellular debris are affected. Another theory states that high hydrostatic pressure technology interrupts the functions responsible for microbial reproduction and survival. Smelt et al. (1994) hypothesized that the ATPase bound to the cell membrane is strongly involved in the mechanism of inactivation by high hydrostatic pressure. These authors showed that microorganisms with a more fluid membrane are more resistant to pressure than cells with a less fluid membrane, by comparing cells of *Lactobacillus plantarum* with different membrane fluidity. In their studies, Smelt et al. (1994) also found that after a high hydrostatic pressure treatment, cells of *L. plantarum* release ATP.

It has also been observed that the growth phase also has an important role in the sensitivity of microorganisms to high hydrostatic pressure; in the stationary phase cells are much more resistant to high hydrostatic pressure treatment than they are in the exponential phase. In the exponential phase, pressure resistance increases as membrane fluidity increases, whereas in the stationary phase a relationship between these two variables has not been observed (Torres and Velazquez 2005). Mañas and Mackey (2004) proposed that cells of *E. coli* J1 in the exponential phase are inactivated under high hydrostatic pressure by irreversible damage to the cell membrane, and cells in the stationary phase have a cytoplasm membrane that is robust enough to withstand very intense pressures, suggesting that the retention of a intact membrane allows cells in the stationary phase to repair gross changes in other cellular structures and to remain viable at pressures that are lethal to cells in the exponential phase.

Table 11.2 Main emerging "nonthermal" inactivation factors (Adapted from Butz 2002; US. FDA 2001; Barbosa-Cánovas 2002)

Factor	Mechanism of inactivation	Critical factors	Potential application/products on the market
High-pressure processing Subjecting liquid and solid foods, with/without packaging, to 100–800 MPa, below 0–100°C, from seconds to about 20 min, instantaneously and uniformly throughout food, independent of size, shape, and food composition	Membrane damage, protein denaturation, leakage of cell contents	Temperature, pressure magnitude, rate of compression and decompression, and holding time at pressure, time to achieve treatment pressure, composition, pH and a_w of the food, product initial temperature, and critical factors of other constraints in combination	Jam, jellies, fruit juices, oysters, guacamole, ham, fruit yogurts, dairy-based fruit smoothie, processed meat products, pressure-treated ready-to-eat rice, sauces (in use since 1990)
Pulsed electric field Application of pulses of high voltage ($\cong$ 20–80 kV/cm) to foods between 2 electrodes, in the form of exponentially decaying, square wave, bipolar, or oscillatory pulses at ambient, subambient, or slightly above ambient temperature, for less than 15 s	Electrical breakdown and electroporation of cell membranes	Electric field intensity, pulse width, treatment time and temperature, and pulse wave shapes, pH, ionic compounds, conductivity and medium ionic strength, and critical factors of other constraints in combination	Only two industrial-scale pulsed electric field systems recently available Fluid foods (fruit juices, liquid eggs, milk, sauces, beverages)
Ultraviolet light Radiation from the ultraviolet region of the electromagnetic spectrum (UV-C 200–280 nm), at least 400 J/m^2	DNA mutations induced by ultraviolet light absorption	Transmissivity of the material, homogeneity of the flow pattern and the radiation field, ultraviolet wavelength, thickness of the radiation path through the food (geometric configuration of the system); product composition, solids content, and starches	Pasteurization of apple and clear juices. Surface and air decontamination

Pulsed light

A few flashes of intense pulses of broad spectrum "white light" (ultraviolet to the near infrared region) applied in a fraction of a second

Chemical modifications and cleavage of the DNA and destruction of cellular components by the high peak power and the broad spectrum of the flash

Light characteristics (wavelength, intensity, duration, and number of pulses), packaging and type, transparency, and color of food are considered to be critical process parameters

Sterilization or reduction of microbial load on surfaces and transparent products

Ultrasound

Energy generated by sound waves of 20 kHz or more

Disruption of cellular structure and functional components and cell lysis attributed to cavitation

Power and amplitude of ultrasonic waves, exposure time, volume and composition of the food to be processed, temperature of treatment, and the critical factors of procedures used in combination

No commercial food products. Limited to product modification and process efficiency improvements (enhancement of mass and heat transfer, degassing of liquids, cleaning of surfaces)

High-pressure homogenization

Pressure ranges from 100 to 350 MPa. The process is affect by the number of passes

Mechanical destruction of the cell integrity, caused by the spatial pressure and velocity gradients, turbulence, impingement, and/or cavitation. Cavitation and viscous shear have been identified as the primary mechanisms of microbial cell disruption during this process

Microbial strain and suspension medium, pressure and temperature of homogenization, and apparatus employed

This process is used in the food, chemical, cosmetic, and pharmaceutical industries primarily to prepare or stabilize emulsions and suspensions

High hydrostatic pressure treatment has a greater effect in inactivating Gram-negative bacteria than Gram-positive bacteria, and in general, vegetative cells are much more sensitive to high pressure than spores. Raso et al. (1998) noted for apple, orange, pineapple, blueberry, and grape juices treated under high hydrostatic pressure that ascospores of *Zygosaccharomyces bailli* exhibited a resistance 5–8 times greater than that exhibited by vegetative cells. The population of vegetative cells decreased by almost 5 log cycles after 5 min of treatment at 300 MPa, whereas the load of ascospores was reduced by only 0.5–1 log cycle.

11.3.2 Pulsed Electric Field

Although many studies have reported the effectiveness of a pulsed electric field to eliminate microorganisms (Álvarez et al. 2000; Martín et al. 1997; Yeom et al. 2000), the mechanisms of action have not been clearly established (Somolinos et al. 2008; Yeom et al. 2002). One of the theories for the inactivation of bacteria by a pulsed electric field is known as "dielectric breakdown," which indicates that the application of an electric field induces an electrical potential across the membrane, causing an electrostatic charge based on the nature of the bipolar molecules forming the membrane.

When the transmembrane potential exceeds a critical value (approximately 1 V), the repulsion between molecules initiates the formation of pores in weak areas of the membrane, or enlarges the existing pores, which may be permanent or temporary depending on the intensity of treatment. This process will increase the permeability of the cell, which causes loss of cellular material or intrusion of medium components into the cell, causing microbial death.

Most microorganisms have a cell wall that surely also interferes in the mechanism of inactivation. Studies of the mechanisms of inactivation of bacteria by a pulsed electric field have shown differences in Gram-positive and Gram-negative bacteria. Bacterial permeabilization in *E. coli* and *Salmonella typhimurium* by a pulsed electric field suggests that differences in structure and composition of the cell wall could be responsible for the different behavior observed under the pulsed electric field treatment (Somolinos et al. 2008).

11.3.3 Ultraviolet Light

When a foodstuff is subjected to radiation of 200–280 nm, the microbial load is reduced because the DNA absorbs the UV-C radiation, resulting in cross-linking between the neighboring pyrimidine nucleoside bases (thymine and cytosine) in the same DNA strand (Miller et al. 1999), affecting both replication and transcription of the nucleic acid (Guerrero Beltrán et al. 2009; Sastry et al. 2009); so microbial functions are damaged and eventually the cell dies (Sastry et al. 2009). The intensity of

UV-C treatment affects the level of cross-linking. The mutation can be reversed, and this depends on the repair systems of each microorganism (Miller et al. 1999).

This treatment has a germicidal effect against bacteria, viruses, protozoa, molds, yeasts, and algae (Bintsis et al. 2000; Guerrero-Beltrán and Barbosa-Cánovas 2005; Morgan 1989; Sizer and Balasubramaniam 1999). For some food products, water is used and for the disinfection of surfaces, a wavelength of 254 nm is used (Guerrero Beltrán et al. 2009; Guerrero-Beltrán and Barbosa-Cánovas 2005). This wavelength and an exposure of at least 400 J/m^2 of UV-C light is necessary to ensure a reduction of 5 log cycles in the microbial load, and therefore safe products can be obtained.

The effectiveness of UV-C treatment depends on many factors. One of these factors is the type of microorganism to be treated. Because of the extensive variety of microorganisms and strains, the doses can differ greatly for each product depending on the desired effect. Another factor influencing the efficiency of treatment is the medium in which the microorganisms are suspended owing to the different penetration capacity of UV-C light through different media. The absorptivity decreases as the concentration of soluble and suspensed solids increases (Guerrero-Beltrán and Barbosa-Cánovas 2005). The geometric configuration of the product is another factor that can affect the treatment with UV-C radiation.

UV-C radiation at a wavelength of 253.7 nm was investigated by Schenk et al. (2008) for its microbicidal effects on pear slices with and without peel. Survival curves of inoculated *L. monocytogenes*, *L. innocua*, *Zygosaccharomyces bailii*, and *Debaryomices hansenii* showed upward concavity and a pronounced tailing effect, indicating that most of the organisms were destroyed in a short time during UV-C exposure and that a fraction of the population survived after the treatment. This tailing effect was attributed among other factors to the heterogeneity in the resistances of the population to UV-C irradiation, and to the presence of suspended solids and of matrix solids that may block the UV irradiation. The Weibull distribution of resistances model described well the inactivation of the microorganisms on pear surfaces which had been inoculated with them.

The importance of conducting kinetic inactivation studies is illustrated by the following example. Gómez et al. (2008) studied the surface color change of fresh-cut apple slices irradiated with UV-C radiation at different doses (0–25 kJ/m^2) and stored in a refrigerator for 7 days. The color parameters were found to be dependent on the UV-C dose and storage time. At the end of storage, samples exposed to UV-C light turned darker and less green when compared with fresh-cut apple slices or with samples before storage, and this browning effect was more pronounced with the greatest UV-C dose. Breakage of cellular membranes in UV-C-treated samples as observed by microscopic observations could explain the increase in browning of irradiated apples. For the surface of cut apples inoculated with *L. monocytogenes* and *Z. bailii*, the survival curves of the microorganisms showed an upward concavity and a pronounced tailing effect, and, in general, for irradiation times longer than approximately 15 min (corresponding dose approximately 8.4 kJ/m^2), the inactivation rate significantly decreased or was null. Thus, the integration of both color and inactivation studies could result in products with better quality.

11.3.4 Ultrasound

It has been observed that the effectiveness of ultrasound treatment depends on the type of organism and the product to be treated. Spores are relatively resistant to ultrasonic waves, and long treatment times are therefore required to obtain safe products. Moreover, when ultrasound is applied to reduce the bacterial load in foods the process is not very effective (Piyasena et al. 2003).

The mechanism of microbial inactivation is due to thinning of cell membranes, localized heating, production of free radicals, and cell lysis attributed to cavitation induced by changes in pressure (there are regions of alternating compression and expansion) created when a sonic wave meets a liquid medium (Butz and Tauscher 2002). The effect of cavitation causes the formation of bubbles that have a larger surface area during the expansion cycle, which increases the diffusion of gas, causing the bubble to expand. When the ultrasonic energy provided is not sufficient to retain the vapor phase in the bubble, a rapid condensation occurs, and the condensed molecules collide quickly, creating shock waves that create regions of very high temperature and pressure, reaching up to 5,500 K and 50,000 kPa. The hot zones can kill some bacteria, but they are localized and do not affect a large area (Piyasena et al. 2003).

The minimum oscillation of pressure that is required to produce cavitation is determined by factors such as dissolved gas, hydrostatic pressure, specific heat of the liquid and the gas in the bubble, and the tensile strength of the liquid. Other factors affecting the effectiveness of microbial inactivation are the amplitude of the ultrasonic waves, the exposure/contact time, the volume of food being processed, the composition of the food, and the treatment temperature. Some microorganisms that have been treated with ultrasound are *L. monocytogenes*, a number of strains of *Salmonella* spp., *E. coli, S. aureus*, and *Bacillus subtilis* (Piyasena et al. 2003).

11.3.5 High-Pressure Homogenization

The capacity of high-pressure homogenization to reduce the microbial load of food and other products has been investigated by several researchers (Diels et al. 2005, 2003). It has been reported that the factors influencing killing of microorganisms are the strains, pressure and temperature of homogenization, the apparatus employed, and the suspension medium (Donsì et al. 2009). High-pressure homogenization kills vegetative bacteria mainly through mechanical destruction of the cell integrity caused by the spatial pressure and velocity gradients, turbulence, impingement, and/or cavitation occurring in liquids during high-pressure homogenization (Donsì et al. 2009). Cavitation and viscous shear have been identified as the primary mechanisms of microbial cell disruption during this process. It has been observed that Gram-negative bacteria are more sensitive than Gram-positive bacteria to this technology (Donsì et al. 2009; Wuytack et al. 2002), and yeasts and fungi have resistance intermediate between that of Gram-negative bacteria and that of Gram-positive bacteria; this is due to their larger size and glucans, mannans, and proteins as basic structural components of their cell walls. Bacterial spores are more resistant than vegetative cells to the high-pressure homogenization process (Wuytack et al. 2002).

11.4 Combined Treatments and Their Effect on Microbial Inactivation

The usefulness of many of the emerging factors described already when used as the only preservation factor is limited because of the following aspects:

1. Most of these emerging factors are effective in inactivating vegetative cells of most microorganisms, but spores are far more tolerant, and very high treatment intensities are required to achieve spore destruction in low-acid foods. These large doses may significantly affect the sensory and/or functional properties of foods. So, these alternative factors for inactivating microorganisms are analogous to thermal pasteurization and their use for the sterilization of low-acid foods at room temperature is not yet possible.
2. The scheduled process for traditional thermal processes is based on knowledge of the thermal inactivation kinetics of the most-heat-resistant pathogen of concern for each specific product and of the heat transfer rate. But alternative nonthermal techniques involve different mechanisms for microbial destruction and the organisms of concern are not the same, nor do the inactivation kinetics follows the exponential relationship that is usually the basis of thermal process design. The most resistant pathogens and the corresponding surrogate organisms for alternative preservation processes are yet to be clearly defined. Moreover, limited information is available about the mechanisms of inactivation of microorganisms by alternative factors and survivor plots commonly exhibit a shoulder and/or a tail. A highly resistant subpopulation usually remains viable over a long period of treatment time, causing a substantial decrease in the efficacy of the processes.

A multifactorial preservation approach can enhance inactivation of highly resistant microbial subpopulations by alternative nonthermal factors, so lower factor intensities can be used. Combining emerging factors with conventional preservation ones or with other novel techniques has been explored in recent years with promising results (Alzamora et al. 2003; Ross et al. 2003; Lado and Yousef 2002; Raso and Barbosa-Cánovas 2003). The effect of combining different preservation technologies can be additive, synergistic, or antagonistic (Raso and Barbosa-Cánovas 2003). According to Leistner (2000), attacking various cellular targets will have a synergistic effect by making the organism use every possible repair mechanism simultaneously, and the activation of stress shock proteins also becomes more difficult. Some successful additive or synergistic combinations are the following:

- High hydrostatic pressure and lowered pH. This combination prevents microbial growth and the germination of spores that can survive high hydrostatic pressure treatment at acidic pH.

- High hydrostatic pressure and heat. This combination increases the lethality of the nonthermal process and decreases the intensity of the nonthermal treatment.
- High hydrostatic pressure and antimicrobial agents. Cells injured sublethally by high hydrostatic pressure become more susceptible to antimicrobials.
- Pulsed electric fields and heat. High temperatures increase the fluidity and the thickness of membranes, increasing the lethality of pulsed electric field treatment.
- Pulsed electric fields and antimicrobials. Antimicrobials which act on the cell membrane may increase the susceptibility of membranes to dielectric breakdown and/or a pulsed electric field may facilitate the access of antimicrobials that cross the membrane and act in the cytoplasm.
- Ultrasound and natural antimicrobials (e.g., cinnamon). The lipophilic antimicrobial can accumulate in the lipid bilayer of the membrane, sensitizing this structure to the action of ultrasonic waves.

11.4.1 High Hydrostatic Pressure Combined with Other Conservation Factors

An antagonistic effect has been observed when high hydrostatic pressure is combined with a low a_w. Palou et al. (1998) found that lethality increased as a_w of the suspension medium increased; therefore, there was a baroprotective effect when a_w decreased.

Low pH has a significant effect on high hydrostatic pressure treatments, because it allows a reduction in the intensity of the nonthermal process and increases its lethality. Low pH prevents microbial growth and the germination of spores that can survive a high hydrostatic pressure process at acidic pH (Raso and Barbosa-Cánovas 2003). Koseki and Yamamoto (2006) found that a reduction of pH greatly affected the high hydrostatic pressure efficiency in the inactivation of *L. monocytogenes*. Hoover et al. (1989) reported that high pressure in combination with pH has a particularly inhibitory effect on membrane ATPase, a very important enzyme in the acid–base physiological processes of cells.

The combined effect of high hydrostatic pressures and antimicrobials has also been studied. The addition of antimicrobials allows both an increase of the lethality and a decrease of the intensity of the nonthermal process; moreover, microbial growth after the treatment is inhibited. High hydrostatic pressure can damage the membrane, increasing both the cell penetrability and the activity of antimicrobial agents. After treatment with high hydrostatic pressure, antimicrobials continue to exert their inhibitory effect, preventing the recovery of any sublethally injured cell (Morgan et al. 2000). It is important to mention that the method of application of a high hydrostatic pressure treatment can also influence the lethal effect of antimicrobials agents (Ross et al. 2003). Palou et al. (1997) obtained a higher inactivation of *Z. bailii* when a high hydrostatic pressure was combined with 1,000 ppm potassium sorbate. Guerrero-Beltran et al. (2006) also obtained a beneficial effect in reducing the microbial load in mango puree when high hydrostatic pressure was combined with 500 ppm ascorbic acid.

The combined effect of high hydrostatic pressure and heat has been widely investigated for the inactivation of several microorganisms suspended in different media and foods. In general, the combination of high hydrostatic pressure with heat has the advantages of increasing the lethality of the nonthermal process and of decreasing the intensity of the thermal treatment (Raso and Barbosa-Cánovas 2003). At higher and lower temperatures, microorganisms are more sensitive to the treatment (Raso and Barbosa-Cánovas 2003), and the resistance among strains is much lower when pressure is combined with moderate temperatures (Alpas et al. 1999). The combined treatment of heat and high hydrostatic pressure has a beneficial effect on the inactivation of spores. The high pressure causes germination of spores and once spores have germinated, high temperature destroys them (Raso and Barbosa-Cánovas 2003). Estrada-Girón et al. (2007) observed an inverse relationship between pressure and temperature and the number of survivors of *Geobacillus stearothermophilus* spores (ATCC 7953) added to soymilk. Palou et al. (1998) found no inactivation of *Byssochlamys nivea* ascospores suspended in apple and cranberry juices after a continuous treatment of pressure and temperature. Nevertheless, in an oscillatory pressurization of three to five cycles, inactivation was observed. Investigations have also been performed with *L. innocua* and *Citrobacter freundii* (Carlez et al. 1993).

11.4.2 Pulsed Electric Field Combined with Other Conservation Factors

It has been observed that the combination of a pulsed electric field with low pH increases the lethality of the nonthermal process and decreases the intensity of the treatment required to achieve a certain microbial reduction; moreover, the low pH inhibits microbial growth after the treatment (Raso and Barbosa-Cánovas 2003). Therefore, the addition of organic acids has a synergistic effect when this is combined with a pulsed electric field; this effect might be because both treatments target the cell membrane, increasing the permeability of the cell wall and membrane, enhancing the entry of undissociated acids into the bacterial cell (Ross et al. 2003). Gómez et al. (2005) found for the inactivation of *L. monocytogenes* in McIlvaine buffer a proportional relationship between microbial load and pH. The same behavior was found by Wouters et al. (1999). However, the opposite effect was found by Álvarez et al. (2000) for the inactivation of *Salmonella senftenberg*. This demonstrates that pH plays an important role in determining the inactivation kinetics under pulsed electric field treatments.

When a pulsed electric field is combined with addition of any antimicrobial, a powerful synergistic effect is expected because antimicrobials can increase the susceptibility of membranes to dielectric breakdown and/or a pulsed electric field can facility the access of such agents to the membranes or cytoplasm. An increase in the inactivation effect of a pulsed electric field has been observed for both Gram-positive and Gram-negative bacteria when the system is treated in the presence of nisin.

This synergistic effect has been found when bacteria are incubated with nisin after pulsed electric field treatment, when nisin is present during the treatment, or when the antimicrobial is added after the pulsed electric field treatment (Raso and Barbosa-Cánovas 2003).

Sobrino-López and Martín-Belloso (2008) found a synergistic effect on the inactivation of *S. aureus* in milk when the addition of nisin and lysozyme ws combined with a pulsed electric field. A synergistic and additive behavior was observed by Mosqueda-Melgar et al. (2008) in the inactivation of *Salmonella enteritidis* and *E. coli* O157:H7 in apple, pear, orange, and strawberry juices which had been inoculated with these microorganisms when pulsed electric field treatment was combined with the addition of citric acid and cinnamon bark oil.

A direct relationship between temperature and pulsed electric field treatment, operated as a batch or as a continuous system, has been observed. The synergistic behavior of heat and a pulsed electric field is related to the temperature effect on membrane fluidity properties. At high temperatures phospholipids are less ordered than at low temperatures, and therefore the membrane has a liquid-crystalline structure, and its thickness is reduced (Raso and Barbosa-Cánovas 2003). Amiali et al. (2007) observed an increase in the inactivation rate of *E. coli* O157:H7 and *S. enteritidis* in liquid egg yolk with increasing temperature.

11.4.3 Ultrasound Combined with Other Conservation Factors

There is not enough information about the effect caused by combining ultrasound with reduced a_w on microbial load, but it appears that the lower the a_w, the greater the resistance of a microorganism to ultrasound, as Valero et al. (2007) observed for orange juice in which the presence of pulp increases the resistance of microorganisms to ultrasound.

The combined effect of ultrasound with low pH has been studied for the inactivation of some microorganism (Pagan et al. 1999; Kinsloe et al. 1954), but no effect was observed in the survival curves with such a combination. However, a beneficial effect on microbial load has been observed when ultrasound treatment is combined with addition of antimicrobials (Ross et al. 2003). As an example, for *L. monocytogenes* it has been observed that the higher the concentration of vanillin, the greater the inactivation of the microorganism obtained, and for the same concentration of vanillin, the microbial load obtained is lower when ultrasound treatment is applied.

It appears that there is a beneficial effect on the microbial inactivation when ultrasound is combined with heat. López-Malo et al. (1999) obtained a beneficial effect of combining ultrasound with heat for the inactivation of *Saccharomyces cerevisiae*. The same effect was observed by Baumann et al. (2005) for the inactivation of *L. monocytogenes* 10403S at lethal and sublethal temperatures combined with ultrasound, and by Stanley et al. (2004). However, D'Amico et al. (2006) observed that ultrasound treatment with or without the effect of heat was effective at reducing microbial levels in raw milk, *L. monocytogenes* levels in ultra-high-temperature milk which had been inoculated with this microorganism, and *E. coli* O157:H7 in apple cider.

11.4.4 High-Pressure Homogenization Combined with Other Conservation Factors

There is not enough information about the combined effect of high-pressure homogenization and reduced a_w, but it appears that there is no influence on microbiological load of such a combination as Diels et al. (2005) found for *E. coli* MG1655 in buffered suspensions.

In contrast to a_w, there is a beneficial effect on the microbial inactivation when high-pressure homogenization is combined with addition of antimicrobials. Tribst et al. (2008) obtained a beneficial effect when high-pressure homogenization and addition of lysozyme (an antimicrobial enzyme) where combined for the inactivation of *Lactobacillus brevis*. Similar results were obtained by Vannini et al. (2004) using lysozyme or lactoperoxidase for the inactivation of some Gram-positive and Gram-negative species in skim milk which had been inoculated with them. They concluded that the interaction of high-pressure homogenization and lysozyme or lactoperoxidase is associated with conformational modifications of the two proteins, with a consequent enhancement of their activity. Pathanibul et al. (2009) used nisin as an antimicrobial combined with high-pressure homogenization for the inactivation of *E. coli* and *L. innocua* in apple or carrot juice which had been inoculated with them, but obtained no effect for such a combination for *E. coli*, but interactions were observed with *L. innocua*.

The effect of high-pressure homogenization combined with heat has also been studied. Diels et al. (2003) observed that temperatures between 5°C and 40°C do not affect inactivation of *S. aureus* by high-pressure homogenization, but *Yesinia enterocolitica* inactivation was affected by temperature over a much wider range.

11.5 Concluding Remarks

The rational design of minimal processes for food preservation using emerging and traditional agents in combination requires, among other factors, a careful analysis of the mechanisms of action of the factors on microorganisms and knowledge of their inactivation kinetics and the characteristics of the interaction of the factors in combination (additive, synergistic, or antagonic), as well as the effect of the preservation system on food quality.

Acknowledgements We acknowledge financial support from Universidad de Buenos Aires, CONICET, and ANPCyT-BID of Argentina, as well as from Instituto Tecnológico y de Estudios Superiores de Monterrey and CONACyT of Mexico.

References

Ahvenainen R. New approaches in improving shelf life of minimally processed fruit and vegetables. Trend Food Sci Technol. 1996;7:179–87.

Ahvenainen R, Mattila-Sandholm T, Ohlsson T. Minimal processing of foods, VTT symposium series, vol. 142. Espoo: Technical Research Center of Finland (VTT); 1994.

Ahvenainen R. Minimal processing in the future: integration across the supply chain. In: Ohlsson T, Bengtsson N, editors. Minimal processing technologies in the food industry. Cambridge: Woodhead; 2002. p. 267–81.

Alpas H, Kalchayanand N, Bozoglu F, Sikes A, Dunne CP, Ray B. Variation in resistance to hydrostatic pressure among strains of food-borne pathogens. Appl Environ Microbiol. 1999;65:4248–51.

Álvarez I, Raso J, Palop A, Sala FJ. Influence of different factors on the inactivation of *Salmonella senftenberg* by pulsed electric fields. Int J Food Microbiol. 2000;55:143–6.

Alzamora SM, Salvatori DM. Minimal processing: fundamental and application. In: Hui YH, editor. Handbook of food science, technology and engineering, vol. 3. Boca Raton: CRC Press/Taylor & Francis; 2006. p. 118–1–6.

Alzamora SM, Guerrero S, López-Malo A, Palou E. Plant antimicrobial combined with convectional preservatives for fruit products. In: Roller S, editor. Natural antimicrobials for the minimal processing of foods. Cornwall: Woodhead; 2003. p. 235–49.

Alzamora SM, Tapia MS, López-Malo A, editors. Minimally processed fruits and vegetables. Fundamentals and applications. Gaithersburg: Aspen Publishers; 2000a.

Alzamora SM, Tapia MS, López-Malo A. Overview. In: Alzamora SM, Tapia MS, López-Malo A, editors. Minimally processed fruits and vegetables. Fundamentals and applications. Gaithersburg: Aspen Publishers; 2000b. p. 1–9.

Alzamora SM, Tapia MS, Welti-Chanes J. New strategies for minimally processed foods. The role of multitarget preservation. Food Sci Technol Int. 1998;4:353–62.

Amiali M, Ngadi MO, Smith JP, Raghavan GSV. Synergistic effect of temperature and pulsed electric field on inactivation of *Escherichia coli* O157:H7 and *Salmonella enteritidis* in liquid egg yolk. J Food Eng. 2007;79:689–94.

Avsaroglu MD, Buzrul S, Alpas H, Akcelik M, Bozoglu F. Use of the Weibull model for lactococcal bacteriophage inactivation by high hydrostatic pressure. Int J Food Microbiol. 2006;108:78–83.

Barbosa-Cánovas GV. Key goals of emerging technologies for inactivating bacteria. Food Safety Mag. 2002;8(4):34–42.

Baumann AR, Martin SE, Feng H. Power ultrasound treatment of *Listeria monocytogenes* in apple cider. J Food Prot. 2005;68(11):2333–40.

Bialka KL, Demirci A, Puri VM. Modeling the inactivation of *Escherichia coli* O157:H7 and *Salmonella enterica* on raspberries and strawberries resulting from exposure to ozone or pulsed UV-light. J Food Eng. 2008;85:444–9.

Bintsis T, Litopoulou-Tzanetaki E, Robinson R. Existing and potential applications of ultraviolet light in the food industry-a critical review. J Sci Food Agric. 2000;80:637–45.

Butz P, Tauscher B. Emerging technologies: chemical aspects. Food Res Int. 2002;35:279–84.

Carlez A, Rosec JP, Richard N, Cheftel J. High pressure inactivation of *Citrobacter freundii*, *Pseudomonas fluorescens*, and *Listeria innocua* in inoculated minced beef. Lebensm Wiss Technol. 1993;26:48–54.

Char C, Guerrero S, Alzamora SM. Survival of *Listeria innocua* in thermally processed orange juice is affected by vanillin addition. Food Control. 2009;20(1):67–74.

Chick H. An investigation of the laws of disinfection. J Hyg Camb. 1908;8:92–158.

D'Amico DJ, Silk TM, Wu J, Guo M. Inactivation of microorganisms in milk and apple cider treated with ultrasound. J Food Prot. 2006;69(3):556–63.

Diels AMJ, Wuytack EY, Michiels CW. Modelling inactivation of *Staphylococcus aureus* and *Yersinia enterocolitica* by high-pressure homogenization at different temperatures. Int J Food Microbiol. 2003;87:55–62.

Diels AMJ, Callewaert L, Wuytack EY, Masschalck B, Michiels CW. Inactivation of *Escherichia coli* by high-pressure homogenization is influenced by fluid viscosity but not by water activity and product composition. Int J Food Microbiol. 2005;101:281–91.

Donsì G, Ferrai G, Lenza E, Maresca P. Main factors regulating microbial inactivation by high-pressure homogenization: operating parameters and scale of operation. Chem Eng Sci. 2009;64:520–32.

Estrada-Girón Y, Guerrero-Beltrán JA, Swanson BG, Barbosa-Cánovas GV. Effect of high hydrostatic pressure on spores of *Geobacillus stearothermophilus* suspended in soymilk. J Food Process Pres. 2007;31:546–58.

Geeraerd AH, Valdramidis VP, Van Impe JF. GInaFit, a free tool to assess non-log-linear microbial survivor curves. Int J Food Microbiol. 2005;102:95–105.

Guerrero-Beltrán JA, Barbosa-Cánovas GV. Reduction of *saccharomyces cerevisiae*, *Escherichia coli* and *Listeria innocua* in apple juice by ultraviolet light. J Food Process Eng. 2005;28:437–52.

Guerrero-Beltrán JA, Barbosa-Cánovas G, Moraga-Ballesteros G, Moraga-Ballesteros MJ, Swanson BG. Effect of pH and ascorbic acid on high hydrostatic pressure-processed mango puree. J Food Process Preserv. 2006;30:582–96.

Guerrero-Beltrán JA, Welti-Chanes J, Barbosa-Cánovas GV. Ultraviolet-c light processing of grape, cranberry and grapefruit juices to inactivate Saccharomyces cerevisiae. J Food Process Eng. 2009;32:916–932.

Gómez PL, Salvatori D. Alzamora SM. Estudio de los cambios de color producidos por la aplicación de tecnologías emergentes de conservación en frutas mínimamente procesadas. 9th Congreso Argentino del Color. Santa Fe, Argentina, 1–3 October 2008.

Gómez N, García D, Álvarez I, Condon S, Raso J. Modelling inactivation of *Listeria monocytogenes* by pulsed electric fields in media of different pH. Int J Food Microbiol. 2005;103:199–206.

Heldman DR, Newsome RI. Kinetic models for microbial survival during processing. Food Technol. 2003;57:40–6. 100.

Hoover DG, Metrick C, Papineau AM, Farkas DF, Knorr D. Biological effects of high hydrostatic pressure on food microorganisms. Food Technol. 1989;43:99–107.

Huang L. Thermal inactivation of *Listeria monocytogenes* in ground beef under isothermal and dynamic temperature conditions. J Food Eng. 2009;90:380–7.

Huxsoll RL, Bolin HR. Processing and distribution alternatives for minimally processed fruits and vegetables. Food Technol. 1989;43(2):132–8.

Juneja VK, Marks HM. Mathematical description of non-linear survival curves of *Listeria monocytogenes* as determined in a beef gravy model. Innov Food Sci Emerg Technol. 2003;4:307–17.

Kinsloe H, Ackerman E, Reid JJ. Exposure of microorganisms to measured sound fields. J Bacteriol. 1954;68:373–80.

Koseki S, Yamamoto K. pH and solute concentration of suspension media affect the outcome of high hydrostatic pressure treatment of *Listeria monocytogenes*. Int J Food Microbiol. 2006;111:175–9.

Lado BH, Yousef AE. Alternative food–preservation technologies: efficacy and mechanisms. Microbes Infect. 2002;4:433–40.

Linton RH, Carter WH, Pierson MD, Hackney CR. Use of a modified Gompertz equation to model nonlinear survival curves for *Listeria monocytogenes* Scott A. J Food Prot. 1995;58:946–54.

Leistner L. Basic aspects of food preservation by hurdle technology. Int J Food Microbiol. 2000;55:181–6.

López-Malo A, Guerrero S, Alzamora SM. *Saccharomyces cerevisiae*, thermal inactivation kinetics combined with ultrasound. J Food Prot. 1999;62(10):1215–7.

Mafart P. Food engineering and predictive microbiology: on the necessity to combine biological and physical kinetics. Int J Food Microbiol. 2005;100:239–51.

Mafart P, Couvert O, Gaillard S, Leguerinel I. On calculating sterility in thermal preservation methods: application of the Weibull frequency distribution model. Int J Food Microbiol. 2002;72:107–13.

Manvell C. Minimal processing of food. Food Sci Technol Today. 1997;11:107–11.

Mañas P, Mackey BM. Morphological and physiological changes induced by high hydrostatic pressure in exponential- and stationary-phase cells of *Escherichia coli*: relationship with cell death. Appl Environ Microbiol. 2004;70(3):1545–54.

Martín O, Qin BL, Chang FJ, Barbosa-Cánovas GV, Swanson BG. Inactivation of *Escherichia coli* in skim milk by high intensity pulsed electric fields. J Food Process Eng. 1997;20:317–36.

McKellar RC, Lu X, editors. Modeling microbial responses in food. Boca Raton: CRC Press; 2004.

McMeekin TA. Predictive microbiology: quantitative science delivering quantifiable benefits to the meat industry and other food industries. Meat Sci. 2007;77:17–27.

McMeekin TA, Olley JN, Ross T, Ratkowsky DA. Predictive microbiology: theory and application. Taunton/New York: Research Studies Press/Wiley; 1993.

Miller R, Jeffrey W, Mitchell D, Elasri M. Bacterial responses to ultraviolet light. Am Soc Microbiol. 1999;65(8):535–41.

Morgan R. UV "green" light disinfection. Dairy Indust Int. 1989;54(11):33–5.

Morgan SM, Ross RP, Beresford T, Hill C. Combination of hydrostatic pressure and lacticin 3147 causes increased killing of *Staphylococcus* and *Listeria*. J Appl Microbiol. 2000;88:414–20.

Mosqueda-Melgar J, Raybaudi-Massilia RM, Martín-Belloso O. Non-thermal pasteurization of fruit juices by combining high-intensity pulsed electric fields with natural antimicrobials. Innov Food Sci Emerg Technol. 2008;9:328–40.

Ohlsson T. Minimal processing–preservation methods of the future: an overview. Trends Food Sci Technol. 1994;5:341–4.

Ohlsson T. New thermal processing methods. Paper presented at the EFFoST Conference on the Minimal Processing of Food, 6–9 Nov 1996.

Ohlsson T, Bengtsson N, editors. Minimal processing technologies in the food industry. Cambridge: Woodhead; 2002.

Oliveira FAR, Oliveira JC, editors. Processing foods: quality optimization and process assessment. Boca Raton: CRC Press; 1999.

Ortega-Rivas E. Processing effect for safety and quality in some non-predominant food technologies. Crit Rev Food Sci Nutr. 2007;47:161–73.

Pagan R, Manas P, Alvarez I, Condon S. Resistance of *Listeria monocytogenes* to ultrasonic waves under pressure at sublethal (manosonication) and lethal (manothermosonication) temperatures. Food Microbiol. 1999;16:139–48.

Palou E, López-Malo A, Barbosa-Cánovas GV, Welti-Chanes J, Swanson BG. High hydrostatic pressure as a hurdle for *Zygosaccharomyces Bailii* inactivation. J Food Sci. 1997;62(4):855–7.

Palou E, López-Malo A, Barbosa-Cánovas GV, Welti-Chanes J, Davidson PM, Swanson BG. Effect of oscillatory high hydrostatic pressure treatments on *Byssochlamys nivea* ascospores suspended in fruit juice concentrates. Lett Appl Microbiol. 1998;27:375–8.

Palou E, López-Malo A, Welti-Chanes J. Innovative fruit preservation methods using high pressure. In: Welti-Chanes J, Barbosa-Cánovas GV, Aguilera JM, editors. Engineering and food for the 21st century. Boca Raton: CRC Press; 2002. p. 715–25.

Pathanibul P, Taylor TM, Davidson PM, Harte F. Inactivation of *Escherichia coli* and *Listeria innocua* in apple and carrot juices using high pressure homogenization and nisin. Int J Food Microbiol. 2009;129:316–320.

Peleg M. On calculating sterility in thermal and non-thermal preservation methods. Food Res Int. 1999;32:271–8.

Peleg M. Advanced quantitative microbiology for foods and biosystems – models for predicting growth and inactivation. Boca Raton: Taylor & Francis, CRC; 2006.

Peleg M, Cole MB. Reinterpretation of microbial survival curves. Crit Rev Food Sci. 1998;38: 353–380.

Peleg M, Cole MB. Estimating the survival of *Clostridium botulinum* spores during heat treatments. J Food Prot. 2000;63:190–5.

Pilavtepe-Çelik M, Buzrul S, Alpas H, Bozoglu F. Development of a new mathematical model for inactivation of *Escherichia coli* O157:H7 and *Staphylococcus aureus* by high hydrostatic pressure in carrot juice and peptone water. J Food Eng. 2009;90:388–94.

Piyasena P, Mohareb E, McKellar RC. Inactivation of microbes using ultrasound: a review. Int J Food Microbiol. 2003;87:207–16.

Raffellini S, Guerrero S, Alzamora SM. Inactivation of *Escherichia coli* in hydrogen peroxide solutions at various concentrations and pHs. J Food Saf. 2008;28:514–533.

Raso J, Barbosa-Cánovas GV. Nonthermal preservation of foods using combined processing techniques. Crit Rev Food Sci Nutr. 2003;43:265–85.

Raso J, Calderón ML, Góngora M, Barbosa-Cánovas G, Swanson BG. Inactivation of *Zygosaccharomyces Bailii* in fruit juices by heat, high hydrostatic pressure and pulsed electric fields. J Food Sci. 1998;63(1):1042–4.

Ross AIV, Griffiths MW, Mittal GS, Deeth HC. Combining nonthermal technologies to control foodborne microorganisms. Int J Food Microbiol. 2003;89:125–38.

Rolle RS, Chism GW. Physiological consequences of minimally processed fruits and vegetables. J Food Quality. 1987;10:187–93.

San Martin MF, Sepulvelda DP, Altunaker B, Gongora-Nieto MM, Sawnson BG, Barbosa-Canovas G. Evaluation of selected mathematical models to predict the inactivation of *Listeria innocua* by pulsed electric fields. Food Sci Technol. 2007;40:1271–9.

Sapru V, Smerage, GH, Teixeira AA, Lindsay JA. Comparison of predictive models for bacterial spore population resources to sterilization temperatures. J Food Sci. 1993;58:223–228.

Sastry SK, Datta AK, Worobo RW. Ultraviolet light. J Food Sci Suppl. 2009;65:90–2.

Saucedo-Reyes D, Marco-Celdrán A, Consuelo Pina-Pérez M, Rodrigo D, Martínez-López A. Modeling survival of high hydrostatic pressure treated stationary- and exponential-phase *Listeria innocua* cells. Innov Food Sci Emerg Technol. 2008;10(2):135–41.

Schenk M, Guerrero SN, Alzamora SM. Response of some microorganisms to ultraviolet treatment on fresh-cut pear. Food Bioprocess Technol. 2008;1:384–92.

Shewfelt RL. Quality of minimally processed fruits and vegetables. J Food Quality. 1987;10:143–56.

Singh RP, Oliveira FAR, editors. Minimal processing of foods and process optimization: an interface. Boca Raton: CRC Press; 1994.

Sizer CE, Balasubramaniam VM. New intervention processes for minimally processed juices. Food Technol. 1999;53:64–7.

Slongo AP, Rosenthal A, Quaresma-Camargo LM, Deliza R, Pereira-Mathias S, Falcão de Aragão GM. Modeling the growth of lactic acid bacteria in sliced ham processed by high hydrostatic pressure. LWT Food Sci Technol. 2009;42(1):303–6.

Smelt JPPM, Rijke AGF, Hayhurst A. Possible mechanism of high pressure inactivation of microorganisms. High Pressure Res. 1994;12(4–6):199–203.

Snyder OP. HACCP and regulations applied to minimally processed foods. In: Novak HC, Sapers GM, Juneja VK, editors. Microbial safety of minimally processed foods. Boca Raton: CRC Press; 2003. p. 127–50.

Sobrino-López A, Martín-Belloso O. Enhancing the lethal effect of high-intensity pulsed eclectic field in milk by antimicrobial compounds as combined hurdles. J Dairy Sci. 2008;91(5):1759–68.

Somolinos M, Mañas P, Condon S, Pagán R, García D. Recovery of *Saccharomyces cerevisiae* sublethally injured cells after pulsed electric fields. Int J Food Microbiol. 2008;125:352–6.

Stanley KD, Golden DA, Williams RC, Weiss J. Inactivation of Escherichia coli O157:H7 by high-intensity ultrasonication in the presence of salts. Foodborne Pathog Dis. 2004;1(4):267–80.

Torres JA, Velazquez G. Commercial opportunities and research challenges in the high pressure processing of foods. J Food Eng. 2005;67:95–112.

Tribst AAL, Franchi MA, Cristianini M. Ultra-high pressure homogenization treatment combined with lysozyme for controlling *Lactobacillus brevis* contamination in model system. Innov Food Sci Emerg Technol. 2008;9(3):265–71.

US. FDA. Report on kinetics of microbial inactivation for alternative food processing technologies. 2001. htp://www.cfsan.gov/~comm/ift–toc.html. Accessed June 2001.

Valero M, Recrosio N, Saura D, Muñoz N, Martí N, Lizama V. Effects of ultrasonic treatments in orange juice processing. J Food Eng. 2007;80:509–16.

Vannini L, Lanciotti R, Baldi D, Guerzoni ME. Interactions between high pressure homogenization and antimicrobial activity of lysozyme and lactoperoxidase. Int J Food Microbiol. 2004;94:123–35.

van Boekel MAJS. On the use of Weibull model to describe thermal inactivation of microbial vegetative cells. Int J Food Microbiol. 2002;74:139–59.

Velázquez G, Vázquez P, Vázquez M, y Torres JA. Avances en el procesado de alimentos. Cienc Tecnol Aliment. 2005;4(5):353–67.

Welti-Chanes J, López-Malo A, Palou E, Bermúdez D, Guerrero-Beltrán JA, Barbosa-Cánovas GV. Fundamentals and applications of high pressure processing to foods. In: Barbosa-Cánovas GV, Tapia MS, Cano P, editors. Novel food processing technologies. Boca Raton FL: CRC Press; 2005. p. 157–81.

Welti-Chanes J, San Martín-González F, Barbosa-Cánovas GV. Water and biological structures at high pressure. In: Buera P, Welti-Chanes J, Llilford P, Corti H, editors. Water properties of food, pharmaceutical, and biological materials. Boca Raton FL: CRC Press; 2006. p. 205–32.

Welti-Chanes J, Vergara F, López-Malo A. Minimally processed foods: state of the art and future. In: Fito P, Ortega-Rodríguez E, Barbosa-Cánovas GW, editors. Food engineering 2000. New York: Chapman and Hall; 1997. p. 181–212.

Wiley R. Introduction to minimally processed refrigerated fruits and vegetables. In: Wiley RC, editor. Minimally processed fruits and vegetables. New York: Chapman and Hall; 1994a. p. 1–14.

Wiley RC, editor. Minimally processed refrigerated fruits and vegetables. New York: Chapman and Hall; 1994b.

Wouters PC, Dutreux N, Smelt JPPM, Lelieveld HLM. Effect of pulsed electric fields on inactivation kinetics of *Listeria innocua*. Appl Environ Microbiol. 1999;65(12):5364–71.

Wuytack EY, Diels AMJ, Michiels CW. Bacterial inactivation by high-pressure homogenisation and high hydrostatic pressure. Int J Food Microbiol. 2002;77:205–12.

Xiong R, Xie G, Edmonson GM, Sheard MA. A mathematical model for bacterial inactivation. Int J Food Microbiol. 1999;46:45–55.

Yeom HW, Streaker CB, Zhang QH, Min DB. Effects of pulsed electric fields on the activities of microorganisms and pectin methyl esterase in orange juice. J Food Sci. 2000;65(8):1359–62.

Yeom HW, Zhang QH, Chism GW. Inactivation of pectin methyl esterase in orange juice by pulsed electric fields. J Food Sci. 2002;67(6):2154–9.

Chapter 12
Emerging Technologies to Improve the Safety and Quality of Fruits and Vegetables

Elisabete M.C. Alexandre, Teresa R.S. Brandão, and Cristina L.M. Silva

12.1 Introduction

Consumers' demand for increased quality standards has led to the development of new and less aggressive processing technologies which permit greater retention of natural taste. As a consequence, minimal processing techniques designed to replace traditional preservation methods have emerged. This will extend shelf life without the detrimental effects attributed to severe heating. Furthermore, the presence of pesticides in food products and the use of synthetic additives to preserve fresh fruits and vegetables have increased the necessity to use other innovative methods that do not leave chemical residues in foods, avoiding some serious diseases (Ohlsson 2002; Bruhn 2005). Nonthermal methods have emerged as attractive alternatives to conventional thermal processing methods. They constitute challenging processes aiming at reducing pernicious effects of thermal methods, by preserving quality and nutritional attributes of fruits and vegetables, and yielding safe and less perishable products. Ozone, UV-C radiation, and ultrasound treatments are promising techniques intended for the fruit and vegetable industry. The application of such technologies may yield products with limited losses of color, flavor, texture, and nutrients, while retaining the desired shelf life and safety. However, the efficiency related to each safety or quality indicator depends on the product/indicator under consideration.

E.M.C. Alexandre • T.R.S. Brandão • C.L.M. Silva (✉)
Escola Superior de Biotecnologia, Universidade Católica Portuguesa,
Rua Dr. António Bernardino de Almeida, 4200-072, Porto, Portugal
e-mail: clsilva@esb.ucp.pt

12.2 Safety and Quality Aspects of Fruits and Vegetables

12.2.1 Safety

Fruits and vegetables are often consumed uncooked; anything that is left on them, after they have come into contact with other inanimate or animate surfaces and entities, will be consumed. Fruits and vegetables often contain a great diversity of microbial flora. Some of those microorganisms are involved with the body's defense; most have no direct impact on human's health, but some are responsible for disease states. Bacteria, parasites, and viruses are the most relevant organisms that can be transported on fruits and vegetables and that may pose a hazard to health (the ones associated with outbreaks are included in Table 12.1). Some of those pathogens are naturally present in the environment, and one of the main sources of contamination is fecal deposits (human and/or animal). From the farmer to the final consumer there are a huge number of occurrences that compromise fresh produce safety. Irrigation and/or washing waters and soils containing such pathogens are subsequent sources of transmission; organic fertilizers or soil conditioners derived from wastewater sludge are also dangerous contaminants; insects and birds are also vehicles that may cause cross-contamination. In a later step, lack of hygiene rules related to manufacturers and procedures may contribute to the presence of undesired microorganisms in the products.

Chemical contaminants such as heavy metals and persistent organic pollutants (e.g., polycyclic aromatic hydrocarbons) are frequently found in fresh fruits and vegetables. They are derived from the application of insecticides, fungicides, herbicides, rodenticides, and fertilizers. Many pesticides, owing to their potent neurotoxic chemicals, attack and disable portions of the nervous system. Compounds containing lead, arsenic, copper, mercury, and zinc are frequently used on annual crops and fruit trees (Alloway 2004). Some studies have reported the presence of unacceptable levels of pesticide residues in fresh fruits and vegetables, which implies that there is need for regular monitoring, especially of imported crops (Torres et al. 1996; Knezevic and Serdar 2009).

Fertilizers are also of concern and are broadly classified in two categories: organic (composed of decayed plant/animal matter) and inorganic (composed of simple chemicals and minerals). These materials may contain significant concentrations of trace metals, including cadmium and zinc (in phosphate fertilizers), and a wide range of metal and organic contaminants (in composts and sewage sludges) (Alloway 2004).

12.2.2 Quality

Food quality is defined by the consumer, buyer, grader, or any other client on the basis of subjective and objective measurements of the food product. The characteristics of the product should satisfy the final consumer, and there are a range of attributes that can be considered to analyze the expected product quality.

Table 12.1 Major pathogens associated with contamination of fruits and vegetables and human illness (Anonymous 2001; Warriner 2005; Zhao 2005; Gorny 2006)

Organism	Species	Transmission source	Food vehicle
Bacteria	Clostridium botulinum	Soil, lakes, streams, coastal waters, intestinal tracts of mammals, decaying vegetation, reptiles	Mushrooms, green pepper, coleslaw, cabbage, garlic, potato, carrot
	Escherichia coli	Animal feces, especially cattle, chickens, deer, sheep, goats, and pigs; cross-contamination from raw meat	Berries, melons, lettuce, spinach, cantaloupe, alfalfa sprouts, cabbage, celery, watercress
	Salmonella spp.	Animal and human feces; cross-contamination from raw meat, poultry, or eggs	Berries, strawberries, melons, raw tomatoes, alfalfa sprouts, cabbage, celery, artichokes, endive, fennel, lettuce
	Shigella spp.	Human feces	Raw vegetables, lettuce, green onions, celery
	Listeria monocytogenes	Soil, food processing environments, surfaces	Lettuce, cabbage, carrots, chicory, potatoes, radish, bean sprouts
Parasites	Cryptosporidium spp.	Water, animal and human feces	Vegetables salads
	Cyclospora spp.	The route of transmission is unclear; water is often mentioned	Raspberries, blackberries, lettuce, basil, and other fresh produce
Viruses	Hepatitis A	Water, human feces and urine	Lettuce, melons, strawberry, watercress, tomatoes, green onions, fresh-cut fruit
	Norovirus	Human feces, vomitus	

In recent years, increased attention has been given to the quality and safety of foods, not only from a consumer's perspective but also from the perspective of governmental entities. To ensure consumer satisfaction, some food standards were designed to provide information about the product and the means to preserve product quality. Food standards were also developed to prevent economic fraud and to establish the market value of products (Anonymous 2002). There are various institutions worldwide that specialize in the preparation of food standards and codes of practice for fresh products and in quality inspection and certification of fresh fruits and vegetables. These measures provide consumer protection in relation to quality parameters that are often difficult to measure when the product is bought. However, there are some quality attributes that consumers can scrutinize themselves, such as the external attributes. Appearance, feel, and defects can be easily evaluated by vision and touch (evaluation of size, shape, state of maturation, general coloration, softening, and defects of the product). Internal quality attributes can be more difficult to evaluate, as characteristic aroma, taste, and texture can only be analyzed if the consumer eats the product before acquiring it. Hidden quality parameters such as nutritive value and wholesomeness are impossible to measure at the time of purchase. The more relevant attributes and some detection methods that can be used for quality evaluation of fresh fruits and vegetables are included in Table 12.2.

The consumption of fresh fruits and vegetables is of significant benefit to human health. Antioxidants, minerals, vitamins, and fibers in produce are good examples of food components that can help in the prevention of diseases. For example, calcium has a significant role in the prevention or control of osteoporosis, folate is associated with the prevention of neural tube defects in infants, and fiber is associated with the prevention of colon cancer. Some phytochemicals such as carotenoids present in carrots and the yellow, red, and orange pigments of plants are highly significant in the prevention of cancer, diabetes, and hypertension. Vitamin A (found in carrots, spinach, and cantaloupe), vitamin B (present in broccoli, strawberries, and tomatoes), and vitamin C (found in significant levels in oranges, kiwifruit, and mango, and in certain types of nuts) are examples of compounds that can help to minimize cellular oxidative damage and have a role in the immune system.

12.3 Traditional Technologies

12.3.1 Thermal Treatments

Conventional thermal processing methods, such as blanching, pasteurization, and sterilization, play a very important role in the food industry, since high temperatures will cause a rapid inactivation of microorganisms and enzymes.

Blanching is one important process that precedes other preservation processes such as freezing and dehydration. It is mainly applied to vegetables with the purpose of reducing enzyme activity, which is responsible for undesirable colors and flavors, texture alteration, and nutritional decay. The process is characterized by the use of

Table 12.2 Quality attributes and some detection methods that can be used for quality evaluation of fresh fruits and vegetables (Anonymous 2002; Rico et al. 2007)

Quality attributes		Based on	Examples	Detection methods
External		Appearance (sight)	Size and shape	Grade standards and differentiate between items
			Color	Visual or mechanical methods (i.e., colorimeters and spectrophotometers)
		Feel (touch)	Firmness (softening of the product)	Mechanical means (i.e., texture analyses)
		Defects	Defects due to production, handling, environment, diseases, and other factors	Measured visually though some mechanical methods (i.e., ultrasound and machine vision)
Internal		Odor/aroma	Aromatic compounds (sum of compounds perceived by nose)	Gas chromatographs, mass spectrometers, or similar mechanisms
		Taste	Sweet, sour, bitter, and astringent (perception of chemical compounds on the tongue and other nerve endings of the mouth)	Spectrophotometric and gravimetric methods and liquid and gas chromatography
		Texture (mouthfeel)	Tenderness, crispness, crunchiness, chewiness, and fibrousness	Mechanical means (i.e., texture analyses)
Hidden		Wholesomeness (freshness)	May be brought about by the food itself or by external factors such as environment or handling	Microscopic, microbiological, and X-ray technologies
		Nutritive value	Vitamins, minerals, fibers, antioxidants, and other phytochemicals (presence and levels of components that support life)	Wet chemistry, chromatographic methods, and other chemical and physical methods
		Safety	Pathogenic microorganisms	Microbiological examination

short bursts of heat lasting about 2–3 min at temperatures usually between 85°C and 100°C. The definition of blanching time/temperature to which the product must be submitted depends on the size, shape, thermal diffusivity, and natural levels of enzymes (Fennema et al. 1973; Knorr 1995).

The negative aspects of blanching are related to the effects that heat has on sensorial and nutritional characteristics of foods. Softening of tissues, vitamin loss, development of undesirable cooked flavors, and loss of soluble solids (to the blanching water) are some examples (Qi et al. 1995; Martins and Silva 2001). These occurrences promoted research into and development of alternative methods as efficient as blanching in the ability to reduce enzyme activity and microbial load of the products, yet allowing them to retain organoleptic and nutritional characteristics (Piyasena et al. 2003; Knorr et al. 2004).

Pasteurization is a thermal treatment that reduces pathogenic bioload to a level that is unlikely to cause disease. The temperatures used are usually below 100°C and the time/temperature combination used depends on the product and microorganism profile in the product. With solid foods, particularly fruits and vegetables, the pasteurization process is often applied to the products' surfaces (immersed in hot water). Annous and Kozempel (2006) reported apples, melons, mangoes, lemons, oranges, cucumbers, pears, tomatoes, and alfalfa seeds as examples of produce negatively affected by surface pasteurization.

Sterilization is a thermal process that kills all microorganisms and vegetative cells. The severity of such a process limits its application to fruits and vegetables since the impact on the quality is considerable.

12.3.2 Chemical Treatments

A number of sanitizing agents may be used to reduce the risk of contamination. Traditionally, fresh fruits and vegetables are rinsed in chlorine, hydrogen peroxide, and acid solutions. However, the impact of those treatments depends on the initial microbial loads; with high bioloads, such treatments may not be sufficient.

Chlorine has been used as an effective disinfectant of drinking water, wastewater, and food processing equipment and surfaces. The chlorination process can be performed by addition of chlorine (Cl_2) or by addition of sodium hypochlorite (NaOCl) or calcium hypochlorite $Ca(OCl)_2$ to wash waters. Such solutions, when applied to the disinfection of fruits and vegetables, are used at concentrations ranging from 50 to 200 ppm with a short treatment time of 1–2 min. The lack of effectiveness of chlorine solutions for disinfection of fresh fruits and vegetables is correlated to the neutralization of chlorine by tissues components, thus limiting its oxidative action on microorganisms (Beuchat 1998; Bachmann and Earles 2000).

In addition to chlorine, iodine is also used to sanitize food processing equipment and surfaces. Iodine formulations called iodophors are water-soluble, less volatile, and less irritating to the skin than other iodine formulations. Iodophors inactivate a wide spectrum of microorganisms. At low temperatures they are less corrosive than chlorine and are minimally affected by the presence of organic compounds.

However, at higher temperatures (above 50°C) they vaporize easily and become very corrosive. The application of iodophors to fruits and vegetables is restricted, as iodine forms a purple complex with starch (Beuchat 1998).

Acetic, citric, succinic, malic, tartaric, benzoic, and sorbic acids are the acids in solution form most commonly used for disinfection of fruits and vegetables. Pathogenic microorganisms present on the surfaces of products are killed or inhibited through washes and sprays with such materials. The use of lemon juice to wash lettuce, or its application during the cutting of fruits and vegetables, can eliminate or prevent microbial growth (Hernandez-Brenes 2002; Parish et al. 2003).

Hydrogen peroxide (H_2O_2) is another well-studied oxidizing agent that is toxic to microorganisms. Hydrogen peroxide has a very low toxicity and is recognized as eco-*friendly* owing to its low impact on the environment. However, considering different types of fruits and vegetables, it has been reported that hydrogen peroxide can negatively impact the quality of some products. Hydrogen peroxide may induce browning (mushrooms, lettuce), may cause bleaching of anthocyanins (strawberries and raspberries), and is not efficient in killing yeasts and molds. However, for some fruits and vegetables (i.e., cantaloupes, grapes, prunes, raisins, walnuts, and pistachios), hydrogen peroxide is an efficient disinfectant with no significant impact on product quality (Sapers and Simmons 1998; Bachmann and Earles 2000).

Sapers (2001) reported that the use of sanitizing agents to wash inoculated apples (on a laboratory scale) can reduce the bioload to the order of 2–3 log units. However, when they were applied as a wash treatment in actual industrial environments, the population reduction was less than 1 log cycle.

The effectiveness of all these technologies depends on microbial sensitivity to the sanitizer agent used and, consequently, variable results are commonly reported by researchers. In part, the lack of efficiency can be attributed to the inaccessibility of locations with structures and tissues that support microbial flora. Investigation of specific combinations of pathogens and produce is lacking.

The information presented in Table 12.3 is related to sanitizer solutions often applied to fruits and vegetables and their impact on safety and quality.

12.4 Emerging Technologies

There is significant interest in the development of new minimal nonthermal processing technologies that do not use severe heat-based methods. These processes are known to preserve quality parameters such as odor, visual appearance, color, texture, nutritive value, and absence of additives. At the same time, they have the added advantage of reduction of the microbial load (Kuldiloke 2002; Piyasena et al. 2003).

Emerging technologies such as ozonation, ultrasonication, and ultraviolet (UV) light treatment are applicable to an extensive variety of food products. Fresh fruits and vegetables have a short shelf life when compared with chilled ingredients and convenience foods (cooked meats and vegetables), which are usually ambient temperature stable foods with a long shelf life (Kuldiloke 2002). The sequential use of innovative preservation methods or their combination with a less intensive heat treatment or with traditional sanitizers, as a form of hurdle technology, is fast gaining interest.

Table 12.3 Overview of the impact of sanitizer solutions on quality and safety of fruits and vegetables

Product	Sanitizer agent/concentration	Safety quality indicator		Reference
Lettuce	Neutral electrolyzed oxidizing water (4.5 and 30 mg/L)	1.6–2.0 log reductions (total aerobic plate counts)	No impact on sensory quality, color, vitamin C, and total phenols; decrease of total antioxidant capacity (sodium hypochlorite, 200 mg/L)	Vandekinderen et al. (2009)
	Sodium hypochlorite (20 and 200 mg/L)	0.5–1.3 log reductions (total aerobic plate counts)		
	Peroxyacetic acid (80 and 250 mg/L)	1.0–2.4 log reductions (total aerobic plate counts)		
	Aqueous chlorine dioxide (10 mg/L)	1.55–3.96 log reductions (*Escherichia coli*)		Singh et al. (2002a)
	Thyme essential oil (0.1%)	1.93–4.05 log reductions (*Escherichia coli*)		
Lettuce and strawberries	Sodium hypochlorite (100, 200, and 300 mg/L) Packaged and stored under refrigerated conditions	2.1, 1.8, and 2.5 log reductions for lettuce (total mesophiles, psychrotrophs, and yeasts and molds, respectively)	Impact on firmness, browning; strawberry color not affected	Wei et al. (2007)
Romaine lettuce and baby carrots	Aqueous chlorine dioxide (5, 10, and 20 mg/L)	*Escherichia coli* O157:H7 0.97–1.72 log reductions for lettuce		Singh et al. (2002b)
	Thyme essential oil (0.1, 1.0, and 10.0 mg/L)	*Escherichia coli* O157:H7 0.43–2.41 log reductions for lettuce 0.64–2.77 log reductions for carrots		
Pear tomatoes and lettuce	Sodium hypochlorite (200 ppm) Stored at 4°C and 22°C	2.6 and 1.3 log reductions for pear tomatoes and lettuce, respectively (*Escherichia coli*) 4.8 and 2.1 log reductions for pear tomatoes and lettuce, respectively (*Yersinia enterocolitica*)		Velázquez et al. (2009)
	Benzalkonium chloride (0.1 mg/mL) Stored at 4°C and 22°C	2.1 and 0.4 log reductions for pear tomatoes and lettuce, respectively (*Escherichia coli*) 4.2 and 1.5 log reductions for pear tomatoes and lettuce, respectively (*Yersinia enterocolitica*)		

Fresh-cut escarole and lettuce	Lactic acid (0.2 and 1%) Stored at 4°C and 22°C	2.2 and 1.7 log reductions for pear tomatoes and lettuce, respectively (*Escherichia coli*) 5.1 and 2.4 log reductions for pear tomatoes and lettuce, respectively (*Yersinia enterocolitica*)	Retention of overall visual quality and firmness; off-flavors not detected (except for acid lactic treatment, 20 mL/L)	Allende et al. (2008)
	Sodium hypochlorite (100 mg/L) Stored at 5°C and 8°C	6.5, 6.6, and 3.9 log reductions for escarole (mesophiles, coliforms, and yeasts and molds, respectively) 6.4, 5.5, and 2.8 log reductions for lettuce (mesophiles, coliforms, and yeasts and molds, respectively)		
	Acidified sodium chlorite (500 and 250 mg/L) Stored at 5°C and 8°C	7.4, 7.5, and 4.7 log reductions for escarole (mesophiles, coliforms, and yeasts and molds, respectively) 6.2–7.1, 5.8–5.7, and 3.5–3.7 log reductions for lettuce (mesophiles, coliforms, and yeasts and molds, respectively)		
	Hydrogen peroxide and peroxyacetic acid (20 and 10 mL/L) Stored at 5°C and 8°C	7.7, 7.5, and 4.1 log reductions for escarole (mesophiles, coliforms and yeasts and molds, respectively) 7.1–6.8, 5.6–5.8 and 2.9–3.1 log reductions for lettuce (mesophiles, coliforms and yeasts and molds, respectively)		
	Peroxyacetic acid and hydrogen peroxide (80 μL/L) Stored at 5°C and 8°C	7.2, 6.9, and 4.8 log reductions for escarole (mesophiles, coliforms, and yeasts and molds, respectively)		

(continued)

Table 12.3 (continued)

Product	Sanitizer agent/concentration	Safety quality indicator	Reference
	Lactic acid (20 and 10 mL/L) Stored at 5°C and 8°C	7.6, 7.5, and 4.0 log reductions for escarole (mesophiles, coliforms, and yeasts and molds, respectively) 6.4–6.6, 5.8–5.9, and 3.4–3.8 log reductions for lettuce (mesophiles, coliforms, and yeasts and molds, respectively)	
	Organic acids and flavonoids (5 and 2.5 mL/L) Stored at 5°C and 8°C	7.8, 7.7, and 4.6 log reductions for escarole (mesophiles, coliforms, and yeasts and molds, respectively) 6.7–6.6, 5.8, and 3.0–3.8 log reductions for fresh-cut lettuce (mesophiles, coliforms, and yeasts and molds, respectively)	
	Lactoperoxidase, hydrogen peroxide, and thiocyanate (40 mg/L) Stored at 5°C and 8°C	7.8, 7.7, and 4.5 log reductions for escarole (mesophiles, coliforms, and yeasts and molds, respectively) 7.0, 5.7, and 3.4 log reductions for lettuce (mesophiles, coliforms, and yeasts and molds, respectively)	
Fresh-cut carrots	Sodium hypochlorite (200 ppm) Packaged and stored at 5°C	2–3 log reductions (*Escherichia coli*; *Salmonella* spp., and *Listeria monocytogenes*) and 0.18 log reductions (total aerobic counts)	Ruiz-Cruz et al. (2007)
	Peroxyacetic acid (40 ppm) Packaged and stored at 5°C	1.24, 2.1, 0.83, and <1 log reductions (*Escherichia coli*; *Salmonella* spp.; *Listeria monocytogenes*, and total aerobic counts)	
	Acidified sodium chlorite (100, 250, and 500 ppm) Packaged and stored at 5°C	4.81, 4.84, 2.5, and 0.86–2.3 log reductions (*Escherichia coli*; *Salmonella* spp., *Listeria monocytogenes*, and total aerobic counts)	

Produce	Treatment	Result	Reference
Apples, tomatoes, and lettuce	Chlorine (10, 50, and 100 μg/mL) Stored at 4 °C, 12 °C, and 25 °C	1.71–4.23, 1.64–3.70, and 1.61–5.46 log reductions for apples, tomatoes, and lettuce, respectively (*Enterobacter sakazakii*)	Kim et al. (2006)
	Chlorine dioxide (10, 50, and 100 μg/mL) Stored at 4 °C, 12 °C, and 25 °C	2.90–4.49, 2.30–3.70, and 1.45–5.46 log reductions for apples, tomatoes, and lettuce, respectively (*Enterobacter sakazakii*)	
	Peroxyacetic acid–base (40, and 80 μg/mL) Stored at 4 °C, 12 °C, and 25 °C	4–4.25, 2.98–3.70, and 2.26–5.46 log reductions for apples, tomatoes, and lettuce, respectively (*Enterobacter sakazakii*)	
Cantaloupe	Chlorine (200 ppm)	2.7, 0.38, 2.7, and 1.8 log reductions (total bacterial counts, *Pseudomonas* spp., yeasts and molds, and lactic acid bacteria, respectively)	Ukuku (2006)
	Hydrogen peroxide (2.5%)	2.6, 0.38, 2.7, and 1.5 log reductions (total bacterial counts, *Pseudomonas* spp., yeasts and molds, and lactic acid bacteria, respectively)	
	Free available chlorine (200, 500, and 1,000 mg/L)	2.29–2.84 and 0.45–0.72 log reductions (mesophilic aerobes and *Salmonella enteriditis*)	Bastos et al. (2005)
	Peracetic acid (60 mg/L)	Complete reduction of coliform group and fecal coliforms 1.15 log reductions (mesophilic aerobes)	
Cantaloupe and honeydew melons	Hydrogen peroxide (2.5%) Stored under refrigerated conditions	3 log reductions for both melons (*Escherichia coli* and *Listeria monocytogenes*) Populations of aerobic mesophilic bacteria and yeast and mold on both products were significantly reduced	Ukuku et al. (2005)
	Hydrogen peroxide (1%), nisin (2.5 μg/mL), sodium lactate (1%), and citric acid (0.5%) Stored under refrigerated conditions	4 log reduction of *Escherichia coli* in cantaloupe; complete reduction of *Escherichia coli* on honeydew melon and *Listeria monocytogenes* in both melons; populations of aerobic mesophilic bacteria and yeast and mold on honeydew were below the detection levels, and on cantaloupe they were significantly reduced	

(continued)

Table 12.3 (continued)

Product	Sanitizer agent/concentration	Safety quality indicator		Reference
	Hydrogen peroxide (2.5 and 5%) Stored under refrigerated conditions	3 log reductions for both melons (aerobic mesophilic bacteria); 1.73–2.56 and ~1.32 to ~3.00 log reductions for cantaloupe and honeydew melons (*Salmonella* spp.)	No impact on overall acceptability of both melons; appearance maintained in honeydew but not in cantaloupe; positive effect on acceptability of fresh-cut melon	Ukuku (2004)
Shredded lettuce and diced tomatoes	Chlorine (120 and 200 μg/mL) Stored at 4°C, 21°C, and 30°C	1.07–1.11 and 1.13–1.17 log reductions on lettuce and tomatoes (*Salmonella baildon*)		Weissinger et al. (2000)
Lettuce and broccoli	Chlorine (50 and 100 mg/L)	1.3–2.8 and 1.7–2.5 log reductions for lettuce and broccoli (*Escherichia coli*)		Behrsing et al. (2000)
Shredded carrots	Chlorine (50 and 200 ppm)	1 log reduction (total aerobic plate counts and yeasts and molds)	Impact on soluble solid content, color, and impairment of aroma perception; low changes in pH	Alegria et al. (2009)
Blueberries	Aqueous chlorine dioxide (1–15 ppm)	4.88, 4.48, 3.32, 4.56, 3.54, and 2.86 log reductions (*Listeria monocytogenes*, *Pseudomonas aeruginosa*, *Salmonella typhimurium*, *Staphylococcus aureus*, *Yersinia enterocolitica*, and yeasts and molds, respectively)		Wu and Kim (2007)
Green bell pepper	Aqueous chlorine dioxide (3 mg/L)	3.7 log reductions (*Listeria monocytogenes*)		Han et al. (2001)
Apples	Aqueous chlorine dioxide (5–40 ppm)	3.1–4.2 and 2.2–3.9 log reductions (*Salmonella* and *Escherichia coli*, respectively)		Huang et al. (2006)
Mungbean sprouts	Aqueous chlorine dioxide (100 ppm) Stored under refrigerated conditions	0.66, 2.96, and 1.45 log reductions (total mesophilic microorganisms, *Salmonella* typhimurium, and *Listeria monocytogenes*, respectively)		Jin and Lee (2007)
Tomatoes	Aqueous chlorine dioxide (5–20 ppm)	5 log reduction (*Salmonella enterica* and *Erwinia carotovora*)		Pao et al. (2007)

12.4.1 Ozone

12.4.1.1 Properties and Microbial Effects

Ozone (O_3) is a powerful oxidizing agent formed in the stratosphere by sunlight (at a wavelength of 185 nm). In nature, ozone is present at very low concentrations which do not satisfy industrial/commercial requirements.

Commercially, this molecule is obtained by exposing oxygen molecules to corona discharges and to UV light; both procedures are similar to what happens in nature (Sharma 2005). The corona discharge is achieved by passing the feed gas between two closely spaced electrodes. When the feed gas (such as dry air, oxygen, or a gaseous mixture) passes between the electrodes, the gas is partially ionized and subsequent atomic collisions produce ozone. The process of ozone generation by UV light is similar. The isolated oxygen atoms are formed by the photodissociation of oxygen caused by short-wavelength UV radiation, around 240 nm. Similarly, oxygen atoms and molecules interact to form ozone (Fig. 12.1).

Ozone can also be produced by water electrolysis, but the cost is very high compared with that for corona discharge (Graham 2000). Electrolysis is predominantly used by the pharmaceutical and electronic industries.

Ozone gas has a blue tint at room temperature; at higher concentrations the color is not noticeable. It has a strong characteristic smell that many authors have described as

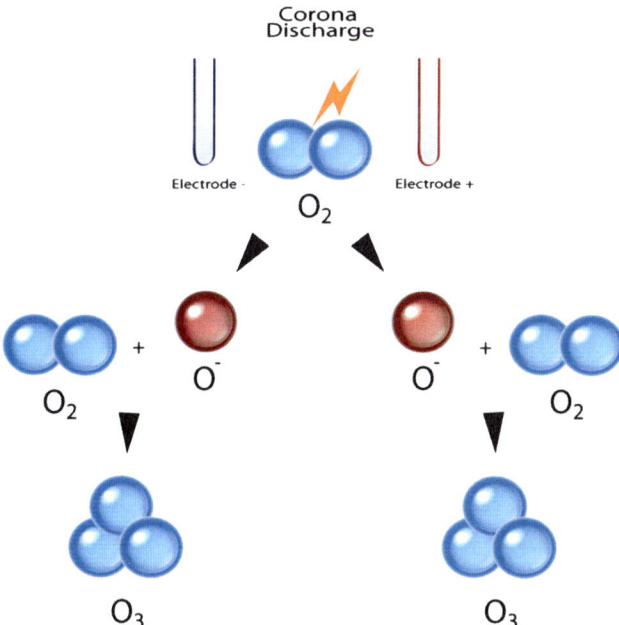

Fig. 12.1 Ozone formation by corona discharge

that of "fresh air after a thunderstorm." Human olfactory senses are able to detect this odor, even when ozone is present at a very low concentration (below 0.01 ppm).

Ozone is thermally unstable and dissociates in a very short time, forming oxygen or reacting with other gases (Butz and Tauscher 2002; Güzel-Seydim et al. 2004a). Ozone is more soluble in water than oxygen, with a half-life in distilled water of 20–30 min before breaking down to form oxygen. However, it is difficult to maintain ozone dissolved in water, especially if temperatures are high. Consequently, to maintain or increase its concentration in aqueous solution, it is necessary to have continuous ozone production and low temperatures (Güzel-Seydim et al. 2004b; Graham 2000). Ozone is considered as a valuable disinfection agent because of its high oxidation potential (of 2.07 mV), with only fluorine having a greater potential (3.06 mV). The efficiency of ozone as an oxidizing agent depends greatly on the balance between ozone concentration and organic matter content (Sharma 2005). In aqueous solution, ozone is slightly more unstable than it is in the gas phase.

Ozone is effective against some Gram-positive bacteria such as *Listeria monocytogenes*, *Staphylococcus aureus*, *Bacillus cereus*, and *Enterococcus faecalis*, as well as against some Gram-negative bacteria such as *Pseudomonas aeruginosa* and *Yersinia enterocolitica*, some yeasts such as *Candida albicans* and *Zygosaccharomyces bacilli*, and some molds such as *Aspergillus niger* (Restaino et al. 1995; Güzel-Seydim et al. 2004b). Ozone's ability to inactivate such a spectrum of microorganisms is due to its strong oxidation power. Figure 12.2 shows the way ozone approaches/destroys the bacterial cell and subsequent cell lysis.

Ozone promotes progressive oxidation of vital cellular components; the cell surface is the initial ozone target. Ozone has the capacity to oxidize the double bonds of unsaturated lipids, resulting in cell disruption and subsequent leakage of cellular contents (Güzel-Seydim et al. 2004b). Kim et al. (1999) reported that lipoprotein and lipopolysaccharide layers of Gram-negative bacteria are the first sites destroyed by ozone, and consequently the cell permeability is very strongly affected, resulting

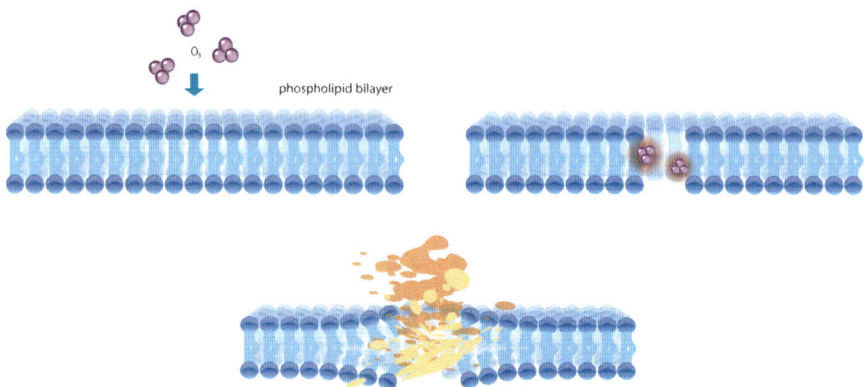

Fig. 12.2 Effect of ozone on the bacterial cell

in cell lysis. The same authors also reported that nucleic acids are significantly damaged or even destroyed by ozone, as is viral RNA, and that ozone alters polypeptide chains in viral protein coats.

12.4.1.2 Applications to Fruits and Vegetables

Ozone was first used as a disinfectant for municipal waters, process waters, bottled drinking waters, and swimming pools. Later, its application was extended wastewater, dairy and swine effluent, cooling towers, hospital water systems and equipment, aquariums and aquaculture, water theme parks, and public spas and spas in the home (Smilanick 2003).

Some applications in the food industry include food preservation, surface hygienization, sanitation, water disinfection, and wastewater reutilization (Sheldon and Brown 1986; Schneider et al. 1991). Even at low ozone concentrations and/or contact times, ozone can be effective in the inactivation of pure cultures of bacteria, molds, yeast, parasites, and viruses in solution. However, the success of the food sanitation process is conditioned by the composition of the food surface, the type of microbial contaminant, and the degree of attachment to or association of microorganisms with food. In such situations, the food itself may compete with the microorganisms for ozone and may limit the accessibility of ozone to the microbial food contaminants (Kim et al. 1999; Sharma 2005). Ozone reacts with a large variety of organic compounds at very different rates, so there are many oxidation by-products that are formed during the ozonation process. However, those by-products and residues are not potentially harmful and, consequently, the ozonation process is considered safe.

Ozone, when dissolved in water, has been tested in postharvest treatments of fruits and vegetables. However, it has greater efficacy when in the gaseous form, associated with refrigerated storage and modified-atmosphere packaging. When ozone is used in water, its concentrations range from 0.03 to 20.0 ppm. When it is used in the gaseous form, the concentration reaches highly elevated doses such as 20,000 ppm (Selma et al. 2008b). Since ozone decomposition is rapid in water, its antimicrobial action may take place only at food surfaces (Aguayo et al. 2006).

The information included in Table 12.4 provides an overview of the impact of ozone on the quality and safety features of fruits and vegetables.

12.4.2 Ultrasound

12.4.2.1 Properties and Microbial Effects

Ultrasound is defined as pressure waves with a frequency of 20 kHz or more, which is a frequency greater than the upper limit of human hearing (Butz and Tauscher 2002). Ultrasound technology has been studied for several years and can be classified as

Table 12.4 Overview of impact of ozone on quality and safety of fruits and vegetables

Product	Processing/storage conditions	Quality/safety indicators	References
Fresh-cut celery	Ozone in water (0.03, 0.08, and 0.18 ppm) Storage under refrigerated conditions	Vitamin C, total sugar, and sensory retention; decrease of enzyme activity 1.69 log reduction (total microbial counts)	Zhang et al. (2005)
Fresh-cut lettuce	Ozone in water (0.5–4.5 ppm) Storage under refrigerated conditions	Vitamin C and β-carotene retention; decrease of enzyme activity; better scores for all sensory parameters 1.5, 1.1, and 1.5 log reduction (aerobic mesophilic bacteria, psychrotrophic bacteria, and *Enterobacteriaceae*, respectively)	Ölmez and Akbas (2009)
	Ozone in water (10 and 20 mg/L) Modified-atmosphere packaging or in air and storage under refrigerated conditions	No impact on respiratory activity, visual appearance, and final phenolic content 1.6 and 3.2 log reduction (aerobic mesophilic and coliforms counts, respectively)	Beltran et al. (2005)
Iceberg lettuce	Ozone in water (4 mg/L)	No impact on color, vitamin C, β-carotene, texture, and moisture content 1.7 and 1.5 log reduction (aerobic mesophilic and psychrotrophic bacteria, respectively)	Akbas and Olmez (2007)
	Ozone in water(0.5–16.5 ppm)	1.8 log reduction (*Shigella sonnei*)	Selma et al. (2007)
	Ozone in water (1 mg/L) Modified-atmosphere packaging and storage under refrigerated conditions	No impact on color, texture, visual appearance, aroma, off-flavor, and off-odor Reductions of aerobic mesophiles were similar to those from water washing but *Enterobacteriaceae* and *Pseudomonades* were strongly reduced	Baur et al. (2004)
Enoki mushrooms	Ozone in water (1, 3, and 5 ppm) Storage at 15°C	<1.0 and 0.5 log reduction (*Escherichia coli* and *Listeria monocytogenes*, respectively)	Yuk et al. (2007)

(continued)

Table 12.4 (continued)

Product	Processing/storage conditions	Quality/safety indicators	References
Peaches and table grapes	Gaseous ozone (0.3 ± 0.05 ppm) Storage under refrigerated conditions	No impact on respiration and ethylene production rates of peaches, incidence and severity of decay, and phytotoxic injuries of fruit tissues; inhibition of aerial mycelial growth and sporulation on peaches and gray mold nesting among grapes; water loss increased in peaches but not in grapes	Palou et al. (2002)
Fresh, prepackaged whole spinach leaves	Gaseous ozone (1.6 and 4.3 mg/L) Storage under refrigerated and nonrefrigerated conditions	No impact on relative humidity in air storage condition but changes for oxygen stored conditions; discoloration of spinach leafs 3–5 log reduction/leaf (*Escherichia coli*)	Klockow and Keener (2009)
Fresh-cut green asparagus	Ozone in water (1 mg/L) Modified-atmosphere packaging and storage under refrigerated conditions	Decrease of enzyme activity; increase of lignin, cellulose and hemicellulose contents of cell wall	An et al. (2007)
Precut or postcut carrots	Ozone in water (1 mg/L)	Losses of soluble solid content, color changes and impairment of aroma perception > 0.4 and 0.6–0.7 log reductions (total aerobic plate and yeasts/molds, respectively)	Alegria et al. (2009)
Date fruits	Gaseous ozone (1, 3 and 5 ppm)	0.26–0.56, 2.65–3.10, 2.67–3.11, and 0.13–0.43 log reductions (total bacterial count, coliforms, *Staphylococcus aureus*, and yeasts/molds, respectively)	Najafi and Khodaparast (2009)
Dried figs	Gaseous ozone (0.1–9.0 ppm)	No impact on color, pH, moisture content, sweetness, rancidity, flavor, appearance, and overall palatability 0.9–3.5, 2.7–3.5, and 1.0–2.0 log reductions (*Escherichia coli, Bacillus cereus*, and *Bacillus cereus* spores, respectively)	Akbas and Ozdemir (2008)

(continued)

Table 12.4 (continued)

Product	Processing/storage conditions	Quality/safety indicators	References
	Ozone in water (1.7 mg/L) Gaseous ozone (13.8 mg/L)	Both forms affect aflatoxin B_1 degradation Aqueous: 1.49–2.42, 1.33, and 1.73 log reductions (aerobic mesophilic bacteria, total yeasts, and total molds, respectively); *Escherichia coli* and coliforms not detected Gaseous: 0.81–1.42, 0.46–1.84, 0.16–2.09, and 0.59 log reductions (aerobic mesophilic bacteria, coliforms, total yeasts, and total molds, respectively); *Escherichia coli* not detected	Zorlugenç et al. (2008)
	Gaseous ozone (1, 5 and 10 ppm)	1.59–2.51, 0.0–0.39, and 0.40–1.30 log reductions (total aerobic mesophilic bacteria, coliforms, and yeast/molds, respectively); *Escherichia coli* not detected	Öztekin et al. (2006)
Fresh-cut cilantro	Ozone in water Modified-atmosphere packaging and storage under refrigerated conditions	Slight decrease of firmness; no influence on color, off-odor, and aroma; overall quality retention 1–1.5 log reduction (total aerobic plate); total *Enterobacteriaceae* increase after 4 days of storage	Wang et al. (2004)
Cherry tomatoes	Gaseous ozone (5, 10, 20, and 30 mg/L) Modified-atmosphere packaging and storage under refrigerated or nonrefrigerated conditions	Color changes for higher concentrations; texture was maintained Complete reduction of *Salmonella enteritidis* The higher the ozone concentration, the less contact time required	Das et al. (2006)

(continued)

Table 12.4 (continued)

Product	Processing/storage conditions	Quality/safety indicators	References
Whole and sliced tomatoes	Gaseous ozone (4 ± 0.5 µL/L) Storage under refrigerated conditions	Decrease of respiration rate during storage; decrease of metabolic activity; increase of contents of sugars and organic acids; tissue firmness maintained; retention of good appearance and overall quality in slices but aroma decayed 1.1–1.2 and 0.5 log reductions (mesophilic and psychrotrophic bacteria and yeast/molds, respectively)	Aguayo et al. (2006)
Longan fruit	Gaseous ozone (200 µL/L) Storage at 25°C	Reduced growth of *Lasidiplodia* sp. and completely inhibited growth of *Cladosporium* sp.	Whangchai et al. (2006)
Blackberries	Gaseous ozone (0.1 and 0.3 ppm) Storage under refrigerated conditions	No impact on total anthocyanin content, color retention; decrease of enzyme activity Fungal decay was not observed	Barth et al. (1995)
Grape berries	Gaseous ozone (16 mg/L)	Induced resistance to postharvest decay; no deleterious effects on appearance; no phytotoxicity; firmness was not adversely affected Total elimination of fungi, yeasts, and bacteria (after 20 min)	Sarig et al. (1996)
Rocket leaves	Ozone in water (10 mg/L) Modified atmosphere packaging and storage under refrigerated conditions	No impact on visual quality, vitamin C, phenolic, and flavonoids contents ~1, ~0.5, and ~0.5 log reductions (total mesophilic bacteria, total coliforms, and yeasts and molds)	Sánchez et al. (2006)
Cantaloupe melon	Gaseous ozone (5,000, 10,000, and 20,000 ppm) Storage under refrigerated conditions	No impact on the visual quality, aroma, firmness, and soluble solid content; increase in translucency and decrease of lightness 1.6, 1.6, 0.7, and 1.1 log reductions/(28.2 ± 5.7g) (total coliforms, *Pseudomonas fluorescens*, yeasts, and lactic acid bacteria, respectively)	Selma et al. (2008b)

(continued)

Table 12.4 (continued)

Product	Processing/storage conditions	Quality/safety indicators	References
	Gaseous ozone (10,000 ppm)	No impact on visual quality, aroma, translucency, firmness, or color 1.1, 1.3, 1.5, and 1.3 log reductions (aerobic mesophilic bacteria, psychrotrophic bacteria, molds, and total coliforms, respectively)	Selma et al. (2008a)
Precut or whole green peppers	Gaseous ozone (2–8 mg/L)	0.69–7.18 log reductions (*Escherichia coli*)	Han et al. (2002)
	Ozone in water (0.30–3.95 ppm) Storage under refrigerated conditions	0.72 log reduction (aerobic plate counts)	Ketteringham et al. (2006)
Strawberries	Gaseous ozone (0.35 ppm) Storage under refrigerated conditions followed by storage at 20°C	Impact on contents of sugars, anthocyanin, and ascorbic acid and aroma Prevention of fungal growth	Pérez et al. (1999)
Fresh strawberries and shredded lettuce	Gaseous ozone (1–10 mg/L) Storage under refrigerated conditions	Little impact on firmness; increase of browning in lettuce; no colur changes in strawberries 0.79–1.28, 1.17–1.51, and 0.99–0.78 log reductions on lettuce–strawberry (mesophiles, psychrotrophs, yeasts, and molds, respectively)	Wei et al. (2007)
Tomatoes, strawberries, table grapes, plums, and clementines	Gaseous ozone (0.005–5.0 µmol/mol ozone) Storage at 13°C	Marked reduction of molds (*Botrytis cinerea*)	Tzortzakis et al. (2007)
Shredded lettuce and baby carrots	Ozone in water (5.2, 9.7 and 16.5 mg/L) Gaseous ozone (2.1, 5.2 and 7.6 mg/L)	Decolorization of lettuce leaves for higher concentrations Water: 5.97–7.32 log reductions (*Escherichia coli*) Gaseous: 5.16–7.21 log reductions (*Escherichia coli*)	Singh et al. (2002b)
Apples, lettuce, strawberries, and cantaloupe	Ozone in water (3 ppm) Storage under refrigerated conditions	~5.6 log reductions (*Escherichia coli* and *Listeria monocytogenes*) No influence on *Escherichia coli* and *Listeria monocytogenes*; increase in mesophilic bacteria and mold and yeast counts	Rodgers et al. (2004)

being of low and high energy (high and low frequency, respectively), having different applications on the basis of this criterion.

Ultrasound is known to cause chemical and physical changes in biological structures (in a liquid medium), owing to the rapid formation and destruction of cavitation bubbles (Fig. 12.3).

When ultrasound waves pass through a liquid, a continuous wave type of motion is created. Longitudinal waves are propagated via a series of compression and rarefaction waves, which are induced in the medium particles. If the wave has sufficiently high amplitude and gas bubbles or cavities are produced in the medium (cavitation process). Growth of these bubbles produces a phenomenon called rectified diffusion, i.e., during the expansion cycle gas diffusion causes bubble expan-

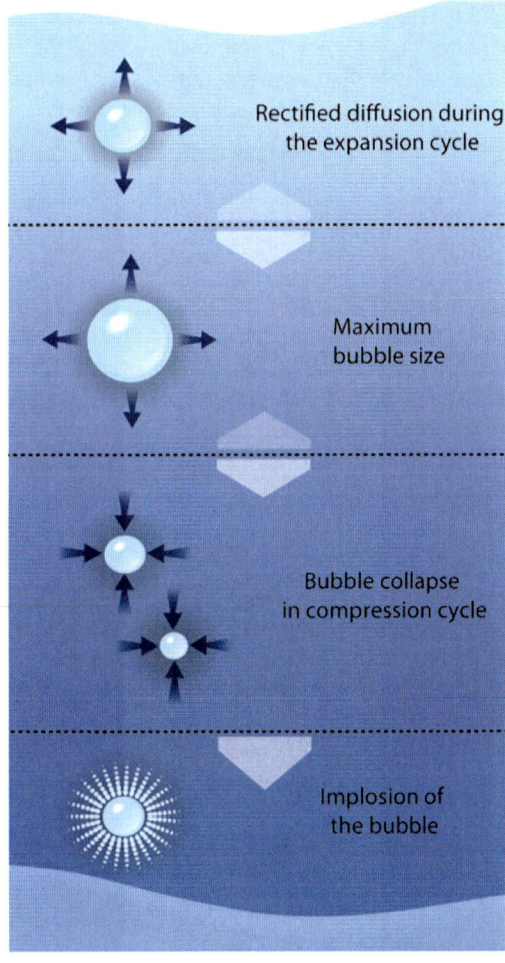

Fig. 12.3 Ultrasonic cavitation and implosion

sion. That gas taken in during the expansion is not completely expelled during the compression cycle and, consequently, the bubbles grow in size. After some compression cycles, rapid condensation occurs, because the ultrasonic energy is not sufficient to retain the vapor phase in the bubble. The condensed bubble implodes violently, creating shock waves and generating the energy for chemical and mechanical effects. In a liquid medium, the collapse of bubbles using a frequency of 20 kHz creates regions of very high temperature, around 5,500°C, and pressure, about 50,000 kPa (Mermelstein 2000; Kuldiloke 2002; Ohlsson 2002).

The destructive potential of ultrasound was discovered in the 1960s when it was observed that sound waves used in antisubmarine warfare killed fish. Since then, studies on how ultrasound disrupts biological structures have been conducted, and numerous theories and mechanisms concerning ultrasound's harmful effects on cells have been proposed (Earnshaw et al. 1995; Piyasena et al. 2003). Intracellular cavitations are responsible for this phenomenon (as previously described) and the pressure changes are the main cause for the bactericidal effect of the ultrasound technology. The micromechanical shocks rapidly form bubbles (that collapse equally fast), which promotes the disruption of cellular structures and functional components, thus causing cellular lysis (Anonymous 2000; Butz and Tauscher 2002; Ross et al. 2003). The effectiveness of ultrasound treatment on microorganisms depends on the type of bacteria and on the treatment time and temperature.

12.4.2.2 Applications to Fruits and Vegetables

Ultrasonication is an innovative technology in food processing that is believed to have potential but still requires further investigation and optimization for use in the food industry application to fruits and vegetables, which have acoustical properties significantly different from those of biological and industrial materials. One of the most important limiting factors is the dispersion of void and pores, which influence the monitoring and interpretation of ultrasound data in those products. However, the development of sensors, microprocessors, and methods of signal analysis are facilitating the use of this nonthermal, nondestructive, and environmentally friendly technology in measurements of fresh products. At the industrial level, it is necessary to study the frequency and power requirements for different fresh products and to choose the best equipment and measurement tests (Mizrach 2008).

Nevertheless, the use of ultrasound technology has increased in the food industry (Knorr et al. 2002). Low-intensity ultrasound, at intensities lower than 1 W/m^2 and a frequency higher than 100 kHz, is used for noninvasive detection (food control), providing information about physical-chemical properties of food materials, and is used in relaxation phenomena studies (Kuldiloke 2002; Knorr et al. 2004). High-intensity ultrasound, at intensities higher than 1 W/m^2 and frequencies between 18 and 100 kHz, is regularly used during physical and chemical changes of food properties, such as extraction, emulsification, homogenization, separation, extrusion, cell disruption, degassing of liquid food, viscosity alteration, chemical reaction promotion (oxidation/reduction), enzyme and protein extraction, meat softening, and modification of crystallization processes (McClements 1995; Villamiel and de Jong 2000; Knorr et al. 2004; Zheng and Sun 2006; Patist and Bates 2008).

Higher-power ultrasound at lower frequencies, associated with heat and pressure, has been reported to successfully inactivate heat-resistant enzymes and cause cavitation and subsequent inactivation of microorganisms (Vercet et al. 1997; Piyasena et al. 2003; Knorr et al. 2004).

The complexity of foods, and their potential protective effect on microorganisms, makes ultrasound treatment ineffective as a preservation technique. However, there are several processes that can be combined with ultrasound treatment: (1) heat and sonication (thermosonication), (2) pressure and sonication (manosonication), and (3) pressure, heat, and sonication (manothermosonication). The combination of such processing technologies may cause additive, antagonist, or synergistic effects, depending on the type of combination and on the safety or quality standards analyzed. In fruits and vegetables, these hurdle combinations are considered improved applications, as positive changes in quality characteristics are attained when compared with results from conventional thermal processes. The amplitude of ultrasonic waves, the contact time, the nature of the microorganism/food, the volume of food to be processed, the food composition, and the treatment temperature are examples of some critical processing factors of concern (Anonymous 2000; Butz and Tauscher 2002; Piyasena et al. 2003; López-Malo et al. 2005). Additionally, ultrasound is responsible for the production of free radicals and localized heating (Earnshaw et al. 1995).

An overview of the impact of ultrasound processing applied to fruits and vegetables is given in Table 12.5.

12.4.3 Ultraviolet Light

12.4.3.1 Properties and Microbial Effects

The electromagnetic spectrum consists of different types of radiation that are classified according to their wavelengths. UV radiation covers a wide range of wavelengths in the nonionizing region of the electromagnetic spectrum, with wavelengths between 100 and 400 nm (Fig. 12.4). UV radiation is categorized as UV-A (315–400 nm), UV-B (280–315 nm), UV-C (200–280 nm) and vacuum UV (100–200 nm). UV-A causes indirect damage to DNA, proteins, and lipids through reactive oxygen intermediates; these are responsible for the changes in human skin that lead to tanning. UV-B and UV-C cause indirect and direct damage to DNA because they are strongly absorbed at wavelengths below 320 nm. UV-B can cause skin burning and eventually lead to skin cancer. UV-C is known to be germicidal; it inactivates bacteria, viruses, protozoa, yeasts, molds, and algae (Anonymous 2000; Hollósy 2002; Matak 2004; Zenoff et al. 2006; Keyser et al. 2008).

The germicidal effect of radiation is mainly due to the photochemical reactions that are induced inside the microorganisms. The major lesion in the microorganisms is due to DNA absorption of UV light (254 nm is the most lethal), taking place by

Table 12.5 Overview of impact of ultrasound on quality and safety of fruits and vegetables

Product	Processing conditions	Quality/safety indicators	References
Precut or postcut carrots	Ultrasonication (45 kHz) and thermosonication (45 kHz and 50°C)	Significant soluble solid content loses, color changes, pH changes did not exceed 0.2 pH units, and impairment of aroma perception	Alegria et al. (2009)
		0.5–1.7 and 0.5–1.2 log reduction (total aerobic plate counts and yeast and molds, respectively)	
Watercress	Thermosonication (20 kHz and 82.5–92.5°C)	No significant differences in color or chlorophyll content	Cruz et al. (2007)
	Thermosonication (20 kHz and 40–92.5°C)	Peroxidase became more heat labile as temperature increased	Cruz et al. (2006)
Strawberries	Sonication (25, 28, 40, or 59 kHz)	Retention of firmness, total soluble solids, total titratable acidity and vitamin C; decay reduction	Cao et al. (2010)
	Storage under refrigerated conditions	0.88 and 1.1 log reductions (aerobic microorganisms and mold and yeast)	
Brine bell pepper	Thermosonication (47 kHz and 25–55°C)	Negative effect on firmness; uptake of solutes increased with exception of Ca^{2+}, Na^+ ions, and acidity; water losses; increase of total and soluble solids	Gabaldón-Leyva et al. (2007)

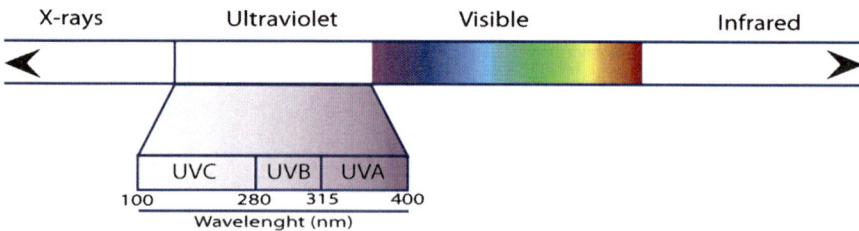

Fig. 12.4 Electromagnetic spectrum

cross-linking pyrimidine nucleoside bases (thymine and cytosine) in the same DNA strand. The formation of thymine dimers in DNA and RNA will compromise the cellular functions, since microorganisms are unable to perform the normal transcription and replication of the nucleic acids, and eventually cell death occurs (Giese and Darby 2000; Ohlsson 2002; Wang et al. 2005; Sharifi-Yazdi and Darghahi 2006; Altic et al. 2007; Unluturk et al. 2008). The germicidal effect of UV-C light has been demonstrated to be effective against some common foodborne photogenes such as

Campylobacter, *Salmonella*, and *Escherichia coli* (Yaun 2002). However, several microorganisms possess repair mechanisms to counteract the effects of UV light exposure. Microorganisms can repair their DNA by light-independent (dark repair) as well as light-dependent (photo reactivation) mechanisms. Dark repair mechanisms occur through nucleotide excision repair, postreplication recombinational repair, or error-prone repair. The induction of those mechanisms is dependent on DNA damage. The photoreactivation can be produced by inactivating microorganisms that recover their activity by repairing pyrimidine dimers in the DNA. This phenomenon is light-dependent because it uses a photolyase enzyme that can be activated in the presence of different wavelengths, such as UV-A radiation and photosynthetically active radiation (400–700 nm) (Yaun 2002; Matak 2004; Zenoff et al. 2006; Sanz et al. 2007).

12.4.3.2 Applications to Fruits and Vegetables

Owing to its strong germicidal effect, UV-C radiation has a great potential in food processing. Industrially, two types of UV lamps are of main interest: (1) the usual low-pressure mercury arc lamps with a monochromatic output at 254 nm (UV-C range), and (2) the high-intensity and medium-pressure mercury arc lamps with a polychromatic output at all UV wavelengths (however concentrated at certain peaks within the UV-C region). Lamps that emit polychromatic light continuously have medium light intensity. For high light intensity, UV light pulses are often used (Giese and Darby 2000; Sassi et al. 2005).

UV-C light is a promising alternative to the traditional sanitizers and thermal treatments. Industrially, it has been used in the disinfection of drinking water, wastewater, and air, as well as in the disinfection of surfaces and food containers (Marquenie et al. 2002; Ohlsson 2002; Matak 2004; Wang et al. 2005; Altic et al. 2007). Some juices, fruits and vegetables, meat, and fish can be processed by UV light with the aim of microbial load reduction (Guerrero-Beltrán and Barbosa-Cánovas 2004; Unluturk et al. 2008). However, water treatment and surface sterilization are the fields where UV-C radiation has been applied with greatest success owing to the difficulty of achieving light deep penetration beyond the opaque surface of several substances (Ohlsson 2002). This may limit the application of UV radiation to solid foods.

The critical parameters to be considered when designing a convenient UV radiation process for reduction of microbial load of foods, extension of shelf life, and maintenance of satisfactory sensory and nutritional quality are the type of food, the type of food surface, the food composition, the intensity of radiation, and the exposure time (Anonymous 2000; Yaun 2002; Guerrero-Beltrán and Barbosa-Cánovas 2004).

The impact of UV radiation on the quality and safety characteristics of fruits and vegetables is shown in Table 12.6. Exposure to low doses of UV-C has been reported to reduce postharvest decay of fruits and vegetables (Erkan et al. 2001; Marquenie et al. 2002; Allende and Artés 2003a; Allende and Artés 2003b), which augurs well

Table 12.6 Overview of impact of UV-C radiation on quality and safety of fruits and vegetables

Product	Processing conditions	Quality/safety indicators	References
Red oak leaf lettuce	UV-C light (1.18, 2.37, and 7.11 kJ/m^2) Modified-atmosphere packaging and storage under refrigerated conditions	Shelf-life extension; higher dose induced tissue softening and browning after 7 days of storage at 5°C Reduction of natural microflora (highest doses showed the greatest microbial inhibitions)	Allende et al. (2006)
Fresh processed Lollo Rosso lettuce	UV-C light (0.4–8.14 kJ/m^2) Package stored under refrigerated conditions	Increase of respiration rate and tissue brightness; decrease of browning (for the highest doses) Reduction of psychrotrophic bacteria, coliforms, and yeasts; growth stimulation of lactic acid bacteria	Allende and Artés (2003b)
Pomegranate arils	UV-C light (0.56–13.62 kJ/m^2) Modified-atmosphere packaging and storage under refrigerated conditions	No impact on respiration rate, total anthocyanin, content and antioxidant activity Reduction of mesophilic, psychrotrophic, and lactic acid bacteria and *Enterobacteriaceae*; yeasts and molds were unaffected	López-Rubira et al. (2005)
Fresh-cut watermelon	UV-C light (1.6–7.2 kJ/m^2) Package stored under refrigerated conditions	Higher UV-C doses induced slightly higher CO_2 production; impact on color parameters; lycopene content was maintained (2.8 kJ/m^2) or it was slightly decreased (1.6 kJ/m^2); no impact on vitamin C; catalase activity and total polyphenols content declined; total antioxidant capacity increased Reduction of mesophilic bacteria, psychrophilic bacteria, and enterobacteria	Hernández et al. (2010)
	UV-C light (1.4–13.7 kJ/m^2)	No impact on juice leakage, color, and overall visual quality >1 log reduction (aerobic plate count)	Fonseca and Rushing (2006)
Pepper	UV-C light (7 kJ/m^2) Storage under refrigerated conditions	No impact on firmness, pH, sugar content, fruit lightness or superficial color; lower carotenoid content; antioxidant capacity increased; reduced decay	Vicente et al. (2005)
Golden papaya	UV-C light (0.2–2.4 kJ/m^2) Storage under controlled atmosphere	Scald of the fruit Inhibition of conidial germination; reduced fungal sporulation; higher doses inhibited mycelial growth	Cia et al. (2007)

(continued)

Table 12.6 (continued)

Product	Processing conditions	Quality/safety indicators	References
Strawberry	UV-C light (0.25 and 1.0 kJ/m^2) Storage under refrigerated conditions	Shelf-life extension; lower respiration rate; higher titratable acidity and anthocyanin content; firmness and free sugars increased; lower electrical conductivity for 0.25 kJ/m^2 (slower rate of senescence) Controlled decay caused by *Botrytis cinerea*	Baka et al. (1999)
	UV-C light (1 kJ/m^2) Storage under refrigerated conditions	Reduction on phenolic and vitamin C contents; no impact on overall visual quality Promoted psychrophilic growth	Allende et al. (2007)
	UV-C light (4.1 kJ/m^2) Storage at 20°C	Reduced decay, softening, and reddening; steady increase in anthocyanin concentration; no impact on total sugar content and titratable acidity; reduction of total phenolic compounds; pH increased Fungal growth was retarded	Pan et al. (2004)
Peach	UV-C light (3.71 and 11,15 kJ/m^2) Storage under room conditions	Impact on total solid soluble content, total titratable acidity, and firmness; no impact on pH	Coutinho et al. (2003)
	UV-C light (0.84–40 kJ/m^2) Storage in the dark at room temperature	Low UV-C doses induced host resistance by controlling latent brown rot infection; photoreversion with visible light resulted in reduction of host resistance to brown rot; increase of phenylalanine ammonia-lyase activity, delayed ripening and suppressed ethylene production Survey of yeast populations; reduction of *Monilinia fructicola* surface lesions	Stevens et al. (1998)
Zucchini squash	UV-C light(0.493–9.86 kJ/m^2) Storage under refrigerated conditions	No impact on the degree of chilling injury, ethylene production, and sugar or malic acid concentrations; higher respiration rates; reddish-brown discoloration for storage at 10°C Retarded microbial growth (mesophilic bacteria and yeast and fungal populations)	Erkan et al. (2001)

(continued)

Table 12.6 (continued)

Product	Processing conditions	Quality/safety indicators	References
Strawberry and cherries	UV-C light (0.05–1.50 J/cm^2)	No impact on firmness of both products, browning and drying of strawberry leaves Fungal growth on strawberries was retarded, but was not retarded in cherries	Marquenie et al. (2002)
Red delicious apples, leaf lettuce, and tomatoes	UV-C light (1.5–24 mW/cm^2)	3.3 and 2.79 log reduction (*Escherichia coli*) on apples and green leaf lettuce 2.19 and 2.65 log reduction (*Salmonella* spp.) on tomatoes and green leaf lettuce	Yaun et al. (2004)
Peaches, sweet potato, tomatoes, and tangerines	UV-C light (1.3–7.5 kJ/m^2)	Reduced the incidence of storage and brown rot of peaches, *Rhizopus* soft rot of tomatoes and sweet potatoes, and green mold of tangerines	Stevens et al. (1997)
Fresh-cut honey, pineapple, *pisang mas* banana, and guava	UV-C light (2.158 J/m^2)	Increase of total phenol (except for pineapple) and flavonoid contents; reduction of vitamin C content; ferric reducing/antioxidant power and free-radical-scavenging activity increase (except for pineapple)	Alothman et al. (2009)

for potential application of UV-C radiation in the postharvest industry. However, low doses may not be sufficient to inactivate some viruses and spores. Reported intensities for efficient UV treatments are around 400 J/m^2 (Anonymous 2000), and it is essential that the product surface is totally exposed to radiation.

Some examples of positive aspects related to the application of UV-C radiation to foods are the absence of nontoxic by-products, removal of some organic contaminants, and low energy requirements (when compared with thermal treatments). However, under particular circumstances, microorganisms may repair injury caused by UV-C light exposure and the safety of the products may be seriously compromised (Solomon et al. 1998; Keyser et al. 2008)

12.4.4 Combination of Processes

Ozone, ultrasound, and UV-C light can be applied individually, with good results in terms of safety and quality of fruits and vegetables. However, the combination of those innovative technologies with a mild thermal treatment, high pressure, or even the traditional disinfectant solutions is promising. The intensity of each technology can be reduced by combining two or more treatments, attaining target microbial inactivation without compromising quality standards.

The application of nonthermal processes may, however, have an induced resistance effect on the microorganisms that survive the treatment. Rames et al. (1997) reported that when UV-C light was applied to *Escherichia coli* in suspension, the surviving bacteria became more resistant to the radiation. The cumulative damage of DNA in microorganisms can lead to a reduction of the number of survivors, but may not result in a complete sterilization. UV-C treatment can be easily combined with other preservation techniques such as chilling disinfection, modified-atmosphere packaging, mild thermal treatments, chemical sanitizers, and even other oxidizing agent (such as ozone).

Hurdle processes have been applied to microorganisms in broth: Burleson et al. (1975) studied the inactivation of selected organisms with public health significance (such as *Staphylococcus aureus*, *Salmonella typhimurium*, and *Escherichia coli*) by ozone and sonication; Ordonez et al. (1984) combined ultrasonic and heat treatments to study the survival of thermoduric streptococci; Pagan et al. (1999) analyzed the resistance of *Listeria monocytogenes* to manosonication and manothermosonication processes; Lopez et al. (1994) also applied manothermosonication to inactivate peroxidase, lipoxygenase, and polyphenol oxidase; López-Malo et al. (2005) combined thermosonication and antimicrobials for fungal inactivation.

Use of ultrasound for disinfection of surfaces of fresh fruits and vegetables requires further investigation. However, ultrasound has been applied with success to liquid foods such as fruit juices (i.e., orange, tomato, and guava juices and apple cider), usually combined with mild thermal treatments and/or pressure (Baumann et al. 2005; Valero et al. 2007; Cheng et al. 2007; Adekunte et al. 2010).

Ozone has been used in the aqueous or gaseous state, combined with most of the treatments mentioned before. Several researchers have studied the synergetic effects of hurdle processes. Allende and Artés (2003a) combined UV-C and modified-atmosphere packaging for reduction of microbial growth of lettuce. López-Rubira et al. (2005) studied the shelf life and overall quality of minimally processed pomegranate arils packaged in a modified atmosphere and treated with UV-C. Pan et al. (2004) and Marquenie et al. (2002) investigated the effect of UV-C and heat treatment on strawberries and sweet cherries throughout storage. Das et al. (2006) analyzed the effect of controlled-atmosphere storage, modified-atmosphere packaging, and gaseous ozone exposure on the survival of *Salmonella enteritidis* on cherry tomatoes. An et al. (2007) reported changes in some quality indexes in fresh-cut green asparagus pretreated with aqueous ozone and subsequent modified-atmosphere packaging. Yuk et al. (2007) combined ozone and organic acid treatment for control of *Escherichia coli* O157:H7 and *Listeria monocytogenes* on enoki mushroom. Allende et al. (2007) combined UV-C light, gaseous ozone, superatmospheric dioxygen and a high level of carbon dioxide and investigated the impact on health-promoting compounds and the shelf life of strawberries. Whangchai et al. (2006) used ozone in combination with some organic acids to control postharvest decay and pericarp browning of longan fruit. Singh et al. (2002b) verified the efficacy of chlorine dioxide, ozone, and thyme essential oil on sequential washings in reducing *Escherichia coli* loads of lettuce and baby carrots.

12.5 Traditional Versus Emerging Technologies and Future Prospects

Heat treatments are efficient in inactivating microbes, thus reducing decay of fruits and vegetables and attaining safety targets. Besides the proven effect of thermal treatments in microbial reduction, they have a significant impact on product quality. Some quality parameters such as texture, color, aroma, flavor, taste, and nutritive value can be severally affected by temperature. The emerging technologies are known to lessen this impact.

The conventional washings of fruits and vegetables with chemical sanitizers have regulatory and mandatory restrictions. Physical decontamination techniques are also the focus of many food regulations (Geysen et al. 2005).

The applications of ozone, ultrasound, and UV-C radiation are examples of nonthermal technologies that may have potential applications in the food industry. These alternative technologies should produce microbiologically safe products, increase the products' shelf life, and reducing the process impact on quality characteristics. Moreover, nonthermal treatments require a significant consumption of energy, and are therefore more economic and environmentally friendly technologies (Piyasena et al. 2003).

All technologies, traditional or innovative, have both advantages and disadvantages. For example, a thermal treatment when correctly applied guarantees safety of fruits and vegetables, however quality attributes may be sacrificed.

Sanitizer agents can be effective against a wide range of pathogens with minimal quality loses, although in some circumstances they may react with natural organic matter, forming halogenated by-products, which can cause severe health problems.

Radiation treatments, such UV-C light processes, do not have problems concerning the formation of by-products but, industrially, it is very difficult to measure the germicidal dose and, in some conditions, the microorganisms can repair their DNA and recover from injury.

Ozone treatments do not produce by-products and are more effective against microorganisms than are most of the sanitizers commonly used. However, as ozone is a strong oxidizing agent, its impact on humans is one of the most important parameters needed to be controlled. Industrially, it is essential to establish safe limits for people who are in contact with ozone during the food processing, and the use of masks with eye protection is recommended for workers.

Ultrasound-based treatments are nondestructive and can be performed online, but they are very expensive to implement industrially.

Probably combinations of technologies are the right road to follow. The combination of the positive aspects of some technologies may conjointly improve safety and quality aspects of fruits and vegetables.

References

Adekunte A, Tiwari BK, Scannell A, Cullen PJ, O'Donnell C. Modelling of yeast inactivation in sonicated tomato juice. Int J Food Microbiol. 2010;137(1):116–20.

Aguayo E, Escalona VH, Artés F. Effect of cyclic exposure to ozone gas on physicochemical, sensorial and microbial quality of whole and sliced tomatoes. Postharvest Biol Tech. 2006;39(2):169–77.

Akbas MY, Olmez H. Effectiveness of organic acids, ozonated water and chlorine dippings on microbial reduction and storage quality of fresh-cut iceberg lettuce. J Sci Food Agric. 2007;87(14):2609–16.

Akbas MY, Ozdemir M. Application of gaseous ozone to control populations of *Escherichia coli*, *Bacillus cereus* and *Bacillus cereus* spores in dried figs. Food Microbiol. 2008;25(2):386–91.

Alegria C, Pinheiro J, Gonçalves EM, Fernandes I, Moldão M, Abreu M. Quality attributes of shredded carrot (*Daucus carota* L. cv. Nantes) as affected by alternative decontamination processes to chlorine. Innov Food Sci Emerg Tech. 2009;10(1):61–9.

Allende A, Artés F. Combined ultraviolet-C and modified atmosphere packaging treatments for reducing microbial growth of fresh processed lettuce. Lebensmittel-Wissenschaft und -Technologie. 2003a;36(8):779–86.

Allende A, Artés F. UV-C radiation as a novel technique for keeping quality of fresh processed 'Lollo Rosso' lettuce. Food Res Int. 2003b;36(7):739–46.

Allende A, McEvoy JL, Luo Y, Artes F, Wang CY. Effectiveness of two-sided UV-C treatments in inhibiting natural microflora and extending the shelf-life of minimally processed 'Red Oak Leaf' lettuce. Food Microbiol. 2006;23(3):241–9.

Allende A, Marín A, Buendía B, Tomás-Barberán F, Gil MI. Impact of combined postharvest treatments (UV-C light, gaseous O_3, superatmospheric O_2 and high CO_2) on health promoting compounds and shelf-life of strawberries. Postharvest Biol Tech. 2007;46(3):201–11.

Allende A, Selma MV, López-Gálvez F, Villaescusa R, Gil MI. Role of commercial sanitizers and washing systems on epiphytic microorganisms and sensory quality of fresh-cut escarole and lettuce. Postharvest Biol Tech. 2008;49(1):155–63.

Alloway BJ. Contamination of soils in domestic gardens and allotments: a brief overview. Land Contam Reclamation. 2004;12(3):179–87.

Alothman M, Bhat R, Karim AA. UV radiation-induced changes of antioxidant capacity of fresh-cut tropical fruits. Innov Food Sci Emerg Tech. 2009;10(4):512–6.

Altic LC, Rowe MT, Grant IR. UV light inactivation of *Mycobacterium avium* subsp. *paratuberculosis* in milk as assessed by FASTPlaque TB phage assay and culture. Appl Environ Microbiol. 2007;73(11):3728–33.

An J, Zhang M, Lu Q. Changes in some quality indexes in fresh-cut green asparagus pretreated with aqueous ozone and subsequent modified atmosphere packaging. J Food Eng. 2007;78(1):340–4.

Annous BA, Kozempel MF. Surface pasteurization with hot water and steam. In: Sapers GM, Gorny JR, Yousef AE, editors. Microbiology of fruits and vegetables. Boca Raton: CRC Press/Taylor & Francis; 2006. p. 479–96.

Anonymous. Kinetics of microbial inactivation for alternative food processing technologies. Washington, DC: Center for food safety and applied nutrition. U.S. Food and Drug Administration; 2000.

Anonymous. Outbreaks associated with fresh produce: incidence, growth and survival of pathogens in fresh and fresh-cut produce, Analysis and evaluation of preventive control measures for the control and reduction/elimination of microbial hazards on fresh and fresh-cut produce Nutrition. U.S. Food and Drug Administration; 2001.

Anonymous. Food safety and quality assurance issues, improving the safety and quality of fresh fruits and vegetables: a training manual for trainers, University of Maryland, Food and Drug Administration; 2002. p. V1–27.

Bachmann J, Earles R. Postharvest handling of fruits and vegetables, horticulture technical note, appropriate technology transfer for rural areas. 2000. p. 1–19.
Baka M, Mercier J, Corcuff R, Castaigne F, Arul J. Photochemical treatment to improve storability of fresh strawberries. J Food Sci. 1999;64(6):1068–72.
Barth MM, Zhou C, Mercier J, Payne AF. Ozone storage effects on anthocyanin content and fungal growth in blackberries. J Food Sci. 1995;60(6):1286–8.
Bastos MSR, Soares NFF, Andrade N, Arruda AC, Alves RE. The effect of the association of sanitizers and surfactant in the microbiota of the Cantaloupe (*Cucumis melo* L.) melon surface. Food Control. 2005;16(4):369–73.
Baumann AR, Martin SE, Feng H. Power ultrasound treatment of *Listeria monocytogenes* in apple cider. J Food Prot. 2005;68(11):2333–40.
Baur S, Klaiber R, Hammes WP, Carle R. Sensory and microbiological quality of shredded packaged iceberg lettuce as affected by pre-washing procedures with chlorinated and ozonated water. Innov Food Sci Emerg Tech. 2004;5(1):45–55.
Behrsing J, Winkler S, Franz P, Premier R. Efficacy of chlorine for inactivation of *Escherichia coli* on vegetables. Postharvest Biol Tech. 2000;19(2):187–92.
Beltran D, Selma MV, Marín A, Gil MI. Ozonated water extends the shelf life of fresh cut lettuce. J Agric Food Chem. 2005;53:5654–63.
Beuchat LR. Surface decontamination of fruits and vegetables eaten raw: a review, food safety issues. Geneva: Food Safety Unit, World Health Organization; 1998. p. 1–42.
Bruhn CM. Explaining the concept of health risk versus hazards to consumers. Food Control. 2005;16(6):487–90.
Burleson GR, Murray TM, Pollard M. Inactivation of viruses and bacteria by ozone, with and without sonication. Appl Microbiol. 1975;29(3):340–4.
Butz P, Tauscher B. Emerging technologies: chemical aspects. Food Res Int. 2002;35(2–3):279–84.
Cao S, Hu Z, Pang B, Wang H, Xie H, Wu F. Effect of ultrasound treatment on fruit decay and quality maintenance in strawberry after harvest. Food Control. 2010;21(4):529–32.
Cheng LH, Soh CY, Liew SC, Teh FF. Effects of sonication and carbonation on guava juice quality. Food Chem. 2007;104(4):1396–401.
Cia P, Pascholati SF, Benato EA, Camili EC, Santos CA. Effects of gamma and UV-C irradiation on the postharvest control of papaya anthracnose. Postharvest Biol Tech. 2007;43(3):366–73.
Coutinho EF, Junior JLS, Haerter JA, Nachtigall GR, Cantillano RFF. Aplicação pós-colheita de luz ultravioleta (UV-C) em pêssegos cultivar Jade, armazenados em condição ambiente. Ciência Rural, Santa Maria. 2003;33(4):663–6.
Cruz RMS, Vieira MC, Silva CLM. Effect of heat and thermosonication treatments on peroxidase inactivation kinetics in watercress (*Nasturtium officinale*). J Food Eng. 2006;72(1):8–15.
Cruz RMS, Vieira MC, Silva CLM. Modelling kinetics of watercress (*Nasturtium officinale*) colour changes due to heat and thermosonication treatments. Innov Food Sci Emerg Tech. 2007;8(2):244–52.
Das E, Gürakan GC, BayIndIrlI A. Effect of controlled atmosphere storage, modified atmosphere packaging and gaseous ozone treatment on the survival of *Salmonella Enteritidis* on cherry tomatoes. Food Microbiol. 2006;23(5):430–8.
Earnshaw RG, Appleyard J, Hurst RM. Understanding physical inactivation processes: combined preservation opportunities using heat, ultrasound and pressure. Int J Food Microbiol. 1995;28(2):197–219.
Erkan M, Wang CY, Krizek DT. UV-C irradiation reduces microbial populations and deterioration in *Cucurbita pepo* fruit tissue. Environ Exp Bot. 2001;45(1):1–9.
Fennema OR, Powrie WD, Marth EH. Low temperature preservation of foods and living matter. New York: Marcel Dekker; 1973.
Fonseca JM, Rushing JW. Effect of ultraviolet-C light on quality and microbial population of fresh-cut watermelon. Postharvest Biol Tech. 2006;40(3):256–61.

Gabaldón-Leyva CA, Quintero-Ramos A, Barnard J, Balandrán-Quintana RR, Talamás-Abbud R, Jiménez-Castro J. Effect of ultrasound on the mass transfer and physical changes in brine bell pepper at different temperatures. J Food Eng. 2007;81(2):374–9.

Geysen S, Verlinden BE, Nicolai BM. Thermal treatments of fresh fruit and vegetables. In: Jongen W, editor. Improving the safety of fresh fruit and vegetables. Cambridge: Woodhead; 2005. p. 429–53.

Giese N, Darby J. Sensitivity of microorganisms to different wavelengths of UV light: implications on modeling of medium pressure UV systems. Water Res. 2000;34(16):4007–13.

Gorny J. Microbial contamination of fresh fruits and vegetables. In: Sapers GM, Gorny JR, Yousef AE, editors. Microbiology of fruits and vegetables. Boca Raton: CRC Press/Taylor & Francis; 2006. p. 3–32.

Graham DM. Ozone as an antimicrobial agent for the treatment, storage and processing of foods in gas and aqueous phases. Direct food additive petition, Palo Alto, CA, Electric Power Research Institute; 2000.

Guerrero-Beltrán JA, Barbosa-Cánovas GV. Advances and limitations on processing foods by UV light. Food Sci Technol Int. 2004;10(3):137–47.

Güzel-Seydim Z, Bever PI, Greene AK. Efficacy of ozone to reduce bacterial populations in the presence of food components. Food Microbiol. 2004a;21(4):475–9.

Güzel-Seydim ZB, Greene AK, Seydim AC. Use of ozone in the food industry. Lebensmittel-Wissenschaft und -Technologie. 2004b;37(4):453–60.

Han Y, Linton RH, Nielsen SS, Nelson PE. Reduction of *Listeria monocytogenes* on green peppers (*Capsicum annuum* L.) by gaseous and aqueous chlorine dioxide and water washing and its growth at 7°C. J Food Prot. 2001;64(11):1730–8.

Han Y, Floros JD, Linton RH, Nielsen SS, Nelson PE. Response surface modelling for the inactivation of *Escherichia coli* O157:H7 on green peppers (*Capsicum annuum*) by ozone gas treatment. Food Microbiol Saf. 2002;67(3):1188–93.

Hernández AF, Robles PA, Gómez PA, Callejas TA, Artés F. Low UV-C illumination for keeping overall quality of fresh-cut watermelon. Postharvest Biol Tech. 2010;55(2):114–20.

Hernandez-Brenes C. Section III-good manufacturing practices for handling, packing, storage and transportation of fresh produce. Improving the safety and quality of fresh fruits and vegetables: a training manual for trainers, University of Maryland, Food and Drug Administration; 2002. p. III1–34.

Hollósy F. Effects of ultraviolet radiation on plant cells. Micron. 2002;33(2):179–97.

Huang TS, Xu CL, Walker K, West P, Zhang SQ, Weese J. Decontamination efficacy of combined chlorine dioxide with ultrasonication on apples and lettuce. J Food Sci. 2006;71(4):M134–9.

Jin HH, Lee SY. Combined effect of aqueous chlorine dioxide and modified atmosphere packaging on inhibiting *Salmonella typhimurium* and *Listeria monocytogenes* in mungbean sprouts. J Food Sci. 2007;72(9):M441–5.

Ketteringham L, Gausseres R, James SJ, James C. Application of aqueous ozone for treating pre-cut green peppers (*Capsicum annuum* L.). J Food Eng. 2006;76(1):104–11.

Keyser M, Muller IA, Cilliers FP, Nel W, Gouws PA. Ultraviolet radiation as a non-thermal treatment for the inactivation of microorganisms in fruit juice. Innov Food Sci Emerg Tech. 2008;9(3):348–54.

Kim JG, Yousef AE, Dave S. Application of ozone for enhancing the microbiological safety and quality of foods: a review. J Food Prot. 1999;62:1071–87.

Kim H, Ryu J-H, Beuchat LR. Survival of *Enterobacter sakazakii* on fresh produce as affected by temperature, and effectiveness of sanitizers for its elimination. Int J Food Microbiol. 2006;111(2):134–43.

Klockow PA, Keener KM. Safety and quality assessment of packaged spinach treated with a novel ozone-generation system. LWT – Food Sci Tech. 2009;42(6):1047–53.

Knezevic Z, Serdar M. Screening of fresh fruit and vegetables for pesticide residues on Croatian market. Food Control. 2009;20(4):419–22.

Knorr D. In: Gould GW, editor. New methods of food preservation. London: Blakie Academic & Prof; 1995.

Knorr D, Ade-Omowaye BIO, Heinz V. Nutritional improvement of plant foods by non-thermal processing. Proc Nutr Soc. 2002;61(2):311–8.

Knorr D, Zenker M, Heinz V, Lee D-U. Applications and potential of ultrasonics in food processing. Trends Food Sci Tech. 2004;15(5):261–6.

Kuldiloke J. Effect of ultrasound, temperature and pressure treatments on enzyme activity and quality indicators of fruits and vegetables juices. Thesis, Technischen Universität Berlin; 2002.

Lopez P, Sala FJ, Fuente JL, Condon S, Raso J, Burgos J. Inactivation of peroxidase, lipoxygenase, and polyphenol oxidase by manothermosonication. J Agric Food Chem. 1994;42(2):252–6.

López-Malo A, Palou E, Jiménez-Fernández M, Alzamora SM, Guerrero S. Multifactorial fungal inactivation combining thermosonication and antimicrobials. J Food Eng. 2005;67(1–2): 87–93.

López-Rubira V, Conesa A, Allende A, Artés F. Shelf life and overall quality of minimally processed pomegranate arils modified atmosphere packaged and treated with UV-C. Postharvest Biol Tech. 2005;37(2):174–85.

Marquenie D, Michiels CW, Geeraerd AH, Schenk A, Soontjen C, Van Impe JF, et al. Using survival analysis to investigate the effect of UVC and heat treatment on storage rot of strawberry and sweet cherry. Int J Food Microbiol. 2002;73(2–3):187–96.

Martins RC, Silva CLM. Modelling colour and chlorophyll loss of frozen green beans (*Phaseolus vulgaris*, L.). Int J Refrig. 2001;25(7):966–74.

Matak KE. Effect of UV irradiation on the reduction of the bacterial pathogens and chemical indicators of milk. Thesis.Blacksburg: Virginia Polytechnic Institute and State University; 2004.

McClements DJ. Advances in the application of ultrasound in food analysis and processing. Trends Food Sci Tech. 1995;6(9):293–9.

Mermelstein NH. Annual meeting papers address. Nonthermal processing methods. Food Tech. 2000;54(5):184–8.

Mizrach A. Ultrasonic technology for quality evaluation of fresh fruit and vegetables in pre- and postharvest processes. Postharvest Biol Tech. 2008;48(3):315–30.

Najafi BHM, Khodaparast MHH. Efficacy of ozone to reduce microbial populations in date fruits. Food Control. 2009;20(1):27–30.

Ohlsson T. Minimal processing of foods with non-thermal methods. In: Ohlsson T, Bengtsson N, editors. Minimal processing technologies in the food industry. Cambridge: Woodhead; 2002. p. 34–60.

Ölmez H, Akbas MY. Optimization of ozone treatment of fresh-cut green leaf lettuce. J Food Eng. 2009;90(4):487–94.

Ordonez JA, Sanz B, Hernandez PE, Lopez-Lorenzo P. A note on the effect of combined ultrasonic and heat treatments on the survival of thermoduric streptococci. J Appl Bacteriol. 1984;56(1):175–7.

Öztekin S, Zorlugenç B, Zorlugenç FK. Effects of ozone treatment on microflora of dried figs. J Food Eng. 2006;75(3):396–9.

Pagan R, Manas P, Alvarez I, Condon S. Resistance of *Listeria monocytogenes* to ultrasonic waves under pressure at sublethal (manosonication) and lethal (manothermosonication) temperatures. Food Microbiol. 1999;16(2):139–48.

Palou L, Crisosto CH, Smilanick JL, Adaskaveg JE, Zoffoli JP. Effects of continuous 0.3 ppm ozone exposure on decay development and physiological responses of peaches and table grapes in cold storage. Postharvest Biol Tech. 2002;24(1):39–48.

Pan J, Vicente AR, Martínez GA, Chaves AR, Civello PM. Combined use of UV-C irradiation and heat treatment to improve postharvest life of strawberry fruit. J Sci Food Agric. 2004;84(14): 1831–8.

Pao S, Kelsey DF, Khalid MF, Ettinger MR. Using aqueous chlorine dioxide to prevent contamination of tomatoes with *Salmonella enterica* and *Erwinia carotovora* during fruit washing. J Food Prot. 2007;70(3):629–34.

Parish ME, Beuchat LR, Suslow TV, Harris LJ, Garret EH, Farber JM, et al. Methods to reduce/eliminate pathogens from produce and fresh-cut produce. Compr Rev Food Sci Food Saf. 2003;2(Suppl):161–73.

Patist A, Bates D. Ultrasonic innovations in the food industry: from the laboratory to commercial production. Innov Food Sci Emerg Tech. 2008;9(2):147–54.

Pérez AG, Sanz C, Ríos JJ, Olías R, Olías JM. Effects of ozone treatment on postharvest strawberry quality. J Agric Food Chem. 1999;47(4):1652–6.

Piyasena P, Mohareb E, McKellar RC. Inactivation of microbes using ultrasound: a review. Int J Food Microbiol. 2003;87(3):207–16.

Qi B, Zang Q, Barbosa-Canovas GV, Pedrow PD. Transactions of the ASAE, Food and process engineering 1995. p. 557–65.

Rames J, Chaloupecky V, Sojkova N, Bencko V. An attempt to demonstrate the increased resistance of selected bacterial strains during repeated exposure to UV radiation at 254 nm. Cent Eur J Public Health. 1997;5(1):30–1.

Restaino L, Frampton EW, Hemphill JB, Palnikar P. Efficacy of ozonated water against various food-related microorganisms. Appl Environ Microbiol. 1995;61(9):3471–5.

Rico D, Martín-Diana AB, Barat JM, Barry-Ryan C. Extending and measuring the quality of fresh-cut fruit and vegetables: a review. Trends Food Sci Tech. 2007;18(7):373–86.

Rodgers SL, Cash JN, Siddiq M, Ryser ET. A comparison of different chemical sanitizers for inactivating *Escherichia coli* O157:H7 and *Listeria monocytogenes* in solution and on apples, lettuce, strawberries, and cantaloupe. J Food Prot. 2004;67(4):721–31.

Ross AIV, Griffiths MW, Mittal GS, Deeth HC. Combining nonthermal technologies to control foodborne microorganisms. Int J Food Microbiol. 2003;89(2–3):125–38.

Ruiz-Cruz S, Acedo-Félix E, Díaz-Cinco M, Islas-Osuna MA, González-Aguilar GA. Efficacy of sanitizers in reducing *Escherichia coli* O157:H7, *Salmonella* spp. and *Listeria monocytogenes* populations on fresh-cut carrots. Food Control. 2007;18(11):1383–90.

Sánchez MA, Allende A, Bennett RN, Ferreres F, Gil MI. Microbial, nutritional and sensory quality of rocket leaves as affected by different sanitizers. Postharvest Biol Tech. 2006;42(1):86–97.

Sanz EN, Dávila IS, Balao JAA, Alonso JMQ. Modelling of reactivation after UV disinfection: Effect of UV-C dose on subsequent photoreactivation and dark repair. Water Res. 2007;41(14):3141–51.

Sapers GM. Efficacy of washing and sanitizing methods for disinfection of fresh fruit and vegetable products. Food Tech Biotechnol. 2001;39(4):305–11.

Sapers GM, Simmons GF. Hydrogen peroxide disinfection of minimally processed fruits and vegetables. Food Technol. 1998;52(2):48–52.

Sarig P, Zahavi T, Zutkhi Y, Yannai S, Lisker N, Ben-Arie R. Ozone for control of post-harvest decay of table grapes caused by *Rhizopus stolonifer*. Physiol Mol Plant Pathol. 1996;48(6):403–15.

Sassi J, Viitasalo S, Rytkonen J, Leppakoski E. Experiments with ultraviolet light, ultrasound and ozone technologies for onboard ballast water treatment. VTT TIEDOTTEITA – research notes 2313. Abo Akademi University; 2005.

Schneider KR, Steslow FS, Sierra FS, Rodrick GE, Noss CI. Ozone depuration of *Vibrio vulnificus* from the southern quahog clam, *Mercenaria campechiensis*. J Invertebr Pathol. 1991;57(2):184–90.

Selma MV, Beltrán D, Allende A, Chacón-Vera E, Gil MI. Elimination by ozone of *Shigella sonnei* in shredded lettuce and water. Food Microbiol. 2007;24(5):492–9.

Selma MV, Ibáñez AM, Allende A, Cantwell M, Suslow T. Effect of gaseous ozone and hot water on microbial and sensory quality of cantaloupe and potential transference of *Escherichia coli* O157:H7 during cutting. Food Microbiol. 2008a;25(1):162–8.

Selma MV, Ibáñez AM, Cantwell M, Suslow T. Reduction by gaseous ozone of *Salmonella* and microbial flora associated with fresh-cut cantaloupe. Food Microbiol. 2008b;25(4):558–65.

Sharifi-Yazdi MK, Darghahi H. Inactivation of pathogenic bacteria using pulsed UV-light and its application in water disinfection and quality control. Acta Med Iran. 2006;44(5):305–8.

Sharma R. Ozone descontamination of fresh fruits and vegetables. In: Jongen W, editor. Improving the safety of fresh fruit and vegetables. Cambridge: Woodhead; 2005. p. 373–86.

Sheldon BW, Brown AL. Efficacy of ozone as a disinfectant for poultry carcasses and chill water. J Food Sci. 1986;51(2):305–9.

Singh N, Singh RK, Bhunia AK, Stroshine RL. Effect of inoculation and washing methods on the efficacy of different sanitizers against *Escherichia coli* O157:H7 on lettuce. Food Microbiol. 2002a;19(2–3):183–93.

Singh N, Singh RK, Bhunia AK, Stroshine RL. Efficacy of chlorine dioxide, ozone, and thyme essential oil or a sequential washing in killing *Escherichia coli* O157:H7 on lettuce and baby carrots. Lebensmittel-Wissenschaft und-Technologie. 2002b;35(8):720–9.

Smilanick JL. Use of ozone in storage and packaging facilities. Proceedings of the Washington tree fruit postharvest conference; Dec 2003, Wenatchee: WSU-TFREC Postharvest Information Network; 2003.

Solomon C, Casey P, Mackey C, Lake A. Ultraviolet disinfection. Environ Tech Initiative 1998; 1–2.

Stevens C, Khan VA, Lu JY, Wilson CL, Pusey PL, Igwegbe ECK, et al. Integration of ultraviolet (UV-C) light with yeast treatment for control of postharvest storage rots of fruits and vegetables. Biol Control. 1997;10(2):98–103.

Stevens C, Khan VA, Lu JY, Wilson CL, Pusey PL, Kabwe MK, et al. The germicidal and hormetic effects of UV-C light on reducing brown rot disease and yeast microflora of peaches. Crop Prot. 1998;17(1):75–84.

Torres CM, Picó Y, Mañes J. Determination of pesticide residues in fruit and vegetables. J Chromatogr A. 1996;754(1–2):301–31.

Tzortzakis N, Singleton I, Barnes J. Deployment of low-level ozone-enrichment for the preservation of chilled fresh produce. Postharvest Biol Tech. 2007;43(2):261–70.

Ukuku DO. Effect of hydrogen peroxide treatment on microbial quality and appearance of whole and fresh-cut melons contaminated with *Salmonella* spp. Int J Food Microbiol. 2004;95(2): 137–46.

Ukuku DO. Effect of sanitizing treatments on removal of bacteria from cantaloupe surface, and re-contamination with *Salmonella*. Food Microbiol. 2006;23(3):289–93.

Ukuku DO, Bari ML, Kawamoto S, Isshiki K. Use of hydrogen peroxide in combination with nisin, sodium lactate and citric acid for reducing transfer of bacterial pathogens from whole melon surfaces to fresh-cut pieces. Int J Food Microbiol. 2005;104(2):225–33.

Unluturk S, AtIlgan MR, Handan Baysal A, TarI C. Use of UV-C radiation as a non-thermal process for liquid egg products (LEP). J Food Eng. 2008;85(4):561–8.

Valero M, Recrosio N, Saura D, Muñoz N, Martí N, Lizama V. Effects of ultrasonic treatments in orange juice processing. J Food Eng. 2007;80(2):509–16.

Vandekinderen I, Camp JV, Meulenaer BD, Veramme K, Bernaert N, Denon Q, et al. Moderate and high doses of sodium hypochlorite, neutral electrolyzed oxidizing water, peroxyacetic acid, and gaseous chlorine dioxide did not affect the nutritional and sensory qualities of fresh-cut iceberg lettuce (*Lactuca sativa* Var. *capitata* L.) after washing. J Agric Food Chem. 2009;57(10): 4195–203.

Velázquez LC, Barbini NB, Escudero ME, Estrada CL, Guzmán AMS. Evaluation of chlorine, benzalkonium chloride and lactic acid as sanitizers for reducing *Escherichia coli* O157:H7 and *Yersinia enterocolitica* on fresh vegetables. Food Control. 2009;20(3):262–8.

Vercet A, Lopez P, Burgos J. Inactivation of heat-resistant lipase and protease from *Pseudomonas fluorescens* by nanothermosonication. J Dairy Sci. 1997;80(1):29–36.

Vicente AR, Pineda C, Lemoine L, Civello PM, Martinez GA, Chaves AR. UV-C treatments reduce decay, retain quality and alleviate chilling injury in pepper. Postharvest Biol Tech. 2005;35(1):69–78.

Villamiel M, de Jong P. Inactivation of *Pseudomonas fluorescens* and *Streptococcus thermophilus* in Trypticase® soy broth and total bacteria in milk by continuous-flow ultrasonic treatment and conventional heating. J Food Eng. 2000;45(3):171–9.

Wang H, Feng H, Luo Y. Microbial reduction and storage quality of fresh-cut cilantro washed with acidic electrolyzed water and aqueous ozone. Food Res Int. 2004;37(10):949–56.

Wang T, MacGregor SJ, Anderson JG, Woolsey GA. Pulsed ultra-violet inactivation spectrum of *Escherichia coli*. Water Res. 2005;39(13):2921–5.

Warriner K. Pathogens in vegetables. In: Jongen W, editor. Improving the safety of fresh fruit and vegetables. Cambridge: Woodhead; 2005. p. 3–43.

Wei K, Zhou H, Zhou T, Gong J. Comparison of aqueous ozone and chlorine as sanitizers in the food processing industry: impact on fresh agricultural produce quality. Ozone: Sci Eng. 2007; 29(2):113–20.

Weissinger WR, Chantarapanont W, Beuchat LR. Survival and growth of *Salmonella baildon* in shredded lettuce and diced tomatoes, and effectiveness of chlorinated water as a sanitizer. Int J Food Microbiol. 2000;62(1–2):123–31.

Whangchai K, Saengnil K, Uthaibutra J. Effect of ozone in combination with some organic acids on the control of postharvest decay and pericarp browning of longan fruit. Crop Prot. 2006;25(8):821–5.

Wu VCH, Kim B. Effect of a simple chlorine dioxide method for controlling five foodborne pathogens, yeasts and molds on blueberries. Food Microbiol. 2007;24(7–8):794–800.

Yaun BR. Efficacy of ultraviolet treatments for the inhibition of pathogens on the surface of fresh fruits and vegetables, Thesis. Blacksburg: Virginia Polytechnic Institute and State University; 2002.

Yaun BR, Sumner SS, Eifert JD, Marcy JE. Inhibition of pathogens on fresh produce by ultraviolet energy. Int J Food Microbiol. 2004;90(1):1–8.

Yuk H-G, Yoo M-Y, Yoon J-W, Marshall DL, Oh D-H. Effect of combined ozone and organic acid treatment for control of *Escherichia coli* O157:H7 and *Listeria monocytogenes* on enoki mushroom. Food Control. 2007;18(5):548–53.

Zenoff VF, Sineriz F, Farías ME. Diverse responses to UV-B radiation and repair mechanisms of bacteria isolated from high-altitude aquatic environments. Appl Environ Microbiol. 2006;72(12):7857–63.

Zhang L, Lu Z, Yu Z, Gao X. Preservation of fresh-cut celery by treatment of ozonated water. Food Control. 2005;16(3):279–83.

Zhao Y. Pathogens in fruit. In: Jongen W, editor. Improving the safety of fresh fruit and vegetables. Cambridge: Woodhead; 2005. p. 44–88.

Zheng L, Sun D-W. Innovative applications of power ultrasound during food freezing processes – a review. Trends Food Sci Tech. 2006;17(1):16–23.

Zorlugenç B, Kiroglu Zorlugenç F, Öztekin S, Evliya IB. The influence of gaseous ozone and ozonated water on microbial flora and degradation of aflatoxin B1 in dried figs. Food Chem Toxicol. 2008;46(12):3593–7.

Chapter 13
Novel Technologies for the Preservation of Chilled Aquatic Food Products

Carmen A. Campos, María F. Gliemmo, Santiago P. Aubourg, and Jorge Barros Velázquez

13.1 Introduction

13.1.1 Aquatic Species: Sources and Composition Characteristics

Most fish and other aquatic species give rise to products of great economic importance in many countries. The demand for such products has been increasing steadily over the last century and shows no signs of lessening, as fishing and farming actually constitute a basic source of food for all populations of the world. Recent FAO reports, 1999–2005, have shown that whereas the volumes of aquatic species caught remain relatively stable (around 90–95 million tons), the volumes of farmed aquatic species have continuously increased in recent years, as shown in Fig. 13.1 (FAO 2007a, b).

This trend in the fishing industry has been attributed to several factors. First, the actual amount of fish caught does not satisfy demand. The utilization of different fish and invertebrate species for the manufacture of various fish or fish-derived

C.A. Campos (✉) • M.F. Gliemmo
Departamento de Industrias, Facultad de Ciencias Exactas y Naturales,
Universidad de Buenos Aires, Intendente Guiraldes 2160, C.A.B.A, 1428, Argentina
e-mail: carmen@di.fcen.uba.ar; mfg@di.fcen.uba.ar

S.P. Aubourg
Instituto de Investigaciones Marinas de Vigo (CSIC), C/Eduardo Cabello 6, Vigo, Pontevedra, E-36208, Spain
e-mail: saubourg@iim.csic.es

J.B. Velázquez
Department of Analytical Chemistry, Nutrition and Food Science, LHICA,
School of Veterinary Sciences, University of Santiago de Compostela, E-27002, Lugo, Spain

Laboratory of Biotechnology, College of Pharmacy, University of Santiago de Compostela,
E-15782, Santiago, Spain
e-mail: jorge.barros@usc.es

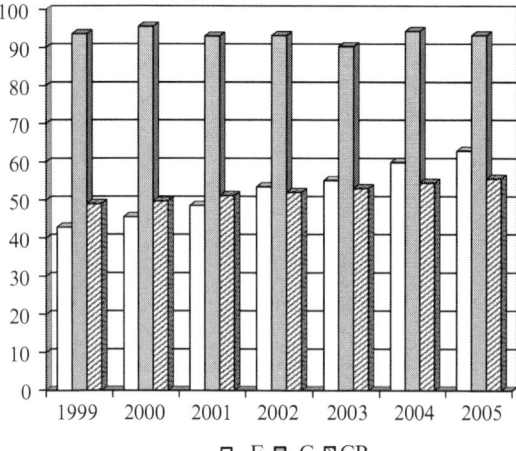

Fig. 13.1 Annual values for farming production (*F*), catching (*C*), and chilled products (*CP*) during the 1999–2005 period (expressed in million tons) (Adapted from FAO 2007a, b)

products is affected by their availability, the processing capacity of the fishing industry, and the size, anatomical structure, and sensory properties of the muscles and other valuable parts of the animals. As a result of dwindling stocks of traditional species, technologists and manufacturers are forced to turn their attention to new and unconventional sources of raw materials. On the other hand, farming provides the advantages of a product of consistent quality, constant supply to consumers, and control of the various production stages, including feeding and slaughtering conditions.

Marine foods especially, and aquatic foods as a whole, are known to provide a high content of important constituents for the human diet, such as nutritional and digestive proteins, including high levels of essential amino acids (lysine and methionine, among others), lipid-soluble vitamins (namely, A and D), microelements (I, F, Ca, Cu, Zn, Fe, and others), and highly unsaturated fatty acids (Simopoulos, 1997). The lipid fraction is now the subject of a great deal of attention owing to its high content of ω-3 polyunsaturated fatty acids, which have been shown to play a positive role in preventing certain human diseases.

Fish and invertebrate constituents deteriorate quickly and have been reported to manifest significant qualitative and quantitative content changes that are termed product spoilage (Kolakowska et al. 1992). The most important composition changes have been shown to occur in lipids. These can be the result of endogenous mechanisms or exogenous agents (Pearson et al. 1977; Bandarra et al. 2001). As regards endogenous mechanisms, lipid matter exhibits a wide content range among species and a heterogeneous distribution throughout the body of aquatic species. This phenomenon is probably influenced by genetic, physiological, and anatomical factors. With reference to exogenous agents, seasonal variation was considered to significantly affect feeding availability and other external factors affect lipid content in different types of species.

13.1.2 Aquatic Species Processing: Chilled Products

Aquatic food products are produced from poikilothermic species that possess a high water and non-protein-nitrogen content, a soft muscular and skin structure, and a low collagen content (Brown, 1986). In addition, the highly unsaturated lipid composition mentioned has been shown to be especially prone to lipid oxidation, this leading to sensory and nutritional quality losses. Such features mean that these products are considered to be among the most perishable foods, thus necessitating rapid and efficient processing as well as refrigerated storage after death.

Aquatic products can be obtained from fresh species through a wide range of technological processes. For years, different traditional strategies have been applied, such as cooling (chilling, freezing, and frozen storage), heating (cooking, canning, smoking), and salt treatment (salting). In the same way, processing technologies can be divided into those that try to retain all the original properties of the fresh specimens (chilling, freezing, and frozen storage) and those that apply a strong treatment so that new and different product can be obtained (canning, smoking).

During recent decades, consumers' choice has been transformed by developments in the production, distribution, and retailing of food, which, with improvements in the design and equipment in the domestic kitchen, have facilitated a major change in our lifestyle. Perhaps the most striking development is the marketing of a wide and expanding range of chilled perishable foods, including aquatic foods (Heap 1992). Convenience, easy preparation, and the "fresh" and "healthy" image are attractive features. These developments have occurred because of the application of advanced technology to the production, packaging, distribution, and retailing of food as a whole, and aquatic food products in particular, the latter exhibiting production levels as high as 50–52 million tons for the 1999–2005 period (FAO 2007c; Fig. 13.1). In developed countries, chilled aquatic foods have taken over the markets, because of increasing demand where high-quality products are offered to the consumer, in spite of the fact that the distance between the catching sites or farms and consumers can be great. Additionally, the fresh market is also of great importance in developing countries as a result of their having fewer alternative methods at their disposal for preservation.

The integrity and safety of chilled aquatic foods is multifactorial. Care is required at every stage in the chain, from primary production of raw materials, through manufacture, distribution, retail, and consumer use. The establishment of the shelf life of chilled aquatic foods requires a full appreciation of the microbial, chemical, physical, and biochemical aspects which influence the sensory acceptability of products, and these have to be considered in relation to the safety and quality of products. Further developments in processing technology will undoubtedly contribute to continued development and innovation in this sector. In this sense, this chapter provides an overview of recent and advanced technologies that may be applied in the manufacture of fresh aquatic products. A brief look at traditional tools on chilling technologies is also included.

13.1.3 Preliminary Steps: Slaughtering and Onboard Handling

It is well known that prior to any technological treatment, steps such as catching, slaughtering, and onboard handling can strongly affect the quality of aquatic species. Such steps, conducted either at the farm, on a vessel or ashore, are intended to (1) prevent physical damage, (2) remove blood, digestive juices, slime, and feces by washing the aquatic species with clean water, and (3) avoid bacterial contamination of cut fillets through proper sanitation and hardware.

One of the factors which for a certain period of time slows down the postmortem autolytic and bacterial decomposition of the flesh and protein is the death-stiffening of the muscle tissue called rigor mortis (Whittle et al. 1990; Ashie et al. 1996). Rigor mortis in aquatic species generally has a shorter duration than in mammals. It starts 1–7 h after death; its peak in slaughtered specimens, kept in ice, is between 5 and 22 h after death. A prolongation of the rigor mortis period is of great economic importance. The period is longer when specimens have exerted less muscular activity prior to death, and when the flesh is refrigerated immediately. In such a period, various physiological reactions occur in the aquatic species, such as the use of energy reserves in the muscle and liver and a change in the acid–base balance. There may also be an increase in blood plasma ion concentration and reduced tissue water content. These reactions affect the enzyme activity in postmortem species and the rate of freshness degradation by changing the conditions in the tissues.

In connection with farm handling, different commercial methods of slaughtering have been shown to cause stress and aversive behavior (Özogul and Özogul 2004; Scherer et al. 2005). Whatever method is used for slaughtering, the important point is it should be efficient. Killing animals requires special consideration in terms of both the welfare of the species and the quality of the final product. As mentioned above, physical activity at the time of or before slaughter negatively affects the flesh quality; as the muscle metabolism is predominantly anaerobic, rapid reduction in adenosine triphosphate (ATP) and pH levels occurs, resulting in lactic acid formation. Stressed specimens may develop stronger rigor mortis, and a softer muscle texture (in comparison with unstressed specimens) that is more susceptible to the undesirable separation of muscle blocks in a raw fillet, a phenomenon referred to as gaping. The handling and processing of aquatic species during rigor mortis can result in a loss of quality and a lower fillet yield.

Concerning fishery handling, a common onboard treatment for white fish is gutting (Tejada and Huidobro 2002). Guts are known to be a reservoir of powerful digestive enzymes and bacteria that can spoil the flesh, resulting in unacceptable odors and flavors. As a result of gutting, many studies have reported an extension of shelf life during chilling of aquatic species. For fatty and small species however, the process of gutting has been shown to be impracticable and may create new cut surfaces for potential bacterial contamination and lipid oxidation, therefore not improving the shelf life.

13.1.4 Damage Pathways During Chilling of Aquatic Species

Wild and farmed aquatic species provide highly perishable products whose quality and freshness rapidly declines after death. Deterioration of such species begins immediately upon catching or slaughtering, and the degree to which it continues depends directly on the iced or refrigerated storage conditions. The rate of alteration has been shown to depend on factors such as the nature of the fish species, size, lipid content, feeding state at the moment of capture, importance and nature of microbial load, and storage temperature.

Great attention has been given to nutritional and sensory changes that aquatic species can undergo during the chilled storage. It has been proved that damage can be produced by the following factors (Whittle et al. 1990; Ashie et al. 1996; Olafsdóttir et al. 1997):

1. *Endogenous enzyme activity*. Different autolysis mechanisms in aquatic species have been described. After death, lactic acid is produced by glycolysis. The drop in the pH value activates proteases, so proteolysis is produced, leading to free amino acid formation. At the same time, ATP follows the nucleotide degradation pathway, this leading to the formation of inosine and hypoxanthine. Besides, lipases and phospholipases present in the muscle of the species can cause lipid hydrolysis, so free fatty acids, partially hydrolyzed lipids (diglycerides and monoglycerides), glycerine, and nitrogenated bases can be produced. This wide variety of relatively small molecules can undergo further interactions with other constituents, leading to quality losses in the product.
2. *Microbial decomposition*. Bacterial enzymes convert trimethylamine oxide into trimethylamine and decompose amino acids and proteins, forming ammonia, hydrogen sulfide, and other undesirable compounds characteristic of microbial spoilage. Biogenic amines are also produced by the action of bacterial enzymes on free amino acids (histidine, tyrosine, tryptophan, lysine, ornithine, and others). Greatest attention has been given to histamine formation, which is reported to be responsible for the "scombrotoxic fish poisoning."
3. *Lipid oxidation*. Since lipids of aquatic species are highly unsaturated, contact with oxygen, especially in the presence of light or other catalysts, can lead to a wide range of lipid oxidation compounds (peroxides, carbonyls, interaction compounds) and to the loss of some essential and beneficial fatty acids for the human diet. Lipid oxidation compounds can interact with protein derivates, leading to a nutritional loss of the product. This damage pathway becomes more relevant as the fat content and both storage temperature and time increase.

As a consequence of these factors, efficient chilling techniques for the refrigeration of aquatic material must be employed to reduce postmortem quality losses before it reaches the consumer or before further processing steps are conducted.

13.2 Chilling Methods

13.2.1 Traditional Chilling Methods

13.2.1.1 Flake Icing

Although an increasing number of factory vessels are processing marine species in various ways at sea, shortly after they are caught, chilling to temperatures close to 0°C by using flake ice is the preservation method most extensively used to cool marine species rapidly and to extend their shelf life (Whittle et al. 1990; Ashie et al. 1996). Under such conditions, the average storage life of cold water species is 1–2 weeks in temperate climates, whereas species from warm tropical waters can be kept somewhat longer at 0°C.

The rate at which heat can be removed from aquatic foods during flake ice chilling depends on several parameters. In the case of packaged products, size and shape will affect the rate of heat transfer, whereas in the case of unpacked products, special attention should be paid to the relative humidity conditions, with a view to prevent dehydration of the food material (Heap 1992). Although pure ice melts at 0°C, the temperature of material stored in crushed ice may be slightly lower. This effect is due to the salts present in the flesh and to the seawater on the surface of the carcasses, which cause some of the ice to melt and extract heat from the surroundings. Ideally, aquatic species should be in contact only with ice and not with each other. Individual carcasses touching one another do not cool as rapidly as when each carcass is buried completely in ice.

Quick and efficient flake ice chilling of the catch to the lowest temperature practicable without actually freezing the flesh is essential if spoilage is to be kept to a minimum. Rapid chilling also aids bleeding and results in more attractive fillets. Besides, contributing to the cooling, the flow or melt water also washes away bacterial slime, spoilage products, and any residual blood, and thus helps to preserve the fresh appearance and smell. Room temperature should be kept above 0°C so that a good flow of melting water is maintained; however, if this temperature is too high, the weight of ice necessary to ensure that the aquatic species will remain covered, and that the specimens' temperature will not rise, will be greater. Finally, much of the care exercised by fishermen in handling and stowing the catch may be wasted unless good standards of cleanliness, hygiene, and sanitation are maintained at all times.

13.2.1.2 Chilled Seawater

Besides use of flake ice, the employment of chilled seawater has also been shown to be popular (Kraus 1992). In this process, aquatic species are stored in chilled seawater, the chilling being effected either by mechanical refrigeration or by the addition of ice. Mechanical refrigeration of ice-chilled seawater has been demonstrated to be an excellent means for chilling and transporting entire specimens while still at sea and for the road transport of bulk quantities; there has been little or no application of fresh species in the retail phase. Salt uptake must be monitored and circulation is crucial to avoid thermal stratification.

"Chilled or refrigerated seawater" is the term commonly used to describe seawater that has been cooled just below 0°C. It has been reported to provide the following advantages compared with traditional flake icing: (1) faster cooling; (2) less stress on the food pieces; (3) the possibility of a lower temperature; (4) faster handling of large quantities of aquatic material; (5) sometimes, a longer storage time.

However, there are also some disadvantages compared with flake ice employment: (1) excess salt intake, (2) water absorption by species of low lipid content; (3) some water-soluble protein loss; (4) presence of anaerobic bacteria responsible for putrefaction; (5) modification of the appearance of gills, skin, and eyes.

13.2.1.3 Salt Addition to Flake Ice

Salt treatment is a frequent practice in certain countries. This can consist of direct addition of sodium chloride to the ice used to cool aquatic specimens or immersion of food material in a brine solution, this being followed by a cooling treatment. The preservative effect of salt is due to a decrease in water activity, which leads to less potential of microbial attack, and enhancement of functional properties, such as water holding capacity, leading to an increase of the shelf life (Huidobro et al. 1990; Toledo-Flores and Zall 1992). Salt is absorbed by the flesh and imparts a desired flavor to the finished product. Brining has an additional beneficial effect as it toughens the skin and prevents its adhesion to any surface. Brining also brightens the appearance of aquatic specimens by removing any remaining slime. However, owing to the lipid pro-oxidant effect attributed to the presence of sodium chloride (Aubourg and Ugliano 2002), some detrimental effects on the lipid composition may be encountered, especially when a fatty fish is concerned.

Salt treatment has mostly been targeted to species that are further processed by another technological treatment. The salt intake by specimens during brining has been shown to depend on several factors, such as the concentration and temperature of the brine, duration, brine to aquatic species ratio, freshness, and fat content. Thus, large specimens require longer treatment times than small ones, fatty specimens require longer treatment times than lean ones, and species that have been filleted or split for kippering require a shorter treatment time than whole or nobbed ones.

13.2.1.4 Quality Changes During Traditional Chilling Processes

Once the aquatic species have been caught and slaughtered, autolytic changes start; the microorganisms associated with the specimens invade the tissues, and after a brief lag period microbial multiplication starts. As previously mentioned, during chilled storage, significant deterioration of sensory quality and loss of nutritional value occur as a result of changes in chemical constituents, which lead to a strong effect on the commercial value (Whittle et al. 1990; Olafsdóttir et al. 1997). This degradation process mostly occurs in the initial stage by the action of muscle enzymes and later by the action of microbial enzymes. The rate of alteration depends on factors such as the nature of the fish species, size, lipid content, state at the moment of capture, importance and nature of the microbial load, and storage temperature.

Inadequate gutting and washing or contamination on the vessel will accelerate spoilage, as will an increase in the activity of the bacteria caused by storage at a temperature higher than 0°C, owing to the use of an inadequate amount of ice or unsatisfactory icing practices, for example.

Sensory and Physical Changes

During the chilled storage, sensory analysis is currently the most important method for freshness evaluation. Characteristic sensory changes occur in appearance, odor, taste, and texture of aquatic species for as long as the chilling time increases (Olafsdóttir et al. 1997); an analysis is normally conducted by a sensory panel of trained professionals, sensory changes are ranked into categories. A good correlation has been found between sensory scores and time in ice for albacore tuna (Pérez-Villarreal and Pozo 1990) and sardine (Nunes et al. 1992; Ababouch et al. 1996). Small species showed relatively short shelf lives: 6 days for herring (Smith et al. 1980), 9 days for sardine (El Marrakchi et al. 1990), and 6–9 days for mackerel depending on the ice-to-fish ratio (Bennour et al. 1991). In contrast, bigger fish species showed longer shelf lives: whole albacore showed a shelf life of 20 days (Pérez-Villarreal and Pozo 1990) and whole gutted salmon showed acceptable quality for 20 days (Whittle et al. 1990).

Changes in physical properties (texture, water holding capacity, electrical properties, and viscosity, among others) have been observed. Torrymeter readings (Nunes et al. 1992) and an increase in pH values (Ababouch et al. 1996) in sardine have shown good correlations with sensory assessment. Torrymeter readings have been shown to be an accurate method for assessing freshness in herring (Damoglou 1980).

Formation of Amines and Degradation of Nucleotides

A sharp content increase of the total volatile bases and trimethylamine has been observed to begin after 9–10 days of storage as a result of the end of the lag phase of microorganisms in sardine (El Marrakchi et al. 1990; Ababouch et al. 1996), mackerel (Bennour et al. 1991), and herring (whole and fillets) (Fernández-Salguero and Mackie 1987). In the case of larger fish, trimethylamine levels in stored albacore tuna increased sharply after 20 days (Pérez-Villarreal and Pozo 1990); in parallel, the sensory acceptability of the specimens decreased.

Among biogenic amines, most efforts have been targeted at histamine detection. An increase in the level of this amine has been observed after 7–9 days of chilled storage in mackerel (Ababouch et al. 1996) and sardine (Gallardo et al. 1997). Fernández-Salguero and Mackie (1987) showed an important increase in histamine and cadaverine contents during chilled storage of whole fish and fillets of herring.

Nucleotide degradation has been observed to occur, leading to a progressive formation of inosine and hypoxanthine with storage time in sardine (Nunes et al. 1992), horse mackerel (Smith et al. 1980), and salmon (Erikson et al. 1997); employment of the k value (ratio of nonphosphorylated ATP metabolites to the total ATP breakdown products) showed a great correlation with both sensory assessment and storage time in albacore (Pérez-Villarreal and Pozo 1990) and salmon (Erikson et al. 1997).

Lipid Damage Analysis

During chilled storage of several fish species, lipid hydrolysis in mackerel, herring, and sardine (Hwang and Regenstein 1993; Smith et al. 1980; Aubourg et al. 1997), formation of peroxides (Smith et al. 1980; Undeland et al. 1999), thiobarbituric acid reactive substances (Nunes et al. 1992; Aubourg et al. 1997), and interaction compounds (Aubourg et al. 1997; Undeland et al. 1999) and loss of endogenous antioxidants (Undeland et al. 1999) have been detected.

It has also been observed that the shelf life in ice is longer for lean fish than for fatty fish; indeed, the shelf life was found to differ within species with capture season according to the lipid content (Whittle et al. 1990; Nunes et al. 1992). Previous chilling was shown to exert a negative effect on the shelf life of frozen herring fillets (Undeland and Lingnert 1999); it was observed that ice storage had a greater effect than frozen storage on the formation of peroxides and fluorescent compounds and the disappearance of endogenous antioxidants (α-tocopherol and ascorbic acid).

A negative effect caused by previous chilling could be observed on canned fish quality; this process of damage was found to have a nonlinear correlation between the previous chilled storage time and the lipid damage found in the canned sardine product (Aubourg and Medina 1997).

13.2.2 *Novel Refrigeration Methods: Subzero Storage in Ice Slurries*

13.2.2.1 General Aspects

Ice slurries can be defined as mixtures of ice particles finely dispersed in an aqueous solution composed of water and other solid components used to decrease its freezing point and achieve temperatures less than 0°C, but always above the initial freezing point of fish material. Two phases coexist in ice slurries: 25–40% of the mixture consists of small spherical ice crystals, although this percentage can be varied according to specific applications, and the remaining 60–75% is water. Spherical microscopic crystals are generated by a fast cooling step, in which the rate of nucleation greatly exceeds the rate of crystal growth, resulting in the generation of numerous small spherical crystals (Hartel 1992).

The application of ice slurries prepared from marine water chilled to subzero temperatures for the storage of fish products has received remarkable attention in recent years (Piñeiro et al. 2004). Such binary systems are also referred to in the literature as flow ice, fluid ice, or liquid ice and provide a significant number of advantages to the handling and storage of aquatic food products. Such systems have a higher chilling rate when compared with flake ice alone. This is due to the higher heat-exchange capacity of the binary mixture. These binary mixtures cause very limited external damage (also as compared with conventional flake ice) to the fish specimens as a consequence of the spherical geometry of the ice crystals. Another advantage is the prevention of oxidation and dehydration mechanisms, owing to the full coverage of the fish surface.

The versatility of ice slurries is also remarkable. These mixtures can be pumped, which allows their hygienic handling, and may also be combined with additives that may act as coadjutants to maintain fish quality during refrigerated storage (Huidobro et al. 2002). Although subzero temperatures can also be achieved through thermal energy storage systems such as ice builders or dynamic ice systems such as ice harvesters or slurry ice makers, a unique feature of ice slurry generators is that the ice being formed does not adhere to any heat transfer surface. As a consequence of this, no defrost cycle is required since no ice is attached to the evaporator (Gladis 1999).

13.2.2.2 Novelty and Main Advantages of Ice Slurries

Aquatic food products have traditionally been stored in ice (Graham et al. 1993) or in hydro-cooling systems. Nevertheless, both techniques have certain limitations, such as the impossibility of being pumped –in the case of flake ice– or storing the fish material below 0°C –in the case of hydro-cooling systems (Jul 1986). As compared with flake ice, hydro-cooling systems offer the advantage that the fish material can be immersed directly in boxes. On the other hand, hydro-cooling systems do not guarantee the fish temperature will be kept at 0°C. Despite these limitations, hydro-cooling systems are considered as a "superchilling" technology (Waterman and Taylor 2001) that permits the extension of the fish shelf life because of the low storage temperatures achieved. Such superchilling techniques have been described as being especially useful when the fishing banks are so distant that storage in flake ice does not guarantee good preservation of the fish quality on landing. Superchilling has also been used for the distribution of live fish (Price et al. 1991) and for the preservation of salmon and tuna (Waterman and Taylor 2001).

With respect to other superchilling techniques, ice slurries represent a well-established advanced alternative for the cooling and storage of fish, from either extractive practices or aquaculture facilities. Although ice slurries were initially applied to dairy products, beverages, vegetables, fruit, meat, and fish, many advantages have been reported as regards the application of this technology to aquatic food products:

1. The storage temperature is slightly below 0°C, but is always above the initial freezing point of the fish material: this implies the slowing down of the chemical and enzymatic breakdown reactions that affect fish quality.
2. The chilling of the fish material is extremely rapid, since the heat exchange rates obtained are about 4 times higher than those provided by flake ice: this permits the fish material to be rapidly placed at subzero temperatures.
3. The geometry of the ice crystals is spherical, in contrast with the sharp edges of flake ice crystals, this limiting the physical damage over the delicate fish surface during handling and storage.
4. The full coverage of the fish surface by the liquid phase prevents the presence of air pockets, thus preventing dehydration during storage.
5. The washing effect of the liquid phase over all the fish skin implies significant reductions of the surface microbial load, this limiting the postmortem diffusion of microorganisms towards the muscle.
6. When marine water is used to prepare the ice slurry, the presence of sodium chloride provides an additional preservation effect, also contributing to stabilize the myofibrillar protein fraction.
7. Ice slurries can be pumped because of their fluid nature, this allowing the automation of the processing and the hygienic distribution of the fish material.
8. Ice slurries can be combined with other preservation agents that may provide additional advantages in terms of microbial inhibition, prevention of lipid oxidation, etc.

There are several ice slurry systems commercially available today. Among them, liquid-Ice™ is characterized by its fluidity, high cooling capacity, and flexibility, also offering a high refrigeration efficiency and a low waste amount of Freon per ton of ice generated. Flo-ice™ is another widely used system, consisting of an ice–water mixture generated in a double-wall heat exchanger, normally by freezing marine water.

13.2.2.3 Pioneer and Recent Applications of Ice Slurries to Seafood Products

The first report of the use of ice slurries with aquatic food products was provided by Chapman (1990), who compared flake ice, hydro-cooling systems, and ice slurry for the onboard storage of finfish. This study demonstrated that the onboard application of ice slurries was the most effective method for maintaining the quality of chilled fish (Chapman 1990). Later, Harada (1991) demonstrated the advantages of ice slurries as a precooling method. Afterwards, other authors successfully applied an ice slurry system to albacore (Price et al. 1991), and reported practical advantages at both microbiological and biochemical levels.

In the case of shellfish, the pioneer use of ice slurries is attributed to Chinivasagam et al. (1998). These authors performed a comparative investigation of the

Table 13.1 Application of subzero storage based on ice slurries to aquatic food products

Aquatic products	References
Finfish	Chapman (1990)
Albacore	Harada (1991), Price et al. (1991)
Tropical prawns	Chinivasagam et al. (1998)
Shrimp	Huidobro et al. (2002)
Sea bass	Martinsdóttir et al. (2002)
Turbot	Piñeiro et al. (2005), Rodríguez et al. (2006) Campos et al. (2006)
European hake	Rodríguez et al. (2004), Losada et al. (2004b)
Sardine	Losada et al. (2004a), Campos et al. (2005)
Horse mackerel	Rodríguez et al. (2005), Losada et al. (2005)
Megrim	Aubourg et al. (2006)
Norway lobster	Losada et al. (2006), Aubourg et al. (2007)
Ray	Múgica et al. (2008), Barros-Velázquez et al. (2008)
Angler	Barros-Velázquez et al. (2008)
Blackspot sea bream	Álvarez et al. (2008)

spoilage patterns of five prawn species stored either in flake ice or in ice slurry. The experimental results showed a significant increase in shelf lives from 10 to 15 days for the flake ice batch and to more than 20 days for the ice slurry batch. Huidobro et al. (2001) suggested the use of ice slurries for the slaughter, chilling, and storage of sea bream in aquaculture facilities. These authors found a significant reduction of the spoilage rate of the sea bream batch processed in ice slurry. They also reported the usefulness of ice slurries for the onboard storage of shrimps (Huidobro et al. 2002) destined to be commercialized as peeled products. With respect to the use of ice slurries as a slaughter method for the killing of farmed fish, only two species –sea bream (*Sparus aurata*) (Huidobro et al. 2001, 2002) and turbot (*Psetta maxima*) (Morzel et al. 2002) – have been investigated.

Over the past 5 years, ice slurries have been optimized and studied in detail. A summary of the most relevant species investigated (Table 13.1) suggests that the subzero temperature of storage significantly slows down the biochemical mechanisms involved in fish spoilage. In addition, the presence of sodium chloride and the washing effect of the liquid phase provide better microbial control. Altogether, these effects are translated into better scores in a sensory analysis and substantial shelf life extensions.

13.2.2.4 Combined Preservation Systems Based on Ice Slurries

The use of a combined system consisting of ozone and ice slurry for the preservation of different fish species such as sardine (Campos et al. 2005) and turbot (Rodríguez et al. 2006) has produced positive results. Ozone has traditionally been used as a water-disinfection agent owing to its strong oxidizing nature, which makes it a useful tool for the inactivation of microorganisms. A combined system successfully installed in a fishing vessel operating in the Grand Sole fishing bank was evaluated as a method for the chilling and storage of megrim (Aubourg et al. 2006), ray, hake, and angler (Barros-Velázquez et al. 2008). In general terms, storage in ozonized slurry ice led to significantly lower counts of several of the following microbial groups: aerobic mesophiles, psychrotrophic bacteria, anaerobes, coliforms, and both lipolytic and proteolytic microorganisms at both skin and muscle levels. The combination of ice slurry and ozone also allowed better control of pH and trimethylamine formation as compared with ice slurry alone. The use of ozonized ice slurries for the storage of fish specimens has been proved to improve the quality and extend the shelf life of these fish species.

Another recent application of combined preservation systems based on ice slurries is the case of Norway lobster (*Nephrops norvegicus*), an aquatic food product of remarkably high commercial value in many European countries. In this case, the use of ice slurries offers advantages both at the microbial and at the biochemical level but, in contrast, it may accelerate the reactions responsible for the enzymatic browning of shellfish carapaces. With a view to avoid this defect, which limits the shelf life of lobster, a combined process using a previous treatment with an antimelanosic agent (sodium hydrogen sulfite, HSO_3Na) followed by storage in ice slurry was evaluated by Aubourg et al. (2007). The combined process provided better maintenance of sensory quality and an extended shelf life of lobster as compared with counterpart untreated batches and control batches stored in flake ice. Remarkably, these practical benefits were achieved without surpassing the 150 mg/kg legal limit established for HSO_3Na. This work opens the way for the development of a combined system consisting of the controlled addition of HSO_3Na to the liquid phase of ice slurries with the aim of extending the shelf life of prawns and shrimps.

13.2.2.5 Prospects of Ice Slurries for the Aquatic Food Sector

The incorporation of slurry ice systems into fishing vessels was initiated in recent years. Besides the demonstrated advantages of such systems for the onboard storage of fish, the introduction of the technology of ice slurries in ports and inland factories is another field that is gaining increasing importance. Thus, the aquatic food product sector, sometimes reticent to incorporate novel technologies and designs, is paying more attention to novel technologies based on ice slurries for the chilling and the subzero storage of aquatic food products (Yamada et al. 2002). Finally, the development of novel tailor-made combined preservation systems based on ice slurries is in its infancy: in this sense the controlled use of antioxidants or biopreservation agents, depending on the specific needs of the species under commercialization, will surely contribute to providing consumers with better and safer food products. For this to happen, the development of prototypes that allow the automatic and hygienic incorporation of certain food-grade additives will surely be one of the most relevant issues for improving the quality and safety of these products.

13.3 Packaging

13.3.1 Modified-Atmosphere Packaging

13.3.1.1 General Aspects

Modified-atmosphere packaging (MAP) is a method widely used to preserve quality and to maintain hygienic, sanitary, and sensory characteristics of perishable products, in particular, those more prone to microbiological and biochemical spoilage such as chilled aquatic food products. It involves the total or partial removal of air contained in the package, followed by reintroduction of an altered or modified atmosphere consisting of other gases, usually carbon dioxide, nitrogen, and oxygen alone, or their combination.

It is widely accepted that storage in MAP instead of air or vacuum packaging increases the shelf life of fish products, including ready-to-eat shrimp (Rutherford et al. 2007), fresh Mediterranean swordfish (*Xiphias gladius*) (Giatrakou et al. 2008; Pantazi et al. 2008), farmed Atlantic halibut (*Hippoglossus hippoglossus*) (Hovda et al. 2007), fillets of rainbow trout (*Oncorhynchus mykiss* W.), (Choubert and Baccaunaud 2006), sardines (*Sardina pilchardus*) (Özogul and Özogul 2006; Özogul et al. 2004), Atlantic herring (*Clupea harengus*) (Özogul et al. 2000), portions of raw whiting, mackerel and salmon (Fagan et al. 2004).

The extension of shelf life by using MAP depends on the fish species, its fat content, the initial microbiological quality, the treatment that the fish undergoes after slaughter, the storage temperature, the gas composition, the packaging materials, the gas-to-product volume ratio, the presence of additives, combination with other hurdle technologies, and even the different analytical methods used for shelf life assessments and criteria for the end point of shelf life (Sivertsvik et al. 2002; Hovda et al. 2007; Torrieri et al. 2006; Randell et al. 1997). Thus, and because of the considerable number of factors implicated, it is difficult to make a general statement on the effects of MAP on fish quality in terms of shelf life extension and fish acceptability.

In relation to the criteria for establishing the shelf life, the moment at which the product quality becomes unacceptable has to be clearly identified. One of the limitations of this technology is that it usually extends the period of moderate-to-low quality (neutral taste period) of fish rather than the initial and preferred fresh quality period (Wang et al. 2008).

13.3.1.2 Effect of the Gas Mixture in Modified-Atmosphere Packaging on the Shelf Life of Seafood Products

Carbon dioxide is nearly always present as the major gas component in the gas used in MAP for fish because of its bacteriostatic and fungistatic action. It is highly soluble in water and fat. Carbon dioxide solubility greatly increases with decreased temperature

(Sivertsvik et al. 2002, 2004; Wang et al. 2008). Therefore, carbon dioxide exists as a dissolved gas and as carbonic acid (2%) in biological tissue, causing a pH decrease and changes in atmosphere composition (Sivertsvik et al. 2004; Torrieri et al. 2006; Randell et al. 1997). This behavior will alter the energetic balance of the microbial cell, affecting different enzymatic activities. Besides, carbon dioxide may alter cell membrane function, this affecting nutrient uptake and absorption. These facts mean carbon dioxide has a major effect on microbial growth. Many examples of this effect can be found in the literature. Sivertsvik (2007) observed a reduction in the microbial population of farmed cod fillets packed in MAP with carbon dioxide and stored at 0°C, Kristinsson et al. (2008) reported similar results during the storage of yellowfin tuna steaks at 4°C, and Rutherford et al. (2007) found a diminution of *Listeria monocytogenes* growth on ready-to-eat shrimp at 12°C.

Sometimes, absorption of carbon dioxide by the fish may cause package deformation as a consequence of a decrease of the gas content in the package; but, in contrast, bacterial and enzymatic activities throughout storage can increase the carbon dioxide level (Lannelongue et al. 1982; Debevere and Boskou 1996). Therefore, because of this behavior, it is necessary to establish the correct value of the gas-to-product volume ratio that is useful to produce an ideal modified atmosphere and to avoid package collapse. Increase of the gas-to-product volume ratio enhanced the amount of carbon dioxide dissolved in the water phase (Devlieghere et al. 1998). The typical commercial gas-to-product volume ratio is usually 2:1 or 3:1 (Sivertsvik 2007; Sivertsvik et al. 2002).

However, with respect to the potential acidification caused by carbon dioxide, it was reported for salmon (*Salmo salar*) and rainbow trout (*Oncorhynchus mykiss* W.) that the pH did not change during storage in a modified atmosphere containing carbon dioxide, suggesting that the strong buffering capacity of fish tissue was the cause of this result (Barnett et al. 1982; Choubert and Baccaunaud 2006; Fey and Regenstein 1982).

Although the use of carbon dioxide enriches modified-atmosphere environments, it has a negative effect on bacterial growth, owing to the acidification produced which may affect the water holding capacity of muscle protein and thus the texture of the product, limiting its shelf life. This trend will depend on the nature of the fish, carbon dioxide exposure, and temperature, which may have a combined effect depending on the specific MAP system. Torrieri et al. (2006) reported a reduction of water loss due to storage by 50% in farmed gutted bass (*Dicentrarchus labrax*) at 3°C by lowering the level of carbon dioxide from 70% to 50%. Similar results were observed in farmed cod (*Gadus morhua*) fillets (Sivertsvik 2007) and in sea bass (*Lates calcalifer*) slices (Masniyom et al. 2002). Randell et al. (1997) observed greater drip loss during storage of Baltic herring (*Clupea harengus membras*) fillets than rainbow trout (*Salmo gairdneri*) fillets packed in 35–40% carbon dioxide atmosphere. However, the possible negative impact on consumers' acceptance produced by the presence of moisture in the package can be avoided by using packs with an absorbent bottom which helps reduce the drip accumulation on the package without significantly affecting the juiciness of the fish flesh (Torrieri et al. 2006).

The presence of oxygen in MAP is useful in some cases to inhibit reduction of trimethylamine oxide to trimethylamine in fresh fish and to maintain the red color of meat fishes such as tunas, yellowtails, and swordfish, reducing and retarding browning caused by formation of metmyoglobin (Sivertsvik et al. 2002; Giatrakou et al. 2008; Kristinsson et al. 2008). As an alternative to oxygen to prevent browning, to stabilize the bright red color and to extend the shelf life, yellowfin tuna steaks were treated with carbon monoxide, present in conventional hickory wood smoke, which displaces the bound oxygen of the heme in hemoglobin and myoglobin and may retard microbial growth by inhibiting the oxidative phosphorylation mechanisms and thus inhibit aerobic respiration and affect microbial survival (Kristinsson et al. 2008).

However, the presence of oxygen in the packages is undesirable in many cases since it can produce oxidative rancidity, discoloration of fish flesh, and/or growth of spoilage aerobic bacteria, such as *Shewanella putrefaciens*. The conjugated double bonds of ketocarotenoids of rainbow trout (*Oncorhynchus mykiss* W.) are oxidized in the presence of air, leading to discoloration of fish tissue (Choubert and Baccaunaud 2006). A drop in the oxygen level of the atmosphere was observed during the storage of farmed cod fillets caused by microbiological and biochemical reactions within the packed product (Sivertsvik 2007).

As an alternative to oxygen in the packages, nitrogen (N_2) is generally used. It is inert and tasteless, with low solubility in water and fat. Moreover, nitrogen delays oxidative rancidity and inhibits the growth of aerobic microorganisms (Sivertsvik et al. 2002). However, it was reported that increasing the level of nitrogen in MAP of farmed cod fillets stored at 0°C enhances the consequent unpleasant raw odor owing to the increase of trimethylamine and total volatile base nitrogen content caused by microbial action. These undesirable characteristics are minimized when using MAP containing pure carbon dioxide or oxygen or combined as reported by Sivertsvik (2007) and Debevere and Boskou (1996) in filleted cod (*Gadus morhua*), by Ruiz-Capillas and Moral (2001) in gutted hake (*Merluccius merluccius* L.), and by Lannelongue et al. (1982) in brown shrimp (*Penaeus aztecus*). It must be highlighted that the inclusion of oxygen in the atmosphere does not prevent *Clostridium botulinum* growth. Therefore, the use of additional stress factors is necessary, involving, for example, the addition of sodium chloride, the decrease of pH, and temperature control (Sivertsvik et al. 2002).

On the other hand, Randell et al. (1997) assayed the presence of argon instead of nitrogen in a gas mixture to preserve rainbow trout and Baltic herring fillets, but the former was not more effective than nitrogen in preserving fish quality. Besides, as argon is much more expensive than nitrogen, the mixture containing argon would not be commercially affordable. As established above, the best combination of gases that maximizes the shelf life of a fish will depend on the species. Hovda et al. (2007) reported that the use of an atmosphere of 50% CO_2 and 50% O_2 is better than use of an atmosphere of 50% CO_2 and 50% N_2 to extend the shelf life of farmed halibut (*Hippoglossus hippoglossus*) and to produce lower bacterial diversity. Torrieri et al. (2006) found that a gas mixture composed of 30% O_2, 50% CO_2, and 20% N_2 guarantees a shelf life of 7–9 days for gutted farmed bass as was measured by physical and sensory attributes. Increasing the carbon dioxide content to 80–100% extended the shelf life of sea bass (*Lates calcalifer*) slices to more than 20 days at 4°C (Masniyom et al. 2002).

It must be taken into account that, as mentioned above, the gas level in MAP can vary with time owing to dissolution of gases into the fish muscle tissue and owing to microbiological and biochemical reactions. Therefore, experiments are usually conducted under a controlled atmosphere to study the real effect of the different gas constituents, serving as a reference for experiments performed in MAP (Choubert and Baccaunaud 2006). Throughout the storage in a controlled atmosphere, a more stable gas composition is ensured and the accumulation of off-odors is avoided. These advantages make the use of a controlled atmosphere of great interest for fish preservation on deep-sea fishing boats, as suggested by Ruiz-Capillas and Moral (2001).

13.3.2 Active Packaging Systems

13.3.2.1 General Aspects

Active packaging involves the incorporation of additives into the packaging with the objective of improving the quality and extending the shelf life of products. According to Coma (2008) and Appendini and Hotchkiss (2002) there are different ways in which the additives can be incorporated:

1. Addition of the active compound to a sachet or pad placed in the package and from which the volatile agent is released
2. Incorporation of the active compound in the packaging film during film manufacture
3. Coating of the packaging with a matrix that acts as a carrier for the active compound
4. Use of an active compound that also possesses film-forming properties or a chemically modified polymer exhibiting bioactive activity
5. Incorporation of the active compound in an edible coating applied onto the food

Oxygen scavengers, carbon dioxide and ethanol generators, and moisture absorbers are active compounds included in sachets or pads. Most of them were developed commercially, being available for processors with different options. Oxygen scavengers are used widely in processed meat products (Kerry et al. 2006).

Incorporation of active compounds into the packaging during film extrusion has been commercially applied by the pharmaceutical and agrochemical industry, but few food applications have been reported. However, a significant number of research reports and patents have been generated. One disadvantage of this technique is that the active compound can be degraded owing to exposure to high temperature and shearing forces of the extrusion process. Different bioactive molecules such as antimicrobials and antioxidants have been included in many films to improve food shelf life (Coma 2008; Appendini and Hotchkiss 2002; Devlieghere et al. 2000; Gennadios et al. 1997).

When the active compound is coated on the surface of the packaging material, any thermal stability problem with the active compound is avoided. Besides, coating can be performed in a later process step so as to minimize the possibility of microbial contamination.

Use of active compounds that also have film-forming properties can be achieved by using chitosan, poly(L-lysine), or chemically modified polymers, which exhibit the active property. The incorporation of an active compound into an edible coating is performed by dipping or spraying on the food. Edible coatings are prepared mainly from polysaccharides, proteins, and lipids. Their use has many advantages: edibility, biodegradability, aesthetic appearance. Moreover, they exhibit barrier properties against oxygen and mechanical stress.

All forms of active packaging have received remarkable attention in recent years because of the important role that they can play in extending food shelf life (Appendini and Hotchkiss 2002; Quintavalla and Vicini 2002). Nevertheless, research studies have been more intense in the field of active edible films and coatings. This trend is due to many factors: the need for new storage techniques, the development of new markets for agricultural commodities and wastes with film-forming properties, etc.

Finally, it must be stressed that the incorporation of preservatives into the packaging instead of their inclusion in the bulk has been proposed to prevent surface microbial growth or lipid oxidation. It must be pointed out that meat from healthy specimens is sterile and microbial growth occurs at the surface. Therefore, active agent in a smaller amount at the place where microbial growth occurs can be more effective. Moreover, a controlled release of the agent from the surface to the inner parts of the food ensures the presence of the agent in the food throughout storage. In contrast, when the preservative is added to the food, it can bind to food components or can react with other additives or food components, losing its activity. The latter trend was observed for nisin in the presence of fat and emulsifiers (Appendini and Hotchkiss 2002).

13.3.2.2 Applications of Active Packaging Systems to Aquatic Products

Numerous active packaging systems have been designed to extend the shelf life of aquatic products. These active systems are used as a single stress factor or as part of a system with multiple hurdles. The active compounds used are antimicrobials and antioxidants in order to prevent microbial growth and lipid oxidation, two deteriorative reactions that determine the end of shelf life in most cases.

Use of active films and coatings for meats in general and in particular for seafood preservation has some advantages:

1. The diminishment of moisture loss throughout refrigerated storage
2. The reduction of lipid oxidation and the sensory spoilage induced by this deteriorative reaction
3. The decrease of the populations of spoilage and pathogenic flora
4. The reduction of volatile flavor loss and foreign odor pickup

Table 13.2 Application of active packaging systems to extend the shelf life of aquatic products

Active component	Carrier	Aquatic product	Deteriorative reaction target	References
Sodium lactate, benzoate, and diacetate, potassium sorbate, nisin	Chitosan-coated plastic films	Cold-smoked salmon	Microbial growth	Ye et al. (2008)
Oyster and hen lysozyme, nisin	Calcium alginate	Smoked salmon	Microbial growth	Datta et al. (2008)
Chitosan	Chitosan	Herring cod	Lipid oxidation, microbial growth	Jeon et al. (2002)
Nisin	Cellulose derivates coated on polyethylene	Cold-smoked salmon	Microbial growth	Neetoo et al. (2008)
Thyme oil, cynamaldehyde	Soy and whey protein, carboxymethylcellulose	Cooked shrimp	Microbial growth	Ouattara et al. (2001)
Oregano and rosemary extracts	Gelatine, gelatine–chitosan	Cold-smoked sardine	Microbial growth, lipid oxidation	Gomez-Estaca et al. (2007)
Butylhydroxytoluene	Polyethylene	Sierra fish	Lipid oxidation, protein quality	Torres-Arreola et al. (2007)
Butylhydroxytoluene	Polyethylene	Tilapia, perch	Lipid oxidation	Huang and Weng (1998)
Chitosan	Chitosan, chitosan–starch	Salmon	Microbial growth	Vásconez et al. (2007)
Butylhydroxyanisole	Rohu proteins	Rohu	Lipid oxidation, color changes	Panchavarnam et al (2003)
Glucose oxidase/catalase	Algin blankets	Winter flounder	Endogenous enzyme activity	Field et al. (1986)
Glucose oxidase/catalase	Glucose solution	Shrimp	Endogenous enzyme activity, microbial growth	Dondero et al. (1993)

Probably, one of the first reports of active packaging was the oxygen scavenging ability of glucose oxidase for extending the shelf life of fish as reported by Field et al. (1986). A summary of the most relevant active packaging systems developed is given in Table 13.2. One of the main objectives is to prevent the contamination by spoilage flora and in many cases the target of this inhibition is *L. monocytogenes* growth, which constitutes the major risk in freshly processed cold-smoked salmon (Ye et al. 2008; Datta et al. 2008; Cortesi et al. 1997. On the other hand, another important objective is to avoid oxidative spoilage by means of oxygen scavengers or by the use of antioxidant agents (Torres-Arreola et al. 2007; Gomez-Estaca et al. 2007; Panchavarnam et al. 2003; Dondero et al. 1993).

13.3.3 Future Trends in Packaging of Aquatic Food Products

Nowadays, MAP and active packaging play an important role in improving the quality and extending the shelf life of seafood. However, these technologies cannot replace good-quality foodstuff and the use of good and hygienic handling and manufacturing practices. Future work will focus on the use of natural preservatives bound to polymers. Some issues yet to be solved for commercial applications include the following: changes in sensory properties of the food induced by the novel technology; the interaction of active agents with the specific food matrix; and the influence of this interaction on the effectiveness of the active agent. Finally, the economic feasibility of the processes and the consumer acceptance are key factors to ensure the widespread use of such novel technologies.

References

Ababouch L, Souibri L, Rhaliby K, Ouahdi O, Battal M, Busta F. Quality changes in sardines (*Sardina pilchardus*) stored in ice and at ambient temperature. Food Microbiol. 1996;13:123–32.

Álvarez A, Feás X, Barros-Velázquez J, Aubourg SP. Biochemical, microbiological and sensory quality of farmed blackspot seabream (*Pagellus bogaraveo*) subjected to slaughtering and storage under advanced chilling technologies. Int J Food Sci Technol. 2008;44:1561–71.

Appendini P, Hotchkiss JH. Review of antimicrobial food packaging. Innov Food Sci Emerg Technol. 2002;3:113–26.

Ashie I, Smith J, Simpson B. Spoilage and shelf-life extension of fresh fish and shellfish. Crit Rev Food Sci Nutr. 1996;36:87–121.

Aubourg S, Medina I. Quality differences assessment in canned sardine (*Sardina pilchardus*) by detection of fluorescent compounds. J Agric Food Chem. 1997;45:3617–21.

Aubourg S, Ugliano M. Effect of brine pre-treatment on lipid stability of frozen horse mackerel (*Trachurus trachurus*). Eur Food Res Technol. 2002;215:91–5.

Aubourg SP, Losada V, Gallardo JM, Miranda JM, Barros-Velázquez J. On-board quality preservation of megrim (*Lepidorhombus whiffiagonis*) by a novel ozonised-slurry ice system. Eur Food Res Technol. 2006;223:232–7.

Aubourg SP, Losada V, Prado M, Miranda JM, Barros-Velázquez J. Improvement of the commercial quality of chilled Norway lobster (*Nephrops norvegicus*) stored in slurry ice: Effects of a preliminary treatment with antimelanosic agent of enzymatic browning. Food Chem. 2007;103:74s1–8.

Aubourg S, Sotelo C, Gallardo J. Quality assessment of sardines during storage by measurement of fluorescent compounds. J Food Sci. 1997;62:295–9.

Bandarra N, Batista I, Nunes M, Empis J. Seasonal variation in the chemical composition of horse mackerel (*Trachurus trachurus*). Eur Food Res Technol. 2001;212:535–9.

Barnett HJ, Stone FE, Roberts GC, Hunter PJ, Nelson RW, Kwok J. A study in the use of a high concentration of CO_2 in a modified atmosphere to preserve fresh salmon. Mar Fish Rev. 1982;44:7–11.

Barros-Velázquez J, Gallardo JM, Calo P, Aubourg SP. Enhanced quality and safety during on-board chilled storage of fish species captured in the Grand Sole North Atlantic fishing bank. Food Chem. 2008;106:493–500.

Bennour M, El Marrakchi A, Bouchriti N, Hamama A, El Ouadaa M. Chemical and microbiological assessments of mackerel (*Scomber scombrus*) stored in ice. J Food Prot. 1991;54:784, 789–792.

Brown W. Fish muscle as food. In: Bechtel P, editor. Muscle as food. Orlando: Academic; 1986. p. 459.

Campos C, Losada V, Rodríguez O, Aubourg S, Barros-Velázquez J. Evaluation of an ozone-slurry ice combined refrigeration system for the storage of farmed turbot (*Psetta maxima*). Food Chem. 2006;97:223–30.

Campos C, Rodríguez Ó, Losada V, Aubourg S, Barros-Velázquez J. Effects of storage in ozonised slurry ice on the sensory and microbial quality of sardine (*Sardina pilchardus*). Int J Food Microbiol. 2005;103:121–30.

Chapman L. Making the grade. Ice slurries get top marks for quality products, Australian Fisheries; July 1990. pp. 16–9.

Chinivasagam HN, Bremner HA, Wood AF, Nottingham SM. Volatile components associated with bacterial spoilage of tropical prawns. Int J Food Microbiol. 1998;42:45–55.

Choubert G, Baccaunaud M. Colour changes of fillets of rainbow trout (*Oncorhynchus mykiss* W.) fed astaxanthin or canthaxanthin during storage under controlled or modified atmosphere. LWT. 2006;39:1203–13.

Coma V. Bioactive packaging technologies for extended shelf life of meat-based products. Meat Sci. 2008;78:90–103.

Cortesi M, Santoro A, Murru N, Pepe T. Distribution and behavior of *Listeria monocytogenes* in three lots of naturally-contaminated vacuum-packed smoked salmon stored at 2 and 10°C. Int J Food Microbiol. 1997;37:209–14.

Damoglou A. A comparison of different methods of freshness assessment of herring. In: Connell J, editor. Advances in fish science and technology. Farnham: Fishing News Books; 1980. pp. 394–9.

Datta S, Janes ME, Xue QG, La Peyre JF. Control of *Listeria monocytogenes* and *Salmonella anatum* on the surface of smoked salmon coated with calcium alginate coating containing oyster lysozyme and nisin. J Food Sci. 2008;73(2):M67–71.

Debevere J, Boskou G. Effect of modified atmosphere packaging on the TVB/TMA-producing microflora of cod fillets. Int J Food Microbiol. 1996;31:221–9.

Devlieghere F, Vermeiren L, Jacobs M, Debevere J. The effectiveness of hexamethylenetetramine-incorporated plastic for the active packaging of foods. Packag Technol Sci. 2000;13:117–21.

Devlieghere F, Debevere J, Van Impeb J. Concentration of carbon dioxide in the water-phase as a parameter to model the effect of a modified atmosphere on microorganisms. Int J Food Microbiol. 1998;43:105–13.

Dondero M, Egaña W, Tarky W, Cifuentes A, Torres A. Glucose oxidase/catalase improves preservation of shrimp (*Heterocarpus reedi*). J Food Sci. 1993;58:774–9.

El Marrakchi A, Bennour M, Bouchriti N, Hamama A, Tagafait H. Sensory, chemical and microbiological assessments of Moroccan sardines (*Sardina pilchardus*) stored on ice. J Food Prot. 1990;53:600–5.

Erikson U, Beyer A, Sigholt T. Muscle high-energy phosphates and stress affect K-values during ice storage of Atlantic salmon (*Salmo salar*). J Food Sci. 1997;62:43–7.

Fagan JD, Gormley TR, Ui Mhuircheartaigh MM. Effect of modified atmosphere packaging with freeze-chilling on some quality parameters of raw whiting, mackerel and salmon portions. Innov Food Sci Emerg Technol. 2004;5:205–14.

FAO. Fishery statistics. Aquaculture production. In: Food and Agriculture Organization of the United Nations, Yearbook 2007, Vol. 100/2. Rome: Food and Agriculture Organization of the United Nations; 2007a. p. 23.

FAO. Fishery statistics. Capture production. In: Food and Agriculture Organization of the United Nations, Yearbook 2007, Vol. 100/1. Rome: Food and Agriculture Organization of the United Nations; 2007b. p. 21.

FAO. Commodities. In: Food and Agriculture Organization of the United Nations, Yearbook 2007, Vol. 101. Rome: Food and Agriculture Organization of the United Nations; 2007c. p. 11.

Fernández-Salguero J, Mackie I. Comparative rates of spoilage of fillets and whole fish during storage of haddock (*Melanogrammus aeglefinus*) and herring (*Clupea harengus*) as determined by the formation of non-volatile and volatile amines. Int J Food Sci Technol. 1987;22:385–90.

Fey MS, Regenstein JM. Extending shelf-life of fresh wet red hake and salmon using CO_2,-O_2 modified atmosphere and potassium sorbate ice at 1°C. J Food Sci. 1982;47:1048–54.

Field CE, Pivarnik LF, Barnett SM, Rand Jr AG. Utilization of glucose oxidase for extending the shelf-life of fish. J Food Sci. 1986;51:66–70.

Gallardo J, Sotelo C, Pérez-Martín R. Determination of histamine by capillary zone electrophoresis using a low-pH phosphate buffer: application in the analysis of fish and marine products. Z Lebensm Unters Forsch. 1997;204:336–40.

Gennadios A, Hanna M, Kurth L. Application of edible coatings on meats, poultry and seafoods: a review. LWT Food Sci Technol. 1997;30:337–50.

Giatrakou V, Kykkidou S, Papavergou A, Kontominas MG, Savvaidis IN. Potential of oregano essential oil and MAP to extend the shelf life of fresh swordfish: a comparative study with ice storage. J Food Sci. 2008;73:167–73.

Gladis SP. Ice slurry thermal energy storage for cheese process cooling. American Society of Heating, Refrigeration and Air-Conditioning Engineers. ASHRAE Transactions 1999; 103, part 2

Gomez-Estaca J, Montero P, Giménez B, Gómez-Guillén MC. Effect of functional edible films and high pressure processing on microbial growth and oxidative spoilage in cold-smoke sardine (*Sardina pilchardus*). Food Chem. 2007;105:511–20.

Graham J, Johnston WA, Nicholson FJ El hielo en las pesquerías, FAO Fisheries Technical Paper 331. 1993. http://www.fao.org.

Harada. How to handle Albacore. Australian Fisheries; February 1991. pp. 28–30.

Hartel RW. Solid-liquid equilibrium: crystallization in foods. In: Schwartzberg HG, Hartel RW, editors. Physical chemistry of foods. New York: Marcel Dekker; 1992. pp. 47–81.

Heap R. Refrigeration of chilled foods. In: Dennis C, Stringer M, editors. Chilled foods. A comprehensive guide. Chichester: Ellis Horwood; 1992. pp. 59–76.

Hovda MB, Sivertsvik M, Lunestad BT, Lorentzen G, Rosnes JT. Characterisation of the dominant bacterial population in modified atmosphere packaged farmed halibut (*Hippoglossus hippoglossus*) based on 16S rDNA-DGGE. Food Microbiol. 2007;24:362–71.

Huang CH, Weng YM. Inhibition of lipid oxidation in fish muscle by antioxidant incorporated polyethylene film. J Food Process Preserv. 1998;22:199–209.

Huidobro A, López-Caballero ME, Mendes R. Onboard processing of deepwater pink shrimp (*Parapenaeus longirostris*) with liquid ice: effect on quality. Eur Food Res Technol. 2002;214:469–75.

Huidobro A, Mendes R, Nunes ML. Slaughtering of gilthead seabream (*Sparus aurata*) in liquid ice: influence on fish quality. Eur Food Res Technol. 2001;213:267–72.

Huidobro A, Montero P, Tejada M, Colmenero F, Borderías J. Changes in protein function of sardines stored in ice. Z Lebensm Unters Forsch. 1990;190:195–8.

Hwang K, Regenstein J. Characteristics of mackerel mince lipid hydrolysis. J Food Sci. 1993;58:79–83.

Jeon YJ, Kamil JYVA, Shahidi F. Chitosan as an edible invisible film for quality preservation of herring and Atlantic cod. J Agric Food Chem. 2002;50:5167–78.

Jul M. Chilling and freezing fishery products: changes in view and usages. Int J Refrig. 1986;9:174–8.

Kerry JP, Grady MN, Hogan SA. Past, current and potential utilization of active and intelligent packaging systems for meat and muscle-based products. A review. Meat Sci. 2006;74:113–30.

Kolakowska A, Kwiatkowska L, Lachowicz K, Gajowiecki L, Bortnowska G. Effect of fishing season on frozen-storage quality of Baltic herring. In: Bligh E, editor. Seafood science and technology. Oxford: Canadian Institute of Fisheries Technology, Fishing News Books; 1992. pp. 269–77.

Kraus L. Refrigerated sea water treatment of herring and mackerel for human consumption. In: Burt J, Hardy R, Whittle K, editors. Pelagic fish. The resource and its exploitation. Aberdeen: Fishing News Books; 1992. pp. 73–81.

Kristinsson HG, Crynen S, Yagiz Y. Effect of a filtered wood smoke treatment compared to various gas treatments on aerobic bacteria in yellowfin tuna steaks. LWT Food Sci Technol. 2008;41:746–50.

Lannelongue M, Finne G, Hanna MO, Nickelson R, Vanderzant G. Storage characteristics of brown shrimp (*Penaeus aztecus*) stored in retail packages containing CO_2-enriched atmospheres. J Food Sci. 1982;47:911–913,923.

Losada V, Barros-Velázquez J, Gallardo J, Aubourg S. Effect of advanced chilling methods on lipid damage during sardine (*Sardina pilchardus*) storage. Eur J Lipid Sci Technol. 2004a;106: 844–50.

Losada V, Piñeiro C, Barros-Velázquez J, Aubourg S. Inhibition of chemical changes related to freshness loss during storage of horse mackerel (*Trachurus trachurus*) in slurry ice. Food Chem. 2005;93:619–25.

Losada V, Piñeiro C, Barros-Velázquez J, Aubourg S. Effect of slurry ice on chemical changes related to quality loss during European hake (*Merluccius merluccius*) chilled storage. Eur Food Res Technol. 2004b;219:27–31.

Losada V, Rodríguez O, Miranda JM, Barros-Velázquez J, Aubourg SP. Development of different damage pathways in Norway lobster (*Nephrops norvegicus*) stored under different chilling systems. J Sci Food Agric. 2006;86:1552–8.

Martinsdóttir E, Valdimarsdóttir P, Porkelsdóttir A, Olafsdóttir G, Tryggvadóttir SV. Shelf life of sea bass (*Dicentrarchus labrax*) in liquid and flake ice studied by quality index method (Q I M) electronic nose and texture. Proceedings of 32nd Annual WEFTA Meeting, Galway; 13–15 May 2002

Masniyom P, Benjakul S, Visessanguan W. Shelf-life extension of refrigerated sea bass slices under modified atmosphere packaging. J Sci Food Agric. 2002;82:873–80.

Morzel M, Sohier D, Van de Vis H. Evaluation of slaughtering method for turbot with respect to animal welfare and flesh quality. J Sci Food Agric. 2002;82:19–28.

Múgica B, Barros-Velázquez J, Miranda JM, Aubourg SP. Evaluation of a slurry ice system for the commercialization of ray (*Raja clavata*): effects on spoilage mechanisms directly affecting quality loss and shelf life. LWT Food Sci Technol. 2008;41:974–81.

Neetoo H, Ye M, Chen H, Joerger RD, Hicks DT, Hoover DG. Use of nisin-coated plastic films to control *Listeria monocytogenes* on vacuum-packaged cold-smoked salmon. Int J Food Microbiol. 2008;122:8–15.

Nunes M, Batista I, Morão de Campos R. Physical, chemical and sensory analysis of sardine (*Sardina pilchardus*) stored in ice. J Sci Food Agric. 1992;59:37–43.

Olafsdóttir G, Martinsdóttir E, Oehlenschläger J, Dalgaard P, Jensen B, Undeland I, et al. Methods to evaluate fish freshness in research and industry. Trends Food Sci Technol. 1997;8:258–65.

Ouattara B, Sabato SF, Lacroix M. Combined effect of antimicrobial coating and gamma irradiation on shelf life extension of pre-cooked shrimp (*Penaeus spp.*). Int J Food Microbiol. 2001;68:1–9.

Özogul F, Özogul Y. Biogenic amine content and biogenic amine quality indices of sardines (*Sardina pilchardus*) stored in modified atmosphere packaging and vacuum packaging. Food Chem. 2006;99:574–8.

Özogul F, Polat A, Özogul Y. The effects of modified atmosphere packaging and vacuum packaging on chemical, sensory and microbiological changes of sardines (*Sardina pilchardus*). Food Chem. 2004;85:49–57.

Özogul F, Taylor KDA, Quantick P, Özogul Y. Chemical, microbiological and sensory evaluation of Atlantic herring (*Clupea harengus*) stored in ice, modified atmosphere and vacuum pack. Food Chem. 2000;71:267–73.

Özogul Y, Özogul F. Effects of slaughtering methods on sensory, chemical and microbiological quality of rainbow trout (*Oncorhynchus mykiss*) stored in ice and MAP. Eur Food Res Technol. 2004;219:211–6.

Panchavarnam S, Basu S, Manisha K, Warrier SB, Venugopal V. Preparation and use of freshwater fish, rohu (Labeo rohita) protein dispersion in shelf-life extension of the fish steaks. LWT Food Sci Technol. 2003;36:433–9.

Pantazi D, Papavergou A, Pournis N, Kontominas MG, Savvaidis IN. Shelf-life of chilled fresh Mediterranean swordfish (*Xiphias gladius*) stored under various packaging conditions: microbiological, biochemical and sensory attributes. Food Microbiol. 2008;25:136–43.

Pearson A, Love J, Shorland F. Warmed-over flavor in meat, poultry and fish. Adv Food Res. 1977;23:2–61.

Pérez-Villarreal B, Pozo R. Chemical composition and ice spoilage of albacore (*Thunnus alalunga*). J Food Sci. 1990;55:678–82.

Piñeiro C, Barros-Velázquez J, Aubourg SP. Effects of newer slurry ice systems on the quality of aquatic food products: a comparative review versus flake ice chilling methods. Trends Food Sci Technol. 2004;15:575–82.

Piñeiro C, Bautista R, Rodríguez O, Losada V, Barros-Velázquez J, Aubourg SP. Quality retention during the chilled distribution of farmed turbot (*Psetta maxima*): effect of a primary slurry ice treatment. Int J Food Sci Technol. 2005;40:817–24.

Price RJ, Melvin EF, Bell JW. Postmortem changes in chilled round, bled and dressed albacore. J Food Sci. 1991;56(2):318–21.

Quintavalla S, Vicini L. Antimicrobial food packaging in meat industry. Meat Sci. 2002;62:373–80.

Randell K, Hattula T, Ahvenainen R. Effect of packaging method on the quality of rainbow trout and Baltic herring fillets. LWT Food Sci Technol. 1997;30:56–61.

Rodríguez Ó, Losada V, Aubourg S, Barros-Velázquez J. Sensory, microbial and chemical effects of a slurry ice system on horse mackerel (*Trachurus trachurus*). J Sci Food Agric. 2005;85:235–42.

Rodríguez Ó, Barros-Velázquez J, Piñeiro C, Gallardo J, Aubourg S. Effects of storage in slurry ice on the microbial, chemical and sensory quality and on the shelf life of farmed turbot (*Psetta maxima*). Food Chem. 2006;95:270–8.

Rodríguez O, Losada V, Aubourg S, Barros-Velázquez J. Enhanced shelf-life of chilled European hake (*Merluccius merluccius*) stored in slurry ice as determined by sensory analysis and assessment of microbiological activity. Food Res Int. 2004;37:749–57.

Ruiz-Capillas C, Moral A. Chilled bulk storage of gutted hake (*Merluccius merluccius* L.) in CO_2 and O_2 enriched controlled atmospheres. Food Chem. 2001;74:317–25.

Rutherford TJ, Marshall DL, Andrews LS, Coggins PC, Schilling MW, Gerard P. Combined effect of packaging atmosphere and storage temperature on growth of *Listeria monocytogenes* on ready-to-eat shrimp. Food Microbiol. 2007;24:703–10.

Scherer R, Augusti P, Staffens C, Bochi V, Hecktheuer L, Lazzari R, et al. Effect of slaughter method on postmortem changes in grass carp (*Ctenopharyngodon idella*) stored in ice. J Food Sci. 2005;70:C348–52.

Simopoulos A. Nutritional aspects of fish. In: Luten J, Börrensem T, Oehlenschläger J, editors. Seafood from producer to consumer, integrated approach to quality. London: Elsevier Science; 1997. pp. 589–607.

Sivertsvik M. "The optimized modified atmosphere for packaging of pre-rigor filleted farmed cod (*Gadus morhua*) is 63 ml/100 ml oxygen and 37 ml/100 ml carbon dioxide. LWT Food Sci Technol. 2007;40:430–8.

Sivertsvik M, Jeksrud WK, Rosnes JT. A review of modified atmosphere packaging of fish and fishery products – significance of microbial growth, activities and safety. Int J Food Sci Technol. 2002;37:107–27.

Sivertsvik M, Rosnes JT, Jeksrud WK. Solubility and absorption rate of carbon dioxide into non-respiring foods. Part 2: raw fish fillets. J Food Eng. 2004;63:451–8.

Smith J, Hardy R, McDonald I, Templeton J. The storage of herring (*Clupea harengus*) in ice, refrigerated sea water and at ambient temperature. Chemical and sensory assessment. J Sci Food Agric. 1980;31:375–85.

Tejada M, Huidobro A. Quality of farmed gilthead seabream (*Sparus aurata*) during ice storage related to the slaughter method and gutting. Eur Food Res Technol. 2002;215:1–7.

Toledo-Flores L, Zall R. Methods for extending the storage life of fresh tropical fish. In: Flick G, Martin R, editors. Advances in seafood biochemistry. Composition and quality. Lancaster: Technomic Publishing Company; 1992. pp. 233–43.

Torres-Arreola W, Soto Valdez H, Peralta E, Cárdenas–Lopez JL, Ezquerra-Brauer JM. Effect of low density polyethylene film containing butylated hydroxytoluene on lipid oxidation and protein quality of Sierra fish (Scomberomorus sierra) muscle during frozen storage. J Agric Food Chem. 2007;55:6140–6.

Torrieri E, Cavella S, Villani F, Masi P. Influence of modified atmosphere packaging on the chilled shelf life of gutted farmed bass (*Dicentrarchus labrax*). J Food Eng. 2006;77:1078–86.

Undeland I, Lingnert H. Lipid oxidation in fillets of herring (*Clupea harengus*) during frozen storage. Influence of prefreezing storage. J Agric Food Chem. 1999;47:2075–81.

Undeland I, Hall G, Lingnert H. Lipid oxidation in fillets of herring (*Clupea harengus*) during ice storage. J Agric Food Chem. 1999;47:524–32.

Vásconez MB, Campos CA, Flores S, Alvarado de Dios J, Gerschenson L. Elaboración de recubrimientos comestibles en base a quitosano y estudio de su efecto antimicrobiano en filetes de salmón. Cienc Tecnol. 2007;16(3)77–9.

Wang T, Sveinsdóttir K, Magnússon H, Martinsdóttir E. Combined application of modified atmosphere packaging and superchilled storage to extend the shelf life of fresh cod (*Gadus morhua*) loins. J Food Sci. 2008;73(1):S11–9.

Waterman JJ, Taylor DH. Superchilling, Torry Research Station Torry Advisory Note No. 32. FAO in partnership with Support Unit for International Fisheries and Aquatic Research, SIFAR; 2001.

Whittle K, Hardy R, Hobbs G. Chilled fish and fishery products. In: Gormley T, editor. Chilled foods. The state of the art. New York: Elsevier; 1990. pp. 87–116.

Yamada M, Fukusako S, Kawanami T. Performance analysis on the liquid-ice thermal storage system for optimum operation. Int J Refrig. 2002;25:267–77.

Ye M, Neetoo H, Chen H. Effectiveness of chitosan-coated plastic films incorporating antimicrobials in inhibition of *Listeria monocytogenes* on cold-smoked salmon. Int J Food Microbiol. 2008; 127:235–40.

Chapter 14
Use of Natural Preservatives in Seafood

Carmen A. Campos, Marcela P. Castro, Santiago P. Aubourg,
and Jorge Barros Velázquez

14.1 Introduction

Seafood products are known to be especially susceptible to both microbiological and biochemical spoilage pathways. Accordingly, efficient and hygienic preservation processes should be applied immediately after capture/slaughter to preserve product freshness and quality. The development of effective processing treatments to extend the shelf life of fresh fish products is a must. Additionally, the consumers' demand for high-quality and minimally processed seafood has recently captivated great attention. However, an increase in foodborne illness outbreaks is concomitant with the increase in consumer demand for less processed foods (Cherry 1999). These trends highlight the importance of studying new microbial stress factors. Among them, natural preservatives are one of the stress factors proposed to improve quality and extend the shelf life of seafoods, since they are perceived by consumers to be safer than chemical or synthetic additives.

C.A. Campos (✉)
Departamento de Indusatrias, Facultad de Ciencias Exactas y Naturales,
Universidad de Buenos Aires, Intendente Guiraldes 2160, C.A.B.A., 1428, Argentina
e-mail: carmen@di.fcen.uba.ar

M.P. Castro
Universidad Nacional del Chaco Austral, Comandante,
Fernández 755, P.R. Sáenz Peña, Chaco, Argentina

S.P. Aubourg
Instituto de Investigaciones Marinas de Vigo (CSIC), C/Eduardo Cabello 6,
Vigo, Pontevedra, E-36208, Spain

J.B. Velázquez
Department of Analytical Chemistry, Nutrition and Food Science, LHICA, School of Veterinary Sciences, University of Santiago de Compostela, E-27002, Lugo, Spain

Laboratory of Biotechnology, College of Pharmacy, University of Santiago de Compostela, ,
E-15782, Santiago, Spain

In recent decades, numerous studies demonstrating the antimicrobial activity of natural compounds against pathogenic and spoilage microorganisms have been conducted. Moreover, it is known that factors such as pH, temperature, and food components can greatly influence the applicability of a food product and it is a must to perform studies on the specific products to test stability throughout the expected shelf life (Roller 2003).

Seafood deterioration occurs owing to microbial growth, oxidation of lipids, and endogenous enzyme activity, which can produce lipid hydrolysis as a consequence of activity of lipases. In the case of crustaceans, where the action of polyphenol oxidases (PPO) induces enzymatic browning, an additional important reaction must be effectively controlled. In this sense, antioxidant and antibrowning agents can be used to prevent these reactions.

Antimicrobial, antioxidant, and antibrowning agents are all additives used for extending the shelf life of a food and all of them are designed as preservatives according to Branen and Haggerty (2002).

14.2 Use of Natural Antimicrobial Preservatives

The use of many natural preservatives has been proposed to increase the shelf life of seafoods. Most of them are antimicrobial agents, but many also act as antioxidants, such is the case of chitosan and some herbs and spices, whereas others, such as many phenolic compounds extracted from plants, also act as antibrowning agents.

14.2.1 Organic Acids

Sodium and potassium salts from low molecular weight organic acids such as acetic acid, lactic acid, and citric acid have been used to control microbial growth, improve sensory quality, and extend the shelf life of many food types, including seafood products. These preservatives are able to control the development of spoilage bacteria and also possess antibacterial activity against various foodborne pathogens such as *Staphylococcus aureus*, *Yersinia enterocolitica*, *Listeria monocytogenes*, *Escherichia coli*, and *Clostridium botulinum*. Moreover, these salts are economical, widely available, and generally recognized as safe (GRAS).

Antimicrobial efficacy depends on pH, water activity, moisture, fat, nitrite and salt content of the product, and storage conditions (temperature, packaging atmosphere) (Chen and Shelef 1992; Shelef 1994; Buncic et al. 1995; Houtsma et al. 1996; Nebrink et al. 1999). The antimicrobial action increases when the additives are combined, even at reduced concentrations (Qvist et al. 1994; Blom et al. 1997; Stekelenburg and Kant-Muermans 2001; Mbandi and Shelef 2001).

Several authors have tested low molecular weight organic acids for their antimicrobial activity against *Listeria* and other spoilage and pathogenic bacteria in aquatic food products.

Sorbic acid and its salts have been used as antimicrobials in a wide variety of food products. It is known that the salts, alone or in combination with other microbial stress factors, are able to inhibit the growth of spoilage flora and pathogens such as *S. aureus* and *C. botulinum* in fish (Sofos 2000). Moreover, it was shown that when sorbates were used at a level not exceeding 0.26% (w/w), no adverse effect on sensory properties was observed (Thakur and Patel 1994). According to the information mentioned, sorbates could be a valuable alternative for the preservation of fish products, but consumers perceive them as chemical preservatives despite their natural origin as extracts from the rowan tree; the compound used in industry is produced by chemical synthesis.

L. monocytogenes contamination is the major risk factor in freshly processed cold-smoked salmon (Joffraud et al. 2006). The inclusion of potassium lactate and sodium diacetate was proposed to solve this problem. Vogel et al. (2006) demonstrated that the addition of potassium lactate (2%) or sodium diacetate (0.14%) to cold-smoked salmon by brine injection prevented the growth of *L. monocytogenes* for 42 days at 10°C in a vacuum-packed product. Moreover, the sensory analysis revealed that the preservatives did not negatively affect flavor or odor. Yoon et al. (2004) also reported that concentrations of potassium lactate and sodium diacetate completely inhibited the growth of *L. monocytogenes* stored at 4°C for 32 days. Sallam (2007) reported that a dipping treatment of 10 min at 4°C in a 2.5% (w/v) aqueous solution of sodium acetate, sodium lactate, or sodium citrate delayed the growth of spoilage flora and diminished lipid oxidation for 15 days at 1°C of salmon fillets packed in polyvinylidene film. A dip treatment for 30 min with a solution containing 2% sodium acetate or potassium sorbate extended the shelf life up to 15 days for vacuum-packed black pomfret or pearlspot under refrigerated storage (Manju et al. 2007).

Consumers generally accept the use of organic acids and their salts in foods since they regard organic acids as food-grade compounds, and recognize their use in households as flavorings or natural food acidulants from ancient times. The only point of concern is that according to the amount added, they can negatively influence sensory quality.

14.2.2 Antimicrobials from Herbs and Spices

Since ancient times, spices and herbs have been added to foods as seasoning additives because of their aromatic properties. They also exhibit useful antimicrobial and antioxidant activity and many of them are classified as GRAS (Kabara 1991). However, their use in foods as preservatives is limited because of flavor considerations, since effective antimicrobial doses may exceed sensorially acceptable levels. Knowledge of the mechanisms and factors determining antimicrobial action are essential to overcome the problems mentioned.

The antimicrobial activity of plant-derived compounds against many different microorganisms tested individually and in vitro is well documented in the literature

(Nychas and Skandamis 2003). They inhibit the microbial growth of Gram-positive and Gram-negative bacteria, yeasts, and molds. The active compounds responsible for the antimicrobial activity of spices are primarily phenolic components of the essential oil fraction (Beuchat 1994). The antimicrobial activity of cinnamon, allspice, and cloves is attributed to eugenol (2-methoxy-4-allylphenol) and cinnamic aldehyde, which are major constituents of the volatile oils of these spices. The essential oil of oregano contains up to 50% thymol; thyme has 43% thymol and 36% *p*-cymene, and savory has 30–45% carvacrol and 30% *p*-cymene (Farag et al. 1989). Antimicrobial activity is influenced by the culture medium, the temperature of incubation, and the inoculum size. Moreover, a synergistic action was found by the joint use with some membrane chelators acting as permeabilizing agents (e.g., ethylenediaminetetraacetic acid) against Gram-negative bacteria (Nychas and Skandamis 2003).

In comparison with the numbers of studies performed in vitro, there have been few studies of the antimicrobial action of essential oils in food systems. In particular, some applied studies were performed with aquatic products. Mejllholm and Dalgaard (2002) reported that oregano oil (0.05%, v/w) extended the shelf life of naturally contaminated modified-atmosphere-packaged cod fillets from 11–12 days to 21–26 days at 2°C. The shelf-life extension resulted from the reduced growth of aerobic microorganisms and *Photobacterium phosphoreum*, a specific spoilage microorganism of modified-atmosphere-packaging. Also, the trimethylamine levels were reduced. From the sensory point of view, the fillets had a distinctive herbal flavor which decreased gradually throughout storage. Neither the texture nor the appearance of the cod fillets was affected by the oregano oil addition. When the oregano oil was added to the modified-atmosphere-packaged salmon fillets, no effect was detected on the development of *P. phosphoreum*. This differential behavior of oregano oil was related to the higher lipid content of salmon. The active compounds of oregano oil, primary thymol and carvacrol, were probably dissolved in the lipid phase of the salmon fillets and therefore had no effects on *P. phosphoreum* growth in the aqueous phase. The lipophilicity of essential oils is a factor that restricted their use as antimicrobials in systems with a lipid phase.

In relation to other essential oils, oregano is more effective on fish than is mint oil, even in fatty fish dishes. This trend was confirmed in two experiments with fish roe salad using both essential oils at the same concentration (0.5–2.0 v/w) (Tassou et al. 1995; Koutsoumanis et al. 1999). Mahmoud et al. (2004) reported that dipping carp (*Cyprinus carpio*) fillets in a solution of 0.5% carvacrol and 0.5% thymol before storage at 5°C reduced the numbers of indigenous flora and decreased the level of volatile base nitrogen. As a consequence, the shelf life of the fillets was extended from 4 to 12 days as determined by sensory evaluation. The incorporation of thyme oil and *trans*-cinnamaldehyde into an edible film prepared from soy protein isolates reduced the microbial growth in precooked shrimp subjected to gamma irradiation. The sensory quality was not affected by essential oil up to 0.9%. However, the incorporation of 1.8% essential oils in the coating solutions significantly decreased the acceptability of the product (Ouattara et al. 2001).

In relation to the activity of essential oils to inhibit foodborne pathogens such as *Salmonella typhimurium*, Kim et al. (1995) found that a dipping treatment of 10 min in a 1% Tween 20 solution containing 0.5%, 1.5%, or 3.0% w/v carvacrol, citral, or geraniol decreased counts of the pathogen in inoculated red grouper fillets (*Epinephelus moriu*), with carvacrol being the most potent bactericidal agent followed by geraniol and citral, and that the refrigeration temperature acted synergistically with essential oils. A concentration of 1.5% was needed for each compound to effectively kill the bacteria. For this level of oil, in carvacrol samples a warmly pungent flavor was detected. Citral induced a yellowish color with a lemon-like flavor. Geraniol gave a strong rose-like smell. In summary, the levels of these essential oils needed to inhibit *Salmonella* are not adequate from the sensory point of view.

The Japanese style of cuisine of eating raw and lightly cooked seafood is increasing nowadays. In accordance with this trend, a high incidence of gastroenteritis caused by *Vibrio parahaemolyticus* was detected (Yano et al. 2006). One possible strategy to enhance the safety of raw seafood lies in the use of spices and herbs. Oregano and rosemary essential oils inhibited the bacteria in a nutrient-rich medium (Yano et al. 2006). Low levels of marjoram (25 ppm) were effective for decreasing bacterial survival at a low temperature in nutrient-poor media. But no data are available for seafood products (Yano et al. 2006). Wasabi (*Wasabi japonica*), a Japanese spice, is added traditionally in sushi preparation. The acrid substance in this spice is allyl isothiocyanate, and it was found that the essential oil of wasabi has antimicrobial effects against *E. coli, Salmonella typhi*, and other bacteria (Hasegawa et al. 1999). The latter authors reported that the addition of an ethanol extract of wasabi or allyl isothyocianate inhibited the growth of *V. parahaemolyticus* in lean and fatty tuna meat suspensions. The effectiveness of preservative extracts was greater in fatty samples than in lean ones. It was proposed that the fatty acids present in tuna suspension, which are mainly *cis*-vaccenic, palmitic, and docosahexaenoic acids, may protect the activity of allyl isothiocyanate.

14.2.3 Chitosan

Chitosan is a modified, natural carbohydrate polymer derived by deacetylation from chitin [poly-β-(1→4)-*N*-acetyl-D-glucosamine]. It is a major component of the shells of crustaceans such as crab, shrimp, and crawfish. Chitosan is produced from shell wastes with different deacetylation grades and molecular weights, and, therefore, different functional properties and biological activities were observed (No et al. 2007). In particular, its antimicrobial activity has been analyzed extensively. It was demonstrated that it inhibits the growth of many spoilage and pathogenic bacteria and also yeast and molds (No et al. 2007; Roller 2003). Moreover, chitosan is a film-making agent, being used in the preparation of edible films and coatings. Therefore, in foods, this preservative has been applied as an additive to a product formulation and also as an agent to develop a film or coating.

In seafood products, chitosan has been successfully used either as an antimicrobial or as an antioxidant. Kamil et al. (2002) tested the antioxidant activity of chitosans of different viscosities (360, 57, and 14 cP; and the corresponding molecular masses of 1,800, 960, and 660 kDa) in cooked, comminuted flesh of herring (*Clupea harengus*). The oxidative stability of fish flesh with added chitosans (50, 100, and 200 ppm) was compared with that of fish flesh to which conventional antioxidants, butylated hydroxyanisole plus butylated hydroxytoluene, and tertiary butyl hydroxyquinone (200 ppm) has been added, throughout storage at 4°C. Among the three chitosans, 14-cP chitosan was the most effective in preventing lipid oxidation. The formation of thiobarbituric acid reactive substances (TBARS) in herring samples containing 200 ppm of 14-cP chitosan was reduced after 8 days of storage by 52% as compared with that of the control. At 200 pm, 14-cP chitosan exerted an antioxidant effect similar to that of commercial antioxidants in reducing TBARS values in comminuted herring flesh.

The results described suggested that the chitosan antioxidant action depended on the molecular weight and concentration of chitosan. A similar trend was reported by Kim and Thomas (2007) when testing the antioxidant action of chitosan with different molecular masses (30, 90, and 120 kDa) on salmon homogenate. The 30-kDa chitosan showed the highest scavenging activity.

Chitosans may retard lipid oxidation by chelating ferrous ions released during storage by proteins such as myoglobin, hemoglobin, and ferritin, thus eliminating the pro-oxidant activity of ferrous ions or preventing their conversion to ferric ion (Kamil et al. 2002) The dependence of antioxidant activity on the viscosities previously mentioned may be attributed to the molecular mass differences, which determine the extent of metal ion chelation. In their charged state, the cationic amino groups of chitosans impart intramolecular electric repulsive forces, which increase the hydrodynamic volume by extended chain conformation. Perhaps this phenomenon may be responsible for lesser chelation by high-viscosity (high molecular mass) chitosans (Kamil et al. 2002).

Almost all the applications of chitosan in seafood products involve chitosan as a film-forming agent (No et al. 2007). Jeon et al. (2002) analyzed the effect of chitosan with different molecular masses and, as a consequence, different viscosities (360, 57, and 14 cP) as coatings for the shelf-life extension of fresh fillets of Atlantic cod (*Gadus morhua*) and herring (*C. harengus*) over 12 storage days at 4°C. A reduction in relative moisture losses was observed for cod samples coated with 360-cP chitosan throughout storage. In addition, the chitosan coating reduced lipid oxidation, chemical spoilage, and growth of microorganisms in both fish model systems compared with uncoated samples. The preservative efficacy and the viscosity of chitosan were interrelated; the efficacy of chitosans with viscosities of 57 and 360 cP was superior to that of chitosan with 14-cP viscosity. Sathivel (2005) evaluated the effects of chitosan as an edible coating on the quality of skinless pink salmon (*Oncorhynchus gorbuscha*) fillets during frozen storage. Coating with chitosan reduced the moisture loss of fillets by 50% compared with the uncoated samples, and also delayed lipid oxidation. In addition, coating with chitosan increased the thaw yield in relation to noncoated fillets. Also, no effects of coating on color parameters for cooked pink salmon fillets were found after 3 months of frozen storage. Similar results were obtained for cooked skinless pink salmon (Sathivel et al. 2007). In addition to the previously mentioned antioxidant action of chitosan, in the case of a coating applied on the surface, it may act as a barrier between the product and its surroundings, thus slowing down the diffusion of oxygen from the surroundings via the surface into the product.

The antimicrobial effectiveness of chitosan added to a seafood formulation or to a coating was demonstrated. Vásconez et al. (2007) reported that dipping salmon fillets in 1% chitosan solution for 3 min reduced aerobic mesophilic and psychrophilic cell counts. Additionally, quality parameters such as pH and loss of weight had acceptable values throughout storage, extending the shelf life of fillet salmon from 3 days (uncoated samples) to 6 days at 2°C. Chen et al. (1998) evaluated chitosan of different deacetylation degree and different derivates. They found that the immersion of oysters in a 2,000-ppm solution of sulfobenzoyl chitosan, a derivate with excellent solubility in water, kept bacterial growth below 10^6 colony-forming units (CFU)/g throughout 13 storage days at 5°C, whereas in untreated samples the bacterial population reached a level of 10^9 CFU/g and developed a putrid odor at the same time. Also, a delay in the growth of coliforms, *Pseudomonas*, and *Aeromonas* and a complete inhibition of *V. parahaemolyticus* was observed. Since *V. parahaemolyticus* is responsible for the foodborne outbreak resulting from eating raw oysters, the treatment proposed showed potential for extending the shelf life of refrigerated oysters.

The antimicrobial action of chitosan is suggested to be due, in part, to the binding of its positively charged molecules with the anionic cell membranes, promoting a disruption of their barrier properties (Sagoo et al. 2002).

14.2.4 *Perspectives in the Use of Natural Antimicrobial Preservatives*

Future research trends regarding organic acids will focus on their uses in novel foods, as well as on the exploitation of novel organic acid derivatives in foods. The significance and determination of synergistic combinations with other natural antimicrobials must be intensified to increase pathogen control and product shelf life. Finally, the fact that organic acid treatments have the potential to create acid-adapted pathogens could be appreciated; the adapted or modified organisms may develop permanent resistance to acid stress (Samelis and Sofos 2003). According to this, research to evaluate the acid tolerance response of bacteria in seafood must be performed.

Essential oils can be added to fish by tumbling or spraying. In relation to other mild preservation procedures such as high-pressure treatment and low-dose irradiation, essential oil addition is inexpensive and easy to perform as a method for extending the shelf life of aquatic products. However, as essential oils influence the sensory characteristics of seafood, they should be used primarily to develop new products with a specific sensory profile. Active compounds and toxicity should also be clearly documented before essential oils are used within the food industry. As demonstrated recently by Dusan et al. (2006), high doses of some essential oils can have detrimental effects on intestinal cells, and therefore the effect on the whole of the intestinal tract needs to be assessed before safe usage can be achieved (Fisher and Philips 2008).

Chitosan has been approved as a food additive in Japan and Korea since 1995 and 1983, respectively. It has been recognized as safe in the USA since 2001. However, more studies must be done to rule out that long-term intake of low levels of chitosan would not selectively inactivate beneficial gut microflora (Roller 2003). Moreover, extensive application in food systems is essential for the full exploitation of chitosan as a preservative.

14.3 Use of Antimicrobial Biopreservatives

14.3.1 Introduction

The term "biopreservation" was defined by Stiles (1996) as the use of a natural or controlled microflora and/or its antimicrobial metabolites to extend the shelf life and improve the safety of food. This term is linked to lactic acid bacteria (LAB), a group of Gram-positive bacteria which are often used in food to ensure its safety, preserve its quality, and develop new flavors. However, and unlike starter cultures, protective cultures used for biopreservation purposes should modify the sensory features of the food product as little as possible.

Certain LAB species and strains have been shown to exert strong antagonistic activity against spoilage and pathogenic microorganisms such as *Listeria, Clostridium, Staphylococcus*, and *Bacillus* spp. Such an antagonistic effect is caused by a decrease in the pH of the food (mainly caused by the production of lactic acid and other organic acids), the competition for nutrients, and/or the production of antimicrobial metabolites (Gibbs 1987; Klaenhammer 1988; Caplice and Fitzgerald 1999). Certain LAB are able to grow at refrigeration temperatures and are tolerant to modified-atmosphere packaging, low pH, high salt concentrations, and the presence of certain additives such as lactic acid, acetic acid, and ethanol.

Because of these benefits, LAB can be used as protective cultures to outgrow competitor microorganisms such as certain spoilage and pathogenic bacteria, with the subsequent benefits in terms of food safety. In addition, several LAB species and strains are able to produce bacteriocins. These molecules are synthesized by the ribosomes, released extracellularly as bioactive peptides or peptide complexes with a bactericidal or bacteriostatic effect on other closely related and non-closely related species, while the producer bacterial cell exhibits specific immunity to the action of its own bacteriocin. Members of the genera *Lactococcus, Lactobacillus, Leuconostoc, Carnobacterium, Enterococcus*, and *Pediococcus* are known to secrete bacteriocins. Accordingly, such LAB and their bacteriocin-containing extracellular extracts may be practically used for the control of spoilage and pathogenic microorganisms in seafood products. Among them, the genera *Carnobacterium* and *Enterococcus*, isolated from fish species, are especially relevant for the biopreservation of raw and processed seafood products.

14.3.2 Natural Presence of Protective Lactic Acid Bacteria in Aquatic Food Products: Types and Applications

LAB are not considered as genuine microflora of the aquatic environment, but certain genera, including *Carnobacterium, Lactobacillus, Enterococcus,* and *Lactococcus,* have been found in fish (Huss 1995; Stiles and Holzapfel 1997; González et al. 2000; Ringø et al. 2000; Campos et al. 2006). *Carnobacterium maltaromaticum* and *Carnobacterium divergens* have been reported to belong to the normal intestinal microbial population of a variety of fish species such as Atlantic salmon (*Salmo salar*), wild pike (*Esox lucius*), and wild brown trout (*Salmo trutta*) (Ringø and Gatesoupe 1998; González et al. 1999). *C. maltaromaticum* and *Lactobacillus* spp., which are able to produce bacteriocins, have also been isolated from chilled fish and cold-smoked fish products (Baya et al. 1991; Leroi et al. 1998). Likewise, enterococci exhibiting antilisterial activity have been isolated from farmed turbot (Campos et al. 2006) and cold-smoked salmon (Tome et al. 2006). Paludan-Müller et al. (1998) identified *Carnobacterium piscicola* as the dominant microorganism isolated from spoiled vacuum-packaged cold-smoked salmon.

Bacteriocins as food preservatives may be used for biopreservation purposes either by using the bacteriocinogenic strain as a protective culture or by using the extracellular extracts of the producing bacteria in the food formulation. LAB bacteriocins or bacteriocin-producing strains can be used either alone or in combination with one another and with other antimicrobials to improve the preservative effect. Some bacteriocins (e.g., lactocin S) have a much slower rate of killing than others and different inhibition spectra against spoilage and pathogenic bacteria. Thus, combining fast- and slow-acting bacteriocins, and broad and narrow inhibition spectra, the food remains safe longer.

Nilsson et al. (2004) demonstrated that bacteriocins are not always the only mechanism by which certain LAB inhibit the growth of certain microbial targets. They conducted a study to evaluate the contribution of *C. piscicola* bacteriocins to the inhibition of *L. monocytogenes* in a cold-smoked salmon system. In their study, a *C. piscicola* strain exhibiting bacteriocin activity (bac$^+$) and an isogenic mutant unable to produce bacteriocin (bac$^-$) were tested against *L. monocytogenes*. The results showed that the bac$^-$ strain was also able to inhibit *L. monocytogenes*, although not to such an extent as the bac$^+$ counterpart, this indicating that bacteriocins are not always the only method for the inhibition of microbial targets.

Duffes et al. (1999) isolated *C. divergens* and *C. maltaromaticum* strains that exhibited listericidal activity in a model experiment with cold-smoked fish. They found that *C. piscicola* V1 inhibited *L. monocytogenes* by the in situ production of bacteriocins in vacuum-packed cold-smoked salmon stored at 4°C and 8°C. In contrast, another related species, namely, *C. divergens* V41 and its divercin V41, only exhibited a bacteriostatic effect on the target microorganism. Silva et al. (2002) used a bacteriocin-producing *Carnobacterium* strain under a spray-dried format. This strain survived the process and retained antilisterial ability, although it lost activity against other Gram-positive targets such as *S. aureus*. The addition of bacteriocin-producing carnobacteria as protective cultures has been reported to inhibit the growth of *L. monocytogenes* in seafood (Alves et al. 2005; Yamazaki et al. 2003). Carnocin UI49, produced by *C. piscicola*, is another potential biopreservative agent that, when combined with nisin Z, provided good control of *Listeria* (Stoffels et al. 1993).

Leroi et al. (1998) also isolated carnobacteria during the first stage of storage of seafood products, whereas *Lactobacillus farciminis*, *Lactobacillus sakei*, and *Lactobacillus alimentarius* were isolated at advanced storage times. Other studies have also confirmed that most bacteria in vacuum-packaged "gravad" fish products stored at refrigeration temperatures are carnobacteria (Leisner 1992; Leisner et al. 1994) and *L. sakei* (Jeppesen and Huss 1993), and to a lesser extent *Leuconostoc* spp., *Lactobacillus curvatus*, and *Weissella viridescens* (Leisner et al. 1994).

Bacteriocin-producing enterococci are widespread in nature, and they have also been isolated from a variety of foods, including fish and seafood products (Dalgaard et al. 2003; Campos et al. 2006; Arlindo et al. 2006). One of the principal bacteriocins produced by enterococci isolated from fish is enterocin P (EntP; Arlindo et al. 2006), which shows good antimicrobial activity against *L. monocytogenes, S. aureus*, and other spoilage and pathogenic bacteria such as *Bacillus* spp. Certain enterococci are able to produce small heat-resistant antilisterial bacteriocins generically known as enterocins. A potential limitation in the use of enterocins is that they may be produced by enterococci harboring antibiotic resistance genes and/or genes coding for potential virulence factors (Eaton and Gasson 2001). In this sense, a preliminary screening of enterococcal strains with potential use in the aquatic food sector should be done to determine their sensitivity to antibiotics. EntP-producing strains of *Enterococcus faecium* (Cintas et al. 1997) have been isolated from fish species such as turbot (Arlindo et al. 2006) and are currently under investigation for their potential practical application as dried cultures for feeding fish in aquaculture facilities. Besides its antimicrobial activity and thermal stability, EntP exhibits other remarkable features, such as protease sensitivity, activity in a broad pH range, and the maintenance of antimicrobial activity after freeze-thawing, lyophilization, or chilled storage (Cintas et al. 1997).

Other bacteriocin-producing genera belonging to the LAB group and with relevance in the seafood sector are *Lactobacillus, Lactococcus*, and *Pediococcus*. Among these, Jeppesen and Huss (1993) studied a *Lactobacillus plantarum* strain with antagonistic effects against *Listeria* in a model fish product containing citric acid or sodium chloride as a curing agent. Nisin, the best studied bacteriocin, has also been evaluated for the inhibition of *Listeria* and other spoilage and pathogenic bacteria in aquatic food products. Al-Holy et al. (2005) reported that nisin, when combined with moderate heat, inhibited the growth of *Listeria innocua* in salmon (*Oncorhynchus keta*) and sturgeon (*Acipenser transmontanus*) caviar. Elotmani and Assobhei (2004) reported the inhibition of the microbial activity in sardine by the combined use of nisin and a lactoperoxidase system. Nilsson et al. (1997) tested a combined treatment with nisin and a modified CO_2-rich atmosphere to control *L. monocytogenes* in cold-smoked salmon. Likewise, Nykanen et al. (2000) also reported good results derived from the combination of nisin and sodium lactate for the control of *L. monocytogenes* in cold-smoked rainbow trout.

The combination of nisin and other bacteriocins has also been evaluated for the biopreservation of aquatic food products. One of the most interesting studies was provided by Bouttefroy and Milliere (2000). They tested combinations of nisin and curvaticin 13 produced by *L. curvatus* SB13 for prevention of the regrowth of bacteriocin-resistant cells of *L. monocytogenes*, and found that this combination was more inhibitory than the use of a single bacteriocin. Chi-Zhang et al. (2004) demonstrated that in the presence of an excess of nisin *L. monocytogenes* rapidly becomes nisin-resistant, whereas the slow addition of nisin only leads to a temporary tolerance before *Listeria* cells become nisin-sensitive. As a consequence of this, the antimicrobial effectiveness of nisin also depends on its concentration and mode of delivery. Wessels and Huss (1996) tested the nisin-producing *Lactococcus lactis* subsp. *lactis* ATCC 11454 as a protective culture for lightly preserved fish products, and found that a sodium chloride concentration of up to 4% allowed growth and nisin production, whereas 5% sodium chloride resulted in very slow growth and no nisin being detected.

When protective cultures are used, special attention should be paid to the method employed to apply them to the food product. In this sense, some authors have evaluated the use of spray-dried cultures as protective cultures (Corcoran et al. 2004; Gardiner et al. 2000). Likewise, Silva et al. (2002) studied bacteriocin production by the spray-dried LAB *C. divergens*, *Lactobacillus salivarius* and *L. sakei* for action against *L. innocua*, *L. monocytogenes*, and *S. aureus*. They concluded that spray-drying is a potentially useful process for large-scale production of dried powders containing viable organisms with antagonistic activity against pathogens.

14.3.3 Biopreservation and Probiotics

An alternative to antimicrobial agents in the prevention and management of fish diseases in aquaculture facilities is the use of probiotics added to animal feeds (Balcazar et al. 2006). Some protective cultures of LAB have been reported to exhibit probiotic activity due to their positive contribution to the healthy microflora of the human gut. Moreover, these probiotic LAB have also been introduced into animal feeds because of both their contribution to the health of farmed fish and their activity for the biological control of fish pathogens in aquaculture facilities (Verschuere et al. 2000). The possible mechanisms of action of probiotics are the production of compounds that are inhibitory towards pathogens, competition with target microorganisms for nutrients and energy, competition with undesirable species for adhesion sites, enhancement of the immune response of the animal, improvement of water quality, and interaction with phytoplankton (Verschuere et al. 2000).

As stated above, in the last decade probiotics have been introduced into aquaculture practice to overgrow and inhibit pathogenic bacteria by the incorporation of beneficial bacteria through the animal feeds. It should be highlighted that the probiotic LAB strain introduced in the food chain through the animal feed must also be safe for humans. The microorganisms used in animal feeds in the European Union territory mainly belong to the genera *Enterococcus, Lactobacillus, Pediococcus,* and *Streptococcus* (Anadon et al. 2006). Since 2000, all microbial strains intended to be used as probiotics in animal feeds must be evaluated by the European Union administration and authorized by a Regulation Commission before being used as ingredients in feedstuffs (Becquet 2003). Council Directive 70/524/EEC (1970) on feed additives is based on the principles of premarket authorization, the positive-list principle, and the thorough risk assessment of the effects of a particular strain on human and animal health as well as on the environment. Before a new probiotic strain is introduced or a new use of an already approved product is promoted, a dossier has to be submitted to the authorities, following the guidelines published in Commission Directive 94/40/EEC. These guidelines contain detailed evaluation methods. The safety requirements refer to (1) the target animal, (2) the consumer and the environment, and (3) the workers (Becquet 2003).

Among the evaluation studies of LAB strains such as probiotics in aquaculture, Nikoskelainen et al. (2001) studied human and dairy-derived probiotics for the prevention of infectious diseases in fish. They studied the potential probiotic properties of six LAB belonging to the *Lactobacillus, Bifidobacterium,* and *Enterococcus* genera. Strains were evaluated on the basis of the following parameters: mucosal adhesion, mucosal penetration, inhibition of pathogen growth and adhesion, and resistance to fish bile. The following fish pathogens were tested: *Vibrio anguillarum, Aeromonas salmonicida,* and *Flavobacterium psychrophilum.* Although most of the LAB strains evaluated had a considerable ability to adhere to different fish mucus types (14–26% of the bacteria added), none of them were able to inhibit the mucus binding of *A. salmonicida.* The results suggested that *Lactobacillus bulgaricus* and *Lactobacillus rhamnosus* ATCC 53103 were the most promising probiotic candidates. The use of probiotics by shrimp farmers has also been found to be effective and environmentally safe (Sambasivam et al. 2003).

14.3.4 Perspectives in the Biopreservation of Aquatic Food Products

Although the evaluation of bacteriocin-producing bacteria and their metabolites for the preservation of aquatic food products has been extensive, only a few reports include bacteriocins produced by LAB isolated from aquatic food products. The use of such bacteriocins and protective cultures may be of great interest since they would be more adapted to the food substrate in which they are destined to exert their effects. In this sense, EntP-producing enterococci isolated from farmed turbot have recently been tested as seafood preservatives under a spray-dried format and exhibited antilisterial, antistaphylococcal, and antibacilli activities. Preliminary results in turbot fillets either vacuum-packaged or subjected to modified-atmosphere packaging suggest that EntP-producing enterococcal strains isolated from fish may be a promising alternative as probiotics in aquaculture technology. However, more work is needed for the isolation and characterization of novel food-grade probiotic LAB that exhibit optimum inhibition activity against pathogenic and spoilage bacteria. Likewise, the adaptation of such protective cultures to new foods, formats, and technological treatments represents another hot topic in the field of seafood technology.

Biopreservation is also gaining increasing importance in combined strategies for the development of minimally processed novel foods as an alternative to traditional and more aggressive preservation techniques. In this sense, the increasing consumers' concern about the use of synthetic chemicals as preservatives in food should be noted, as well as the negative perception of intense technological treatments that may negatively affect the nutritional quality of foods. This scenario opens the way for novel applications of food-grade LAB and/or their metabolites for the biopreservation of minimally processed foods subjected to slight combined preservation treatments, such as pasteurization and/or refrigeration.

14.4 Antioxidant Preservatives from Natural Sources

14.4.1 Introduction

Seafood products deteriorate through several degradation reactions both on short-term and on long-term storage. The main deterioration processes are oxidation reactions and the decomposition of oxidation products, which result in decreased nutritional value and sensory quality. The retardation of these oxidation processes is important for the seafood producer/manager and, indeed, for all persons involved in the entire food chain from the sea to the consumer. Oxidation may be inhibited by various methods, including prevention of oxygen access, use of lower temperature, inactivation of enzymes that promote oxidation, reduction of oxygen pressure, and the use of suitable packaging. Another method of protection against oxidation is to use specific additives which inhibit oxidation. Antioxidant activity depends on many factors, such as the lipid composition, antioxidant concentration, temperature, oxygen pressure, and the presence of other antioxidants and many common food components, e.g., proteins and water (Pokorny 2001). Various antioxidants exhibit substantial differences in effectiveness when used with different types of oils or fat-containing foods, and when used under different handling and processing conditions. These differences have been attributed to their molecular structure. Several factors should be considered when choosing an antioxidant. These include knowing the potency of an antioxidant in a particular application, the ease of incorporation into the food, availability, and cost. The problem of selecting the best antioxidant is also further complicated by the difficulty of predicting how the added antioxidant will interact with pro-oxidants and antioxidants already present in the food item (Nawar 1996).

Seafoods possess high nutritional value, and also functional properties, thanks to their readily digested protein. As such, they are a good source of vitamins and minerals, and fatty fish furthermore contain high concentrations of polyunsaturated fatty acids, eicosapentaenoic acid and docosahexaenoic acid (Ackman 1999). However, because of this high unsaturated lipid content, fish products are very susceptible to loss of quality through lipid oxidation. The onset of rancidity is fast, particularly in fatty and semifatty species such as horse mackerel (*Trachurus trachurus*), in whose muscle large amounts of hemoglobin (a well-known activator of lipid oxidation) and lipids coexist (Richards and Hultin 2002). The use of antioxidants is an effective way to minimize or prevent lipid oxidation in food products (Boyd et al. 1993; Frankel 1998), retarding the formation of toxic oxidation products, maintaining nutritional quality, and prolonging the shelf life of foods (Jadhav et al. 1996). Synthetic antioxidants have been widely used to retard lipid oxidation in foods (Ahmad 1996), but nowadays there is growing interest in finding naturally occurring antioxidants for use in foods (Löliger 1983), and for possible in vivo use. There has been increasing interest in identifying plant extracts which minimize or retard lipid oxidation in lipid-based food products; for example, some natural phenolic compounds that are effective in preventing rancidity in many lipid systems such as fish muscle (Ikawa 1998; Medina et al. 1999; Medina et al. 2003).

Another oxidation process involved in the deterioration of aquatic foods is an enzymatic reaction mediated by polyphenol oxidases (PPO): browning or melanosis, which occurs primarily in crustaceans. These highly prized and economically valuable products are extremely vulnerable to enzymatic browning. Although the products of melanosis are not harmful and do not influence flavor or aroma, consumers will not select these products since their brown discoloration connotes spoilage. Severe melanosis on these products can cause tremendous economic losses because of the high value commanded by these aquatic products in the marketplace (Marshall et al. 2000). Black spots form on fresh shrimp and other shellfish within a few hours after harvest if they are not refrigerated. The reaction involves the oxidation of phenols to quinines by PPO (Taoukis et al. 1990). The endogenous shrimp enzyme PPO catalyzes the initial step in black spot formation and remains active throughout postharvest processing unless the shrimps are frozen or cooked (McEvily et al. 1991; Lu and Foo 1999). The black spots on shrimp begin to form on the head and proceed down the shrimp, forming black lines just under the shell that outline the sections of the tails. Refrigeration or storage on ice slows down but does not prevent this reaction because PPO enzyme systems remain active under these conditions. The rate of spread of melanosis differs among the various species (Montero et al. 2001).

Many studies have focused on either inhibiting or preventing PPO activity in foods. Various techniques and mechanisms have been developed over the years for the control of these undesirable enzyme activities. These techniques attempt to eliminate one or more of the essential components (oxygen, enzyme, copper, or substrate) from the reaction (Lee et al. 1988). Many techniques are applied in the prevention of enzymatic browning. Relatively new techniques, such as the use of killer enzymes, naturally occurring enzyme inhibitors, and ionizing radiation, have been explored and exploited as alternatives to heat treatment and certain chemical treatments because of the health risks associated with the latter (Marshall et al. 2000). Reducing agents play a role in the prevention of enzymatic browning either by reducing *o*-quinines to colorless diphenols or by reacting irreversibly with *o*-quinines to form stable colorless products (Green 1976). Reducing compounds are very effective in the control of browning. Sulfating agents are the most widely applied reagents for the control of browning in the food industry. Melanosis in crustaceans is normally controlled by means of certain sulfite derivatives (Ferrer et al. 1989). These compounds are known to produce allergic reactions and serious disturbances in asthmatic subjects (Taylor and Bush 1986; DeWitt 1998). Increased regulatory attention and heightened consumer awareness of the risks associated with sulfated foods have created a need for a safe, effective sulfite alternative for use in foods (Taylor and Bush 1986; McEvily et al. 1991). Several synthetic antioxidants, such as butylated hydroxyanisole, butylated hydroxytoluene, tertiary butyl hydroxyquinone, and propyl gallate, are suspected to be carcinogenic (Madhavi and Salunkhe 1995). Plant phenolic compounds such as tocopherols, flavinoid compounds, cinnamic acid derivatives, and coumarins are naturally occurring compounds which have an antioxidant effect that renders them inhibitory to PPO and thus browning (Jayaprakasha et al. 2001). Therefore, control of the deteriorative effects of browning poses a major challenge to the food scientist. In this sense, the search for natural antioxidants, especially of plant origin, has greatly increased in recent years.

14.4.2 *Preserving Seafood from Lipid Oxidation Reactions*

The direct application of antioxidants to fish is the most widely studied method for reducing off-flavor development in fish. Fish muscle can be sprayed with or dipped in solutions or suspensions of antioxidants, or it can be packed in films containing antioxidants (see Table 14.1). Other methods include (1) "glazing" of the previously frozen fish fillets with a chilled antioxidant solution in order to form a thin glazing layer and (2) washing the fillets with seawater before spraying them with the corresponding antioxidant solution. Rinsing with water or washing has been employed in chilled and frozen fish for removing heme proteins and metallic traces (Richards et al. 1998; Undeland et al. 1998). However, endogenous hydrophilic antioxidants can be lost in these treatments, leading to a higher susceptibility of the muscle tissue (Undeland et al. 1998). Therefore, the balance between endogenous pro-oxidants and antioxidants must be carefully controlled. Fish tissues contain endogenous antioxidants that are able to stabilize their high content of unsaturated lipids in vivo. In postmortem conditions, the endogenous antioxidants are consumed sequentially (Petillo et al. 1998; Undeland et al. 1999), and some studies have related the loss of endogenous antioxidants to the development of oxidative rancidity (Erickson 1993; Jia et al. 1996; Pazos et al. 2005).

Even though a number of herbs and spices have been identified as having substantial antioxidant activity, a general problem encountered with these sources of antioxidants is their characteristic odor and taste. This may make them unsuitable for use in many foods, particularly in seafood. The antioxidant activity of these ingredients is also less than that of the corresponding extracts or purified compounds. Moreover, the antioxidant accessibility to the fish tissue may also be limited because much of the active component(s) may be held within the herb or spice used. Ramanathan and Das (1993) tested a number of dried and fresh herbs for their

Table 14.1 Hurdle technology comprising natural preservatives applied to seafood

Preservative	Other preservative factors	Seafood product	Deteriorative reaction target	References
Sodium lactate, sodium benzoate, sodium diacetate, potassium sorbate, nisin	Chitosan-coated plastic films, refrigeration	Cold-smoked salmon	Microbial growth	Ye et al. (2008)
Oyster and hen lysozyme, nisin	Calcium alginate coating, refrigeration	Smoked salmon	Microbial growth	Datta et al. (2008)
Chitosan	Refrigeration	Herring and cod	Lipid oxidation, microbial growth	Jeon et al. (2002)
Nisin coated on cellulose derivates	Polyethylene packaging, refrigeration	Cold-smoked salmon	Microbial growth	Neetoo et al. (2008)
Thyme oil, cinnamaldehyde included in an edible coating	Low-dose gamma irradiation	Cooked shrimp	Microbial growth	Ouattara et al. (2001)
Oregano and rosemary extracts included in an edible coating	High pressure	Cold-smoked sardine	Microbial growth, lipid oxidation	Gomez-Estaca et al. (2007)
Bacteriocin-producing lactic acid bacteria	Vacuum packaging, refrigeration	Cold-smoked salmon	Microbial growth	Tomé et al. (2008), Ghalfi et al. (2006)
NaCl	MAP, refrigeration	Hake	Microbial growth, endogenous enzyme activity	Pastoriza et al. (1998)
Chitosan coating	Refrigeration	Salmon	Microbial growth	Vásconez et al. (2007)
Thyme and oregano essential oils	Refrigeration	Asian sea bass	Microbial growth	Harpaz et al. (2003)
Glucose oxidase/catalase	Refrigeration	Winter flounder	Endogenous enzyme activity	Field et al. (1986)
Glucose oxidase/catalase	Refrigeration	Shrimp	Endogenous enzyme activity, microbial growth	Dondero et al. (1993)

Oregano and cranberry extracts and lactic acid	Refrigeration	Cod fish fillets and shrimp	Microbial growth	Lin et al. (2005)
Lysozyme	Whey protein coating	Smoked salmon	Microbial growth	Min et al. (2008)
NaCl and clove powder	Refrigeration	Mackerel muscle extract	Microbial growth, endogenous enzyme activity (histamine production)	Wendakoon and Sakaguchi (1993)
Thyme essential oil	MAP and refrigeration	Swordfish fillets	Microbial growth	Kykkidou et al. (2008)
Oregano essential oil, MAP, and light salting	Refrigeration	Sea bream fillets	Lipid oxidation	Goulas and Kontominas (2007)
Sodium lactate and rosemary extract	Smoking	Blue catfish	Microbial growth, lipid oxidation	da Silva (2002)

MAP modified-atmosphere packaging

antioxidant effect in salted minced mackerel as assessed by the TBARS assay. The antioxidant efficacy for the dried herbs was cloves>cinnamon>cumin=black pepper>fennel=fenugreek. The dried spices were more effective than the fresh ones. Several polyphenols were also assessed, being the most potent group of natural antioxidants tested. At 0.01% the order was ellagic acid>tannic acid>myricetin> quercetin.

In recent years, the addition of natural bioactive compounds obtained from plants and vegetables has been employed in an attempt to improve oxidative stability in fish muscle. Tea catechins (He and Shahidi 1997; Tang et al. 2001) and rosemary (Vareltzis et al. 1997), olive oil (Medina et al. 1999), and ginger (Fagbenro and Jauncey 1994) extracts, among many others, have successfully inhibited rancidity of different seafoods.

14.4.2.1 Plant Extracts

Rosemary (*Rosmarinus officinalis*), a member of the *Labiatae* plant family, has been recognized for its medicinal properties since antiquity. Its antioxidant properties have been attributed to the presence of various phenolic diterpenes, carnosol, and carnosinic acid as well as rosmanol, epirosmanol, and isorosmanol acids and flavonoids (Sewalt et al. 2005), which can terminate free-radical reactions and scavenge reactive oxygen species (Sánchez-Escalante et al. 2001). The antioxidant activity of this seasoning herb has been compared with that of other known antioxidant substances. Sewalt et al. (2005) found that low doses of rosemary extract (100–500 ppm) resulted in induction times by the Rancimat method similar to those obtained with the maximum level of the synthetic antioxidants butylated hydroxyanisole and butylated hydroxytoluene (200 ppm) allowed in animal fat and feeds containing vegetable oil.

As regards fish products, many species were found to be effectively protected by the antioxidant activity of rosemary extract. Sardine (*Sardina pilchardus*) mince was treated with rosemary extract (300 ppm) and onion juice (1 mL/100 g) and then stored at −20°C for 5 months. Thiobarbituric acid level, peroxide value, and free fatty acid levels increased in all experimental groups because of the lipid oxidation. Rosemary extract showed an antioxidative effect on sardine mince during frozen storage as indicated by the thiobarbituric acid level, peroxide value, and free fatty acid levels. Oxidation was delayed for 3 months by onion juice treatment. At the end of the 5-month storage period, the thiobarbituric acid values for onion juice treatment and control treatment were outside consumable limits; however, for those samples treated with rosemary extract the values were within consumable limits. After frozen storage for 5 months, the polyunsaturated fatty acid level decreased and the saturated fatty acid level increased in the control treatment, which denotes enzymatic hydrolysis of sardine lipids (Serdaroğlu and Felekoğlu 2005).

Medina et al. (2003) tested the activity of rosemary extract and extra virgin olive oil for preserving fish emulsions and fish oils *n*-3 polyunsaturated fatty acids. Horse mackerel (*T. trachurus*), an underutilized fatty fish species, was selected for stabilizing its quality and obtaining high-value-added products. In fish oil-in-water emulsion, both natural extracts were more effective than butylated hydroxytoluene in preventing the formation and decomposition of hydroperoxides. Extra virgin olive oil phenolics were more effective than rosemary extracts during the first steps of oxidation, probably owing to the higher solubility of hydrophilic rosemary compounds in water, resulting in their having only little effectiveness at the oil–water interface. As the mechanism of phenolic antioxidative action is also related to their bonding with proteins, Medina et al. (2003) determined the protein–oxidized lipid interactions and the partition coefficients of the bioactive compounds in the different phases of the fish emulsion. Polyphenols can act as multidentate ligands bonding with protein surfaces and provoking conformational changes. The antioxidative activity of the system can be reinforced or reduced by the protein fraction. Thus, the synergism between fish proteins and rosemary extract was attributed to a protector effect of fish proteins for inhibiting carnosoic acid oxidation, whereas the lower synergistic activity of extra virgin olive oil phenolics in emulsions was suggested to be due to the partition of hydroxytyrosol and tyrosol into the aqueous phase, where they may antagonize any synergistic effects with water-soluble proteins. In conclusion, the study showed that the antioxidative properties attributed to phenolic natural antioxidants in combination with bioactive lipids such as n-3 polyunsaturated fatty acids could result in well-accepted healthy and nutritional foods.

A protective activity of rosemary extract used as a pretreatment against lipid oxidation in dehydrated salted fish was found by Da Silva Afonso and Sant'Ana (2008). Brine-salted tilapia (*Oreochromis niloticus*) fillets were treated (brine salting with filtered rosemary extract for 3 h) or pretreated (1 h with filtered rosemary extract followed by brine salting for 3 h) with the extract and stored for 240 days at −18°C. Those samples subjected to the pretreatment showed higher a_w and moisture values, whereas the TBARS values in the treatment (3.31 ± 0.79) and pretreatment (3.39 ± 0.53) were half the value in the control treatment (6.14 ± 1.21) at the end of storage. The pretreatment showed a more protective effect in protein oxidation, which could be observed in trichloroacetic acid soluble nitrogen values in 180 days (0.112 ± 0.020) and 240 days (0.132 ± 0.017).

It must be emphasized that the scientific research cited above was intended to test the herb extract on fish products where rosemary would play a significant role in the flavor and odor of the original product. Up to now, no work based on the use of rosemary extract as an antioxidant for preserving fish muscle, intended for wholesale and retail, has been found in the literature.

Catechins belong to the flavonoid family of naturally occurring antioxidants in food. In particular, green tea leaves contain relatively large amounts of (−)-epicatechin, (−)-epicatechin gallate, (−)-epigallocatechin, and (−)-epigallocatechin gallate. They have been recognized as efficient antioxidants by scavenging oxygen radicals and chelating metal ions (Shahidi and Wanasundara 1992). Tea catechins were found to be more efficient than α-tocopherol (both applied at 300 mg/kg) in inhibiting minced muscle lipid oxidation in fresh meats, poultry, and fish (Tang et al. 2001). In a conclusive study by He and Shahidi (1997), individual catechins, crude tea extracts, and ground green tea exhibited excellent antioxidant properties as evidenced by the

TBARS values and total volatile aldehyde and propanal contents of the treated fish meat samples, which were cooked at 75°C and stored at 4°C for 7 days. The order of potency of catechins in the mackerel meat system was (−)-epigallocatechin gallate ≈ (−)-epicatechin gallate > (−)-epigallocatechin >> (−)-epicatechin. Commonly used antioxidants such as α-tocopherol and butylated hydroxytoluene were not very effective in controlling the oxidative deterioration of fish meat. (−)-Epigallocatechin gallate showed slightly better antioxidative activity than tertiary butyl hydroxyquinone. He and Shahidi (1997) postulated that extensive hydroxylation of catechin molecules such as (−)-epigallocatechin gallate and (−)-epicatechin gallate and changes in the hydrophilicity/hydrophobicity of the molecules involved were perhaps the main reasons for their strong antioxidative properties as compared with other catechins tested. These findings also demonstrated that tea extracts and ground green tea leaves containing different amounts of (−)-epigallocatechin gallate, (−)-epicatechin gallate, (−)-epigallocatechin, and (−)-epicatechin may be considered as potential natural antioxidants for the stabilization of lipid-containing foods. However, dechlorophyllization of crude extracts may be necessary to use them for application to substrates in which the original color of the crude extracts might be of concern or when chlorophyll might act as a pro-oxidant.

The rate of oxidation of three species of fish showed a significant difference between species. Rainbow trout (*Oncorhynchus mykiss*), arctic char (*Salvellnus alpinus*) and tilapia (*O. niloticus*), stored at three temperatures (10°C, 5°C, and 0°C), were subjected to different treatments in order to determine their oxidative profiles. Trout oxidized the most, whereas tilapia did not oxidize at an appreciable rate. Several antioxidants (tertiary butyl hydroxyquinone, propyl gallate, rosemary, catechin, and α-tocopherol) were found to be equally effective in slowing oxidation. Catechin was chosen over the other natural antioxidants because of its good antioxidant activity, its nutraceutical properties, and its antimicrobial activity. The trout fillets did not seem to be as greatly affected by the presence of catechin as were the char and tilapia fillets. Ethanol (70%) was added to catechins to avoid microbial growth during the experiment. This treatment (catechin plus 70% ethanol) had the best antioxidant effect on the shelf life of trout and char, but it did not have a significant effect on tilapia fillets (Tozer 2001).

An original study by Lin and Lin (2005) presented the relative efficacies of glazing solutions of green tea, pouchong tea, and black tea as regards preserving the freshness of frozen bonito flesh compared with that of ice glazing. They compared the level of phenolics in the three tea extracts prepared from a single species of Taiwanese tea, and found that green tea and pouchong tea had almost the same values, whereas black tea had half the level of the other teas. Catechins were also found to be present in all of the three tea extracts; pouchong tea exhibited a level of antioxidative catechins similar to that determined for green tea. The extent of fermentation that had occurred for pouchong tea was similar to that which had occurred for green tea. Black tea contained the lowest level of catechins because they had undergone further fermentation in this tea. Green tea and pouchong tea exhibited better antioxidant activity than did black tea as regards lipid oxidation. In the context of protein deterioration, the antioxidant activity of lightly fermented tea (pouchong tea) was more substantial than the corresponding antioxidant activity of nonfermented tea (green tea) and completely fermented tea (black tea). The glazing treatments described and the incorporation of tea extracts in such a process could greatly enhance the shelf life of frozen bonito fillets.

According to Alghazeer et al. (2008), the role of whole instant green tea as an antioxidant in the fatty fish Atlantic mackerel (*Scomber scombrus*) depends on its concentration. They found that 250 ppm of tea extract was more effective than 500 ppm in the protection against lipid peroxidation during frozen storage of the fish. It has been reported that a higher concentration of green tea may act as a pro-oxidant (Honglian and Etsuo 2001). It was proposed that 250 ppm of green tea is sufficient to scavenge the free radicals until 16 weeks, whereas 500 ppm was excessive. After 16 weeks of storage, 500 ppm produced lower levels of lipid oxidation products, particularly hexanal, which was the final product. After 26 weeks, inhibition of lipid oxidation was minimal, as all the antioxidants were probably used up. Nevertheless, it was shown that whole instant green tea, used at the right concentration, may delay or inhibit lipid oxidation and can be used as a natural antioxidant in raw minced fatty fish tissue to enhance preservation.

The results mentioned above as well as the fact that teas are natural, cheap, and widely commercially available products make the application of tea as a stabilizer with antioxidative properties feasible for frozen fish meat.

14.4.2.2 Agro-industrial By-products

Considering the low economic value and the high content of phenolic compounds from agro-industrial processing wastes, by-products could be a valuable source of natural antioxidants for use in the food industry. Many of those applied to the preservation of seafoods are mentioned in this text as examples of the potential use they could soon have as natural antioxidants.

Grape polyphenolics are well-known antioxidants. Grape skins and seeds are a rich source of these compounds, including flavonoids with a different degree of polymerization known as proanthocyanidins (Souquet et al. 1996). The antioxidant activity of grape polyphenols may depend on the degree of polymerization (Yamaguchi et al. 1999). Although oligomers (roughly two to seven residues) are considered more efficient than monomers, materials of a higher degree of polymerization, which are also active, may be mucosal irritants and show astringency effects (Bravo 1998). Pazos et al. (2006) reported the effective antioxidant activity of grape flavonoids obtained from wine industry by-products in food systems containing fish oils and frozen fatty fish. Different monomeric and oligomeric procyanidin mixtures were tested as antioxidants in frozen mackerel, fish oils, and fish oil-in-water emulsions. Oligomers and monomers showed different 1,1-diphenyl-2-picrylhydrazyl free radical scavenging efficiency, with oligomers being more effective than monomers. Partition coefficients and the reversed-phase high-performance liquid chromatography retention revealed that the components of grape fractions showed different physical properties, such as lipophilicity, solubility, and partition. In bulk fish oils, propyl gallate and monomeric flavanols showed the highest efficiency. In oil-in-water emulsions, grape oligomers were more efficient than monomers. Propyl gallate was also highly effective in emulsions, in agreement with the observation that hydrophobic compounds are efficient in emulsions because they are largely accumulated in the oily–aqueous interface (Frankel 1998). Grape flavonoids were totally adsorbed within the fish muscle. This is an important result considering the development of technological strategies to minimize oxidation, e.g., glazing of whole fish or fillets with antioxidant solutions or their immersion in antioxidant solutions. When an aqueous solution of grape phenolics is used, the compounds will migrate from water to the flesh surface. Propyl gallate was also highly adsorbed within the fish muscle. However, propyl gallate is not soluble in aqueous solutions and will consequently not be suitable for immersion techniques. In both oils and emulsions, monomeric flavanols were more effective than monomeric glycosylated flavonols. The effectiveness of inhibiting oxidation in fish oils seemed to be more dependent on the physicochemical properties rather than the intrinsic redox capacity of the antioxidant, related to the number of *ortho*-dihydroxy groups and gallates. An optimal combination of procyanidin degree of polymerization and percentage galloylation may be related to the highest antioxidant efficacy of grape polyphenols in emulsions and in frozen fish muscle.

Wine by-products are also rich in dietary fiber (Bravo and Saura-Calixto 1998). It was recently proposed to define antioxidant dietary fiber as a natural product that combines the beneficial effects of dietary fiber and natural antioxidants, such as polyphenol compounds (Saura-Calixto 1998). Dietary fiber can also be an effective tool in seafood processing for improving functional properties, such as water binding, gelling, etc. Sánchez-Alonso et al. (2007) studied the effect of grape antioxidant dietary fiber addition to minced fish muscle on lipid stability during frozen storage, and found that the addition of red grape fiber considerably inhibited oxidation in horse mackerel minced muscle during the first 3 months of frozen storage. These results indicated that grape antioxidant dietary fiber could be used as an ingredient to prevent oxidation in minced fish during frozen storage. The reason for this effect could be attributed either to the chelating action of fiber on some pro-oxidant metals or to the action of polyphenols associated with dietary fiber. Grape antioxidant dietary fiber was shown to be a good antioxidant which could not only preserve minced fish muscle from oxidation during storage but could also produce health benefits for the consumer, thanks to its bioactive compound and dietary fiber content. It should be noted in this connection that in vitro activities can only be considered as potentially relevant in biological systems and that in vivo activities also depend on bioavailability and biotransformation. It is known that polyphenols such as gallic acid, caffeic acid, malvidin, catechin, and rutin, which are contained in red grapes, can be partly bioavailable (Manach et al. 2005), and consequently their in vivo activities could be significant.

Mango (*Mangifera indica* L.), peach (*Prunus persica*), and passion fruit (*Passiflora* sp.) seeds, as by-products from the industrialization of fruits, were studied in order to find a formulation for enhancing the production of fish burgers from silver catfish (*Rhamdia quelen*) filleting wastes (Bochi 2007). The three fruit seeds were examined for their total phenolic content, radical scavenging capacity against 1,1-diphenyl-2-picrylhydrazyl radicals, ferric-reducing antioxidant power, and antioxidant activity against lipid oxidation in a fish model system (0.05, 0.1, 0.15, and 0.3 mg phenolic compounds/4.4 ml silver catfish homogenate). Mango seed extract showed the highest phenolic content and total antioxidant activity by radical scavenging capacity against 1,1-diphenyl-2-picrylhydrazyl and ferric-reducing antioxidant power assays. Lipid oxidation in a fish model system (at 37°C for 90 min) was retarded by the three seed extracts at all concentrations tested. Mango seed extract, which had the highest antioxidant activity in vitro, was also evaluated for its antioxidant effect against lipid oxidation in fish burgers produced from silver catfish filleting wastes. Burger formulations with 50% pulp from filleting wastes were prepared containing mango seed extract (0, 30, and 90 ppm phenolic compounds) and the oxidation products during frozen storage at −10°C and −20°C were measured. All lipid damage measurements were affected by the storage time and temperature. However, after 120 days at both temperatures, TBARS levels of fish burgers did not reach threshold limit values for human consumption. Mango seed extract had no antioxidant effect against lipid oxidation in silver catfish burgers.

In an attempt to prevent color deterioration in rockfish species (*Sebastolobus alascanus*, *Sebastes ruberriumus*, and *Sebastes alutus*), a crude extract from shrimp shell wastes was prepared from by-products of the fishery industry. Different mechanisms are involved in color or pigment degradation between poultry/meat muscle and rockfish skin. The degradation of meat color is due to the oxidation of myoglobin to metmyoglobin, but color degradation in rockfish skin is also due to carotenoid oxidation. The antioxidant activity of crude and partially purified extracts was evaluated in a β-carotene–linoleic acid emulsion. The ability of the antioxidant to prevent the oxidative destruction of the β-carotene and linoleic acid emulsion was expressed as the decrease in absorbance at 470 nm. The crude extract o shrimp shell wastes in water solution was used as a solution dip for rockfish treatment. Shrimp shell wastes had lower antioxidant activity compared with a commercial antioxidant mixture which contained butylated hydroxytoluene, butylated hydroxyanisole, and citric acid. Antioxidant extracts from shrimp waste improved red color stability for species of rockfish. The effectiveness was lower than that of sodium erythorbate (a commercial antioxidant), which has been commercially used to stabilize the red skin color. However, the crude extract at a higher concentration (0.5% w/v) prevented color degradation in rockfish, whereas only 0.05% w/v sodium erythorbate was requited to prevent color degradation (Li et al. 1998).

Fish proteins from marine species are good sources of high-quality proteins, and there is an opportunity to use more fish processing by-products as protein ingredients for use in foods, feeds, and industrial applications. Pollock (*Theragra chalcogramma*) skin is an abundant and underutilized resource that can be used as a unique protein source to make fish protein hydrolysates. Utilizing proteolytic enzymes, one can prepare pollock skin protein hydrolysates that have new and/or improved chemical and functional properties. The natural antioxidant and water-soluble properties together with the film-forming ability of pollock skin protein hydrolysates make them ideal for coating material to suppress lipid oxidation in fish fillets during frozen storage. The edible hydrolysate coating could also provide some degree of protection against damage during transportation and handling of fish fillets. Shativel et al. (2008) found that glazing salmon (*O. gorbuscha*) fillets with pollock skin protein hydrolysates resulted a higher yield and thaw yield than for control nonglazed salmon fillets. After 4 months of frozen storage, the thiobarbituric acid value of nonglazed fillets was approximately 16 times higher than that of fresh raw pink salmon fillets, and the thiobarbituric acid values of glazed fillets were significantly lower than for the nonglazed fillets. In conclusion, edible coatings prepared from pollock skin hydrolysates have potential applications for enhancing the storage stability and quality of frozen fillets. They have many desirable properties, including emulsifying stability and fat adsorption capacity, which support their use as natural food ingredients.

14.4.3 Preserving Crustaceans from the Development of Melanosis: Natural Antibrowning Agents

Most of the natural antioxidants described in Sect. 14.2 are found in the literature as being also applied to inhibit enzymatic browning in crustaceans. Dipping in or spraying with antioxidant solutions are the main processes employed for the preservation of these seafood products. Afterwards, refrigerated or frozen storage is the key to extending shelf life.

Gokoglu and Yerlikaya (2008) found that grape seed extracts delayed melanosis in shrimp (*Parapenaeus longirostris*). The best results were obtained with a crude grape seed extract concentration of 15 g/l. The dipping solutions were prepared by dissolving grape seed extract in distilled water. Fresh shrimps, which were obtained directly from fishing boats, packed in ice, and transferred to the laboratory within 1 h, were dipped into the solutions at 15°C (1:4 shrimp per solution) for 1 min. After they had been dipped, the shrimps were drained at ambient temperature, packed by overwrapping with poly(vinylidene chloride) film, and stored at 4°C. During the storage, black spot formation on the shrimp shell was determined by sensory analysis, according to the melanosis scale proposed by Otwell and Marshall (1986), and color was measured by an instrumental method. Melanosis scores significantly increased during the storage, with the control samples being the ones that developed black spots at the highest rate. The control group exhibited heavy melanosis which was considered unacceptable at the end of storage for 3 days, whereas the samples treated with grape seed extract of 15 g/l kept their acceptability for 3 days. These findings suggest that grape seed extract can be used as an alternative to sulfating agents, helping in the improvement of food safety control for the shrimp industry. The use of grape seed extracts will also allow the utilization of by-products of wine and grape juice processing, giving added value to the economy of a given region. Maillard reaction products (MRPs) are known for their ability to retard lipid oxidation, as well as for their ability to inhibit certain oxidoreductases, such as PPO (Tan and Harris 1995). The effectiveness of such agents depends on several mechanisms of action in preventing the oxidation reaction, including free-radical-scavenging action; metal-ion-chelating ability, and/or reducing activity (Tan and Harris 1995). The use of MRPs in inhibiting PPO from crustaceans was reported by Matmaroh et al. (2006). They studied the effect of reactant concentrations in a fructose–glycine system on the formation of MRPs and the inhibitory activity of MRPs towards PPO from black tiger shrimp. The enzyme was isolated and treated with a solution of the MRPs prepared by heating an equimolar mixture of fructose and glycine at various concentrations (0.75–30 mM) at 100°C for 12 h. The development of MRPs was associated with a decrease in pH and a loss of reducing sugar and free amino groups, with a coincidental increase in reducing power and copper-chelating ability. MRPs with a reactant concentration of 4.5–30 mM were able to chelate the copper ion. Therefore, the inhibitory activity of MRPs towards browning, induced by PPO, was most likely due to their copper-chelating ability as well as their reducing power. It has to be highlighted that these constitute preliminary results derived from an experiment carried out in vitro, and that further analyses have to be performed to test whether the antioxidant activity of MRPs keeps on acting among PPO in situ on the shrimp.

Rosemary extract has antioxidant activity on marinated deepwater pink shrimp (*Parapenaeus longirostris* Lucas, 1846) stored at 1°C. A marinade solution contained 2% citric acid, 4% sodium chloride and preservatives, 0.1% sorbic acid, and 0.1% benzoic acid. A concentration of 300 ppm of rosemary extract was added to the marinade solution for the experimental group. Boiled shrimp and marinade solution were filled into plastic containers at a ratio of 1:1 (w/v), the containers were sealed with their lids, and the containers were stored at 1 ± 0.5°C for 75 days. The appropriate analyses were performed to investigate the quality changes and to determine the shelf life of marinated shrimps. There was no significant difference between the sensory analysis of control and experimental groups on storage days 0, 15, 30, 45, and 60, whereas rancidity was noted by the panelists in the control group on day 75. Although the bacterial loads of both groups were lower than the consumption limits on storage day 75, the thiobarbituric acid value limited the shelf life of the control group, whereas the quality of the experimental group was still good enough for consumption after 75 days. It was determined that the addition of rosemary extract did not render any additional antibacterial activity on the microbial load of the experimental group (Cadun et al. 2008).

Latex derived from various botanical sources is an effective inhibitor of certain enzymatic reactions. The latex derived from the fig tree, *Ficus* sp., contains the protease ficin. Extracts containing ficin prepared from the fig latex have been shown to be effective inhibitors of the enzymatic browning reaction (Labuza et al. 1990). However, ficin treatment is detrimental to the texture and quality of foods because of ficin's proteolytic activity. McEvily (1991) showed that protease-free extracts from natural latex can inhibit browning of foods at least as effectively as the protease-containing extract. Pink shrimps (*Penaeus duorarum*) were treated with a ficin-free extract and this treatment was compared with treatment of shrimps with ficin and sodium bisulfite, and no treatment of shrimps. The ficin-free extract was as effective as ficin and was more effective than sodium bisulfite in inhibiting browning in shrimp. A solution containing 0.25% by weight of the ficin-free extract can be effective in inhibiting shrimp melanosis. The fact that the treatment with a protease preparation (e.g., crude ficin) for inhibiting the browning reaction does not require an active form of the protease or require protease at all led to the hypothesis that ficin itself was not involved in the mechanism for inhibiting the browning reaction. McEvily (1991) attributed this fact to a low molecular mass component (less than 5,000 Da) in crude latex which exists in the protease-free extract that either inhibits the activity of PPO by binding with the enzyme to form an enzyme–inhibitor complex or binds with or otherwise alters its phenolic substrates or quinoid products. Thus, the browning reaction can be stopped if the substrate is bound to a protein molecule instead of the enzyme, PPO, or if PPO is bound to an inhibitor instead of its substrate. Undoubtedly, these findings need to be correlated with scientific studies to determine what mechanism is involved in the antioxidant activity exerted by this protease-free extract, and to find out what substance is responsible for this achievement.

Llamas Marcos et al. (2006) claimed that LAB could be used as melanosis-inhibiting agents in crustaceans. The LAB used in the procedure were *Lactobacillus helveticus*, *L. lactis*, *C. piscicola*, and *Pediococcus acidilacti*. The microorganism concentration, the immersion times, and the temperatures of the whole treatment did not seem to be fixed appropriately since their ranges were extremely wide. Moreover, the effectiveness of the method is not attributed to any metabolite or substance excreted by LAB; it is only mentioned that the contact between both the crustaceans and the LAB suspension is the reason why melanosis is being inhibited.

14.4.4 Legal Issues Relating to the Addition of Antioxidants to Seafood

The most detailed information about permitted and recommended antioxidants may be seen in Codex Alimentarius, a reference source that is regularly updated by the World Health Organization. However, various countries regulate the use of antioxidants with their own national legislation. The divergences result from tradition, the composition of local diet, and the boarding custom practices – procedures made onboard. The use of antioxidants in fish in the European Union is strictly controlled by legislation. Fish blocks are considered to be unprocessed and therefore antioxidants, apart from ascorbic acid and citric acid, cannot be used in their manufacture. Any other antioxidants would have to be cleared for use and this can take between 2 and 3 years. Antioxidant blends can be used, which contributes to flavoring, but the addition would need to have a discernible flavor impact for valid declaration. Relevant governing legislation is European Parliament and Council Directive No. 95/2/EC on food additives other than odors and sweeteners (Ashton 2002).

If the final product is to be a composite product, then the antioxidants allowed for use under legislation include those allowed for general food use. Nevertheless, the issue is always one of the additives being perceived as "natural" (by the consumers) and label declaration. Customers prefer foods with a minimum of additives as indicated by E numbers, and so the trend is directed towards mixtures of spices that contain antioxidants that do not need to be declared (Miková 2002). In terms of further guidance on the use of antioxidants in composite fish products in the European Union, information on the exact type of product is needed because all of them have nuances in terms of European Union legislation.

14.4.5 Future Trends

The most recent investigations have been targeted towards the identification of novel antioxidants from natural sources. Plant phenolic compounds such as flavonoids, sterols, lignin phenols, and various terpene-related compounds are potent antioxidants. Since the antioxidant activities of natural extracts and compounds have been determined by a wide range of methods and various end points, it has also become increasingly difficult to make a realistic assessment of the efficacy of various natural antioxidants. There is an urgent need for standardization of evaluation methods in order to obtain meaningful information (Madhavi and Salunkhe 1995). In this sense, Decker et al. (2005) outlined model systems for the evaluation of antioxidants in three types of food: bulk oil, oil-in-water emulsions, and muscle foods. The model systems suggested are not intended to be inclusive of all possible methods to measure lipid oxidation and antioxidant activity but they would allow researchers to compare more easily research results from one article to another.

The optimized use of natural antioxidants in food is still in its infancy and needs a lot of research and development. Factors affecting antioxidative efficiency must be carefully studied. The physical properties of the system, the partitioning of active molecules into the different phases, the real effectiveness in each food, the effects of food processes, the molecular structure of the antioxidant, and the effect of other intrinsic components in the food system should be determined (Medina et al. 2003).

On the one hand, the lower effectiveness of natural antioxidants compared with synthetic antioxidants and the requirement of low-cost production are important limitations for the practical use of many natural compounds as antioxidant additives in foods. Furthermore, many natural substances might also have health-associated risks. For example, it has been pointed out that a high concentration of tea induces apoptosis in many types of cells (Chen et al. 2002; Razat and John 2005). Thus, further studies on ascertaining optimum levels of green tea, including polyphenol fractions, for food preservation and their effect on safety using cultured human cells need to be performed.

Considering the new features of the seafood preservation challenge, it has been reported recently that PPO should not be considered as the unique enzyme responsible for melanosis in crustaceans. Martínez-Alvarez et al. (2008a) reported the presence of hemocyanin with diphenol oxidase activity in deepwater pink shrimp (*Parapenaeus longirostris*) after death. PPO and hemocyanin are two proteins which, although they are very similar, perform different physiological functions in crustaceans. In freshly caught deepwater pink shrimp there is an active form of hemocyanin which has diphenol oxidase activity. This activity appeared to be present in the live shrimp, and in vitro activation was not necessary. During storage, a form of hemocyanin with a higher molecular weight also acquires such activity through a mechanism which is still unidentified. The two oligomeric forms of hemocyanin with diphenol oxidase activity seem likely to be chiefly responsible, more than PPO, for the intensity of the melanosis observed in this species. The fact that hemocyanin acquires the ability to oxidize phenols could be of considerable economic importance given the large amount of this protein found in crustaceans, about 1,000 times more than PPO (Adachi et al. 2005), because this could accelerate the onset of melanosis and cause severe loss of quality. Future research should concentrate on characterizing hemocyanin, examining the kinetics of the effect of PPO inhibitors on hemocyanin, and searching for new antibrowning agents to add to commercial formulae. Furthermore, Martínez-Alvarez et al. (2008b) also demonstrated the presence of an active laccase-like enzyme from deepwater pink shrimp. This enzyme was found in all anatomical parts of the deepwater pink shrimp, but particularly in the cephalothorax, and became active during the course of storage. The enzyme was inhibited by a specific inhibitor, but not by 4-hexylresorcinol, a specific inhibitor of PPO. Nevertheless, low concentrations of the antioxidants ascorbic acid and sodium metabisulfite were sufficient to inhibit the laccase-like enzyme. In conclusion, given the difficulties involved in inhibiting melanosis in deepwater pink shrimp by means of authorized concentrations of sulfites, there is a need to find effective alternatives to inhibit the onset of melanosis in this species and prolong storage life without the risk of health problems for consumers. These new melanosis-inhibiting compounds should complement specific inhibitors of PPO activity that are already applied.

14.5 Natural Preservatives Used in Combination with Other Microbial Stress Factors

Preservatives alone have never provided the "magic bullet" for the prevention of food spoilage and poisoning, and it is accepted that the combination of preservatives and other stress factors is one of the possible pathways proposed to improve food safety.

As previously mentioned, natural preservatives are used together with other preservation factors such as traditional or novel refrigerated storage, nonthermal technologies such as gamma irradiation, high pressure, active packaging materials, and modified-atmosphere packaging. A summary of the most relevant combinations of preservatives with other preservation factors is given in Table 14.1.

References

Ackman RG. In: Ackman RG, editor. Marine biogenic lipids, fats and oils. Boca Raton: CRC Press; 1999.

Adachi K, Endo H, Watanabe T, Nishioka T, Hirata T. Hemocyanin in the exoskeleton of crustaceans: enzymatic properties and immunolocalization. Pigment Cell Res. 2005;18:136–43.

Ahmad JI. Free radicals and health: is vitamin E the answer? Food Sci Technol. 1996;10(3): 147–52.
Alghazeer R, Saeed S, Howell NK. Aldehyde formation in frozen mackerel (*Scomber scombrus*) in the presence and absence of instant green tea. Food Chem. 2008;108:801–10.
Al-Holy M, Lin M, Rasco B. Destruction of *Listeria monocytogenes* in sturgeon (*Acipenser transmontanus*) caviar by a combination of nisin with chemical antimicrobials or moderate heat. J Food Prot. 2005;68:512–20.
Alves VF, De Martinis EC, Destro MT, Vogel BF, Gram L. Antilisterial activity of a *Carnobacterium piscicola* isolated from Brazilian smoked fish (surubim [*Pseudoplatystoma* sp.]) and its activity against a persistent strain of *Listeria monocytogenes* isolated from surubim. J Food Prot. 2005;68:2068–77.
Anadon A, Martinez-Larrañaga MR, Aranzazu Martinez M. Probiotics for animal nutrition in the European Union. Regulation and safety assessment. Regul Toxicol Pharmacol. 2006;45:91–5.
Arlindo S, Calo P, Franco C, Prado M, Cepeda A, Barros-Velazquez J. Single nucleotide polymorphism analysis of the enterocin P structural gene of *Enterococcus faecium* strains isolated from nonfermented animal foods. Mol Nutr Food Res. 2006;50:1229–38.
Ashton IP. Understanding lipid oxidation in fish. In: Bremner HA, editor. Safety and quality issues in fish processing. Cambridge/Boca Raton: CRC Press/Woodhead; 2002.
Balcazar JL, Decamp O, Vendrell D, De Blas I, Ruiz-Zarzuela I. Health and nutritional properties of probiotics in fish and shellfish. Microb Ecol Health Dis. 2006;18:65–70.
Baya AM, Toranzo AE, Lupiani B, Li T, Robertson BS, Hetrick FM. Biochemical and serological characterization of *Carnobacterium* spp. isolated from farmed and natural populations of stripped bass and catfish. Appl Environ Microbiol. 1991;57:3114–20.
Becquet P. EU assessment of enterococci as feed additives. Int J Food Microbiol. 2003;88: 247–54.
Beuchat LR. Antimicrobial properties of spices and their essential oils. In: Dillon, Y.M., Board, R.G. editor. Natural Antimicrobial Systems and Food Preservation. CAB International, Oxon;1993. pp. 167–79.
Blom H, Nerbrink E, Dainty R, Hagtvedt T, Borch E, Nissen H, et al. Addition of 2.5% lactate and 0.25% acetate controls growth of *Listeria monocytogenes* in vacuum-packed, sensory-acceptable servelat sausage and cooked ham stored at 4°C. Int J Food Microbiol. 1997;38:71–6.
Bochi VC. Otimização de uma formulação de fish burgers de jundiá (*Rhamdia quelen*) visando o aproveitamento de subprodutos da filetagem e do processamento de frutas. Santa Maria: Universidade Federal de Santa Maria; 2007.
Bouttefroy A, Milliere JB. Nisin-curvaticin 13 combinations for avoiding the regrowth of bacteriocin resistant cells of *Listeria monocytogenes* ATCC 15313. Int J Food Microbiol. 2000;62:65–75.
Boyd LC, Green DP, Giesbrecht FB, King MF. Inhibition of oxidative rancidity in frozen cooked fish flakes by tertbutylhydroquinone and rosemary extract. J Sci Food Agric. 1993;61:87–93.
Branen AL, Haggerty R.J. Introduction to food additives. In: Branen AL, Salminen S, Thorngate JH III, editors. Food additives. Marcel Dekker; 2002. p. 1–9.
Bravo L. Polyphenols: chemistry, dietary sources, metabolism, and nutritional significance. Nutr Rev. 1998;56:317–33.
Bravo L, Saura-Calixto F. Characterization of dietary fiber and the in vitro indigestible fraction of grape pomace. Am J Enol Viticult. 1998;49:135–41.
Buncic S, Fitzgerald CM, Bell RG, Hudson JA. Individual and combined listericidal effects of sodium lactate, potassium sorbate, nisin and curing salts at refrigeration temperatures. J Food Saf. 1995;15:247–64.
Cadun A, Kişla D, Çakli S. Marination of deep-water pink shrimp with rosemary extract and the determination of its shelf-life. Food Chem. 2008;109:81–7.
Campos C, Rodríguez O, Calo-Mata P, Prado M, Barros-Velazquez J. Preliminary characterization of bacteriocins from *Lactococcus lactis*, *Enterococcus faecium* and *Enterococcus mundtii* strains isolated from turbot (*Psetta maxima*). Food Res Int. 2006;39:356–64.
Caplice E, Fitzgerald GF. Food fermentations: role of microorganisms in food production and preservation. Int J Food Microbiol. 1999;50:131–49.

Chen N, Shelef LA. Relationship between water activity, salts of lactic acid and growth of *Listeria monocytogenes* in a meat model system. J Food Prot. 1992;55:574–8.

Chen CS, Liau WY, Tsau GJ. Antibacterial effects of N-sulfonated and N-sulfobenzoyl chitosan and application to oyster preservation. J Food Prot. 1998;61:1124–8.

Chen L, Yang X, Jiao H, Zhao B. Tea catechins protect against lead-induced cytotoxicity, lipid peroxidation, and membrane fluidity in HepG2 cells. Toxicol Sci. 2002;69:149–56.

Cherry JP. Improving the safety of fresh produce with antimicrobials. Food Technol. 1999;53 (11):54–9.

Chi-Zhang Y, Yam KL, Chikindas ML. Effective control of *Listeria monocytogenes* by combination of nisin formulated and slowly released into a broth system. Int J Food Microbiol. 2004;90:15–22.

Cintas LM, Casaus P, Havarstein LS, Hernández PE, Nes IF. Biochemical and genetic characterization of enterocin P, a novel sec-dependent bacteriocin from *Enterococcus faecium* P13 with a broad antimicrobial spectrum. Appl Environ Microbiol. 1997;63:4321–30.

Commission Directive. 94/40/EC of 22 July 1994 amending Council Directive 87/53/EEC fixing guidelines for the assessment of additives in animal nutrition.

Corcoran BM, Ross RP, Fitzgerald GF, Stanton C. Comparative survival of probiotic lactobacilli spray-dried in the presence of prebiotic substances. J Appl Microbiol. 2004;96:1024–39.

Council Directive. 70/524/EEC of 23 November 1970 concerning additives on feeding-stuffs.

Da Silva LVA. Hazard analysis critical control point (HACCP), microbial safety, and shelf life of smoked blue catfish (*Ictalurus furcatus*). Tashkent State University; 2002 (Thesis).

Da Silva AM, Sant'Ana LS. Effects of pretreatment with rosemary (*Rosmarius officinalis* L.) in the prevention of lipid oxidation in salted tilapia fillets. J Food Qual. 2008;31:586–95.

Dalgaard P, Vancanneyt M, Euras Vilalta N, Swings J, Fruekilde P, Leisner JJ. Identification of lactic acid bacteria from spoilage associations of cooked and brined shrimps stored under modified atmosphere between 0°C and 25°C. J Appl Microbiol. 2003;94:80–9.

Datta S, Janes ME, Xue QG, La Peyre JF. Control of *Listeria monocytogenes* and *Salmonella anatum* on the surface of smoked salmon coated with calcium alginate coating containing oyster lysozyme and nisin. J Food Sci. 2008;73(2):M67–71.

De Witt BJ. An improved methodology for the estimation of sulphur dioxide in shrimp. Resumen Internacional. 1998;48:3463–4.

Decker E, Warner K, Richards M, Shahidi F. Measuring antioxidant effectiveness in food. J Agric Food Chem. 2005;53:4303–10.

Dondero M, Egaña W, Tarky W, Cifuentes A, Torres A. Glucose oxidase/catalase improves preservation of shrimp (*Heterocarpus reedi*). J Food Sci. 1993;58:774–9.

Duffes F, Corre C, Leroi F, Dousset X, Boyaval P. Inhibition of *Listeria monocytogenes* by in situ produced and semipurified bacteriocins of *Carnobacterium* spp. on vacuum-packed, refrigerated cold-smoked salmon. J Food Prot. 1999;62:1394–403.

Dusan F, Marian S, Katarina D, Dobroslava B. Essential oils – their antimicrobial activity against *Escherichia coli* and effect on intestinal cell viability. Toxicol In Vitro. 2006;20(8):1435–45.

Eaton TJ, Gasson MJ. Molecular screening of *Enterococcus* virulence determinants and potential for genetic exchange between food and medical isolates. Appl Environ Microbiol. 2001;67:1628–35.

Elotmani F, Assobhei O. In vitro inhibition of microbial flora of fish by nisin and lactoperoxidase system. Lett Appl Microbiol. 2004;38:60–5.

Erickson MC. Compositional parameters and their relationship to oxidative stability of channel catfish. J Agric Food Chem. 1993;41:1213–8.

Fagbenro O, Jauncey K. Chemical and nutritional quality of fermented fish silage containing potato extracts, formalin or ginger extracts. Food Chem. 1994;50:383–8.

Farag RS, Daw ZY, Hewedi FM, El-Baroty GSA. Antimicrobial activity of some Egyptian spice essential oils. J Food Prot. 1989;52:665–7.

Ferrer OJ, Koburger JA, Otwell WS, Gleeson RA, Simpson BK, Marshall MR. Phenoloxidase from the cuticle of Florida spiny lobster (*Panulirus argus*): mode of activation and characterization. J Food Sci. 1989;54:63–7.

Field CE, Pivarnik LF, Barnett SM, Jr Rand AG. Utilization of glucose oxidase for extending the shelf-life of fish. J Food Sci. 1986;51:66–70.

Fisher K, Philips C. Potential antimicrobial uses of essential oils in food: is citrus the answer? Trends Food Sci Technol. 2008;19:156–64.

Frankel EN. Antioxidants. In lipid oxidation. Dundee: The Oily Press; 1998.

Gardiner GE, O'Sullivan E, Kelly J, Auty MA, Fitzgerald GF, Collins JK, et al. Comparative survival rates of human-derived probiotic *Lactobacillus paracasei* and *L. salivarius* strains during heat treatment and spray drying. Appl Environ Microbiol. 2000;66:2605–12.

Ghalfi H, Allaoui A, Destain J, Benkerroum N, Thonart P. Bacteriocin activity by *Lactobacillus curvatus* CWBI-B28 to inactivate *Listeria monocytogenes* in cold-smoked salmon during 4°C storage. J Food Prot. 2006;69(5):1066–71.

Gibbs PA. Novel uses of lactic acid fermentation in food preservation. J Appl Bacteriol Symp Suppl. 1987;63:51S–8.

Gokoglu N, Yerlikaya P. Inhibition effects of grape seed extracts on melanosis formation in shrimp (*Parapenaeus longirostris*). Int J Food Sci Technol. 2008;43:1004–8.

Gómez-Estaca J, Montero P, Giménez B, Gómez-Guillén MC. Effect of functional edible films and high pressure processing on microbial growth and oxidative spoilage in cold-smoked sardine (*Sardina pilchardus*). Food Chem. 2007;105:511–20.

González CJ, Lopez-Diaz TM, García-López ML, Prieto M, Otero A. Bacterial microflora of wild brown trout (*Salmo trutta*), wild pike (*Esox hucius*) and aquacultured rainbow trout (*Oncorhynchus mykiss*). J Food Prot. 1999;62:1270–7.

González CJ, Encinas JP, García-López ML, Otero A. Characterisation and identification of lactic acid bacteria from freshwater fishes. Food Microbiol. 2000;17:383–91.

Goulas AE, Kontominas MG. Combined effect of light salting, modified atmosphere packaging and oregano essential oil on the shelf-life of sea bream (*Sparus aurata*): biochemical and sensory attributes. Food Chem. 2007;100:287–96.

Green LF. Sulphur dioxide and food preservation – a review. Food Chem. 1976;1:103–24.

Harpaz S, Glatman L, Drabkin V, Gelman A. Effects of herbal essential oils used to extend the shelf life of freshwater – reared Asian sea bass fish (*Lates calcarifer*). J Food Prot. 2003;66(3):410–7.

Hasegawa N, Matsumoto Y, Hoshino A, Iwashita K. Comparison of effects of Wasabia japonica and allyl isothiocyanate on the growth of four strains of *Vibrio parahaemolyticus* in lean and fatty tuna meat suspensions. Int J Food Microbiol. 1999;49:27–34.

He Y, Shahidi F. Antioxidant activity of green tea and its catechins in a fish meat model system. J Agric Food Chem. 1997;45(11):4262–6.

Honglian N, Etsuo N. Introducing natural antioxidants. In: Pokorny J, Yanishlieva N, Gordon M, editors. Antioxidants in food. Practical applications. Cambridge: Woodhead; 2001. p. 147–55.

Houtsma PC, Kant-Muermansm L, Rombouts FM, Zwietering MH. Model for the combined effects of temperature, pH and sodium lactate on growth rates of *Listeria innocua* in broth and Bologna-type sausages. Appl Environ Microbiol. 1996;62:1616–22.

Huss HH. Quality and quality changes in fresh fish. In: Huss HH, editor. FAO Fishing Technical Paper 348. Rome/Italy: FAO; 1995. p. 51.

Ikawa Y. Use of tea extracts (sanfood) in fish paste products. New Food Industries. 1998;40: 33–9.

Jadhav SJ, Nimbalkar SS, Kulkarni AD, Madhavi DL. Lipid oxidation in biological and foods systems. In: Madhavi DL, Deshpande SS, Salunkhe DK, editors. Food antioxidants. New York: Marcel Dekker; 1996. p. 1–18.

Jayaprakasha GK, Singh RP, Sakariah KK. Antioxidant activity of grape seed (*Vitis vinifera*) extracts on peroxidation models in vitro. Food Chem. 2001;73:285–90.

Jeon YJ, Kamil JYVA, Shahidi F. Chitosan as an edible invisible film for quality preservation of herring and Atlantic cod. J Agric Food Chem. 2002;50:5167–578.

Jeppesen VT, Huss HH. Characteristic and antagonistic activity of lactic acid bacteria isolated from chilled fish products. Int J Food Microbiol. 1993;18:305–20.

Jia T-D, Kelleher SD, Hultin HO, Petillo D, Maney R, Krzynowek J. Comparison of quality loss and changes in the glutathione antioxidant system in stored mackerel and bluefish muscle. J Agric Food Chem. 1996;44:1195–201.

Joffraud JJ, Cardinal M, Cornet J, Chasles JS, León S, Gigout F, et al. Effect of bacterial interaction on the spoilage of cold-smoked salmon. Int J Food Microbiol. 2006;112:51–61.

Kabara JJ. Phenols and chelators. In: Russell NJ, Gould GW, editors. Food preservatives. Glasgow: Blackie; 1991. p. 200–14.

Kamil JYVA, Jeon YJ, Sahidi F. Antioxidative activity of chitosans of different viscosity in cooked comminuted flesh of herring (*Clupea harengus*). Food Chem. 2002;79:69–77.

Kim KW, Thomas RL. Antioxidative activity of chitosans with varying molecular weights. Food Chem. 2007;101:308–13.

Kim JM, Marshall MR, Cornell JA, Preston III JF, Wei CI. Antibacterial activity of Carvacrol, citral, and geranium against *Salmonella typhymurium* in culture medium and fish cubes. J Food Sci. 1995;60:1364–8.

Klaenhammer TR. Bacteriocins of lactic acid bacteria. Biochimie. 1988;70:337–49.

Koutsoumanis K, Lambropoulou KA, Nychas GJE. A predictive model for the non-thermal inactivation of *Salmonella enteritidis* in a food model system supplemented with a natural antimicrobial. Int J Food Microbiol. 1999;49:67–74.

Kykkidou S, Giatrakou V, Papavergou A, Kontominas MG, Savvaidis IN. Effect of thyme essential oil and packaging treatments on fresh Mediterranean swordfish fillets during storage at 4°C. Food Chem. 2008. doi:10.1016/j.foodchem.2008.11.083.

Labuza TP, Lin S, Lillemo J, Taoukis PS. Inhibition of black spot formation in shrimp by ficin. LebensmittlenWissenshaft-und-Technologie. 1990;23:52–4.

Lee CY, Smith NL, Hawbecker DE. Enzyme activity and quality of frozen green beans as affected by blanching and storage. J Food Qual. 1988;11:279–87.

Leisner JJ. Characterisation of lactic acid bacteria isolated from lightly preserved fish products and their ability to metabolise various carbohydrates and amino acids. Ph.D. Thesis. Denmark: Royal Veterinary and Agricultural University; 1992.

Leisner JJ, Millan JC, Huss HH, Larsen CM. Production of histamine and tyramine by lactic acid bacteria isolated from vacuum-packaged sugar-salted fish. J Appl Bacteriol. 1994;76:417–23.

Leroi F, Joffraud JJ, Chevalier F, Cardinal M. Study of the microbial ecology of cold-smoked salmon during storage at 8°C. Int J Food Microbiol. 1998;39:111–21.

Li SJ, Seymour TA, King AJ, Morrissey MT. Color stability and lipid oxidation of rockfish as affected by antioxidant from shrimp shell waste. J Food Sci. 1998;63:438–41.

Lin C-C, Lin C-S. Enhancement of the storage quality of frozen bonito fillets by glazing with tea extracts. Food Control. 2005;16:169–75.

Lin YT, Labbe RG, Shetty K. Inhibition of *Vibrio parahaemolyticus* in seafood systems using oregano and cranberry phytochemical synergies and lactic acid. Innov Food Sci Emerg Technol. 2005;6:453–8.

Llamas Marcos A, Llamas Galilea P, Vargas Jiménez JM, Navarro Roldán F, Córdoba García F, Borrero Romero MJ. WO 2006/082267. 2006.

Löliger J. Natural antioxidants. In: Allen JC, Hamilton RJ, editors. Rancidity in foods. London: Applied Science; 1983. p. 89–107.

Lu Y, Foo YL. The polyphenol constituents of grape pomace. Food Chem. 1999;65:1–8.

Madhavi DL, Salunkhe DK. Toxicological aspects of food antioxidants. In: Madhavi DL, Deshpande SS, Salunkhe DK, editors. Food antioxidants. New York: Marcel Dekker; 1995. p. 267.

Mahmoud BSM, Yamazakia K, Miyashitab K, Il-Shikc S, Dong-Sukd C, Suzuki T. Bacterial microflora of carp (*Cyprinus carpio*) and its shelf-life extension by essential oil compounds. Food Microbiol. 2004;21(2004):657–66.

Manach C, Wiliamson G, Morand C, Scalbert A, Remesy C. Bioavailability and bioefficacy of polyphenols in humans. I. Review of 97 bioavailability studies. Am J Clin Nutr. 2005;81:230s–42.

Manju S, Srinivasa Gopla TK, José L, Ravinshankar CN, Ashok Kumar K. Nucleotide degradation of sodium acetate and potassium sorbate dip treated and vacuum packed black pomfret (*Parastromateus niger*) and pearlspot (*Etroplus Suratensis*) during chill storage. Food Chem. 2007;102:699–706.

Marshall MR, Kim J, Wei C. Enzymatic browning in fruits, vegetables and seafoods. Project report. FAO; 2000.
Martínez-Álvarez O, Gómez-Guillén C, Montero P. Presence of hemocyanin with diphenol oxidase activity in deepwater pink shrimp (*Parapenaeus longirostris*) post mortem. Food Chem. 2008a;107:1450–60.
Martínez-Álvarez O, Montero P, Gómez-Guillén C. Evidence of an active laccase-like enzyme in deep water pink shrimp (*Parapenaeus longirostris*). Food Chem. 2008b;108:624–32.
Matmaroh K, Benjakul S, Tanaka M. Effect of reactant concentrations on the Maillard reaction in a fructose–glycine model system and the inhibition of black tiger shrimp polyphenoloxidase. Food Chem. 2006;98:1–8.
Mbandi E, Shelef LA. Enhanced inhibition of *Listeria monocytogenes* and *Salmonella Enteritidis* in meat by combinations of sodium lactate and diacetate. J Food Prot. 2001;64:640–4.
McEvily AJ. US Patent 4,981,708; 1991.
McEvily AJ, Iyengar R, Otwell S. Sulfite alternative prevents shrimp melanosis. Food Technol. 1991;45:80–6.
Medina I, Satué-Gracía MT, German JB, Frankel EN. Comparison of natural polyphenol antioxidants from extra virgin olive oil with synthetic antioxidants in tuna lipids during thermal oxidation. J Agric Food Chem. 1999;47:4873–9.
Medina I, González MJ, Pazos M, Medaglia DD, Sacchi R, Gallardo JM. Activity of plant extracts for preserving functional food containing n _ 3 PUFA. Eur Food Res Technol. 2003;217:301–7.
Mejlholm O, Dalgaard P. Antimicrobial effect of essential oils on the seafood spoilage microorganism *Photobacterium phosphoreum* in liquid media and fish products. Lett Appl Microbiol. 2002;34:27–31.
Miková K. The regulation of antioxidants in food. In: Watson DH, editor. Food chemical safety, Additives, vol. 2. Boca Raton/Cambridge: CRC Press/Woodhead; 2002.
Min S, Rumsey TR, Krochta JM. Diffusion of the antimicrobial lysozyme from a whey protein coating on smoked salmon. J Food Eng. 2008;84:39–47.
Montero P, Ávalos A, Pérez-Mateos M. Characterization of polyphenoloxidase of prawns (*Penaeus japonicus*). Alternatives to inhibition additives and high-pressure treatment. Food Chem. 2001;75:317–24.
Nawar WW. Lipids. In: Fennema O, editor. Food chemistry. 3rd ed. New York/Basel: Marcel Dekker; 1996. p. 280–3.
Nebrink E, Borch E, Blom H, Nesbakken T. A model based on absorbance data on the growth rate of *Listeria monocytogenes* and including the effects of pH, NaCl, Na-lactate and Na-acetate. Int J Food Microbiol. 1999;47:99–109.
Neetoo H, Ye M, Chen H, Joerger RD, Hicks DT, Hoover DG. Use of nisin-coated plastic films to control *Listeria monocytogenes* on vacuum-packaged cold-smoked salmon. Int J Food Microbiol. 2008;122:8–15.
Nikoskelainen S, Salminen S, Bylund G, Ouwehand AC. Characterization of the properties of human- and dairy-derived probiotics for prevention of infectious diseases in fish. Appl Environ Microbiol. 2001;67:2430–5.
Nilsson L, Huss HH, Gram L. Inhibition of *Listeria monocytogenes* on cold-smoked salmon by nisin and carbon dioxide atmosphere. Int J Food Microbiol. 1997;38:217–27.
Nilsson L, Ng YY, Christiansen JN, Jorgensen BL, Grotinum D, Gram L. The contribution of bacteriocin to inhibition of *Listeria monocytogenes* by *Carnobacterium piscicola* strains in cold-smoked salmon systems. J Appl Microbiol. 2004;96:133–43.
No HK, Meyres SP, Prinyawiwatkull W, Xu Z. Applications of chitosan for improvement of quality and shelf life of foods: a review. J Food Sci. 2007;72:R87–100.
Nychas G-JE, Skandamis PN. Antimicrobials from herbs and spices. In: Roller S, editor. Natural antimicrobials for the minimal processing of foods. Cambridge: Woodhead; 2003.
Nykanen A, Weckman K, Lapvetelainen A. Synergistic inhibition of *Listeria monocytogenes* on cold-smoked rainbow trout by nisin and sodium lactate. Int J Food Microbiol. 2000;61: 63–72.

Otwell WS, Marshall MR. Screening alternatives to sulphating agents to control shrimp melanosis (black spot). Florida Sea Grant Technical Paper, vol 46; 1986. p. 1–20.
Ouattara B, Sabato SF, Lacroix M. Combined effect of antimicrobial coating and gamma irradiation on shelf life extension of pre-cooked shrimp (*Penaeus spp*). Int J Food Microbiol. 2001;68:1–9.
Paludan-Muller C, Dalgaard P, Huss HH, Gram L. Evaluation of the role of *Carnobacterium piscicola* in spoilage of vacuum and modified-atmosphere-packed cold-smoked salmon stored at 5°C. Int J Food Microbiol. 1998;39:155–66.
Pastoriza L, Sampedro G, Herrera JJ, Cabo ML. Influence of sodium chloride and modified atmosphere packaging on microbiological, chemical and sensorial properties in ice storage of slices of hake (*Merluccius merluccius*). Food Chem. 1998;61(1/2):23–8.
Pazos M, González MJ, Gallardo JM, Torres JL, Medina I. Preservation of the endogenous antioxidant system of fish muscle by grape polyphenols during frozen storage. Eur Food Res Technol. 2005;220:514–9.
Pazos M, Alonso A, Fernández-Bolaños J, Torres JL, Medina I. Physicochemical properties of natural phenolics from grapes and olive oil byproducts and their antioxidant activity in frozen horse mackerel fillets. J Agric Food Chem. 2006;54:366–73.
Petillo D, Hultin HO, Krzynowek J, Autio WR. Kinetics of antioxidant loss in mackerel light and dark muscle. J Agric Food Chem. 1998;46:4128–37.
Pokorny J. In: Pokorny J, Yanishlieva N, Gordon M, editors. Antioxidants in food. Practical applications. Boca Raton/Cambridge: CRC Press/Woodhead; 2001.
Qvist S, Sehested K, Zeuthen P. Growth suppression of *Listeria monocytogenes* in a meat product. Int J Food Microbiol. 1994;24:283–93.
Ramanathan L, Das NP. Natural products inhibit rancidity in salted cooked ground fish. J Food Sci. 1993;58:318–20.
Razat H, John A. Green tea polyphenol epigallocatechin-3-gallate differentially modulates oxidative stress in PC12 cell compartments. Toxicol Appl Pharmacol. 2005;207:212–20.
Richards MP, Hultin HO. Contributions of blood and blood components to lipid oxidation in fish muscle. J Agric Food Chem. 2002;50(3):555–64.
Richards MP, Kelleher SD, Hultin HO. Effect of washing with or without antioxidants on quality retention of mackerel fillets during refrigerated and frozen storage. J Agric Food Chem. 1998;46:4363–71.
Ringø E, Gatesoupe F-J. Lactic acid bacteria in fish: a review. Aquaculture. 1998;160:177–203.
Ringø E, Bendiksen HR, Wesmajervi MS, Olsen RE, Jansen PA, Mikkelsen H. Lactic acid bacteria associated with the digestive tract of Atlantic salmon (*Salmo salar L.*). J Appl Microbiol. 2000;89:317–22.
Roller S. Introduction. In: Roller S, editor. Natural antimicrobials for the minimal processing of foods. Boca Raton: CRC Press; 2003. p. 1–10.
Sagoo S, Board R, Roller S. Chitosan inhibits growth of spoilage microorganism in chilled pork products. 2002.
Sallam KI. Antimicrobial and antioxidant effects of sodium acetate, sodium lactate, and sodium citrate in refrigerated sliced salmon. Food Control. 2007;18:566–75.
Sambasivam S, Chandran R, Khan SA. Role of probiotics on the environment of shrimp pond. J Environ Biol. 2003;24:103–6.
Samelis J, Sofos JN. Organic acids. In: Roller S, editor. Natural antimicrobials for the minimal processing of foods. Boca Raton: CRC Press; 2003.
Sánchez-Alonso I, Jiménez-Escrig A, Saura-Calixto F, Borderías AJ. Effect of grape antioxidant dietary fibre on the prevention of lipid oxidation in minced fish: Evaluation by different methodologies. Food Chem. 2007;101:372–8.
Sánchez-Escalante A, Djenane D, Torrescano G, Beltran JA, Roncales P. The effects of ascorbic acid, taurine, carnosine and rosemary powder on colour and lipid stability of beef patties packaged in modified atmosphere. Meat Sci. 2001;58:421–9.
Sathivel S. Chitosan and protein coatings affect yield, moisture loss, and lipid oxidation of pink salmon (*Oncorhynchus gorbuscha*) fillets during frozen storage. J Food Sci. 2005;70:E455–459459.

Sathivel S, Liu Q, Huang J, Prinyawiwatkul W. The influence of chitosan glazing on the quality of skinless pink salmon (*Oncorhynchus gorbuscha*) fillets during frozen storage. J Food Eng. 2007;83:366–73.

Saura-Calixto F. Antioxidant dietary fiber product: a new concept and a potential food ingredient. J Agric Food Chem. 1998;48:4303–6.

Serdaroğlu M, Felekoğlu E. Effects of using rosemary extract and onion juice on oxidative stability of sardine (*Sardina pilchardus*) mince. J Food Qual. 2005;28(2):109–20.

Sewalt V, Robbins KL, Gamble W. Lipid-soluble antioxidant components of *Rosmarinus officinalis*. Lipid Technol. 2005;17:106–11.

Shahidi F, Wanasundara PKJPD. Phenolic antioxidants. CRC Crit Rev Food Sci Nutr. 1992;32: 67–103.

Shativel S, Huang J, Bechtel PJ. Properties of pollock (*Theragra chalcogramma*) skin hydrolysates and effects on lipid oxidation of skinless pink salmon (*Oncorhynchus gorbuscha*) fillets during 4 months of frozen storage. J Food Biochem. 2008;32:247–63.

Shelef LA. Antimicrobial effects of lactates: a review. J Food Prot. 1994;57:445–50.

Silva J, Carvalho AS, Teixeira P, Gibbs PA. Bacteriocin production by spray-dried lactic acid bacteria. Lett Appl Microbiol. 2002;34:77–81.

Sofos JN. Sorbic acid, chapter 23. In: Naidu AS, editor. Natural food antimicrobial systems. Boca Raton: CRC Press LLC; 2000.

Souquet J-M, Cheynier V, Brossaud F, Moutounet M. Polymeric proanthocyanidins from grape skins. Phytochemistry. 1996;43:509–12.

Stekelenburg FK, Kant-Muermans MLT. Effect of sodium lactate and other additives in a cooked ham product on sensory quality and development of a strain of *Lactobacillus curvatus* and *Listeria monocytogenes*. Int J Food Microbiol. 2001;66:197–203.

Stiles ME. Biopreservation by lactic acid bacteria. Antonie Van Leeuwenhoek. 1996;70:331–45.

Stiles ME, Holzapfel WH. Lactic acid bacteria of foods and their current taxonomy. Int J Food Microbiol. 1997;36:1–29.

Stoffels G, Sahl HG, Gudmundsdóttir A. Carnocin UI49, a potential biopreservative produced by *Carnobacterium piscicola*: large scale purification and activity against various gram-positive bacteria including *Listeria* sp. Int J Food Microbiol. 1993;20:199–210.

Tan BK, Harris ND. Maillard reaction products inhibit apple polyphenoloxidase. Food Chem. 1995;53:267–73.

Tang SZ, Sheehan D, Buckley DJ, Morrissey PA, Kerry JP. Anti-oxidant activity of added tea catechins on lipid oxidation of raw minced red meat, poultry and fish muscle. Int J Food Sci Technol. 2001;36:685–92.

Taoukis PS, Labuza TP, Lillemo JH, Lin SW. Inhibition of shrimp melanosis (black spot) by ficin. J Food Sci Technol. 1990;23:52–4.

Tassou CC, Drosinos EH, Nychas G-JE. Effects of essential oil from mint (*Mentha piperita*) on *Salmonella enteritidis* and *Listeria monocytogenes* in model food systems at 4 and 10°C. J Appl Bacteriol. 1995;78:593–600.

Taylor SL, Bush RK. Sulphite as food ingredients. Food Technol. 1986;40:47–52.

Thakur BR, Patel TR. Sorbates in fish and fish products-a review. Food Rev Int. 1994;10(1): 93–107.

Tome E, Teixeira P, Gibbs PA. Antilisterial inhibitory lactic acid bacteria isolated from commercial cold smoked salmon. Food Microbiol. 2006;23:399–405.

Tomé E, Pereira VL, Lopes CI, Gibbs PA, Teixeira PC. In vitro tests of suitability of bacteriocin-producing lactic acid bacteria, as potential biopreservation cultures in vacuum- packaged cold-smoked salmon. Food Control. 2008;19:535–43.

Tozer KN. Quality improvement and shelf-life extension of fish fillets from three aquaculture species. Thesis. University of Guelph, National Library of Canada; 2001.

Undeland I, Ekstrand B, Lingnert H. Lipid oxidation in minced herring (*Clupea harengus*) during frozen storage. Effect of washing and precooking. J Agric Food Chem. 1998;46:2319–28.

Undeland I, Hall G, Lingnert H. Lipid oxidation in fillets of herring (*Clupea harengus*) during ice storage. J Agric Food Chem. 1999;47:524–32.

Vareltzis K, Koufidis D, Gavriilidou E, Papavergou E, Vasiliadou S. Effectiveness of a natural rosemary (*Rosmarinus officinalis*) extract on the stability of filleted and minced fish during frozen storage. Food Res Technol. 1997;205:93–6.

Vásconez MB, Campos C A, Flores S, Alvarado de Dios, J y Gerschenson, L. Elaboración de recubrimientos comestibles en base a quitosano y estudio de su efecto antimicrobiano en filetes de salmón. Ciencia y Tecnología 2007; 16(3):77–79.

Verschuere L, Rombaut G, Sorgeloos P, Verstraete W. Probiotic bacteria as biological control agents in aquaculture. Microbiol Mol Biol Rev. 2000;64:655–71.

Vogel BF, Yin NG Y, Hyldig G, Mohr M, Gram L. Potassium lactate combined with sodium diacetate can inhibit growth of *Listeria monocytogenes* in vacuum-packed cold smoked salmon and has no adverse sensory effects. J Food Prot. 2006;68:2134–42.

Wendakoon CN, Sakaguchi M. Combined effect of sodium chloride and clove on growth and biogenic amine formation of *Enterobacter aerogenes* in mackerel muscle extract. J Food Prot. 1993;56(5):410–3.

Wessels S, Huss HH. Suitability of *Lactococcus lactis* subsp. *lactis* ATCC 11454 as a protective culture for lightly preserved fish products. Food Microbiol. 1996;13:323–32.

Yamaguchi F, Yoshimura Y, Nakazawa H, Ariga T. Free radical scavenging activity of grape seed extract and antioxidants by electron spin resonance spectrometry in an H_2O_2/NaOH/DMSO system. J Agric Food Chem. 1999;47:2544–8.

Yamazaki K, Suzuki M, Kawai Y, Inoue N, Montville TJ. Inhibition of *Listeria monocytogenes* in cold-smoked salmon by *Carnobacterium piscicola* CS526 isolated from frozen surimi. J Food Prot. 2003;66:1420–5.

Yano Y, Satomi M, Oikawa H. Antimicrobial effect spices and herbs on *Vibrio parahaemolyticus*. IntJ Food Microbiol. 2006;111:6–11.

Ye M, Neetoo H, Chen H. Effectiveness of chitosan-coated plastic films incorporating antimicrobials in inhibition of *Listeria monocytogenes* on cold-smoked salmon. Int J Food Microbiol. 2008;127:235–40.

Yoon KS, Burnette CN, Abou-Zeid KA, Whiting RC. Control of growth and survival of *Listeria monocytogenes* on smoked salmon by combined potassium lactate and sodium diacetate and freezing stress during refrigeration and frozen storage. J Food Prot. 2004;67:2465–71.

Chapter 15
Edible Films: Use of Lycopene as Optical Properties Enhancer

Rosemary A. de Carvalho, Carmen Silvia Fávaro-Trindade, and Paulo J.A. Sobral

15.1 Introduction

Edible films are thin layers of biopolymers sourced from renewable sources which contain colorless, low molecular weight components called plasticizers. Color in edible films is generally derived from the color of the biopolymer. Gelatin-based films, for example, are almost transparent and colorless. Products thus retain their characteristics, and when colored the films may also contribute to the product's visual appeal and may enhance the appearance of the package. Edible gelatin-based films may contain lycopene or artificial colorants that possess interesting physical properties, such as light barrier properties. The number of studies involving the production and characterization of edible and/or biodegradable films from diverse sources has increased in recent decades as a consequence of the increased awareness of the impact of synthetic packaging on the environment (Gontard and Guilbert 1996; Krochta and De-Mulder-Johnston 1997; Van de Velde and Kiekens 2002; Tharanathan 2003). Edible films are normally produced from macromolecules of biological origin, such as polysaccharides and proteins, capable of forming a continuous matrix (Gennadios et al. 1994; Nisperos-Carriedo 1994; Torres 1994). All additives used should be food grade and the films produced are classified as biodegradable (Gontard and Guilbert 1996).

Other synthetic biopolymers have also been studied for the development of biodegradable films (Van de Velde and Kiekens 2002). Among natural biopolymers, and specifically among proteins, gelatin was one of the first macromolecules used

R.A. de Carvalho • C.S. Fávaro-Trindade • P.J.A. Sobral (✉)
Department of Food Engineering, Faculty of Animal Science and Food Engineering,
University of São Paulo, Pirassununga, SP, Brazil
e-mail: pjsobral@usp.br

in biodegradable film production and continues to be widely studied because of its excellent film-forming properties, large-scale production, and low market prices (Arvanitoyannis 2002). In general, gelatin-based films have excellent gas barrier and good mechanical properties (Arvanitoyannis 2002), and can additionally serve as base for active compounds (antioxidants, antibiotics, etc.) (Gómez-Guillén et al. 2007; López-Carballo et al. 2008; Gómez-Estaca et al. 2009).

In general, gelatin films are colorless and transparent (Sobral et al. 2001; Vanin et al. 2005). For this reason, the incorporation of natural pigments in gelatin films may render them more attractive owing to coloring, without causing loss of edible and/or biodegradable character. Thus, gelatin films continue to be an environmentally friendly material, increasingly in demand by modern consumers. Apart from the coloration of films, the pigment may, eventually, alter physical and functional properties, including the ultraviolet (UV) barrier property (Corat et al. 2007). Among natural pigments, lycopene is thought to have potential because of its antioxidant and nutritional characteristics.

Studies involving the incorporation of natural pigments, such as chlorophyll-derived pigments chlorophyllide and chlorophyllin, in biopolymer-based films are not so commonly encountered (Corat et al. 2007; López-Carballo et al. 2008). Corat et al. (2007) and López-Carballo et al. (2008) were concerned only with the development of a gelatin film containing chlorophyllide that could affect some physical properties of the material or with the production of active films in which chlorophyllin would promote some light-induced antibiotic activity. Biopolymer-based films containing lycopene are novel.

15.2 Protein-Based Edible and/or Biodegradable Films

Proteins are macromolecules, normally water-insoluble, which may be solubilized by, among other techniques, the control of the pH of the solution (Cuq et al. 1995). The drying of such protein dispersions under controlled conditions, with the addition of a plasticizer, may result in the formation of a flexible film (Gennadios et al. 1994; Torres 1994; Cuq et al. 1995). Plasticizers are molecules of low molecular weight which penetrate biopolymer matrices, increasing the volume of empty spaces between the chains of macromolecules, causing a decrease in the strength of intermolecular forces along the matrix, resulting in improved functionality (Banker 1966; Krochta and De-Mulder-Johnston 1997).

Proteins from diverse sources have been studied in edible and/or biodegradable film technology. Among vegetable proteins, cottonseed proteins (Marquié et al. 1997), peanut proteins (Jangchud and Chinnan 1999), sunflower-seed protein isolate (Orliac et al. 2003), soy protein isolate (Ghorpade et al. 1995; Denavi et al. 2009), and wheat gluten (Gontard et al. 1993; Cherian et al. 1995), among others, have been studied.

There have also been numerous studies involving proteins from animal sources: these include pure whey protein isolate (Yoshida and Antunes 2004; Coupland et al.

2000) or in blends with sodium caseinate (Longares et al. 2005), fish myofibrillar proteins (Cuq et al. 1996; Monterrey-Quintero and Sobral 1999, 2000), fish-muscle proteins (Sobral et al. 2005; García and Sobral 2005; Paschoalick et al. 2003), soluble fish-muscle proteins (Iwata et al. 2000; Bourtoom et al.; 2006), and gelatin (Lim et al. 1999; Sobral et al. 2001; Carvalho and Grosso 2004; Thomazine et al. 2005; Vanin et al. 2005; Cao et al. 2009; Gómez-Estaca et al. 2009).

The properties of protein-based films are determined by protein–protein and protein–water interactions, which may be controlled by the preparation process for the film-forming solution and also by the addition of plasticizers (Gennadios et al. 1994; Torres 1994; Gontard and Guilbert 1996; Arvanitoyannis 2002). Additionally to proteins and plasticizers, other functional compounds have been proposed for the development of active films, i.e., for active packaging utilization. Some examples are presented in Table 15.1. By definition, a packaging may be considered active when it is capable of selecting, progressively releasing, absorbing, or transforming a compound, and therefore interacts with the packed food (Gontard 2000). However, the additional functional components may affect the main properties of the material.

Among the examples presented in Table 15.1, gelatin is most frequently used. This material is a protein biopolymer of animal origin produced by either acid or base hydrolysis of collagen obtained from bone of cattle or pigs, their skin, and connective tissues The molecular mass of gelatin depends on the type of raw material and on the process conditions, and ranges from 300 to 200,000 Da (Gennadios et al. 1994).

Gelatin is water-soluble at temperatures over 30°C. When subjected to temperatures above its melting point, it swells and dissolves, and when it cools to room temperature, it forms thermoreversible gels (Slade and Levine 1987). At the molecular level, gelatin-gel formation involves protein restructuring that involves transition from a disordered state to a more ordered stated formed by triple-helix structures typical of collagen in the native state. The structure and physical properties of those gels are due to the degree of formation of microcrystalline bonds (Slade and Levine 1987; Achet and He 1995; Ziegler and Foegeding 1990).

Gelatin sourced from either mammalian or fish sources has been used in many studies on edible and/or biodegradable films, in addition to those on active films (Table 15.1). Owing to the hydrophilic character of gelatin, these films are very sensitive to environmental conditions, which restricts their use as packaging materials. To improve these characteristics, various alternatives have been studied, such as enzymatic modifications (Lim et al. 1999; Carvalho and Grosso 2004), chemical modifications (Bigi et al. 2001; Carvalho and Grosso 2004, 2006; Cao et al. 2007; Carvalho et al. 2008a), incorporation of different plasticizers (Vanin et al. 2005), use of plasticizer blends (Thomazine et al. 2005), and lipid addition (Bertan et al. 2005). In general, significant improvement was not observed in water vapor barrier and mechanical properties.

Another alternative to improve the mechanical characteristics of these materials is the blending of biopolymers with synthetic polymers (Tharanathan 2003). An interesting synthetic polymer for this is poly(vinyl alcohol), which, despite being synthetic, is hydrophilic and biodegradable (Matsumura et al. 1999). Some films

Table 15.1 Examples of studies on protein-based active films found in the literature

Macromolecules	Additives	Expected functional activity	References
Soy protein isolate, spray-dried wheat gluten, egg albumen protein, and whey protein isolate	Nisin	Inhibition of *Listeria monocytogenes*	Ko et al. (2001)
Soy protein	Organic acids (citric, lactic, malic, or tartaric), nisin	Antimicrobial activities (*L. monocytogenes*; *Escherichia coli* O157:H7	Eswaranandam et al. (2004)
Whey protein isolate	*p*-Aminobenzoic acid or sorbic acid	Inhibition of *L. monocytogenes*, *E. coli* O157:H7, and *Salmonella typhimurium* DT104	Cagri et al. (2001)
Whey protein	Lactoferrin, lactoferrin hydrolysate, lactoperoxidase	Inhibition of *Salmonella enterica* and *E. coli* O157:H7	Min et al. (2005)
Alginate	Garlic oil	Antibacterial activity (*E. coli*, *Salmonella typhimurium*, *Staphylococcus aureus*, *Bacillus cereus*)	Pranoto et al. (2005)
Starch–casein	Neem (*Melia azardirachta*) extract	Antibacterial activity (*E. coli*, *S. aureus*, *B. cereus*, *L. monocytogenes*, *Pseudomonas* spp., and *Salmonella*)	Jagannath et al. (2006)
Soy protein	Nisin, grape seed extract, green tea extract	Inhibition of *L. monocytogenes*	Theivendran et al. (2006)
Tuna fish gelatin	Aqueous extract of murta	Antioxidant	Gómez-Guillén et al. (2007)
Alginate and gellan	*N*-Acetylcysteine and glutathione	Antibrowning	Rojas-Graü et al. (2007)
Soy protein	Grape seed extract, nisin, EDTA	Antimicrobial activites (*L. monocytogenes*, *E. coli* O157:H, and *S. typhimurium*)	Sivarooban et al. (2008)
Soy protein	Soluble extracts from Mexican oregano	Antioxidant	Pruneda et al. (2008)
Fish skin gelatin	Butylhydroxytoluene and α-tocopherol	Antioxidant	Jongjareonrak et al. (2008)
Giant squid gelatin	Gelatin hydrolysates	Antioxidant	Giménez et al. (2009)
Gelatin	Chlorophyllin	Antimicrobial activities	López-Carballo et al. (2008)

based on blends of gelatin and poly(vinyl alcohol) have been developed (Chiellini et al. 2001; Bergo et al. 2006; Mendieta-Taboada et al. 2008; Maria et al. 2008; Silva et al. 2008; Carvalho et al. 2009). However, all those films were colorless, i.e., the studies did not include a pigment to impart attractiveness.

In the study conducted by López-Carballo et al. (2008), a pigment was used (Table 15.1). In that study, López-Carballo et al. (2008) made gelatin-based films with sodium magnesium chlorophyllin (E140) and sodium copper chlorophyllin (E141) with the aim of developing antimicrobial photosensitizer-containing edible films and coatings. These authors observed that it was possible to reduce microorganism growth in cooked frankfurters inoculated with *Staphylococcus aureus* and *Listeria monocytogenes* by covering them with sodium magnesium chlorophyllin–gelatin films and coatings. According to these authors, the films had a vivid green-yellow color.

Corat et al. (2007) had previously produced gelatin-based films with different concentrations of chlorophyllide (0, 2, 4, 6, 8, and 10 g of 10% chlorophyllide solution/100 g gelatin), and observed that the pigment significantly altered the color of the films without affecting the other properties studied, such as mechanical properties and solubility.

15.3 Lycopene

Lycopene is a chromoplast carotenoid responsible for the brilliant yellow, orange, and red color of plants and animals (Nguyen and Schwartz 1999). This pigment is a cyclic open-chain polyene hydrocarbon with 13 double bonds, of which 11 are conjugated in a linear array (Shi et al. 2003) (Fig. 15.1). The highly conjugated double-bond system (chromophore) is responsible for the molecular shape, chemical reactivity,, and light absorption in the visible region of the spectrum, and hence the color (Ötles and Çagindi 2008; Rodriguez-Amaya 2002). Furthermore, owing to its chemical structure, lycopene is a lipid-soluble phytochemical (Nguyen and Schwartz 1999).

In a raw food, lycopene is generally present as the all-*trans* geometric isomer; the most thermodynamically stable form. However, light, heat, or chemical reaction may introduce a kink along the conjugated double bonds and transform all-*trans* lycopene into various *cis* isomers (Shi et al. 2004). Also because of that characteristic, lycopene acts as an antioxidant and a free-radical quenching agent in foods. It interacts rapidly with free radicals and with oxygen, thereby inhibiting the propagation step of lipid peroxidation (Britton 1995; Ötles and Çagindi 2008). However, the lycopene structure (long-chain conjugated hydrocarbon) suggests that it will be easily oxidized in the presence of oxygen (Rodriguez-Amaya 2002). The isomerization and oxidation reactions cause, in this material, loss of color and biological activity.

Reports of dietary lycopene decreasing the risk of a number of health conditions have generated new interest in the addition of lycopene to functional foods (Boon et al. 2008). The functional properties of this pigment have caused increased consumer

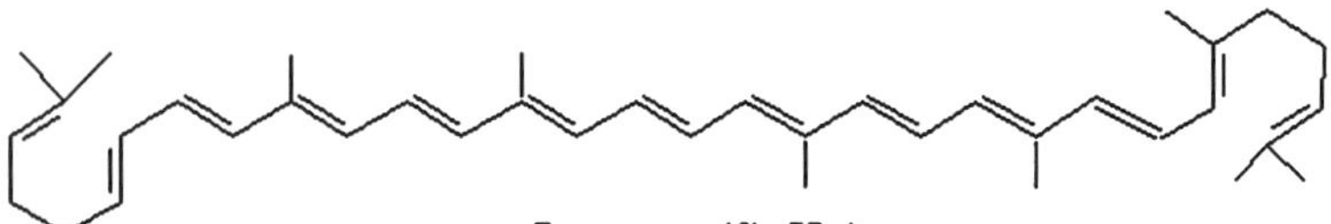

Fig. 15.1 Chemical structure of lycopene

interest in lycopene-containing food products and have generated new applications for lycopene-containing food ingredients (Boon et al. 2008). Because of these new applications, the market for carotenoids has grown. It was estimated that in 2009 the sales of lycopene could reach US $26 million.

Carotenoids, like lycopene, are largely used as colorants in foods, drinks, cosmetics, and animal (mainly poultry and fish) feed (Nunes and Mercadante 2007). The main industrial sources of carotenoids are through chemical synthesis and extraction from plants (Nunes and Mercadante 2007). As a result of the instability of lycopene, its storage and handling of should be performed under controlled environmental conditions to minimize oxidative degradation (Nguyen and Schwartz 1999). Additionally, the use of lycopene as an additive faces technological difficulties owing to its liposoluble character, as the absolute majority of food products are water-rich. Therefore, encapsulation may be useful to render lycopene water-dispersible and protect it from environmental conditions (Matioli and Rodriguez-Amaya 2003; Nunes and Mercadante 2007).

15.4 Experimental Considerations

Edible films were produced using a pigskin gelatin supplied by Gelita South America (São Paulo, Brazil), and sorbitol (Synth) as a plasticizer. To color this material, a sample of purified lycopene was encapsulated in maltodextrin (LycoVit® 10 CWD), supplied by BASF (São Paulo, Brazil).

The films were made by casting (Sobral et al. 2001; Vanin et al. 2005) a 2 g gelatin/100 g film-forming solution; 25 g sorbitol/100 g gelatin, and 0.0, 0.2, 0.4, 0.6, 0.8, and 1.0 g lycopene/100 g gelatin. Gelatin was first hydrated for 30 min at room temperature, followed by solubilization at 55°C for 15 min, in a thermostat-controlled water bath (TE 184; TECNAL, Brazil). The plasticizer was then added and the solution was cooled to 38°C, lycopene was added at this point, and the whole mix was stirred with a magnetic stirrer for 15 min. The films were formed after drying at 30°C for 2 4 h in an air-convection oven. The thickness of the films was kept constant (0.800 ± 0.008 mm) by controlling the ratio dry mass to support area, was and determined with a digital micrometer (Mitutoyo, ± 0.001 mm).

Before characterization, the films were kept in desiccators containing sodium bromide saturated solution (relative humidity of 58%) at 25°C for at least 7 days.

15 Edible Films: Use of Lycopene as Optical Properties Enhancer

The characterization of the films was conducted in an acclimatized room with a temperature range of 22±3°C and relative humidity of approximately 55%.

The moisture content of the films was determined in an oven at 105°C using the drying to constant weight method. The solubility in water of the films was calculated after 24 h immersion, according to the method of Gontard et al. (1993), and was expressed in terms of solubilized dry matter.

Water vapor permeability was determined according to the method proposed by Gontard et al. (1993). The films were fixed in cells containing silica gel. The cells were kept inside desiccators containing distilled water and maintained at 25°C in an oven (BOD TE 390; TECNAL) (±0.2°C). The weight gain of the system was followed at 24 h intervals for 120 h. Water vapor permeability was calculated with Eq. 15.1 (Gontard et al. 1993):

$$\text{WVP} = \frac{w}{tA}\frac{x}{\Delta P}, \qquad (15.1)$$

where x is the average thickness of the film, A is the permeation area, ΔP is the partial vapor pressure difference between the interior of the cell (silica gel; $P=0$ Pa) and the outside (distilled water; $P=3,166$ Pa), and w/t corresponds to the angular coefficient of the weight versus time plot. The mechanical properties of the films were determined by tensile tests (tensile strength, elongation at break, and elastic modulus) and perforation tests (puncture force and puncture deformation) using a TA.XT2i texture analyzer (Stable Micro Systems) with the tensile grips probe moving at 0.9 mm/s (Thomazine et al. 2005) and with a cylindrical 3-mm probe moving at 1 mm/s (Sobral et al. 2001), respectively.

The color parameters (a^*, b^*, L^*), the total color difference (ΔE^*), and the opacity of the films were determined according to the method of Sobral (1999), using a portable color measurement instrument (Miniscan XE; HunterLab). The parameters a^*, b^*, and L^* were determined by placing the films over the white standard, and the total difference was calculated according to Eq. 15.2 (Gennadios et al. 1996):

$$\Delta E^* = [(\Delta L^*)^2 + (\Delta a^*)^2 + (\Delta b^*)^2]^{0.5} \qquad (15.2)$$

where $\Delta L^* = L^*_{\text{standard}} - L^*_{\text{sample}}$, $\Delta a^* = a^*_{\text{standard}} - a^*_{\text{sample}}$, and $\Delta b^* = b^*_{\text{standard}} - b^*_{\text{sample}}$. The opacity was determined according to the method of Sobral (2000) using the same measuring instrument and software used for the color measurements. The opacity was calculated as the relationship ($Y = Y_p/Y_b$) between the opacity of the film placed over the black standard (Y_p) and the opacity of the film paced over the white standard (Y_b).

The UV and visible light barrier properties were determined with a UV/visible spectrophotometer (Libra S22; Biochrom) within the range from 190 to 800 nm by transmittance measurements of the films (Fang et al. 2002).

The data were statistically analyzed for means comparison by the Duncan test ($p<0.05$), using the software program SAS (version 6.8; SAS, Cary, NC, USA).

Table 15.2 Moisture content (MC), water vapor permeability (WVP), and solubility in water (S) of gelatin-based films with different concentrations of lycopene (C_L)

C_L (g lycopene/ 100 g gelatin)	MC (g water/ 100 g film)	WVP ($\times 10^{-8}$ g mm/ h cm^2 Pa)	S (g solubilized dry solids/100 g dried film)
0	11.2 ± 0.6^c	4.1 ± 0.3^a	37.5 ± 1.2^a
0.2	12.5 ± 0.5^a	3.3 ± 0.3^b	37.7 ± 0.8^a
0.4	12.5 ± 0.5^{a,b}	4.4 ± 0.4^a	44.3 ± 4.5^b
0.6	11.4 ± 0.5^{b,c}	4.7 ± 0.6^a	43.8 ± 0.9^b
0.8	11.3 ± 0.6^c	4.8 ± 0.3^a	44.3 ± 1.1^b
1.0	13.1 ± 1.2^{a,b}	4.1 ± 0.4^b	44.1 ± 3.5^a

Different *letters* in a column denote significant difference ($p < 0.05$) between averages obtained through the Duncan test

15.5 Properties of the Lycopene-Added Gelatin Films

The lycopene-colored films were most affected by the concentration of lycopene, and were homogeneous (absence of insoluble particles). The films had a uniform red color, attributed to the complete lycopene solubilization in the film-forming solution, which was aided by the fact that the colorant was microencapsulated (Vertzoni et al. 2006; Fávaro-Trindade et al. 2008). This result is a very important milestone, as this form of solubilization is difficult to accomplish in synthetic plastics technology. According to Durston (2006), there are limits on the quantity of colored pigment that can be added to the synthetic films without affecting the physical properties of these films.

Significant differences were observed between the moisture content of films (Table 15.2); these ranged from 11% to 13%. This is a very low variation, and is thus insufficient to interfere with the results for the other properties (Gontard et al. 1993; Sobral et al. 2001).

The water vapor permeability of the films was not affected by the added lycopene (Table 15.2). Therefore, lycopene addition, to improve the UV/visible light barrier of the gelatin-based films may be technologically beneficial, as the water vapor barrier is not significantly affected. The water vapor permeability of the lycopene-added films was lower than that observed by Thomazine et al. (2005) in gelatin-based films plasticized with the same amount of glycerol and sorbitol: 7×10^{-8} g mm/h cm^2 Pa and 5×10^{-8} g mm/h cm^2 Pa, respectively, and of fish-skin gelatin-based films (12×10^{-8} g–mm/h cm^2 Pa) plasticized with 30% sorbitol (Carvalho et al. 2008b).

The presence of the lycopene in the polymeric matrix caused a slight increase of the solubility in water of the films (Table 15.2). As lycopene is a hydrophobic molecule, a reduction of the water solubility of the films could be expected. In fact, it is quite possible that the existence of hydrophobic components in the film may have hindered the perfect formation of the polymeric matrix, thus facilitating the film solubilization. According to the literature, solubility in water of gelatin-based films may vary mainly as a function of the formulation (plasticizer concentration, type of plasticizer, hydrophobic compound addition, and macromolecule concentration) and of the type of treatment of the gelatin (chemical or enzymatic modification, irradiation, etc.). Carvalho and Grosso (2004, 2006) determined the solubility in water to be between 25% and 30% in films of chemically and enzymatically modified gelatin (10 g/100 g film-forming solution) films plasticized with glycerol (4.5 g glycerol/100 g film-forming solution). Bertan et al. (2005) found values of solubility in water from 30% to 36% in films of gelatin and hydrophobic compounds (triacetine, stearic acid, palmitic acid, and pitch).

Recently, Pérez-Mateos et al. (2009) determined values of solubility in water near 88% for fish-skin gelatin films containing sunflower oil and plasticized with sorbitol and glycerol, whereas Carvalho et al. (2008b) observed the complete solubilization of films based in halibut or tuna fish-skin gelatin after 24 h immersion in water. Regarding active films, Goméz-Estaca et al. (2009) determined solubility values ranging between 34% and 84% when studying gelatin films from two sources (bovine hide and tuna skin) with aqueous oregano and rosemary extracts added in different concentrations. Giménez et al. (2009) observed solubility values above 90% in films based on giant squid (*Dosidicus gigas*) gelatin.

Concerning the mechanical properties of the films, the lycopene did not significantly affect the strength, the stiffness, or the deformability of the films (Table 15.3). It was observed that 1 g lycopene/100 g gelatin concentration caused a greater tensile strength, but this was almost the same as that observed for films with 0.2 g lycopene/100 g gelatin, thus not proving a monotonic behavior. These results can be explained by the hydrophobic character of lycopene, which must have hardly interacted with gelatin polypeptides, thus not having any plasticizer-type or antiplasticizer-type effect.

In general, the films with and without lycopene showed similar strength (Table 15.3) compared with other gelatin films, either in strain tests (21–48 MPa obtained by Thomazine et.al. 2005; 35–50 MPa obtained by Cao et al. 2009) or in perforation tests (8–16 N obtained by Sobral et al. 2001; 16–20 N obtained by Vanin et al. 2005). Besides, the films were more deformable in strain tests than films made from bovine-bone gelatin with 20–30 g sorbitol/100 g gelatin (4–8% obtained by Cao et al. 2009), but were much less deformable than films based on fish gelatin, where elongation during strain tests was around 300% Carvalho et al. (2008b) In general, the puncture deformation was typical of gelatin-based films (Sobral et al. 2001; Thomazine et al. 2005; Vanin et al. 2005).

Increasing concentration of lycopene in films has been shown to improve the UV barrier property (Fig. 15.2). The films had no transmittance at 200 nm and values between 1 and 5 at 280 nm. Gelatin also has excellent UV barrier properties owing to the presence of amino acid residues containing aromatic rings (Sobral et al. 2001). Similar occurrences have been reported with other materials; similar values were reported in films of fish protein muscle at wavelengths of 200–280 nm (Artharn et al. 2007). Fang et al. (2002) determined transmittance of 6–7% in the range from 200 to 280 nm in films based in whey proteins, which are also rich in amino acid residues containing aromatic rings.

Table 15.3 Tensile strength (*TS*), elongation at break (*EB*), elastic modulus (*EM*), puncture force (*PF*), and puncture deformation (*PD*) of gelatin-based films with different concentrations of lycopene (C_L)

C_L (g lycopene/ 100 g of gelatin)	TS (MPa)	EB (%)	EM (MPa)	PF (N)	PD (%)
0	42.3 ± 5.7[b]	16.2 ± 4.3[a]	12.0 ± 1.5[a]	30.0 ± 1.9[a,b,c]	2.4 ± 0.2[a]
0.2	46.0 ± 5.8[a,b]	18.7 ± 1.9[a]	12.0 ± 1.2[a]	30.7 ± 1.8[a,b]	1.9 ± 0.1[b]
0.4	45.7 ± 2.3[b]	16.2 ± 2.5[a]	12.3 ± 1.5[a]	32.9 ± 1.1[a]	2.6 ± 0.4[a]
0.6	44.2 ± 4.8[a, b]	16.2 ± 4.4[a]	12.8 ± 1.9[a]	28.4 ± 2.5[b,c]	2.7 ± 0.3[a]
0.8	41.4 ± 4.7[b]	16.5 ± 3.8[a]	11.9 ± 1.1[a]	33.3 ± 2.6[a]	2.7 ± 0.2[a]
1.0	49.75 ± 8.0[a]	16.2 ± 3.6[a]	13.5 ± 0.6[a]	26.4 ± 2.0[c]	2.3 ± 0.1[a]

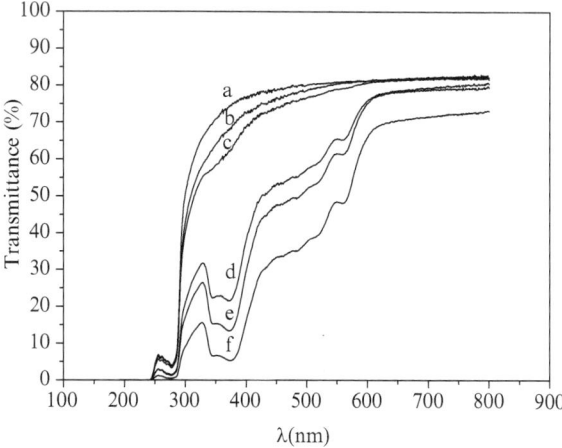

Fig. 15.2 Transmittance of gelatin-based films with different concentrations of lycopene: *a* 0.0 g lycopene/100 g gelatin; *b* 0.2 g lycopene/100 g gelatin; *c* 0.4 g lycopene/100 g gelatin; *d* 0.6 g lycopene/100 g of gelatin; *e* 0.8 g lycopene/100 g gelatin; *f* 1.0 g lycopene/100 g gelatin

Lycopene possesses 11 chromophore conjugated double bonds and is capable of absorbing light in a wide wavelength spectrum (Ötles and Çagindi 2008; Rodriguez-Amaya 2002). Thus, lycopene absorbs UV radiation and protects both the material and the foodstuff from oxidation (Hanlon et al. 1998). This structural configuration reportedly produces a protective effect against the degradation of the material and of the packaged foodstuff (Hanlon et al. 1998; Coltro et al. 2003). These results suggest that this type of material is suitable for packaging lipid-rich foodstuffs and/or vitamins susceptible to oxidation (Bekbolet 1990; Fang et al. 2002; Artharn et al. 2007). The most pronounced effect of light-catalyzed reactions is observed with light in the lower wavelengths of the visible spectrum and in the UV spectrum (Bekbolet 1990). This is of special significance as light is known to have a damaging effect on foodstuffs such as milk and other dairy products owing to the presence of light-sensitive vitamins, such as vitamins A and B_2 (Saffert et al. 2009).

However, lycopene also showed an excellent visible light barrier property, mainly between 350 and 570 nm (Fig. 15.2). The linear reduction (Table 15.4) of the transmittance of the film as a function of lycopene concentration in a wide wavelength range denoted improved light barrier property of the material (Coltro et al. 2003).

Table 15.4 Parameters and correlation coefficients (R) of the linear model used to fit transmittance as a function of lycopene concentration (C_L) at different wavelengths (λ)

λ (nm)	Equation	R
280	$T = 3.9 - 3.4 C_L$	0.980
350	$T = 65.7 - 65.6 C_L$	0.981
400	$T = 75.3 - 63.4 C_L$	0.993
500	$T = 81.5 - 43.1 C_L$	0.997
600	$T = 82.3 - 8.3 C_L$	0.919
700	$T = 82.6 - 2.7 C_L$	0.739
800	$T = 82.7 - 2.2 C_L$	0.728

Addition of lycopene to films caused significant changes in color parameters (Figs. 15.3, 15.4) as expected for a pigment. Even in food products, the intensity of color depends on the lycopene concentration (Rodriguez-Amaya 2002).

The color parameter $a*$, which varies from green (−) to red (+), and the parameter $b*$, which varies from blue (−) to yellow (+), increased linearly ($R^2 = 0.983$ and $R^2 = 0.992$, respectively) with the pigment concentration (C_L) (Eqs. 15.3, 15.4), indicating that effectively the material became more red-yellowish as the pigment concentration increased:

$$a* = -1.2 + 19.1 C_L \qquad (15.3)$$

$$b* = 3.3 + 24.5 C_L \qquad (15.4)$$

The color parameter $L*$ decreased linearly ($R^2 = 0.990$) as a result of lycopene concentration increase (Eq. 15.5), indicating that the films became darker as the lycopene concentration increased:

$$L* = 90.1 - 15.9 C_L \qquad (15.5)$$

From the results observed for $L*$, $a*$, and $b*$, increased chlorophyllide concentration also caused a linear increase ($R^2 = 0.990$) of the total color difference ($\Delta E*$) (Fig. 15.4), according to Eq. 15.6:

$$\Delta E* = 3.2 + 33.8 C_L \qquad (15.6)$$

Similar values can be found in the specialized literature for $b*$ and $L*$ for films without pigment addition (Table 15.5). For example, Hernandez-Muñoz et al. (2004) determined $b* = 23$ for gliadin-rich films containing 33 g glycerol/100 g protein and heat treated at 115°C for 24 h, and Rhim et al. (2000) obtained $b* = 26$ when working with modified soy protein isolate films plasticized with 50 g glycerol/100 g soy protein isolate. Moreover, almost all values of $L*$ presented in Table 15.5 are similar to those for films containing lycopene (Fig. 15.3).

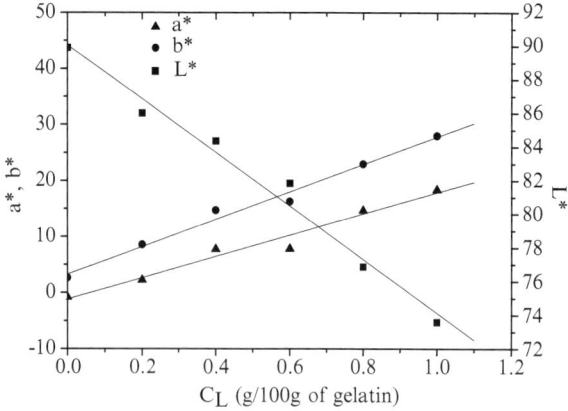

Fig. 15.3 Effect of lycopene concentration (C_L) variation on color parameters a^*, b^*, and L^* of gelatin-based films

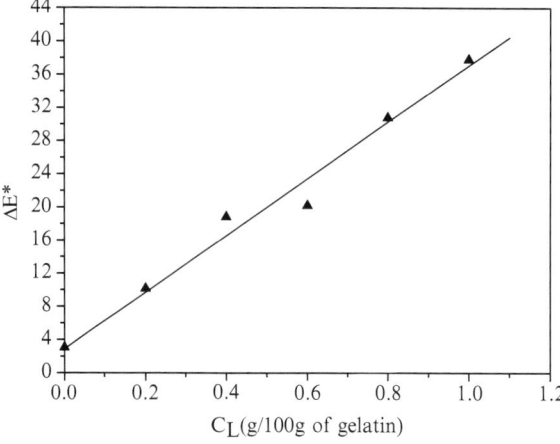

Fig. 15.4 Effect of lycopene concentration (C_L) variation on total color difference (ΔE^*) of gelatin-based films

Nevertheless, the a^* and ΔE^* parameters of the films containing lycopene were, in general, higher than the values presented in Table 15.5, excepted for the heat-cured (140°C) wheat gluten films with 3 g glycerol/100 mL solution, for which $a^* = 11$ (Micard et al. 2000), and the chemically modified gelatin films with 4.5 g glycerol/100 g solution, for which $\Delta E^* = 20$ (Carvalho and Grosso 2006). In the first case, the thermal treatment must have caused film browning, whereas in the second case, the elevated ΔE^* was caused by Schiff base formation as a consequence of bifunctional aldehyde cross-linking (Carvalho and Grosso 2006).

The films containing lycopene exhibited higher values of color parameters as they are normally quite colorless (Sobral 1999; Vanin et al. 2005; Carvalho and Grosso 2006), even in the case of addition of chlorophyll water-soluble derivatives (Table 15.5).

Table 15.5 Examples of color parameters of edible films based on proteins

Films	ΔE^*	a^*	b^*	L^*	References
Gliadin-rich films; 33 g glycerol/100 g protein; without heat treatment	–	−0.2	4.5	97.3	Hernandez-Muñoz et al. (2004)
Gliadin-rich films; 33 g glycerol/100 g protein; heat treated at 115°C for 24 h	–	0.9	22.8	91.3	Hernandez-Muñoz et al. (2004)
Heat-cured wheat gluten films, $T=20$°C; 12.8 g wheat gluten/100 mL solution; 3 g glycerol/100 mL solution	–	−0.8	11.1	94.1	Micard et al. (2000)
Heat-cured wheat gluten films, $T=140$°C; 12.8 g wheat gluten/100 mL solution; 3 g glycerol/100 mL solution	–	10.5	45.4	75.4	Micard et al. (2000)
Modified SPI films, 50 g glycerol/100 g SPI	–	−2.4	14.4	93	Rhim et al. (2000)
Modified SPI films, 50 g glycerol/100 g SPI; 10 g dialdehyde starch/100 g SPI	–	−1.0	26.0	90	Rhim et al. (2000)
Protein-based film from round scad washed mince of an ordinary muscle, 50 g glycerol/100 g protein	–	−1.9	4.9	87	Artharn et al. (2007)
Protein-based film from round scad washed mince of a dark muscle, 50 g glycerol/100 g protein	–	3.0	22.9	82	Artharn et al. (2007)
Films based on bigeye snapper skin gelatin; 50 g glycerol/100 g protein; and 25% palmitic acid instead of glycerol	–	−1.4	2.7	89.9	Jongjareonrak et al. (2006)
Films based on brownstripe red snapper skin gelatin; 50 g glycerol/100 g protein; and 25% palmitic acid instead of glycerol	–	−1.7	8.9	89.3	Jongjareonrak et al. (2006)
Egg white protein films; 60 g poly(ethylene glycol) 400/100 g protein; 10 g oleic acid/100 g protein	–	−1.4	6.0	96	Handa et al. (1999)
Egg white protein films; 50 g sorbitol/100 g protein	2.0	−0.9	4.9	96	Gennadios et al. (1996)
Laboratory-prepared SPI films; 30 g glycerol/100 g protein	8.5	−2.3	9.9	95	Kunte et al. (1997)
Films based on amaranth flour; 22.5 g glycerol/100 g flour	8.9	−1.2	8.1	90	Tapia-Blácido et al. (2007)
Films based on muscle protein of Nile tilapia; 25 g glycerol/100 g protein; heat treated at 40°C for 30 min	8.5	−1.6	8.9	94	Paschoalick et al. (2003)
Water-soluble fish protein films; 25 g sorbitol/100 g protein	–	−0.8	8	33	Bourtoom et al. (2006)
Chemically modified gelatin films; 4.5 g glycerol/100 g solution; 8.8 mmol formaldehyde/100 mL solution	12.1	−1.2	14.1	84.3	Carvalho and Grosso (2006)

(continued)

Table 15.5 (continued)

Films	ΔE^*	a^*	b^*	L^*	References
Chemically modified gelatin films; 4.5 g glycerol/100 g solution; 26.3 mmol glyoxal/100 mL solution	20.2	−0.8	25.2	81.6	Carvalho and Grosso (2006)
Gelatin-based film; 25 g sorbitol/100 g gelatin; 1 g of 10% solution chlorophyllide/100 g gelatin	5.3	−2.5	4.6	91	Corat et al. (2007)
Gelatin films; 10 g gelatin/100 mL; 2.5 g glycerol/100 mL and 80 μg/mL chlorophyllin (E140)	–	−3.2	7.2	85	López-Carballo et al. (2008)

SPI soy protein isolate

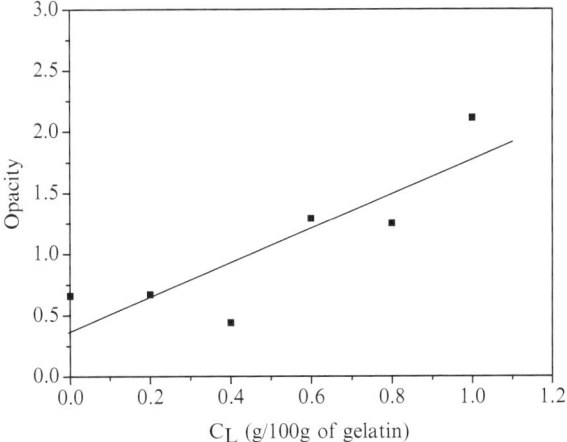

Fig. 15.5 Effect of lycopene concentration (C_L) variation on opacity of gelatin-based films

In fact, according to Nassau, K. (1996), the appearance of color as detected by the eye and interpreted by the brain depends significantly on the exact viewing circumstances, including transparency of samples. And also the color of a pigmented medium, such as a plastic, depends of both transmittance and reflectance, which, in turn, are different if the material is opaque or transparent (Saunderson 1942). Hence, gelatin films (Table 15.5) normally exhibit low color parameter values owing to the added effects of transparency and thickness.

The films containing lycopene also exhibited low opacity, despite the linear increase of this property (Eq. 15.7; $R^2 = 0.856$) as a function of lycopene concentration (Fig. 15.5);

$$Y = 0.4 + 1.4C_L \tag{15.7}$$

Nevertheless, these films could be considered as quite transparent, which suggests that lycopene was perfectly dissolved in the polymeric matrix as a consequence of encapsulation, as previously discussed (Matioli and Rodriguez-Amaya 2003; Nunes and Mercadante 2007; Fávero-Trindade et al. 2008). According to Villalobos et al. (2005), the presence of a disperse, nonmiscible phase promotes opacity as a function of the differences in the refractive index of the phases and the concentration and particle size of the dispersed phase. Gelatin-based films normally exhibit high transparency ($Y < 3$) (Sobral 1999; Vanin et al. 2005; Carvalho et al. 2008a).

15.6 Conclusions

Encapsulation in maltodextrin facilitated lycopene incorporation into gelatin films. Besides delivering a reddish color to the material, which may potentially improve attractiveness, the presence of lycopene as an additive improved UV/visible light barrier properties, which made such films very appropriate for use as packaging material for products susceptible to lipid oxidation. The presence of lycopene to a concentration of 1% in gelatin does not affect the other physical and/or functional properties of the films.

Acknowledgements FAPESP and CNPq are thanked for financial support.

References

Achet D, He XW. Determination of the renaturation level in gelatin films. Polymer. 1995;36: 787–91.
Artharn A, Benjakul S, Prodpran T, Tanaka M. Properties of a protein-based film round scad (*Decapterus maruadsi*) as affected by muscle types and washing. Food Chem. 2007;103: 867–74.
Arvanitoyannis IS, Nakayama A, Aiba S. Edible films made from hydroxypropyl starch and gelatin and plasticized by polyols and water. Carbohyd Polym. 1998a;36:105–19.
Arvanitoyannis IS, Nakayama A, Aiba S. Chitosan and gelatin based edible films: state diagrams, mechanical and permeation properties. Carbohyd Polym. 1998b;37:371–82.
Arvanitoyannis IS, Psomiadou E, Nakayama A, Aiba S, Yamamoto N. Edible films made from gelatin soluble starch and polyols part 3. Food Chem. 1997;60:593–604.
Arvanitoyannis IS. Formation and properties of collagen and gelatin films and coatings. In: Gennadios A, editor. Protein-based films and coatings. Boca Raton: CRC Press; 2002. p. 275–304.
Banker GS. Film coating theory and practice. J Pharmacol Sci. 1966;55:81–9.
Bekbolet M. Light effects on food. J Food Prot. 1990;53:430–40.
Bergo PVA, Carvalho RA, Sobral PJA, Silva FBR, Pinto JKC, Souza JP. Microwave insertion loss measurements in gelatin-based films. Meas Sci Technol. 2006;17:3261–4.
Bertan LC, Tanada-Palmu PS, Siani AC, Grosso CRF. Effect of fatty acids and "Brazilian elemi" on composite films based on gelatin. Food Hydrocolloid. 2005;19:73–82.
Bigi A, Cojazzi G, Panzavolta S, Rubini K, Roveri N. Mechanical and thermal properties of gelatin films at differents degrees of glutaraldehyde crosslinking. Biomaterials. 2001;22:763–8.
Boon CS, Xu Z, Yue X, Mcclements J, Weiss J, Decker EA. Factors affecting lycopene oxidation in oil-in-water emulsions. J Agric Food Chem. 2008;56:1408–14.
Bourtoom T, Chinnan MS, Jantawat P, Sanguandeekul R. Effect of select parameters on the properties of edible films from water-soluble fish proteins in surimi wash-water. Lebensmittel-Wissenschaft und -Technologie. 2006;39:406–19.
Britton G. Structure and properties of carotenoids in relation to function. FASEB J. 1995;9:1551–8.
Cagri A, Ustunol Z, Ryser ET. Antimicrobial mechanical and moisture barrier properties of low ph whey protein-based edible films containing p-aminobenzoic or sorbic acids. J Food Sci. 2001;66:865–70.
Cao N, Fu Y, He J. Mechanical properties of gelatin films cross-linked, respectively, by ferulic acid and tannin acid. Food Hydrocolloid. 2007;21:575–84.
Cao N, Yang X, Fu Y. Effects of various plasticizers on mechanical and water vapor barrier properties of gelatin films. Food Hydrocolloid. 2009;23:729–35.
Carvalho RA, Grosso CRF. Characterization of gelatin based films modified with transglutaminase glyoxal and formaldehyde. Food Hydrocolloid. 2004;18:717–26.
Carvalho RA, Grosso CRF. Properties of chemically modified gelatin films. Braz J Chem Eng. 2006;23:45–53.
Carvalho RA, Grosso CRF, Sobral PJA. Effect of chemical treatment on the mechanical properties water vapour permeability and sorption isotherms of gelatin-based films. Packag Technol Sci. 2008a;21:165–9.

Carvalho RA, Sobral PJA, Thomazine M, Habitante AMQB, Gimenez B, Gómez-Guillén MC, et al. Development of edible films based on differently processed Atlantic halibut (*Hippoglossus hippoglossus*) skin gelatin. Food Hydrocolloid. 2008b;22:1117–23.

Carvalho RA, Moraes ICF, Bergo PVA, Kamimura ES, Habitante AMQB, Sobral PJA. Study of some physical properties of biodegradable films based on blends of gelatin and poly(vinyl alcohol) using a response-surface methodology. Mater Sci Eng C Biomim Mater Sens Syst. 2009;29:485–91.

Cherian G, Gennadios A, Weller C, Chinachoti P. Thermomechanical behavior of wheat gluten films: effect of sucrose glycerin and sorbitol. Cereal Chem. 1995;72:1–6.

Chiellini E, Cinelli P, Corti A, Kenawy ER. Composite films based on waste gelatin: thermal–mechanical properties and biodegradation testing. Polym Degrad Stabil. 2001;73:549–55.

Coltro L, Padula M, Saron ES, Borghetti J, Buratin AEP. Evaluation of a UV absorber added to PET bottles for edible oil packaging. Packag Technol Sci. 2003;16:15–20.

Corat M, Carvalho RA, Favaro-Trindade CS, Sobral PJA. Produção e caracterização de filmes a base de gelatina coloridos com clorofila. Alimentos: Ciencia e Ingeniería. 2007;16:94–6.

Coupland JN, Shaw NB, Monahan FJ, O'Riordan ED, O'Sullivan M. Modeling the effect of glycerol on the moisture sorption behavior of whey protein edible films. J Food Eng. 2000;43:25–30.

Cuq B, Aymard C, Cuq JL, Guilbert S. Edible packaging films based on fish myofibrillar proteins: formulation and functional properties. J Food Sci. 1995;60:1369–74.

Cuq B, Gontard N, Cuq JL, Guilbert S. Functional properties of myofibrilar protein-based biopackaging as affected by films thickness. J Food Sci. 1996;61:580–4.

Denavi G, Tapia-Blacido DR, Añon MC, Sobral PJA, Mauri AN, Menegalli FC. Effects of drying conditions on some physical properties of soy protein films. J Food Eng. 2009;90:341–9.

Durston J. Flexible package closures and sealing systems. In: Theobald N, Winder B, editors. Package closures and sealing systems. Boca Raton: CRC Press; 2006. p. 204–30.

Eswaranandam S, Hettiarachchy NS, Johnson MG. Antimicrobial activity of citric lactic malic or tartaric acids and nisin-incorporated soy protein film against *Listeria monocytogenes Escherichia coli* 0157:H7 and *Salmonella gaminara*. J Food Sci. 2004;69:79–84.

Fang Y, Tung MA, Britt IJ, Yada S, Dalgleish DG. Tensile and barrier properties of edible films made from whey proteins. J Food Sci. 2002;67:188–93.

Fávaro-Trindade CS, Pinho SC, Rocha GA. Microencapsulação de ingredientes alimentícios. Braz J Food Technol. 2008;11:103–12.

García FT, Sobral PJA. Effect of the thermal treatment of the filmogenic solution on the mechanical properties color and opacity of films based on muscle proteins of two varieties of tilapia. Lebensmittel-Wissenschaft und -Technologie. 2005;38:289–96.

Gennadios A, McHugh TH, Weller CL, Krochta JM. Edible coating and films based on proteins. In: Krochta JM, Baldwin EA, Nisperos-Carriedo MO, editors. Edible coatings and to improve food quality. Lancaster: Technomic Publishing Company; 1994. p. 201–77.

Gennadios A, Weller CL, Hanna MA, Froning GW. Mechanical and barrier properties of egg albumen films. J Food Sci. 1996;61:585–9.

Ghorpade VM, Gennadios A, Hanna MA, Weller CL. Soy protein isolate/poly(ethylene oxide) films. Cereal Chem. 1995;72:559–63.

Giménez B, Gómez-Estaca J, Alemán A, Gómez-Guillén MC. Physico-chemical and film forming properties of giant squid (*Dosidicus gigas*) gelatin. Food Hydrocolloid. 2009;23:585–92.

Gómez-Estaca J, Montero P, Fernández-Martín F, Alemán A, Gómez-Guillén MC. Physical and chemical properties of tuna-skin and bovine-hide gelatin films with added aqueous oregano and rosemary extracts. Food Hydrocolloid. 2009;23:1334–41.

Gómez-Guillén MC, Ihl M, Bifani V, Silva A, Montero P. Edible films made from tuna-fish gelatin with antioxidant extracts of two different murta ecotypes leaves (Ugni molinae Turcz). Food Hydrocolloid. 2007;21:1133–43.

Gontard N, Guilbert S. Bio-packing: technology and properties of edible and/or biodegradable material of agricultural origin. Boletim da SBCTA. 1996;30:3–15.

Gontard N, Guilbert S, Cuq JL. Water and glycerol as plasticizer affect mechanical and water vapor barrier properties of an edible wheat gluten film. J Food Sci. 1993;58:206–11.

Gontard N. (2000) Panorama des emballages alimentaires actifs. In: Gontard N. (Cord.) Les emballages actifs. Tec & Doc, p.1.

Handa A, Gennadios MA, Weller CL, Kuroda N. Physical and molecular properties of egg-white lipids films. J Food Sci. 1999;64:860–4.

Hanlon JF, Kelsey RJ, Forcinio HE. Handbook of package engineering. 3rd ed. Lancaster: Technomic Publishing Company; 1998.

Hernandez-Muñoz P, Villalobos R, Chiralt A. Effect of thermal treatments on functional properties of edible films made from wheat gluten fractions. Food Hydrocolloid. 2004;18:647–54.

Iwata K, Ishizaki S, Handa A, Tanaka M. Preparation and characterization of edible films from fish water-soluble proteins. Fisheries Sci. 2000;66:372–8.

Jagannath JH, Radhika M, Nanjappa C, Murali HS, Bawa AS. Antimicrobial mechanical barrier and thermal properties of starch–casein based neem (*Melia azardirachta*) extract containing film. J Appl Polym Sci. 2006;101:3948–54.

Jangchud A, Chinnan MS. Properties of peanut protein film: sorption isotherm and plasticizer effect. Lebensmittel-Wissenschaft und -Technologie. 1999;32:84–9.

Jongjareonrak S, Benjakul W, Visessanguan W, Tanaka M. Effects of plasticizers on the properties of edible films from skin gelatin of bigeye snapper and brownstripe red snapper. Eur Food Res Technol. 2006;222:229–35.

Jongjareonrak A, Benjakul S, Visessanguan W, Tanaka M. Antioxidative activity and properties of fish skin gelatin films incorporated with BHT and a-tocopherol. Food Hydrocolloids. 2008;22: 449–58.

Ko S, Janes ME, Hettiarachchy NS, Johnson MG. physical and chemical properties of edible films containing nisin and their action against *Listeria Monocytogenes*. J Food Sci. 2001;66:1158–62.

Krochta JM, De-Mulder-Johnston CD. Edible and biodegradable polymer films: challenges and opportunities. Food Technol. 1997;51:61–74.

Kunte LA, Gennadios A, Cuppett SL, Hanna MA, Weller CL. Cast films from soy protein isolates and fractions. Cereal Chem. 1997;74:115–8.

Lim LT, Mine Y, Tung A. Barrier and tensile properties of transglutaminase cross-linked gelatin films as affect by relative humidity temperature and glycerol content. J Food Sci. 1999;64:616–22.

Longares A, Monahan FJ, O'Riordan ED, O'Sullivan M. Physical properties of edible films made from mixtures of sodium caseinate and WPI. Int Dairy J. 2005;15:1255–60.

López-Carballo G, Hernández-Muñoz P, Gavara R, Ocio MJ. Photoactivated chlorophyllin-based gelatin films and coatings to prevent microbial contamination of food products. Int J Food Microbiol. 2008;126:65–70.

Maria TMC, Carvalho RA, Sobral PJA, Habitante AMBQ, Feria JS. The effect of the degree of hydrolysis of the PVA and the plasticizer concentration on the color opacity and thermal and mechanical properties of films based on PVA and gelatin blends. J Food Eng. 2008;87:191–9.

Marquié C, Tessier A-M, Aymard C, Guilbert S. HPLC Determination of reactive lysine content of cottonseed protein films to monitor the extent of cross-linking by formaldehyde glutaraldehyde e glioxal. J Agric Food Chem. 1997;45:922–6.

Matioli G, Rodriguez-Amaya DB. Microencapsulação do licopeno com ciclodextrinas. Ciência e Tecnologia de Alimentos. 2003;23:102–5.

Matsumura S, Tomizawa N, Toki A, Nishikawa K, Toshima K. Novel poly(vinyl alcohol)-degrading enzyme and the degradation mechanism. Macromolecules. 1999;32:7753–61.

Mendieta-Taboada O, Sobral PJA, Carvalhob RA, Habitante AMBQ. Thermomechanical properties of biodegradable films based on blends of gelatin and poly(vinyl alcohol). Food Hydrocolloid. 2008;22:1485–92.

Menegalli FC, Sobral PJA, Roques MA, Laurent S. Characteristics of gelatin biofilms in relation to drying process conditions near melting. Dry Technol. 1999;17:1697–706.

Micard V, Balamri R, Morel M-H, Guilbert S. Properties of chemically and physically treated wheat gluten films. J Agric Food Chem. 2000;48:2948–53.

Min S, Harris LJ, Krochta JM. Antimicrobial effects of Lactoferrin lysozyme and the lactoperoxidase system and edible whey protein films incorporating the lactoperoxidase system against *Salmonella enterica* and *Escherichia coli* O157:H7. J Food Sci. 2005;70:332–8.

Monterrey-Quintero ES, Sobral PJA. Caracterização de propriedades mecânicas e óticas de biofilmes à base de proteínas miofibrilares de tilápia do nilo usando uma metodologia de superfície-resposta. Ciência e Tecnologia de Alimentos. 1999;19:294–301.

Monterrey-Quintero ES, Sobral PJA. Preparo e caracterização de proteínas miofibrilares de tilápia do nilo (Oreochromis niloticus) para elaboração de biofilmes. Pesquisa Agropecuária Brasileira. 2000;35:179–89.

Nassau K. Fundamentals of color science. In: Color for Science, Art and Technology, ed. K. Nassau. Lebanon (NJ): North Holland; 1996. p. 1–30.

Nguyen ML, Schwartz SJ. Lycopene: chemical and biological properties. Food Technol. 1999;53:38–45.

Nisperos-Carriedo MO. Edible coatings and films based on polysaccharides. In: Krochta JM, Baldwin EA, Nisperos-Carriedo M, editors. Edible coatings and films to improve food quality. Lancaster: Technomic; 1994. p. 305–36.

Nunes IL, Mercadante AZ. Encapsulation of lycopene using spray-drying and molecular inclusion processes. Braz Arch Biol Techn. 2007;50:893–900.

Orliac O, Rouilly A, Silvestre F, Rigal L. Effects of various plasticizers on the mechanical properties water resistance and aging of thermo-moulded films made from sunflower proteins. Ind Crop Prod. 2003;18:91–100.

Ötles S, Çagindi Ö. Carotenoids as natural colorants. In: Socaciu C, editor. Food colorants chemical and functional properties. New York: CRC Press; 2008. p. 51–70.

Paschoalick TM, Garcia FT, Sobral PJA, Habitante AMQB. Characterization of some functional properties of edible films based on muscle proteins of Nile tilápia. Food Hydrocolloid. 2003;17:419–27.

Pérez-Mateos M, Montero P, Gómez-Guillén MC. Formulation and stability of biodegradable films made from cod gelatin and sunflower oil blends. Food Hydrocolloid. 2009;23:53–61.

Pranoto Y, Salokhe VM, Rakshit SK. Physical and antibacterial properties of alginate-based edible film incorporated with garlic oil. Food Res Int. 2005;38:267–72.

Pruneda E, Peralta-Hernández JM, Esquivel K, Lee SY, Godínez LA, Mendoza S. Mechanical properties and antioxidant effect of Mexican oregano–soy based edible films. J Food Sci. 2008;73:C488–93.

Rhim JW, Gennadios A, Handa A, Weller CL, Handa MA. Solubility tensile and color properties of modified soy protein isolated films. J Agric Food Chem. 2000;48:4937–41.

Rodriguez-Amaya DB. Effects of processing and storage on food carotenoids. Sight Life Newslett. 2002;3:25–35.

Rojas-Graü MA, Tapia MS, Rodríguez FJ, Carmona AJ, Martin-Belloso O. Alginate and gellan-based edible coatings as carriers of antibrowning agents applied on fresh-cut Fuji apples. Food Hydrocolloid. 2007;21:118–27.

Saffert A, Pieper G, Jetten J. Effect of package light transmittance on the vitamin content of milk part 3: fortified UHT low-fat milk. Packag Technol Sci. 2009;22:31–7.

Saunderson JL. Calculation of the color of pigmented plastics. J Opt Soc Am. 1942;32(12):727–9.

Shi J, Maguer ML, Bryan M, Kakuda Y. Kinetics of lycopene degradation in tomato puree by heat and light irradiation. J Food Process Eng. 2003;25:485–98.

Shi J, Qu Q, Kakuda Y, Yeung D, Jiang Y. Stability and synergistic effect of antioxidative properties of lycopene and other active components. Cr Rev Food Sci. 2004;44:559–73.

Silva GGD, Sobral PJA, Carvalho RA, Bergo PVA, Taboada OWM, Habitante AMQB. Biodegradable films based on blends of gelatin and poly(vinyl alcohol): effect of PVA type or concentration on some physical properties of films. J Polym Environ. 2008;16:276–85.

Sivarooban T, Hettiarachchy NS, Johnson MG. Physical and antimicrobial properties of grape seed extract nisin and EDTA incorporated soy protein edible films. Food Res Int. 2008;41:781–6.

Slade L, Levine H. Polymer-chemical properties of gelatin in foods. In: Pearson AM, Dutson TR, Bailey AJ, editors. Advances in meat research collagen as a food. London: Elsevier Applied Science; 1987. p. 251–66.

Sobral PJA. Propriedades funcionais de biofilmes de gelatina em função da espessura. Ciência & Engenharia. 1999;8:60–7.

Sobral PJA. Influência da espessura sobre certas propriedades de biofilmes à base de proteínas miofibrilares. Pesquisa Agropecuária Brasileira. 2000;35:1251–9.

Sobral PJA, Menegalli FC, Hubinger MD, Roques MA. Mechanical water vapor barrier and thermal properties of gelatin based edible films. Food Hydrocolloid. 2001;15:423–32.

Sobral PJA, Santos JS, García FT. Effect of protein and plasticizer concentrations in films forming solutions on physical properties of edible and films based on muscle proteins of a Thai Tilapia. J Food Eng. 2005;70:93–100.

Tallon MJ (2006) The power of lycopene. Functional Foods & Nutraceuticals January: 22–24.

Tapia-Blácido DT, Mauri A, Menegalli FC, Sobral PJA, Añon MC. Contribution of the starch protein and lipid fractions to the physical thermal and structural properties of amaranth (*Amaranthus caudatus*) flour films. J Food Sci. 2007;72:E293–300.

Tharanathan RN. Biodegradable films and composite coatings: past present and future. Trends Food Sci Tech. 2003;14:71–8.

Theivendran S, Hettiarachchy NS, Johnson MG. Inhibition of *Listeria monocytogenes* by nisin combined with grape seed extract or green tea extract in soy protein film coated on turkey frankfurters. J Food Sci. 2006;71:39–44.

Thomazine M, Carvalho RA, Sobral PJA. Physical properties of gelatin films plasticized by blends of glycerol and sorbitol. J Food Sci. 2005;70:172–6.

Torres JA. Edible films and coatings from proteins. In: Hettiarachy NS, Ziegler GR, editors. Protein functionality in food systems. New York: Marcel Dekker; 1994. p. 467–507.

Van de Velde K, Kiekens P. Biopolymers: overview of several properties and consequences on their applications. Polym Test. 2002;21:433–42.

Vanin FM, Sobral PJA, Menegalli FM, Carvalho RA, Habitante AMQB. Effects of plasticizers and their concentrations on thermal and functional properties of gelatin based films. Food Hydrocolloid. 2005;19:899–907.

Vertzoni M, Kartezini T, Reppas C, Archontaki H, Valsami G. Solubilization and quantification of lycopene in aqueous media in the form of cyclodextrin binary systems. Int J Pharm. 2006; 309:115–22.

Villalobos R, Chanona J, Hernández P, Gutiérrez G, Chiralt A. Gloss and transparency of hydroxypropyl methylcellulose films containing surfactants as affected by their microstructure. Food Hydrocolloid. 2005;19:53–61.

Yoshida CMP, Antunes AJ. Characterization of whey protein emulsion films. Braz J Chem Eng. 2004;2:247–52.

Zhang S, Wang Y, Herring JL, Oh J-H. Characterization of edible film fabricated with channel catfish (*Ictalurus punctatus*) gelatin extract using selected pretreatment methods. J Food Sci. 2007;72:C498–503.

Ziegler CR, Foegeding EA. The gelation of proteins. In: Kinsella JE, editor. Advances in food and nutrition research. San Diego: Academic Press; 1990. p. 203–29.

Chapter 16
Clean Strategies for the Management of Residues in Dairy Industries

Giovana Tommaso, Rogers Ribeiro, Carlos Augusto Fernandes de Oliveira, Katerina Stamatelatou, Georgia Antonopoulou, Gerasimos Lyberatos, Cecilia Hodúr, and József Csanádi

16.1 Introduction

In the food industry, the generation of residues is related to the difference between the balance of mass and energy in the processes of transformation of raw materials into food products. Appropriate management of residues generated by or during manufacture of food products usually includes storage and treatment before discharge or reuse as raw material for other processes.

Discharge of treated residues implies costs to the company and for the environment. This cost is mostly related to the use of energy, water, chemical products, and the cost of the discharge process per se. For example, residual waters from the dairy industry contain whey and traces of milk, cheese, yogurt, dairy drinks, and butter. Treatment of these residues involves large amounts of water and produces large volumes of effluent that still have high concentrations of organic material that should be adequately treated before being disposed of in aquatic environments.

The transformation of raw milk into dairy products is one of the most important activities of the dairy industry worldwide. Along with the production of pasteurized and sterilized milk for direct consumption, a variety of process are used for the production of cheese, butter, cream, ice cream, yogurt, and other fermented milk products.

G. Tommaso (✉) • R. Ribeiro • C.A.F. de Oliveira
Department of Food Engineering, School of Animal Science and Food Engineering,
University of São Paulo, Pirassununga, Brazil
e-mail: tommaso@usp.br

K. Stamatelatou • G. Antonopoulou • G. Lyberatos
School of Chemical Engineering, National Technical University of Athens, Athens 15780, Greece

Institute of Chemical Engineering and High Temperature Chemical Processes, Patras, Greece

C. Hodúr • J. Csanádi
Faculty of Engineering, University of Szeged, Mars sq. 7, 6724 Szeged, Hungary

Table 16.1 Annual production of cheese (×1,000 tons)

Year	European Union	North America	South America	Total world production
2000	5,861	4,208	890	12,345
2001	5,865	4,216	900	12,450
2002	5,993	4,372	840	12,870
2003	6,205	4,350	785	13,054
2004	6,481	4,504	840	13,643
2005	6,625	4,645	895	14,043
2006	6,801	4,761	1,008	14,413
2007	6,870	4,833	1,055	14,187
2008[a]	6,975	4,911	1,155	14,501

[a]Predicted

This processing of milk generates huge amounts of waste products, especially whey, which is considered the most important, high-nutrient residue in the dairy industry.

Whey is the serum or liquid part of milk that is separated from curd when making cheese; it is generated at a mean rate of 10 L/kg of cheese produced. The whey usually contains nearly 50% of the total solids present in the original milk. The components of whey are proteins, lactose, minerals, vitamins, and traces of milk fat. Whey is a by-product with high nutrient content, high economic value, and interesting functional properties. Whey can be used as an ingredient for other food products, such as flavored milk beverages, chocolates, candies, cookies, processed cheese, and ice cream, and in the composition of animal feeds.

The world production of cheese increased in the period of 2000–2008 as shown in Table 16.1. The most important cheese manufacturers are located in European Union member states, North America, and South America, which are collectively responsible for nearly 90% of the total cheese production. If it is assumed that generation of each unit of cheese produces ten units of whey, the calculated amount of whey produced in the world is around 145 million tons per year. This figure represents nearly seven million tons of lactose and one million tons of whey proteins that can potentially be incorporated into processes by the food industry.

16.2 Cleaner Production for the Dairy Industry

Cleaner production is a forward-looking, "anticipate and prevent" philosophy. The definition of "cleaner production" adopted by the United Nations Environment Programme (UNEP 2004) is as follows:

> Cleaner Production is the continuous application of an integrated preventive environmental strategy to processes, product and services to increase overall efficiency, and reduce risks to humans and the environment. For production processes, Cleaner Production results from one or a combination of conservation processes of raw materials, water and energy; elimination of toxic and dangerous raw material and reduction of the quantity and toxicity of all emissions and wastes at source during the production process.

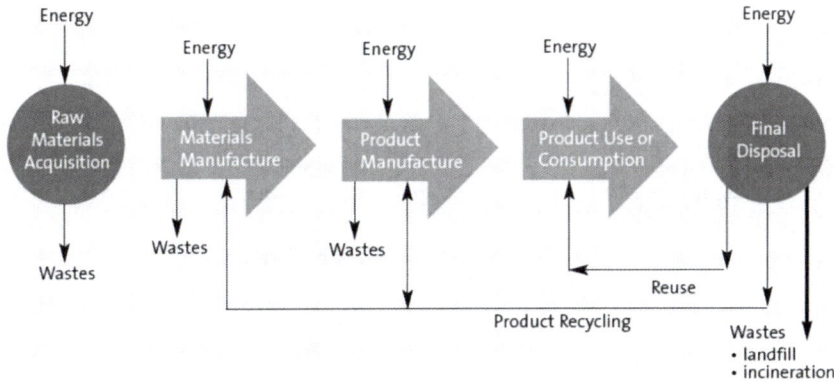

Fig. 16.1 Conventional production line (From Thorpe, B 1999)

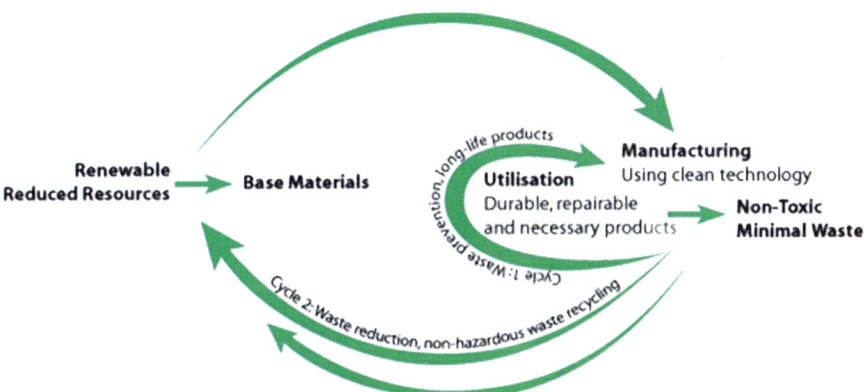

Fig. 16.2 Clean production cycle (From http://www.cleanproduction.org)

For a production process, cleaner production includes conservation of raw materials and energy, elimination of toxic raw materials, and reduction of the quantity and toxicity of all emissions and wastes before they leave the process.

Clean production can be implemented in the manufacture of products and the growth of food. Sometimes called "clean technology," or in the case of food production "organic agriculture," it is a way of protecting workers and communities, and avoiding the generation of hazardous releases into the environment.

Conventional industrial production is linear (Fig. 16.1), but clean production (Fig. 16.2) is rooted within the concepts of product life cycles.

Clean production is not just about cleaner production in factories. Instead, it is a holistic way of looking at how design and consumption of products causes severe ecological problems. Clean production offers ways to reverse trends of the nonsustainable use of materials and energy. Clean production promotes renewable energy and materials and sustainable product design, which results in the production of nontoxic products and processes. More importantly, it protects biological and cultural diversity while encouraging an approach to production and consumption that is precautionary and preventive.

Clean production ultimately means the use of renewable energy and materials, the minimal use of resources, the design of sustainable products, the production of food in a sustainable way, and the generation of waste that is benign and returnable to the production processes.

Clean production begins with a comprehensive look at the way materials flow in society. In particular, it looks at the product chain: where raw materials come from, how and where they are processed, what wastes are generated along the product chain, what products are made from the materials, and what happens to these products during their use and at the end of their commercial life.

According to various definitions developed over the years, four main elements make up the concept of clean production:

1. Precautionary principle. The burden is on proponents of an activity to prove there is no safer way to proceed, rather than on victims or potential victims of the activity to prove it will be harmful.
2. Preventive principle. It is cheaper and more effective to prevent environmental damage than to attempt to manage or cure it.
3. Democratic principle. Clean production can only be implemented with the full involvement of workers and consumers within the product chain.
4. Holistic principle. For each product we buy, we need to have access to information about the materials, energy, and people involved in making it. Access to this information would help build alliances for sustainable production and consumption.

Clean production systems for food and manufactured products are nontoxic and energy-efficient, nonpolluting throughout their entire life cycle, preserve diversity in nature and culture, and support the ability of future generations to meet their own needs.

The cleaner production techniques can be categorized into the following areas: improved housekeeping, process optimization, new process technology, and new product design.

Improved housekeeping is based on improvements to work processes and proper maintenance, including preventing spills and wastage, and improved monitoring, training, and inventory. Examples of improved housekeeping are:

- Collect the first flush of product changeovers and where possible blend these back into the next product or dispose of them as animal feed.
- Elimination of rinsing between different flavored batches of yoghurt.
- Insulation of pipes and repair of steam leaks and steam traps; e.g., a 1-mm hole in a steam line at 700 kPa will lead to a loss of 3,000 L fuel oil/year (Prasad et al. 2004)
- Process optimization reduces resource consumption by optimizing existing processes. The following are changes that are easily achievable in the dairy industry:

- Optimizing start-up/shutdown procedures and changeovers by fine-tuning timers and accurately detecting product interfaces to reduce product waste
- Minimizing waste during separator desludging by optimizing bowl opening frequency
- Process technology changes involve the installation of new processes and technology to minimize waste and increase efficiency. Examples of installations of new technology include:
- Replacing batch pasteurizers with continuous process pasteurizers incorporating plate heat exchangers with less waste and countercurrent heat regeneration reducing energy consumption
- A fully automated cleaning system in a milk processing plant to reduce water and energy usage
- Reliable instrumentation, such as turbidity meters, to identify the concentration of milk solids, and a combination of temperature, pH, and conductivity meters to monitor clean in place frequency, effectiveness, and chemical loss

New product design should result in benefits throughout the life cycle of the product, with reduced waste, reduced energy and water consumption, and higher yields. The new process often requires new process equipment and marketing efforts to successfully establish the new product in the marketplace. Ultrafiltration of milk for cheese manufacture is an example of new design intended to increase the recovery of whey proteins in the cheese and reduces whey volumes.

16.3 Main Elements of Dairy Wastes and Coproducts

The main environmental problem caused by dairy processing is the discharge of large quantities of (liquid) effluent (Table 16.2). The effluent loads from dairy processing depend on the type of product being produced, the scale of the operation, and whether a plant uses batch or continuous processes. Small batch processes require more frequent cleaning, with increased losses from product changeover, drainage, and cleaning losses. Evaporation processes have the potential to create very high organic loads owing to losses during start-up and shutdown. Cheese whey, if not used as a by-product (Table 16.3), is discharged along with other wastewaters and can have a considerable impact on the organic load of the waste from the plant.

Dairy processing effluent generally exhibits the following properties:

- High organic load owing to the presence of milk components
- Fluctuation in pH owing to the presence of caustic and acidic cleaning agents and other chemicals
- High levels of nitrogen and phosphorus and fluctuations in temperature (UNEP 2000)

Whey or milk plasma is the liquid remaining after milk has been curdled and strained; it is a by-product of the manufacture of cheese or casein and has several commercial uses. Sweet whey is manufactured during the manufacture of types of

Table 16.2 Sources of wastage in dairy processing (From Hale et al. 2003)

Milk processing	Cheese processing	Milk powder
Overfill	Whey	Flushing at start-up and shutdown
Spillage	Cheese fines	Deposit on heating surfaces
Purge of product prior to clean in place	Curd losses from curd transfer system	Exhaust loss of fines from driers
Product retained in poorly drained pipelines	Wastewater from flushing pipelines	
Heat-deposited waste	Cheese brine-salt whey	
Product defects/returns	Separator desludge	
Laboratory samples	UF permeate	

UF ultrafiltration

Table 16.3 World annual whey balance and utilization in 2001 (From Zadow 2005)

	Total whey		Industrially utilized whey		Whey protein and lactose		Demineralized whey	
Region	(%)	(MT)	(%)[a]	(MT)	(%)[b]	(MT)	(%)[b]	(MT)
USA	24.8	36	80	28.8	50	14.4	10	2.9
European Union	41.5	60	60	36	60	21.6	10	3.6
Canada	2.8	4	80	3.2	50	1.6	10	0.3
Argentina/Brazil	6.2	9	40	3.6	70	2.5	5	0.2
Australia/New Zealand	5.9	8.5	90	7,7	40	3.1	10	0.8
Rest of world	19	27.5	25	6.9	75	5.2	5	0.3
Total	100	145	59	86.1	56	48.3	9	8.1

MT million tons
[a] Percentage of total whey volume
[b] Percentage of whey utilized

hard rennet cheese such as cheddar and Swiss cheese. Acid whey (also known as sour whey) is obtained during the manufacture of acid types of cheese such as cottage cheese.

Whey contains 95% of the original water, 5% solids, most of them lactose, 20% of the milk protein, and traces of fat (Table 16.4). The composition of whey depends upon the type of cheese produced. Factors such as the season, location, and type and health of dairy cattle also affect whey composition. There are two main types of whey: sweet whey and acid whey. Sweet whey (pH > 5.6) is produced from the manufacture of rennet cheese such as cheddar or mozzarella. Acid whey (pH < 5.1) is produced by lactic acid fermentation to produced fresh cheese such as cottage cheese or cream cheese or by hydrochloric acid casein production. Acid whey contains higher levels of calcium phosphate than does sweet whey.

The main alternatives for processing of whey and milk are either concentration and spray drying for production of powder, or treating the milk/whey with ultrafiltration for the production of proteins followed by transformation into protein concentrate

Table 16.4 Composition of sweet and acid whey, and ultrafiltration (*UF*) permeate (From Durham 2000)

Composition	Sweet whey cheddar	Acid whey HCl	Acid whey Lactic acid	UF permeate cheddar
Solids (%)	6.6	5.1	6	5.5
pH	6.1	4.7	4	6.1
Lactose (%)	4.8	3.7	3.9	4.7
Protein (%)	0.9	0.73	0.72	0.01
Ash (%)	0.59	0.6	0.72	0.53
Lactic acid (%)	0.13	0	0.6	0
Fat (%)	0.06	0.05	0.003	0
Calcium (ppm)	430	1,200	1,140	375
Phosphorus (ppm)	440	680	900	275
Potassium (ppm)	1,460	1,200	1,530	1,450
Sodium (ppm)	430	270	400	430
Chloride (ppm)	970	2,600	910	940

powder; this process is not entirely green as the permeate produced is a major by-product.

Production of whey protein concentrate/milk powder is an attractive alternative for processing whey/milk, as the product has a high value on the market. It is used mainly as a component for baby food and also for protein fortification in various food formulations.

Whey protein/milk concentrates, obtained by the ultrafiltration of whey, are also valuable products and therefore application of spray-drying technology for dehydration is fully justified. On the other hand, the permeate by-product of the ultrafiltration of whey is a material of low value and, from the point of view of the manufacturer, the most convenient way to handle its disposal was, until recently, dumping. This is no longer permitted, and therefore a spray-drying process is currently applied to the permeate even though such a process is not cost-effective.

In the food industry, spray drying is a primary process for formation of particulate matter. Powders that contain fat, protein, and sugar are potentially explosive. The self-heating of milk powders can become a source of a fire or even dust explosion in dairy powder plants. The stickiness of milk powder causes deposits to form on parts of the drying chamber wall during the spray drying of milk. Because of the high costs involved in frequent washing of the drier, the chamber is washed only when deposits have built up to an unacceptable level. The problems caused by buildup are fire or explosion hazards and deterioration of product quality. Fire or explosion hazards are created by the tendency of the deposit to spontaneously combust in the presence of critical conditions (Chong et al. 1999).

Hot surfaces are capable, if the temperature is sufficiently high, of igniting surrounding dust accumulations. The layer ignition temperature is measured in a standard test for a depth of 5 mm, but because of the insulating effect of dusts, thicker deposits can ignite at lower temperatures (Gummer and Lunn 2003). The practical dangers are that a smoldering or burning layer can act either directly as an ignition source for a dust cloud or by means of agglomerations or nests of burning material, which break away from deposits and ignite a dust cloud in another part of the plant. Harper et al. (1997) discussed the burning behavior of powder accumulations on hot surfaces. The stages of ignition and the form of the combustion zone can be complex. Some powders burn directly in the solid phase either with a flame or by smoldering, others melt and burn as a liquid, whereas some burn with a large amount of flame. Some dust can evolve large amounts of flammable gas when subjected to heat.

To predict heating behavior in milk powder deposits, Zockoll (1999) investigated milk powders as both the nest material and the explosive dust cloud. At 1,200°C, nests ignited the dust clouds; this, however, only occurred after impact on the floor of the explosion vessel. When dust was dropped over smoldering nests on the vessel floor, cloud explosions could occur at about 860°C. A flaming nest could, however, be practically extinguished by the dispersal of milk powder around it in the explosion vessel. Tests on the development of smoldering in nests under the influence of a 0.5 m/s air stream showed that smoldering developed differently depending on the type of milk powder. At higher air speeds, open fires occurred in relatively large quantities of skim milk powder. The transition from smoldering to open fire occurs around 800–850°C, depending on the type of milk.

Since milk is an organic product it may be spoilt afterwards as a consequence of its microbial bioload. These microorganisms either survived the heat treatment – pasteurization – or were later introduced into the product. Spoilage is evident when some perceptible (sensorial, color, mold) changes occur; but an expired product might not apparently differ from one which is edible. The practice adopted in most countries until the 1980s was to recall the products and then reutilize them as a feed.

European Union legislation requires the destruction of expired dairy products. This regulation guarantees both product safety and consumers' safety. The regulation is very rigorous, i.e., raw and partially manufactured products have to be destroyed together with the items stored at the manufacturing facility. This regulation is enforced even if the item was not visibly spoilt, or had passed all quality control tests. Repasteurization is allowed by legislation, provided that the product was stored by the dairy company; only then can it be used as a raw material.

16.4 Utilization of By-products

Whey is first clarified to remove cheese solids and then pasteurized to stop starter culture activity. Options for processing whey are summarized in Fig. 16.3.

Whey contains many high-value biologically active protein and nutritional components that are attractive to an increasingly sophisticated market. Milk and whey proteins have been found to have different physiological functions because of the numerous bioactive peptides that are carried within the intact protein. Whey peptides contain 3–20 amino acid residues in the amino acid sequence which defines

Milk processing	_Cheese processing_	_Milk powder_
Overfill	Whey	Flushing at start-up and shut down
Spillage	Cheese fines	Deposit on heating surfaces
Purge of product prior to CIP	Curd losses from curd transfer system	Exhaust loss of fines from driers
Retained product in poorly drained pipelines	Waste water from flushing pipelines	
Heat deposited waste	Cheese brine-salt whey	
Product defects/returns	Separator de-sludge	
Laboratory samples	UF permeate	

Fig. 16.3 Process alternatives for whey. *Evap* evaporation, *UF* ultrafiltration (From Waldron 2007)

function. Whey peptides have been found to exhibit various properties, such as antioxidative, antihypertensive, antimicrobial, cytomodulatory, and immunoregulatory properties, angiotensin-converting enzyme inhibition, and mineral-carrying capacity (Korhonan and Pihlanto 2003). Bioactive peptides are commonly produced by enzymatic hydrolysis of the whole protein, in a batch or continuously with immobilized enzyme membrane reactors. Hydrolysis is coupled with various separation procedures depending on the target peptide. A crossflow filtration process allows the transfer of low molecular weight species, water, and/or solvents through a membrane without changing the solution volume. This process is used for purifying retained large molecular weight species, increasing the recovery of low molecular weight species, buffer exchange, and simply changing the properties of a given solution. Stepwise ultrafiltration (with diafiltration) can be used to fractionate the peptides on the basis of size, whereas ion exchange/size-exclusion chromatography is used for separations based on size and charge.

A new technique reported by Groleau et al. (2004) involves peptide fractionation by electronanofiltration, which involves inserting a cathode into the permeate compartment of the nanofiltration unit to increase separation of neutral and basic peptides.

Evaporation and spray drying are the generally accepted methods for the disposal of large quantities of whey, and a whey drying plant is often an integral part of a cheese factory (Caric 1994).

Whey powder is used in a wide range of foods, mostly as a lower-cost skim milk replacer, and in animal food. The properties of the whey powder are mainly governed by the amount of lactose converted to α-monohydrate crystals before drying, creating a nonhygroscopic free-flowing product. If the lactose in the liquid whey is not crystallized, the resulting high levels of amorphous lactose in the dried whey will lead to caking and poor storage stability (Listiohadi et al. 2005). New technology, known as the Paradry process, has been developed to improve the drying of hygroscopic liquids. This process has been designed for evaporation of liquids with high total solids contents, high viscosity, or tendency to foul. It is continuous, has low operating cost, and produces a noncaking power.

Lactose is recovered from whey permeate as edible-grade lactose in most dairy manufacturing countries. Edible-grade lactose is used in infant formula, chocolate, confectionery, and baked goods and as a flavor/color carrier. Pharmaceutical-grade lactose is further refined and converted into a range of products: milled lactose crystals, spray-dried lactose, anhydrous β-lactose, and micronized lactose. It is mainly used in pharmaceuticals as a tabletting excipient, but is also used as a high-price, dry-powder inhaler. Pharmaceutical-grade lactose and edible-grade lactose are also used as a raw material for lactose derivatives such as lactose, lactobionic acid, lactitol, and galacto-oligosaccharides, or are hydrolyzed to liquid lactose syrup.

The process of lactose manufacture typically involves concentration of the ultrafiltration permeate to 60% solids in a multiple-effect evaporator. The concentrate is transferred to large crystallizer vats and slowly cooled over 20 h. The crystallized lactose is then separated, creating mother liquor containing 20–30% of the lactose plus 90% of the ash. The lactose crystals are washed in a refiner washer with water in a ratio of 1:1 water to lactose, then dried, milled, and bagged.

Lactose can be enzymatically hydrolyzed to glucose and galactose with β-galactosidase, producing a syrup that is more easily digested by those who are lactose-intolerant. Hydrolyzed lactose is sweeter and more soluble than lactose and can be used for sweetening syrups in ice cream, yoghurts, and drinks without lactose crystallization problems. Galactose, formed from hydrolyzed lactose, is also finding a market as an endurance sports drink additive (G-Push). However, cost and difficulty of storage limit the production and use of hydrolyzed lactose syrup.

Whey fractionation imposes greater processing demands, but results in a wide range of products with unique functionalities and high value in the marketplace. Fractionated whey products include whey protein concentrate; whey protein isolate; α and β fractions; bioactive proteins such as immunoglobulins, lactoferrin, lactoperoxidase, and glycomacropeptide; and hydrolyzed bioactive peptides. By-products from the ultrafiltration permeate include edible-grade lactose, pharmaceutical-grade lactose, hydrolyzed lactose, lactose derivatives, and milk salts. Applications of whey by-products are summarized in Table 16.5.

Ultrafiltration membranes with a molecular mass cutoff in the range 10,000–30,000 Da are used to concentrate the whey proteins, whereas the lactose and minerals readily pass through the membrane into the ultrafiltration permeate. The ultrafilter concentrates the whey protein concentrate to 15–20% total solids. The spray-dried whey protein concentrate contains approximately 34% protein, dry weight; the remainder consists of lactose and minerals. Whey protein concentrate with approximately 34% protein has a gross composition similar to that of skim milk powder and can be used in food as a skim milk replacement.

Nanofiltration is a membrane process used to separate small molecules such as mineral ions and water salts but to retain larger molecules (including lactose and proteins), coupling demineralization with concentration. Nanofiltration using 300-Da membranes preferentially removes up to 65% of the monovalent ions (Na^+, K^+, Cl^-), with further reductions achieved by diafiltration. Removal of monovalent ions improves lactose crystallization from whey permeate. Guu and Zall (1992) reported that lactose recovery improved by 8–10% after nanofiltration. Nanofiltration is also

Table 16.5 Properties and uses of whey and permeate coproducts (From Durham et al. 1997)

Product	Properties	Usage
Whey powder	Low-cost milk solids	Skim milk replacement
Demineralized whey powder	High-quality protein	Infant formula
Whey protein concentrate (35–85% protein)	High-quality protein, gelation, adhesion, emulsification, foaming	Infant formula, sports diets, nonfat milk replacement, processed meats, desserts
Whey protein isolate (90%+)	High-quality protein	Infant formula, sports diets
α-Lactalbumin (α fraction)	High-quality protein	Infant formula
β-Lactoglobulin (β fraction)	Gelling solubility and nutrition	Restructured meat or fish, sports and dietetic beverages
Lactoferrin	Antibacterial	Infant formula
Lactoperoxidase	Anticaries	Toothpaste
Glycomacropeptide	Bifidobacteria, enhanced immunity	Infant formula
Immunoglobulins	Anticancer	Cancer prevention, cancer treatment
Peptides (hydrolyzates)	Bioactive	Convalescent diets, athletes
Edible lactose	Carrier, filler, free-flow agent	Color, flavor carrier, instant powdered foods
Pharmaceutical lactose	Bulking agent, binder, flavor	Milk standardization Tabletting excipient

advantageous for the production of demineralized whey powder for infant formula, as calcium phosphate is retained, but monovalent ions are removed. The nanofiltration permeate contains about 0.3–0.5% solids including potassium, sodium, nonprotein nitrogen, and lactose. The permeate can be cleaned, by reverse osmosis, to produce clean water, with the remaining solution being a "dairy salt" concentrate. This monovalent salt mixture could be a useful by-product as a natural low-sodium table salt substitute or it could be used in sports and health beverages.

This salt has also been reported to be recovered and used to regenerate ion-exchange resins (Durhan et al. 2004). Whey is supersaturated with calcium phosphate and contains high levels of potassium and sodium. Calcium phosphate precipitation can cause problems such as fouling of the membrane or evaporator. High levels of minerals can inhibit lactose crystallization, which can adversely affect nonhygroscopic whey production and reduce yields and purity for lactose manufacture. The presence of high mineral levels limits the use of whey powder in infant formula and adversely affects the flavor and range of applications of whey products. A range of demineralization techniques have been developed, including precipitation, ion exchange, electrodialysis, and nanofiltration. These processes can be combined to remove a greater proportion of minerals.

Calcium phosphate precipitation is enhanced by raised pH, heat, and concentration. This leads to evaporators fouling, shortened process runs, and additional cleaning losses. Removal of protein further destabilizes calcium phosphate, causing more fouling to occur during the evaporation of ultrafiltration permeate (Schmidt and

Both 1987). Calcium phosphate precipitation is also responsible for fouling of ultrafiltration membranes (Ramachandra et al. 1994). Rapid removal of calcium and phosphate prior to heating and concentration is the preferred option, but this is not always possible when the whey protein needs to be protected.

Many dairy manufacturers are now recovering calcium phosphate from whey to produce dairy calcium supplements marketed to combat osteoporosis. The precipitated calcium phosphate can be recovered using centrifugal separators, followed by washing, drying, and milling.

Ion exchange is capable of removing up to 95% of the minerals that are present in whey, with the dematerialized whey mostly used in infant formula. There are many difficulties with the ion-exchange process, such as short running times between regeneration, high consumption of regenerant chemicals and associated waste problems, high water requirements to remove excess regenerant, losses of whey protein due to irreversible adsorption, and loss of protein functionality due to pH fluctuations during processing Jonsson and Arph, 1987). Ion-exchange resins can also be used just to decalcify whey and permeates, by employing cationic resins in the sodium or potassium from; thereafter, the decalcified whey or permeate is nanofiltered to remove the excess sodium and potassium. The ion-exchange resin can be regenerated using the concentrated permeate from the nanofilter, thereby recycling the salt from within the process, and avoiding the cost and pollution associated with purchasing salt to regenerate the resin (Durham et al. 2004).

16.4.1 Hydrogen Production from Cheese Whey

Cheese whey contains nutrients, such as lactose (4.5–5% w/v), soluble proteins (0.6–0.8% w/v), lipids (0.4–0.5% w/v), and mineral salts (8–10% dried extract). The high organic content of cheese whey can be converted to a mixture of methane and carbon dioxide (biogas). Moreover, under specific operating conditions, hydrogen can be formed from the conversion of the soluble carbohydrates such as lactose, and this is mainly responsible for the high chemical oxygen demand (COD) values (60–80 g COD/L).

Biogas production takes place during anaerobic digestion. Anaerobic digestion is a complex process involving the synergy of several groups of microorganisms that break down the organic matter into a gas mixture consisting of methane and carbon dioxide (biogas). It is applicable to stabilize wastes of high organic load, such as sewage sludge and agro-industrial wastes (such as manure and cheese whey), producing low volumes of sludge (compared with aerobic treatment) and large volumes of methane-rich biogas suitable for energy production. Cheese whey is an easily biodegradable organic source, consisting mainly of soluble carbohydrates (lactose). During anaerobic digestion (Fig. 16.4), the solids contained in the whey (mainly carbohydrates, proteins, and lipids) are hydrolyzed by bacterial enzymes to their soluble derivatives (i.e., sugars, amino acids, long-chain fatty acids). The acidogenic bacteria can take up these monomers into their cells and

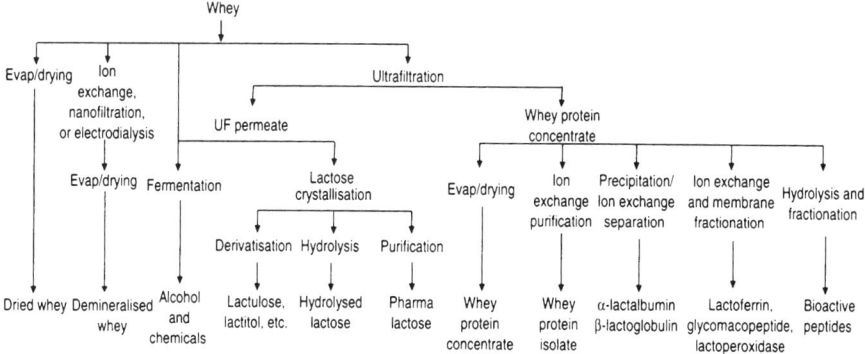

Fig. 16.4 Organic matter breakdown in the anaerobic digestion process

convert them into a mixture organic acids (lactic, acetic, propionic, and butyric acids), alcohols, and hydrogen. This is the fastest step in anaerobic digestion, resulting in the production of acids that tend to decrease the pH. Under conditions of low hydrogen partial pressure (lower than 10^{-6} atm.) and pH above 6, the intermediate products of acidogenesis are converted to acetic acid by acetogenic bacteria. Finally, the acetic acid and hydrogen are converted to methane and carbon dioxide by methanogenic bacteria under conditions of neutral pH.

The organic matter contained in the bulk liquid volume of cheese whey is transformed into gaseous products, leaving the wastewater with a reduced organic load. The energy stored in the chemical bonds of the organic compounds in cheese whey is converted to methane during anaerobic digestion. This is both an energy conservation process and a means of pollution control.

Methanogenesis is performed in the secondary treatment of cheese whey and other effluents. There are several configurations developed in one-stage or two-stage systems intended to improve the performance of the process in terms of COD reduction and methane yield. In two-stage systems, the acidogenesis and methanogenesis phases are separated in two bioreactors combined in series. As cheese whey contains soluble carbohydrates at high concentrations (36 g/L; Antonopoulou et al. 2008), it is possible to obtain biogas rich in hydrogen in the acidogenic stage.

Much attention has been given to hydrogen gas and its potential use as a fuel for transport purposes and electricity generation. Hydrogen is considered to be an alternative energy candidate because it is a clean and environmentally friendly fuel, oxidized to harmless water as a combustion product, instead of greenhouse gases. Biological methods for hydrogen production are environmentally friendly and may be viable for exploiting these wastes, as they can be combined with other biological methods for further treatment to reduce the COD. Biohydrogen may be produced by cyanobacteria and algae through biophotolysis of water or by photosynthetic and chemosynthetic fermentative bacteria. The latter process seems to be the most promising because it is carried out without light energy, and a variety of renewable feedstocks can be used as substrates. Carbohydrates are the main source of hydrogen during fermentative processes and, therefore, wastes/wastewaters or agricultural residues rich in carbohydrates (such as dairy wastewaters) can be considered as potential sources of hydrogen.

Degradation of glucose (or its isomers or its polymers, starch, and cellulose) during anaerobic conditions is accompanied by the production of hydrogen and various metabolic products, mainly volatile fatty acids (acetic, propionic, and butyric acids), lactic acid, and alcohols (butanol and ethanol), depending on the microbial species present and the prevailing conditions. The hydrogen yield can be correlated stoichiometrically with the final metabolic products through the reactions describing the individual processes of acidogenesis (Eqs. 16.1 and 16.2). The production of acetic and butyric acids favors the simultaneous production of hydrogen. The fermentation of glucose to acetic acid produces a maximum theoretical yield of 4 mol hydrogen per mole of glucose (reaction in Eq. 16.1) and the conversion to butyric acid results in 2 mol hydrogen per mole of glucose (reaction in Eq. 16.2):

$$C_6H_{12}O_6 + 2H_2O \rightarrow 2CH_3COOH + 2CO_2 + 4H_2 \qquad (16.1)$$

$$C_6H_{12}O_6 \rightarrow CH_3CH_2CH_2COOH + 2CO_2 + 2H_2 \qquad (16.2)$$

In practice, the production of multiple metabolic products, such as lactic acid, propionic acid, or ethanol, produces a lower net yield of hydrogen. Moreover, acetate-producing metabolism may occur via different, non-hydrogen-yielding pathways. In mixed fermentation processes, the microorganisms may select different pathways, while converting sugars, as a response to changes in their environment (pH, sugar concentration, etc.). The absence or presence of hydrogen-consuming microorganisms in the microbial consortium also affects the microbial metabolic balance and, consequently, the fermentation end products. To improve hydrogen production, all of these aspects should be taken into account and assessed. The following are factors considered crucial for hydrogen production from cheese whey:

- *Regulation of pH*. The optimum pH range for hydrogen production is considered to be 5–6. The high production of acids from cheese whey, if not controlled, will cause the pH to drop below 5. Antonopoulou et al. (2008) found that addition of 20 g/L $NaHCO_3$ was sufficient to keep the pH of the hydrogen reactor at 5.2; this is the pH that is required to obtain a stable operation of a continuous stirred tank reactor at a hydraulic retention time of 24 h. The yield in terms of liters of hydrogen produced per liter of whey added was 2.49. Davila-Vazquez et al. (2008) studied the feasibility to produce hydrogen from cheese whey powder in batch reactors and found that 1 mmol H_2/L/h could be attained at pH 7.5 from 25 g/L cheese whey powder, indicating that hydrogen production may occur at pH values higher than 6. Ferchichi et al. (2005) studied the influence of initial pH on hydrogen production from cheese whey. The initial pH varied between 5 and 10 in batch experiments where the crude cheese whey (87.5% v/v) was fermented by *Clostridium saccharoperbutylacetonicum*. The maximum hydrogen production

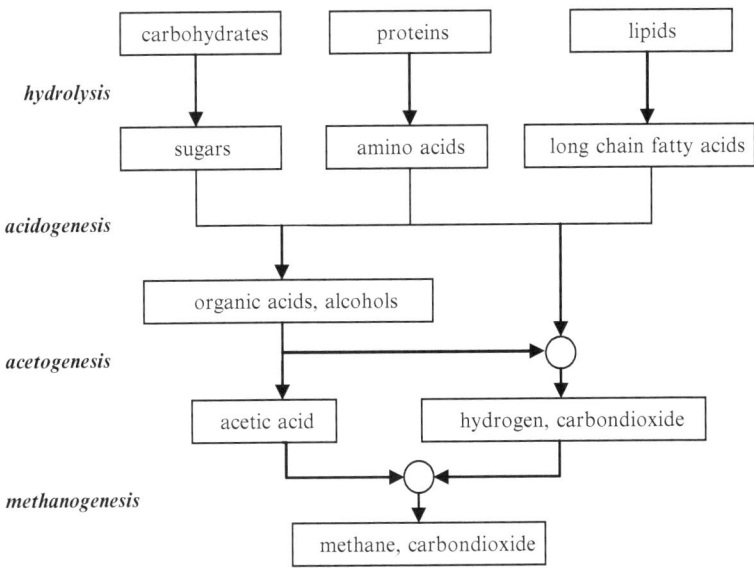

Fig. 16.5 The evolution of biogas and hydrogen in the hydrogenogenic reactor throughout the experimental period (Reprinted with permission from Antonopoulou et al. 2008)

rate and yield were observed at an initial pH 6 and then steadily decreased as the pH increased. The highest rate and yield were 28.3 mL/h and 7.89 mmol/g lactose, respectively. It was also found that lag phase times were much longer at acidic pH than at alkaline pH. Yang et al. (2007) performed experiments in a completely mixed reactor (continuous stirred tank reactor). The pH of the bioreactor was controlled in a range of 4.0–5.0 by addition of carbonate to the feed material consisting of simulated cheese processing wastewater. Maximum hydrogen yields were between 1.8 and 2.3 mmol/g COD fed for the loading rates tested with a hydraulic retention time of 24 h.

- *Inoculation.* The use of heat-treated sludge is a common practice used to seed a bioreactor to produce hydrogen. This process favors the presence of the spore-forming, hydrogen-producing *Clostridium*. This practice was followed by Davila-Vazquez et al. (2008), who treated the inoculum as follows: granular sludge was used as inoculum and washed with three volumes of tap water and then boiled for 40 min to inactivate methanogenic microflora and was stored at 4°C before use. However, Antonopoulou et al. (2008) followed a different strategy: the indigenous microorganisms of the cheese whey were used for the start-up of the bioreactor by filling it with cheese whey and letting it operate in batch mode for 24 h. After this inoculation time, the mode was switched to continuous and hydrogen continued to be produced at a constant rate (Fig. 16.5). The inoculation procedure followed by Antonopoulou et al. (2008) indicated that no spore-forming bacteria able to produce hydrogen prevailed. The fact that the raw cheese whey can be directly used for the production of hydrogen is very important for the operation

of a full-scale plant, suggesting that no extra energy will be required either for the start-up of the reactor or for the pasteurization or sterilization of the influent.
- *Organic acid production*. In general, hydrogen yields are correlated to the final metabolic products. The production of acetic and butyric acids is accompanied by the production of hydrogen, whereas the production of propionic acid indicates that hydrogen is consumed. Moreover, lactic acid and ethanol production is accompanied by no hydrogen generation. In the cases studied by Davila-Vazquez et al. (2008) and Yang et al. (2007), acetic and butyric acids were detected at higher concentrations than propionic acid. The same was also found by Antonopoulou et al. (2008), who also monitored the lactic acid concentration in a hydrogenotrophic bioreactor. In fact, lactic acid was produced at significant concentrations. It was also observed that when the hydrogen production decreased, the lactic acid concentration increased and the pH fell below 5. Therefore, these three parameters seem to be closely correlated and reveal the dynamic behavior of the microbial community towards factors that affect the pH of the medium, and result in a shift of the metabolism towards to more lactate and less hydrogen or vice versa.

In conclusion, the microbial production of hydrogen from cheese whey has not been extensively studied so far. Fermentative hydrogen production processes do not reduce the organic content of the feed. A total COD removal below 5% can be achieved and, therefore, another, subsequent stage is required for COD reduction.

16.5 Wastewater Treatment

16.5.1 *Characterization of Dairy Wastewaters*

According to Janczukowicz et al. (2008), the key pollutants in dairy wastewater are organic compounds [defined as the 5-day biological oxygen demand (BOD_5) and COD] that occur in abundance, suspended solids, and biogenic elements. The main contributors to organic load in dairy wastewater are lactose, fats, and proteins originating from the milk (Perle et al. 1995). The correlation between fats, proteins and carbohydrates might be varied in the case of dairy wastewater, which is extremely significant as far as its biodegradability is concerned. All dairy production effluents can be treated together, with the exception of whey. The biodegradation demands are complex and may cause too much burden on wastewater treatment technology systems. Whey by-product should therefore be managed within a separate installation (Janczukowicz et al. 2008).

Orthon et al. (1993) investigated the biological treatability of integrated dairy plant wastewater containing a small fraction of whey wash waters. The results showed that the wastewater tested had practically no initially inert fraction, but generated residual microbial products accounting for 6–7% of the initial degradable COD; thus, it was noted that effluent COD could not be biologically reduced below 85 mg/L. Danalewich et al. (1998b) evaluated effluents from 15 different dairy plants and, on the basis of the mean total BOD_5 and total COD values (1,856 mg/L and 2,855 mg/L), confirmed that milk processing wastewaters often have a relatively high organic content, which varied greatly within and among plants. For the great majority of the plants evaluated, the BOD_5 to COD ratio was above 0.5, with an average of 0.6320, showing that most of the organic compounds in dairy wastewaters should be easily biodegradable. On average, about 76% of the soluble solids were volatile, even though the ratios varied over a wide range. Total solids and volatile levels also varied significantly. On average, 52% of the total solids were found to be volatile, indicating that soluble inorganic constituents were important in these waste streams. The pH values of composite samples collected over a 24-h period were generally near neutrality or basic, indicating that the large quantities of caustic soda used for cleaning, apparently had a greater impact on overall wastewater pH than the acids used for cleaning. As regards nutrient requirements, Danalewich et al. (1998) found that phosphorus was present at much greater levels than the amount needed for aerobic biological treatment, and in the most of the cases nitrogen levels were sufficient.

16.5.2 Preliminary Processes

Preliminary treatment is the first step in dairy wastewater treatment plants and removes large objects such sticks, bottles, cans, stone, grit, and rags; this process protects the equipment in the plant from physical damage or interference with downstream equipment and treatment processes, including pumps, pump inlets, and pipelines (Nielsen 1996). Treatment equipment such as bar screens and grit chambers are used as the wastewater first enters a treatment plant. There are several types of screens used in wastewater treatment, including static or stationary, rotary drum, brushed, and vibrating screens. Mesh sizes in such equipment can range from 0.5 in. in a static screen to 40 mesh in a more finely spaced mechanically brushed system (Britz et al. 2006). Screening systems may be used in combination (i.e., prescreen – polish screen) to achieve the desired solids removal efficiency (USEPA 2002). To prevent the settling of coarse matter in the wastewater before it is filtered, the ratio of the depth to the width of the approach channel of the screen should be 1:2, and the velocity of water should not be less than 0.6 m/s (Droste 1997). Screens can be cleaned manually or mechanically and the material collected can be sent for composting and used as a soil amendment or as a medium to grow plants. The screened material can also used to produce animal food or it can be disposed of in a landfill. In dairy wastewater plants it is recommended that the retained material from the screening process should be removed a quickly as possible to prevent an increase in the COD concentration as a result of solubilization of the solids (Britz et al. 2006).

Most dairy factories operate intermittently, with discontinuous wastewater production, and where weekly variation of wastewater flow is also a common occurrence. In addition, there are significant differences in flow during processing and cleanup periods, producing a substantial diurnal variation in flow and organic load on processing days. To minimize the need for sizing subsequent treatment units to handle peak flows and loads, flow equalization tanks may be installed to provide a constant 24-h flow rate on processing days, but sizing may also be used to provide a constant daily flow rate, including on nonprocessing days. Large variations in pH wastewater are encountered in a dairy plant's discharge, which may be attributed to the different cleaning strategies employed. Alkaline detergents generally used for organic matter removal would have a pH above 10, whereas acid-based sanitizers and acidic cleaners used for the removal of mineral deposits have a pH of 1.5–6.0 (Britz et al. 2006). The typical pretreatment of dairy wastewaters consists in the modification of the flow rate and pH of the dairy processing waste stream by a flow equalization – neutralization system. Such processes even out the wide variations encountered in a dairy plant's discharge, and also provide a means of isolating accidental causes of acid spills which would shift the pH of the wastewater above 9.0 or below 6.0.

The presence of equalization basins, in which the effluent remains for at least 6–12 h, promotes better operation conditions of secondary wastewater treatment systems as they are not subjected to shock loads and variations of feed and pH values. In these facilities, residual oxidants can react completely with solid particles, neutralizing cleaning solutions. Maintaining the pH between 6.5 and 7.5, minimizes or even prevents the corrosion as this is a process which may occur at pH values below 6.5 and above 10. To avoid settling of solids and odor generation, aeration and mixing of flow equalization basins are required. Methods of aeration and mixing include diffused air, diffused air with mechanical mixing, and mechanical aeration (Tchobanoglous and Burton 2003). The best localization for equalization facilities differs with the characteristics of the collection system, the wastewater to be handled, land availability, and the type of treatment required.

16.5.3 Primary Treatment

Primary treatment is the second step in treatment and removes settable and floatable materials from the dairy wastewater. Wastewater is, often first, held in a quiet tank for several hours to allow particles to settle to the bottom and the greases to float to the top. The solids drawn off the bottom and skimmed off the top are then moved and undergo further treatment through a solids handling processes. The clarified wastewater flows to the next stage of wastewater treatment. The presence of fats, oil, and grease in dairy wastewater can cause a variety of problems in biological wastewater treatment systems.

Gravity traps are extremely effective, self-operating, and easily constructed systems, where wastewater flows through a series of cells, and fats, oil, and grease, which usually float on top, are removed by retention within the cells. Drawbacks include frequent monitoring and cleaning to prevent buildup of fats, oil, and grease, and decreased removal efficiency at pH above 8 (IDJ 1997).

Most fats, oils, and grease have low density and cannot be separated efficiently by sedimentation from water streams. Thus, flotation is a preferred technique for the removal of oil and particles with low density from water during or after de-emulsification; flotation may be natural or forced depending on technical/economic considerations. Dissolved air flotation (DAF) is used extensively in the primary treatment of dairy industry wastewaters to remove suspended solids. The principal advantage of DAF over gravity settling is its ability to totally remove very small or light particles (including grease) in a shorter period of time. Once particles have reached the surface, they are removed by skimming (Tchobanoglous and Burton 2003).

In DAF, either the entire influent (incoming material), some fraction of the influent, or some fraction of the recycled DAF effluent is saturated with air at a pressure of 40–50 psi (250–300 kPa), and is then introduced into the flotation tank. The dissolved air is converted to minute air bubbles under the normal atmospheric pressure in the tank. These air bubbles attach to the fat particles, and the scum formed is skimmed off at the top of tank. Chemicals (e.g., lime, iron and aluminum salts, polymers, and flocculants) are often added prior to the DAF to improve the DAF performance. A DAF system has many advantages, including its low installation costs, compact design, ability to accept variable loading rates, and low level of maintenance (Nielsen 1996). The mechanical equipment used in the DAF system is fairly simple, requiring limited maintenance attention for things such as pumps and mechanical drives (USEPA 2002). The resulting scum is formed by an unstable waste material that should preferably not be mixed with sludge from biological and chemical treatment process since it is very difficult to dewater (Britz et al. 2006).

16.5.4 Secondary Treatment

Secondary treatment is the stage responsible for the consumption of the organic matter. The process in this treatment stage can be physicochemical or biological, or also a combination of both. The dairy effluents selected for a specific treatment process are dependent on the effluent's characteristic composition. Those effluents treated by a biological process can be divided in two major groups, those treated via aerobic or anaerobic processes. Both aerobic and anaerobic technologies have their characteristic advantages and disadvantages, and many authors have recommended their combination as the best option for the treatment of wastewaters with high organic matter concentrations. Through combined processes, optimized for dairy wastewater treatment, effluent discharge limits for agro-industry wastewaters are adequately reached (Demirel et al. 2005).

The aerobic technologies adopted by many dairy industries for processing wastewaters, when applied as the sole solution, are usually highly energy intensive and may lead to uncertainty regarding stabilized performance, owing to factors such as overloading and bulking sludge. On the other hand, anaerobic technologies are simpler, require a lower budget to operate, and have the potential of producing energy from the utilization of the main process product, biogas with a high content of methane (Arvanitoyannis and Giakoundis 2006). It is also known that alone anaerobic processes are not capable of providing an acceptable quality of effluent, so the application of anaerobic processes followed by aerobic processes is the best option for any high-strength wastewater, with the additional possibility of promoting nitrogen removal.

In the conventional anaerobic treatment processes often used for treating dairy wastewaters, anaerobic filters and upflow anaerobic sludge blanket (UASB) reactors are the most common reactor configurations used. The UASB reactors are preferred, as they can treat large volumes of dairy wastewaters in a relatively short time. Lipid degradation and inhibition in single-phase anaerobic systems is frequently discussed in the literature, as lipids are potential inhibitors in anaerobic systems. Moreover, high concentrations of suspended solids in dairy waste streams can also adversely affect the performance of conventional anaerobic treatment processes, particularly the most commonly used upflow anaerobic filters. Thus, two-phase anaerobic digestion processes should be considered more often to overcome problems that may be experienced in conventional single-phase design applications (Demirel et al. 2005).

Alkalinity must be recognized as one of the major factors affecting anaerobic treatment exploratory studies: the selection of the type, amount, and the process configuration itself has considerable impact on proper design and operation (Speece 1996). According to Demirel et al. (2005), adequate levels of alkalinity are considered to be most important for controlling process reliability during anaerobic treatment of dairy wastewater. This statement was also made by Ratusznei et al. (2003a), Mockaitis et al. (2006), Bezerra et al. (2007), and Fuzatto et al. (2009). These authors insist that the operating stability of anaerobic batch reactors is strongly dependent on the alkalinity supplementation strategy.

One of the main problems related to dairy effluents is associated with lipid content. Although lipids are easily hydrolyzed to long-chain fat acids (Hanaki et al. 1981), they can cause many problems related to the biomass present in anaerobic reactors. These problems are always related to biomass washout, short circuit formation, and microorganism inhibition due to exposure to long-chain fatty acids (Hwu and Lettinga, 1997; Vidal et al. 2000; Alves et al. 2000; Demirel et al. 2005).

Some alternatives, such as the inclusion of filters in the upper part of UASB reactors, are indicated for the minimization of the problem of expulsion of the baiomass. Córdoba et al. (1995) compared two anaerobic systems for diluted dairy wastewater treatment, a biological filter with polyurethane foam as a support and a UASB reactor, in which a gas/liquid/solid separator was substituted by a polyurethane foam stream bed. With organic loading rates from 1 to 8 kg/m^3/day, the hybrid configuration showed better results, reaching 89.9% of COD removal at the highest load.

The search for solutions for problems related to the presence of lipids has motivated researchers to use lipases in the pretreatment of wastewaters. Rigo et al. (2008) studied the use of enzymatic compounds in the prehydrolysis of oils and fats present in effluents from slaughterhouses. An enzymatic "pool" produced by solid-state fermentation using *Penicillium restrictum* was compared with a commercial lipase (Lipolase® 1000 T from Novozymes). The conclusion was that both conditions tested were adequate to perform the prehydrolysis of oils and fats present in floating residues from slaughterhouse effluents. Although the specific activity of the commercial enzyme was higher than that found in the enzymatic "pool," similar effluent hydrolysis levels were obtained. Using the same enzymatic pool, Leal et al. (2002) studied the effect of enzymatic pretreatment in the efficiency of a UASB-type reactor in pilot-scale treatment of wastewaters from the dairy industry. It was observed that enzymatic pretreatment increased the overall efficiency of the process and the initial organic matter degradation rate. The authors concluded that the results confirm the potential viability of enzymatic systems combined with anaerobic reactors in the degradation of wastewaters rich in lipids. Promising results have also been reported by Mendes et al. (2006), with the use of pancreatin in the hydrolysis of dairy wastewaters. They detected biogas production of the order of 445 ± 29 mL when the effluent was submitted to hydrolytic pretreatment and 209 ± 47 mL without pretreatment in reactors of total volume 500 mL, with 50 mL of anaerobic inoculum previously adapted. In addition, Gomes et al. (2011) found that enzymatic treatment with porcine pancreas lipase was harmful to the granular structure, promoting biomass disruption and expulsion, and loss of efficiency and stability.

The following biochemical and physiologically based inhibitory mechanisms were recognized in the anaerobic digestion of dairy effluents:

- Lack of a developed proteolytic enzymatic system capable of casein solubilization in the nonacclimatized sludge. The result is a very slow rate of degradation and consumption of casein and hence low process efficiency. This could be overcome by a long period of careful acclimatization to casein, as when effluent is fed to casein-acclimatized reactors the degradation of casein is very fast and its degradation products are biologically available and noninhibitory to the anaerobic process.
- Inhibition of the process by milk fat and by the long-chain fatty acids resulting from milk fat lipolysis. The inhibition is expressed both by reduction of the methanogenic activity and by reduction of the ATP level, which indicate a general reduction in the physiological activity of the cells. In relation to glycerol, this compound was found to be a noninhibitory, readily available by-product of milk fat biodegradation (Perle et al. 1995).

Concerning the inhibition aspect, Alves et al. (2000) suggested that the long-chain fatty acids could be potentially deleterious to methane formation, especially the fraction related to acetate consumption, which represents 70% of the overall production. They also indicated that both tolerance to toxicity and biodegradability of oleic acid were improved by acclimatization with lipids. In this way, Vidal et al. (2000) studied the anaerobic biodegradability of two synthetic wastewaters: one rich in fats (COD ratio; fats/proteins/carbohydrates 1.7:0.57:1) and the other with a low fat content (COD ratio; fats/proteins/carbohydrates 0.05:0.54:1) were studied in samples with total COD ranging from 0.4 to 20 g/L. There were no problems of sludge flotation and the maximum biodegradability and methanation were obtained when operating with wastewaters in the range 3 ± 5 g COD/L. The anaerobic biodegradation rate of fat-rich wastewaters was slower than that of fat-poor wastewaters owing to the slower rate of the fat hydrolysis step. However, this fact avoided the accumulation of volatile fatty acids and the overall process was facilitated. The presence of fats in the wastewater specifically prevented the periodic production of high concentrations of volatile fatty acids, which may adversely affect the process. Under the actual experimental conditions, the intermediates of fat degradation (glycerol and long-chain fatty acids) do not seem to reach sufficient concentrations to affect the anaerobic process.

Raw whey is known to be quite problematic when subjected to anaerobic processes; this condition is precipitated by prevailing low bicarbonate alkalinity, its high COD concentration, and its tendency to rapidly become acidified. During anaerobic digestion, it is important to keep the pH at levels at which the methanogens are active; this is the most crucial factor during the operation of anaerobic digesters. Several authors have pointed out that it is difficult for an anaerobic digester to self-sustain the pH in the case of cheese whey treatment (Mah 1983; Ghaly and Pyke 1991; Ghaly 1989; Yan et al. 1989; Clanton et al. 1987; Lo and Liao 1986; Wolfe 1983). The indigenous bacterial population of cheese whey is rich in microbes that convert lactose rapidly to lactic acid. The acute accumulation of organic acids combined with the low bicarbonate alkalinity causes the pH to fall below 5. Moreover, a massive growth of filamentous bacteria with low specific activity and poor stability may occur in an acidic environment, resulting in low treatment efficiency (Ghaly and Pyke 1991; Yan et al. 1989; Schroder and De Haast 1989; Nordstedt and Thomas 1985).

Therefore, the addition of an alkalinity source (sodium bicarbonate) is necessary to avoid anaerobic process failure (Lo and Liao 1986; Wildenauer and Winter 1985). The alkalinity supplementation may be moderated through proper manipulation of the operating conditions, such as operating with high hydraulic residence times or diluting the influent (Kalyuzhnyi et al. 1997; Yan et al. 1988; Kato et al. 1994). The recirculation of the anaerobic effluent may be an alternative to increase alkalinity and dilute the influent (Malaspina et al. 1996). Another strategy is the separation of the acidogenic and methanogenic phases in two bioreactors in series (Garcia et al. 1991; Germerli et al. 1993; Yilmazer and Yenigu"n 1999; Antonopoulou et al. 2008). In this configuration, the methanogenic effluent can be reused to dilute the influent and enhance alkalinity. The increase in alkalinity can also be achieved by mixing the cheese whey with manure, which is a waste with high alkalinity (Gavala et al. 1999; Lyberatos et al. 1997).

Moreover, the methanogenic bacteria are slow-growing microorganisms affected adversely by low pH values. To increase the low rates of organic acid consumption and the low biogas production rate, the anaerobic digestion of cheese whey has been studied in bioreactors that enhance the concentration of microorganisms by preventing them from leaving the bioreactors with the effluent. Typical bioreactors designed with this concept are those that allow:

- The granulation of biomass as in UASB reactors (Gavala et al. 1999; Erguder et al. 2001).
- The formation of biofilms through the colonization of certain materials which increase the useable surface for bacterial growth, as in fluidized bed reactors, anaerobic upflow fixed bed reactors, or anaerobic downflow fixed bed reactors. Borja et al. (1994) have shown that immobilization of microorganisms on sepiolite,

saponite, and bentonite supports in an anaerobic upflow fixed bed reactor increases the anaerobic digestion rate of diluted cheese whey. Patel et al. (1995) studied the biomethanation of undiluted cheese whey using different support material such as charcoal, gravel, brick pieces, PVC pieces, and pumice stones, with the charcoal resulting in the best performance.

- The settling of biomass after stirring as in anaerobic sequencing batch reactors (Mockaitis et al. 2006; Ratusznei et al. 2003b). The settling capacity of the bacteria is increased through attachment to materials such as polyurethane foam cubes; this type of anaerobic sequencing batch reactor is called an anaerobic sequencing batch biofilm reactor. Damasceno et al. (2007) studied the performance of this type of bioreactor under various organic loading rates and feed strategies.
- The accumulation of bacteria as in the periodic anaerobic baffled reactor, which is modification of the simple anaerobic baffled reactor configuration (Antonopoulou et al. 2008)

Another strategy is to use a membrane-based technology for the removal of biosolids from the anaerobic liquid effluent; the biosolids carried in the effluent consist mainly of bacteria. The biosolids retained on the membrane are recycled back to the bioreactor (Saddoud et al. 2007).

Table 16.6 summarizes the performance of several bioreactor configurations (in terms of COD removal) under various operating conditions, such as the COD of the influent, the hydraulic retention time, and the organic loading rate expressed in grams of COD added per liter of reactor volume per day. It can be seen that high COD removals can be achieved, but attention should be paid to the bioreactor start-up and pH maintenance in the methanogenic bioreactor. The anaerobic digestion of dairy wastewaters generally (not only cheese whey) in a single-stage or two-stage configuration was reviewed by Demirel et al. (2005) and Siso (1996). In the review of Siso (1996), other methods of treatment and/or exploitation of dairy wastewaters were also discussed.

16.5.5 Tertiary Treatment

Tertiary treatment is the part of the treatment plant related to nutrient (nitrogen and phosphorous) removal, since discharging wastewater with high levels of phosphorus and nitrogen can result mainly in eutrophication of receiving waters, particularly lakes and slow-moving rivers.

16.5.5.1 Nitrogen Removal

In dairy wastewater treatment plants, a fraction of the total nitrogen is removed in biological reactors designed remove the carbon-based organic matter that is vital for growth and maintenance of microorganisms. The amount of nonremovable fraction

Table 16.6 Overview of the operating conditions and performance of various anaerobic bioreactor configurations treating dairy cheese whey wastewater

Reactor	Type of waste water	COD (g/L)	HRT (days)	OLR (g COD/L/day)	COD removal (%)	References
Downflow, upflow anaerobic hybrid	Raw cheese whey	69	–	10	98.4	Malaspina et al. (1995)
Upflow anaerobic fixed bed	Raw cheese whey	70	2		81	Patel et al. (1995)
2-stage unmixed	Raw cheese whey	73	10–20		46–65	Ghaly (1996)
Upflow anaerobic sludge blanket	Deproteinated cheese whey		6	2–7.3	85–99	Gavala et al. (1999)
Single-fed and multifed upflow anaerobic fixed bed	Raw cheese whey			20	90–92	Punal et al. (1999)
Upflow filter	Dairy	0.5		21	80	Ince et al. (2000)
2-stage upflow anaerobic sludge blanket	Raw cheese whey	43–55	2–3		95–97	Erguder et al. (2001)
Sequencing batch	Raw cheese whey	0.5–4		0.81–5.7	96	Ratusznei et al. (2003b)
Upflow anaerobic sludge blanket	Synthetic dairy		0.125	13.5	83–86	Ramasamy et al. (2004)
Anaerobic sequencing batch	Raw cheese whey	0.5–4		0.6–4.8	90	Mockaitis et al. (2006)
2-stage anaerobic stirred tank with membrane separator	Raw cheese whey	68.6	4	19.7	98.5	Saddoud et al. (2007)
anaerobic sequencing batch biofilm	Raw cheese whey	1–6		2–12	72–87	Damasceno et al. (2007)
Upflow anaerobic filter	Clarified cheese whey	5–20	2–5	4	98	Gannoun et al. (2008)
Periodic anaerobic baffled	Acidified cheese whey	58	4.4–20	2.9–13.2	94.2–98.9	Antonopoulou et al. (2008)

COD chemical oxygen demand, *HRT* hydraulic retention time, *OLR* organic loading rate

must be reduced before the disposal of the effluent. This process can be performed by biological or by physicochemical means. In a conventional biological tertiary treatment plant, nitrogen is removed in two stages, nitrification and denitrification, which are performed under different conditions and by different microorganisms. The removal has to be sequential to be performed well as the various processes have to be separated in time or space.

In the aerobic nitrification stage, performed by autotrophic bacterial species, ammonium (NH_4^+) is firstly oxidized to nitrite (NO_2^-) by *Nitrosomonas*. Nitrite is then oxidized by *Nitrobacter* to nitrate (NO_3^-) (Weijers et al. 1996). The denitrification process involves the biological reduction of nitrate and/or nitrite to nitrogen gas in the absence of dissolved oxygen. Unlike nitrification, denitrification is performed by a wide range of heterotrophic bacterial species, many of which are commonly found in secondary tanks, since they are facultative, and they can use oxygen, nitrate, or nitrite as their terminal electron acceptor. Heterotrophic bacteria need carbon sources to accomplish the denitrification process; hence, a carbon source must be added to the medium to enable the process to proceed. The substrates can be organics found in the wastewater, ethanol, methanol, or acetic acid (Brown and Koch 2006).

Grady et al. (1999) summarized general guidelines for organic matter to nitrogen ratios for wastewaters that are amenable to complete biological nitrogen removal. COD to total Kjeldahl nitrogen, BOD_5 to total Kjeldahl nitrogen, and BOD_5 to ammonia nitrogen ratios greater than 9, 5, and 8, respectively, should result in excellent nitrogen removal. Danalewich et al. (1998) studied 15 milk processing plants in the Upper Midwest of the USA to obtain information on general process operation, waste generation and treatment practices, chemical usage, and wastewater characteristics. All plants had organic matter to nitrogen ratios exceeding the minimum values necessary to accomplish excellent biological nitrogen removal process.

The conventional nitrification reaction consumes a large amount of oxygen, requiring 4.2 g of oxygen for each gram of nitrified ammonium nitrogen. During denitrification, the requirement of organic carbon is significant. To optimize these processes, various novel and sustainable technologies have been proposed. Simultaneous nitrification and denitrification is the process in which, given the correct conditions, nitrification and denitrification occur simultaneously. This process is based on the presence of anoxic zones and the availability of electron donors for the denitrification process where the nitrification was already happening. In the Anammox (anaerobic ammonium oxidation) process, nitrite and ammonium are converted into nitrogen gas under anaerobic conditions without the need to add an external carbon source. In the SHARON (single reactor system for high ammonium removal over nitrite) process, ammonium is oxidized in a reactor system under aerobic conditions to nitrite, which in turn is reduced to nitrogen gas under anoxic conditions by using an external carbon source.

Ammonia nitrogen can be removed from wastewater by physicochemical means such as volatilization of gaseous ammonia. The process is simple in concept, but it has serious drawbacks that make it expensive to operate and maintain (Tchobanoglous and Burton 2003). The volatilization can be achieved by raising the pH of the liquor so that the aqueous ammonia exists in equilibrium with its gaseous counterpart in accordance with Henry's law. When the pH exceeds 9.5, unionized ammonia prevails and can be stripped from the side stream liquid. The process requires a large amount of air and the gas stream must be captured and treated (Brown and Koch 2006). As Henry's constant varies with temperature, the ammonia solubility is also variable, and the costs of the process can seriously rise with temperature reduction.

Another physicochemical method is breakpoint chlorination, which involves the addition of chlorine to wastewater to oxidize the ammonia nitrogen in solution to gas and other stable compounds. With proper control, all ammonia can be oxidized, but there are serious considerations involved with the process, such as the formation of high chlorine residuals, trihalomethanes, or trichloride gas (which can avoided by the careful control of the pH of the process, which requires high-quality skilled operators). The process can be used following other nitrogen removal processes for fine-tuning of the process (Tchobanoglous and Burton 2003).

Ion exchange can also be used where climatic conditions inhibit biological nitrification and where strict effluent standards are required. This process can meet total nitrogen standards, but has high capital and operational costs and requires high-quality skilled operators (Tchobanoglous and Burton 2003).

16.5.5.2 Phosphorous Removal

Microbes usually have low phosphorous requirements; thus, to achieve low effluent concentrations it is necessary to have additional uptake beyond that needed for normal cell maintenance and synthesis. This situation is achieved when microorganisms store phosphorous for later utilization. This is possible when specific organisms are successively exposed to different environments (aerobic and anaerobic or anoxic). This situation provides a favorable growth condition for microorganisms capable of storing phosphorous, in comparison with other microorganisms (Brown and Koch 2006). When an anaerobic zone is followed by an aerobic zone, the microorganisms exhibit phosphorous uptake above normal levels. On the other hand, in anoxic conditions, phosphorous may be released by microorganisms. Many processes, such as the five-stage Bardenpho process and the PhoStrip process are based on these sequential exposures (Tchobanoglous and Burton 2003).

The presence of nitrate in dairy waste streams adds a level of complexity not commonly encountered in domestic wastewaters. Dairy plants have several options to compensate for the presence of nitrate in their wastewaters. One option is to reduce or eliminate the use of nitric acid as a cleaning agent, decreasing the effect of nitrate on enhanced biological phosphorus removal. Several plants use nitric acid only because of phosphorus surcharges and prefer phosphoric acid based cleaners from a cleaning perspective. Thus, eliminating nitrate from their waste streams may result in improved enhanced biological phosphorus removal, while resulting in better cleaning performance at a lower cost (Danalewich et al. 1998).

Only physicochemical treatment (precipitation and subsequent removal) can be applied, and the most common chemicals used in the process are metal salts (ferric chloride and aluminum sulfate), lime, and polymers. (Tchobanoglous and Burton 2003). The lime precipitation clarification process is primarily used for removal of soluble phosphates by precipitating the phosphate with the calcium oxide to produce insoluble calcium phosphate. It is postulated that orthophosphates are precipitated as calcium phosphate, and polyphosphates are removed primarily by adsorption on calcium floc.

Enhanced biological phosphorus removal can be more cost-effective than chemical precipitation strategies. Therefore, it is important for the dairy industry to evaluate enhanced biological phosphorus removal, combined with nitrification and denitrification (to remove nitrogen), as a treatment option for nutrient removal. Another important aspect is that volatile fatty acids or fermentable organics need to be added to some dairy wastewaters to accomplish complete phosphorus removal (Danalewich et al. 1998).

16.6 Summary and Conclusions

The transformation of raw milk into dairy products is one of the most important activities of the dairy industry worldwide. Along with the production of pasteurized and sterilized milk for direct consumption, a variety of processes are used for the production of cheese, butter, cream, ice-cream, yogurts, and other fermented milk products. As a consequence, the processing of milk usually generates huge amounts of waste products, such as whey and buttermilk. These products have high nutrient and economic values; therefore, they can be used as ingredients for others milk products, such as flavored milk beverages, chocolates, candies, cookies, processed cheese, ice cream, and powdered milk. Because of their intrinsic composition, dairy wastewaters cannot be discharged into septic systems or state water systems or directly dumped; appropriate treatment is required before their release into the environment. There are many alternatives for the treatment of this type of effluent, all based on a sequence of processes and operations presented as functions of physicochemical characteristics of dairy waste effluent. The selection of some processes require careful consideration and must be based on technical, economic, and social aspects.

References

Alves MM, Mota Vieira JA, Álvares Pereira RM, Pereira MA, Mota M. Efects on lipids and oleic acid on biomass development in anaerobic fixed-bed reactors. Part II: oleic acid toxicity and biodegradability. Water Res. 2000;35(1):264.

Antonopoulou G, Stamatelatou K, Venetsaneas N, Kornaros M, Lyberatos G. Biohydrogen and methane production from cheese whey in a two-stage anaerobic process. Ind Eng Chem Res. 2008;47(15):5227–33.

Arvanitoyannis S, Giakoundis A. Current strategies for dairy waste management: a review. Crit Rev Food Sci Nutr. 2006;46:379–90.

Bezerra Jr RA, Rodrigues JAD, Ratuszney SM, Zaiat M, Foresti E. Whey treatment by AnSBBR with circulation: effects of organic loading, shock loads, and alkalinity supplementation. Appl Biochem Biotechnol. 2007;143:257.

Borja R, Duran MM, Martin A, Luque M, Alonso V. Influence of immobilization supports on the kinetic constants of anaerobic digestion of cheese whey. Resour Conserv Recy. 1994;10:329–39.

Britz TJ, van Shalkwyk C, Hung YT. Treatment of dairy processing wastewater. In: Wang LK, Hung Y-T, Lo HH, Yapijakis YapijakisC, editors. Waste treatment in the food processing industry. 2nd ed. Boca Raton: CRC Press; 2006.

Brown JA, Koch CM. Biological nutrient removal operation in wastewater treatment plants. Maidenhead: New York; 2006.

Caric M. Concentrated and dried airy products. New York: VCH Publishers; 1994.

Chong LV, Dong CX, Mackereth AR. Effect or ageing and composition on the ignition tendency of dairy powders. J Food Eng. 1999;39:269–76.

Clanton CJ, Backus BD, Goodrich PR, Fox EJ, Morris HA. Anaerobic digestion of cheese whey. Proceedings 1987 Food Processing Waste Conference, Atlanta; 1987.

Cordoba PR, Francese A P, Sineriz, F. Improved performance of a hybrid design over an anaerobic filter for treatment of dairy industry wastewater at laboratory scale. Journal of Fermentation and Bioengineering, 1995;79:270–272.

Damasceno LHS, Rodrigues JAD, Ratusznei SM, Zaiat M, Foresti E. Effects of feeding time and organic loading in an anaerobic sequencing batch biofilm reactor (ASBBR) treating diluted whey. J Environ Manage. 2007;85:927–35.

Danalewich JR, Papagiannis TG, Belyea RL, Tumbleson ME, Raskin L. Characterization of dairy waste streams, current treatment practices, and potential for biological nutrient removal. Water Res. 1998a;32(12):3555–68.

Danalewich JR, Papagiannism TG, Belyea RL, Tumbleson ME, Raskin L. Characterization of dairy waste streams, current treatment practices, and potential for biological nutrient removal. Water Res. 1998b;32(12):3555–68.

Davila-Vazquez G, Alatriste-Mondragon F, Leon-Rodriguez A, Razo-Flores A. Fermentative hydrogen production in batch experiments using lactose, cheese whey and glucose: Influence of initial substrate concentration and pH. Int J Hydrogen Energy. 2008;3:4989–97.

Demirel B, Yenigun O, Onay TT. Anaerobic treatment of dairy wastewaters: a review. Process Biochem. 2005;40:2583–95.

Droste RL, editor. Theory and practice of water and wastewater treatment. New York: Wiley; 1997.

Durham RJ, Hourigan JA, Sleigh RW, Johnson RL. Whey fractionation: wheying up the consequences. Food Aust. 1997;49(10):460–5.

Durham RJ, Sleigh RW, Hourigan JA. Pharmaceutical lactose: a new whey with no waste. Aust J Dairy Technol. 2004;59(2):138–41.

Durham RJ. Development of a process for the purification of lactose from whey. PhD Thesis, University of Western Sydney, Sydney; 2000.

Erguder TH, Tezel U, Guven E, Demirer GN. Anaerobic biotransformation and methane generation potential of cheese whey in batch and UASB reactors. Waste Manag. 2001;21:643–50.

Ferchichi M, Crabbe E, Gil G-H, Hintz W, Almadidy A. Influence of initial pH on hydrogen production from cheese whey. J Biotechnol. 2005;120:402–9.

Fuzatto MC, Adorno MAT, Pinho SC, Ribeiro R, Tommaso G. Simplified mathematical model for an anaerobic sequencing batch biofilm reactor (asbbr) treating lipid-rich wastewater subject to rising organic loading rates. Environ Eng Sci. 2009;26(7):1197–205.

Gannoun H, Khelifi E, Bouallagui H, Touhami Y, Hamdi M. Ecological clarification of cheese whey prior to anaerobic digestion in upflow anaerobic filter. Bioresour Technol. 2008;99:6105–11.

Garcia PA, Rico JL, Polanco F. Anaerobic treatment of cheese whey in a 2-phase UASB reactor. Environ Technol. 1991;12(4):355–62.

Gavala HN, Skiadas IV, Lyberatos G. On the performance of a centralised digestion facility receiving seasonal agroindustrial wastewaters. Water Sci Technol. 1999;40(1):339–46.

Germerli F, Orhon D, Artan N, Ubay E, Gorgun E. Effect of two-stage treatment on the biological treatability of strong industrial wastes. Water Sci Technol. 1993;28(2):145–54.

Ghaly AE. A comparative study of anaerobic digestion of acid cheese whey and dairy manure in a two-stage reactor. Bioresour Technol. 1996;58:61–72.

Ghaly AE, Pyke JB. Amelioration of methane yield in cheese whey by controlling the pH of the methanogenic stage. Appl Biochem Biotechnol J. 1991;27(3):217–37.

Ghaly AE. Biogas production from acid cheese whey using a two-stage digester. Biochem Biotechnol J. 1989;11(4):237–50.

Gomes DRS, Papa LG, Cichello GC V, Belançon D, Pozzi EG, Balieiro, JCC, Monterrey-Quintero, ES, Tommaso G. Effect of enzymatic pretreatment and increasing the organic loading rate of lipid-rich wastewater treated in a hybrid UASB reactor. Desalination. 2011;279:96–103.

Grady CPL, Daigger GT, Lim HC. Biological wastewater treatment. New York: Marcel Dekker; 1999.

Groleau PE, Lapoinet JF, Gauthier SF, Pouliot Y. Fractionation of whey protein hydrolysate using nanofiltration membranes. IDF Bull. 2004;389:85–91.

Gummer J, Lunn GA. Ignitions of explosive dust clouds by smouldering and flaming agglomerates. J Loss Prevent Process Indust. 2003;16:27–32.

Guu MYK, Zall RR. Nanofiltration concentration effect on the efficacy of lactose crystallization. J Food Sci. 1992;57(3):735–9.

Hale N, Bertsch R, Barnet J, Duddleston WL. Sources of wastage in the dairy industry. IDF Bull. 2003;382:7–30.

Hanaki K, Matsuo T, Nagase M. Mechanisms of inhibition caused by long chain fatty acids in anaerobic digestion process. Biotechnol Bioeng. 1981;23:1591–610.

Harper DJ, Plain G. Use of intrinsically safe circuits and enclosures to control ignition risk form equipment in powder handling plant. IChemE Symposium Series No.141; 1997. pp.143 Available on http://www.icheme.org/communities/subject_groups/safety%20and%20loss%20prevention/resources/hazards%20archive/hazards%20xiii.aspx, Acessed in 09/12/2011.

Hwu C-S, Lettinga G. Acute toxicity of oleate to acetate-utilizing methanogens in mesophilic and thermophilic anaerobic sludges. Enzyme Microb Technol. 1997;21:297.

Ince O, Ince BK, Donnelly T. Attachment, strength and performance of a porous media in an upflow anaerobic filter treating dairy wastewater. Water Sci Technol. 2000;41:261–70.

International Dairy Journal. Removal of fats, oils and grease in the pretreatment of dairy wastewaters. Bull Inter Dairy Fed. Doc. No. 327; 1997.

Janczukowicz W, Zielinski M, Debowski M. Biodegradability evaluation of dairy effluents originated in selected sections of dairy production. Bioresour Technol. 2008;99:4199–205.

Jonsson H, Arph SO. Ion exchange for the demineralization of cheese whey. IDF Bull. 1987;212(4):91–8.

Kalyuzhnyi SV, Martinez EP, Martinez JR. Anaerobic treatment of high-strength cheese-whey wastewaters in laboratory and pilot UASB-reactors. Bioresour Technol. 1997;60:59–65.

Kato MT, Field JA, Kleerebezem R, Lettinga G. Treatment of low strength soluble wastewaters in UASB reactors. J Ferment Bioeng. 1994;77(6):679–86.

Korhonan H, Pihlanto A. Bioactive peptides: new challenges and opportunities for the dairy industry. Aust J Dairy Technol. 2003;58(2):129–33.

Listiohadi YD, Hourigan JA, Sleigh RW, Steele RJ. Role of amorphorous lactose in caking behaviour of α-lactose monohydrate powders. Aust J Dairy Technol. 2005;60(1):33–52.

Lo KV, Liao PH. Digestion of cheese whey with anaerobic rotating biological contact reactor. Biomass. 1986;10:243–52.

Lyberatos G, Gavala HN, Stamatelatou A. An integrated approach for management of agricultural industries wastewaters. Nonlinear Anal Theory Methods Appl. 1997;30(4):2341–51.

Gonzfilez Siso MI. The biotechnological utilization of cheese whey: a review. Bioresour Technol. 1996;57:1–11.

Mah RA. Interaction of methanogens and nonmethanogens in microbial ecology. Proceedings of the Third International Symposium on Anaerobic Digestion, Boston; 1983. pp. 11–22.

Malaspina F, Cellamare CM, Stante L, Tilche A. Anaerobic treatment of cheese whey with a downflow-upflow hybrid reactor. Bioresour Technol. 1996;55:131–9.

Malaspina F, Sante L, Cellamare CM, Tilche A. Cheese whey and cheese factory wastewater treatment with a biological anaerobic – aerobic process. Water Sci Technol. 1995;32(12):59–72.

Mendes AA, Pereira, EB, Castro, HF Effect of the enzymatic hydrolysis pretreatment of lipids-rich wastewater on the anaerobic biodigestion. Biochem. Eng. Jour. 2006;32:185–190.

Mockaitis G, Ratusznei SM, Rodrigues JAD, Zaiat M, Foresti E. Anaerobic whey treatment by a stirred sequencing batch reactor (ASBR): effects of organic loading and supplemented alkalinity. J Environ Manage. 2006;79:198–206.

Nielsen VC. Treatment and disposal of processing wastes. In: Mead GC, editor. Processing of poultry. New York: Chapman and Hall; 1996.

Nordstedt RA, Thomas MV. Start-up characteristics of anaerobic fixed-bed reactors. Trans ASAE. 1985;28(4):1242–8.

Orhon D, Gorgon E, Germirli F, Artan N. Biological treatability of dairy wastewaters. Water Res. 1993;27(4):625–33.

Patel P, Desai M, Madamwar D. Biomethanation of cheese whey using anaerobic upflow fixed film reactor. J Ferment Bioeng. 1995;79(4):398–9.

Perle M, Kimchie S, Shelef G. Some biochemical aspects of the anaerobic degradation of dairy wastewater. Water Res. 1995;29(6):1549–54.

Prasad R, Pagan R, Kauter M, Price N. Eco-efficiency for the dairy processing industry; Dairy Australia 2004.

Punal A, Mendez-Pampin RJ, Lema JM. Characterization and comparison of biomasses from single and multi fed upflow anaerobic filters. Bioresour Technol. 1999;68:293–300.

Ramachandra Rao HG, Lewis MJ, Grandison AS. Effect of soluble calcium of milk on fouling of ultrafiltration membranes. J Sci Food Agric. 1994;65:249–56.

Ramasamy EV, Gajalakshmi S, Sanjeevi R, Jithesh MN, Abbasi SA. Feasibility studies on the treatment of dairy wastewaters with upflow anaerobic sludge blanket reactors. Bioresour Technol. 2004;93:209–12.

Ratusznei SM, Rodrigues JAD, Camargo EFM, And Zaiat M. Operating feasibility of anaerobic sequencing batch reactor containing immobilized biomass. Water Sci Technol. 2003a;48:179–84.

Ratusznei SM, Rodrigues JAD, Zaiat M. Operating feasibility of anaerobic whey treatment in a stirred sequencing batch reactor containing immobilized biomass. Water Sci Technol. 2003b;48(6):179–86.

Rigo E, Rigoni RE, Lodea P, De Oliveira D, Freire DMG, Treichel H, and Di Luccio M. Comparison of Two Lipases in the Hydrolysis of Oil and Grease in Wastewater of the Swine Meat Industry. Ind. Eng. Chem. Res. 2008;47:1760–65.

Saddoud A, Hassairi I, Sayadi S. Anaerobic membrane reactor with phase separation for the treatment of cheese whey. Bioresour Technol. 2007;98:2102–8.

Schmidt DG, Both P. Studies on the precipitation of calcium phosphate. Neth Milk Dairy J. 1987;41:105–20.

Schroder EW, De Haast J. Anaerobic digestion of deproteinated cheese whey in an upflow sludge-blanket reactor. J Dairy Res. 1989;56:129–39.

Speece RE. Anaerobic biotechnology for industrial wastewater. Nashville: Archae Press; 1996.

Tchobanoglous, G. and Burton, F. L. Wastewater engineering: treatment, disposal and reuse. 4th ed. New York: McGraw Hill; 2003.

Thorpe, B. Citizen's Guide to Clean Production. Produced for the Clean Production Network, by the Lowell Center for Sustainable Production, University of Massachusetts Lowell. 1999.

UNEP. Cleaner production assessment in dairy processing, prepared by COWI Consulting Engineers and Planners, Denmark. Paris: UNEP; 2000.

UNEP. Cleaner production activities. United Nations Environment Programme; 2004 http://www.unep.org/Documents.multilingual/Default.asp?DocumentID=67&ArticleID=4717&l=en, Acessed in 09/12/2011.

USEPA (2002) - U.S. Environmental Protection Agency - EPA. 2002. Development Document for the Proposed Effluent Limitations Guidelines and Standards for the Meat and Poultry Products Industry Point Source Category. United States Environmental Protection. Agency 40 CFR 432, available on http://water.epa.gov/scitech/wastetech/guide/mpp/upload/2008_05_13_guide_mpp_proposed_technicaldev.pdf, accessed in 09/12/2011.

Vidal G, Carvalho A, Méndez R, Lema JM. Influence of the content in fats and proteins on the anaerobic biodegradability of dairy wastewaters. Bioresour Technol. 2000;74:231–9.

Waldron KW, editor. Handbook of waste management and co-product recovery in food processing, vol. 1. Abington: Woodhead; 2007.

Weijers SR, Kok, JJ, Preisig, HA, Buunen, A, Wouda, TWM. Identifiability and estimation of parameters in the IAWQ Model No. 1 for modelling activated sludge plants for enhanced nitrogen removal, Computers Chem. Engng. 1996;20:S1455–S1460.

Wildenauer FX, Winter J. Anaerobic digestion of high strength acidic whey in a pH-controlled up-flow fixed-film loop reactor. Appl Microbiol Biotechnol. 1985;22(5):367–72.

Wolfe RS. Fermentation and anaerobic respiration in anaerobic digestion. Proceedings of the Third International Symposium on Anaerobic Digestion, Boston; 1983. pp. 3–10.

Yan JQ, Lo KV, Liao PH. Anaerobic digestion of cheese whey using upflow anaerobic sludge blanket reactor. Biol Waste. 1989;27:289–305.

Yan JQ, Liao PH, Lo KV. Methane production from cheese whey. Biomass. 1988;17:185–202.

Yang P, Zhang R, McGarvey JA, Benemann JR. Biohydrogen production from cheese processing wastewater by anaerobic fermentation using mixed microbial communities. Int J Hydrogen Energy. 2007;32:4761–71.

Yilmazer G, Yenigün O. Two-phase anaerobic treatment of cheese whey. Water Sci Technol. 1999;40(1):289–95.

Zadow JG. Review and report on whey utilisation. Melbourne: Dairy Australia; 2005.

Zockoll C. Ignition effect of smouldering pockets in dust air mixtures. VDI-Beriche. 1999;701:295.

Index

A
Acetogenesis, 84
Acidogenesis, 83–84
Active packaging systems, aquatic food products
 applications, 316–318
 future trends, 318
 general aspects of, 315–316
Adenocalymma alliaceum, 209
Adsorption, 73–74
Advisory Committee on Novel Foods and Processes, 151
Aerobic treatment, 87–89
Aflatoxins
 antifungal potential, 211–212
 chemical structure of, 222
 contamination level, 221
Agro-industrial wastes and by-products
 fish proteins, 347–348
 grape polyphenolics, 345
 rockfish species, 347
 wine by-products, 346
Algae fuel, 48
Aloe vera, 209
Alternaria disease control, tomatoes, 214–217
Anaerobic digestion, 83–87
Antibrowning agents, 348–350
Antifungal agents
 chemical additives, 205–206
 contaminaton, fungi, 205
 food preservation techniques, 205
 fungicides, 206
 hurdle technology
  fruits and vegetables preservation, 213–214
  physical and chemical methods, 212–213
 plant extracts
  Alternaria disease control, 213–217
  crude extracts, 207–208
  essential oils and oleoresins, 209–210
  medicinal plants, 208–209
  spices and herbs, 209
  against toxigenic fungi, 210–211
Antimicrobial metabolites, 192–193
Antimicrobial preservatives
 biopreservatives
  definition, 332
  evaluation of, 336–337
  lactic acid bacteria protection, 333–335
  and probiotics, 335–336
  species and strains, 332
 chitosan, 329–331
 organic acids, 326–327
 plant origin, 193–195
 spices and herbs, 327–329
 uses, 331–332
Antioxidant preservatives, seafood
 crustaceans preservation, melanosis development, 348–350
 deterioration process, 337
 enzymatic reaction, 338
 evaluation of, 351
 legal issues, 350–351
 lipid oxidation, 337, 339–348
 oxidation, 337
Aquatic food products preservation
 active packaging systems
  additives incorporation, 315
  applications, 316–318
  preservatives incorporation, 316

Aquatic food products preservation (*cont.*)
  aquatic species
    chilled products, processing of, 301
    damage pathways, 303
    slaughtering and onboard handling, 302
    sources and composition, 299, 300
  modified-atmosphere packaging
    argon, 314–315
    carbon dioxide, 312–313
    nitrogen, 314
    oxygen, 314
    shelf life extension, 312
  refrigeration methods, ice slurries storage
    application, 308
    combined preservation systems, 310–311
    food product sector, 311
    novelty and advantages, 308–309
    seafood products, 309–310
    versatility, 308
  traditional chilling methods
    amines formation and nucleotide degradation, 306–307
    chilled seawater, 304–305
    flake icing, 304
    lipid damage analysis, 307
    salt treatment, 305
    sensory and physical changes, 306
Aspergillus flavus, 211
Aspergillus parasiticus, 211

B
Bacteria. *See also* Foodborne bacteria, biocontrol of
  Gram-negative, 188, 189, 191, 195, 274, 328
  Gram-positive, 195, 274, 332
Bacterial predators. *SeeBdellovibrio bacteriovorus*
Bacteriocin, 192, 193, 333
Bacteriophages
  applications, 188–189
  examples, 186–187
  host specificity, 187–188
  inactivation, bacteria, 185, 187
  lytic cycle, 184–185
  properties, 184
  T4 phage, 184
Bdellovibrio bacteriovorus, 189–190
Bioalcohols
  anhydrous ethanol, 40
  ethyl tertiary butyl ether (ETBE), 40
  gasohol, 40
  methanol, 40–41
  source of, 41–42
  technologies, process description, 42–43

Biocontrol, foodborne bacteria. *See* Foodborne bacteria, biocontrol of
Biodiesel
  bio-oils, 47
  definition, 46
  from olive oil, 98
  production of, 49
  sources of, 47–48
  technologies, process description, 48–49
  vegetable oils, 46–47
Biofuels
  advantages and disadvantages of, 61
  bioalcohols
    anhydrous ethanol, 40
    ethyl tertiary butyl ether (ETBE), 40
    gasohol, 40
    methanol, 40–41
    source of, 41–42
    technologies, process description, 42–43
  biogas
    composition, 44
    generation, 50–52
    landfill gas, 44
    production of, 44
    purification, 52–54
    sources of, 45
    storage, 54–55
    technologies, process description, 45–46
    utilization, 52
  case studies and examples, 60–61
  definition, 39
  lipid
    bio-oils, 47
    definition, 46
    sources of, 47–48
    technologies, process description, 48–49
    vegetable oils, 46
  from OMWW, 97–98
  raw material for
    lignocellulosic biomass, 49–50
    solid by-products, 55–60
    wastewater, 50–55
  traditional, 39
  types, 39
  wastewater for, 50–55
  wastewater sludge for, 58–60
Biogas
  composition, 44
  generation, 50–52
  landfill gas, 44
  production of, 44
  purification, 52–54
  sources of, 45
  storage, 54–55

Index

technologies, process description, 45–46
utilization, 52
Biological treatment
    aerobic treatment, 87–89
    agricultural land irrigation, 91–92
    anaerobic digestion, 83–87
    composting, 92
    phytoremediation, 89–90
Biomethanol, 43–44
Bio-oils, 47
Biopolymers
    microencapsulation techniques, 168–169
    from OMWW, 98–99
Biopreservatives, 332–337
Biotechnology, 137
Blanching, 264–265
Brassica napus L., 47
Brine-salted tilapia *(Oreochromis niloticus)*, 343

C

Campylobacter phage Cj6, 187–188
Cellulolysis, 43
Centrifugation, 71
Chemical oxidation
    advanced oxidation processes, 74–77
    electrochemical oxidation, 78–79
    wet oxidation, 77–78
Chemical treatments
    acid solutions, 267
    chlorine, 266
    hydrogen peroxide, 267
    iodine, 266–267
Chilled aquatic food products. *See* Aquatic food products preservation
Chilling methods
    refrigeration methods, ice slurries storage
        application, 308
        combined preservation systems, 310–311
        food product sector, 311
        novelty and advantages, 308–309
        seafood products, 309–310
        versatility, 308
    traditional methods
        amines formation and nucleotide degradation, 306–307
        chilled seawater, 304–305
        flake icing, 304
        lipid damage analysis, 307
        salt treatment, 305
        sensory and physical changes, 306
Chitosan, 329–331
Chlorine, 266

Clostridium botulinum, 263
Combustion, 82–83
Composting, olive mill wastes, 92
Consumer acceptance, 140, 144–149
Consumer behavior
    emerging trends, 153–154
    and food choice
        consumer acceptance, 144–149
        determinants of, 140–142
        food policies, 149–152
        new product buying, 138–140
        novel technologies, overview of, 143–144
    future research, recomendations, 154–155
    novel food, 137
Corrective action, 22, 27
Critical control point (CCP), 22
Critical limits, 22, 25
Cronobacter sakazakii, 186
Crude extracts, 207–208
Crustaceans preservation, melanosis development, 348–350
Cryptosporidium spp., 263
Cyclospora spp., 263

D

Decision tree, 22
Dehydration, 43
Denaturing, 43
Deoxynivalenol, 222
Disintegration, 83
Distillation, 43
Drying, 82

E

Electrochemical oxidation, 78–79
Emerging preservation technologies
    fruits and vegetables
        combination processes, 288–289
        ozone, 273–280
        ultrasound, 275, 281–284
        ultraviolet light, 283–288
    novel food
        consumer acceptance, 144–149
        food policies, 149–152
        overview of, 143–144
    vs. traditional, fruits and vegetables, 290
Emulsification, 170–173
Encapsulation, 159
Endogenous enzyme activity, 303
Endolysins, 188
Environmental impact, 11

Esherichia coli O157, 185, 186, 191
Essential oils
　antifungals activity, 209
　antimicrobial activity, 328–329
　foodborne bacteria, biocontrol of, 194–195
Ethanol production, 42–43
Ethyl tertiary butyl ether (ETBE), 40
Evaporation/distillation, 80–82
Extended producer responsibility, 15–16
Extracellular polymers (EPS), 99
Extrusion, 167–168

F

Fermentation, 42
Filtration, 71
Flocculation, 72–73
Flotation, 71
Fluidized bed incinerators, 13
Food and Drug Administration (FDA), 27, 132
Foodborne bacteria, biocontrol of
　antimicrobial metabolites, 192–193
　antimicrobials, plant extracted oils
　　action mechanism, 194
　　animal feed, 195
　　delivery method, 194
　　essential oils incorporation, 194–195
　　modified-atmosphere packaging, 194
　bacteriophages, viral predators
　　applications, 188–189
　　examples, 186–187
　　host specificity, 187–188
　　inactivation, bacteria, 185, 187
　　lytic cycle, 184–185
　　properties, 184
　　T4 phage, 184
　Bdellovibrio, bacterial predators
　　applications, 190
　　characterization, 189
　　replication, 189–190
　berry extract, 195
　competitive enhancement techniques, 190–192
　plant extracts, 195
　protective cultures, 192–193
　quorum sensing
　　enzymes, 197
　　gene expression regulation, 196
　　Yersinia enterocolitica, 197
Food chemicals codex (FCC), 127–128
Food industry, HACCP system in. *See* Hazard analysis and critical control points (HACCP) system
Food irradiation, 138

Food policies, 149–152
Food preservation technologies. *See also* Antifungal agents
　with conservation factors
　　high hydrostatic pressure, 252–253
　　high-pressure homogenization, 255
　　pulsed electric field, 253–254
　　ultrasound, 254–255
　food irradiation, 143
　hurdle technology, 212–213
　linear microbial inactivation response, 238–239
　minimal processing concept, 235–237
　nonlinear microbial inactivation response
　　comparison model, 242–243
　　mechanistic and vitalistic approach, 240
　　modified Gompertz model, 241–242
　　Weibull distribution, resistances model, 240–241
　nonthermal, 137
　preservation factors
　　high hydrostatic pressure, 244–245, 248
　　high-pressure homogenization, 250–251
　　nonthermal inactivation factors, 246–247
　　pulsed electric field, 248
　　ultrasound, 250
　　ultraviolet light, 248–249
　traditional techniques, 237
Food quality, 262, 264
Food safety
　fruits and vegetables, 262
　HACCP on, 30
　of nutraceuticals and functional foods (*see* Functional foods and nutraceuticals)
Foods for specified health use (FOSHU)
　consumer awareness, factors responsible, 124–125
　functional food development strategy, 124
　functional food universe, 123
　scope of nutraceuticals, 124
Fruits and vegetables
　chemical treatments
　　acid solutions, 267
　　chlorine, 266
　　hydrogen peroxide, 267
　　iodine, 266–267
　combination of processes, 288–289
　emerging technologies, 267, 273–288
　hurdle technology, 213–214
　ozone
　　applications, 275
　　impacts of, 276–280
　　microbial effects, 274–275
　　properties, 273–274

Index 417

pathogens, 263
quality, 262, 264
safety, 262
thermal treatments
  blanching, 264–265
  pasteurization, 265
  sanitizer solutions, impacts of, 268–272
  sterilization, 265
traditional technologies, 264–272
traditional *versus* emerging technologies, 290
ultrasound
  applications, 282–283
  impacts of, 284
  microbial effects, 282
  properties, 275, 281–282
ultraviolet light
  applications, 285, 288
  impacts of, 286–288
  microbial effects, 283–285
  properties, 283
Fumonisins, 221, 222, 229
Functional foods and nutraceuticals
  definition, 122
  FOSHU, 122–125
  regulation of, 131–134
  safety of
    benefits and risks, 128–129
    biologically active components, 128
    clinical data, 126
    evaluation of, 129–131
    GAO recommendations, 127
    ingredients, 127–128
    requirements, 125–126
    risk–benefit relationships, 127
    specific questions, 126
Fungicides, 206
Fusarium sambucinum, 207

G

General Accounting Office (GAO 2000), 127
Genetically modified foods, 149–151
Glycine max, 47
Gompertz model, modified, 241–242
Gram-negative bacteria, 188, 189, 191, 195, 274, 328
Gram-positive bacteria, 195, 274, 332

H

Hazard, 22
Hazard analysis and critical control points (HACCP) system
  application in food industry, 27–30
  basic conditions for, 25
  contamination prevention, 21
  cost/benefit analysis, 31–33
  definitions, 22–23
  and environment, 33–35
  food safety, 30
  general principles, 22–23
  implementation of
    preliminary procedures, 23–24
    prerequisites for, 23
    principles of application, 24–27
Hazardous wastes, 4, 6
Heliantus annus L., 47
Hepatitis A virus, 263
Herbs, 209, 327–329
High-added-value product, 92–93
High hydrostatic pressure
  with conservation factors, 252–253
  food preservation method, 244–245, 248
High-pressure homogenization
  with conservation factors, 255
  food preservation method, 250–251
Household waste, 8
Hurdle technology
  application to fruits and vegetables, 213–214
  natural seafood preservatives, 340–341
  physical and chemical methods, 212–213
Hydrogen peroxide (H_2O_2), 267
Hydrolysis, 83

I

Ice slurries
  application, 308
  aquatic food sector, 311
  combined preservation systems, 310–311
  definition, 307
  novelty and advantages, 308–309
  in seafood products, 309–310
  versatility, 308
Inactivation kinetics, microbial
  linear response, 238–239
  nonlinear response
    linear, Weibull and Gompertz models, comparison of, 242–245
    modified Gompertz model, 241–243
    Weibull distribution, 240–241
  semilogarithmic curves, 238, 239
Incineration, 12–13. *See also* Combustion
Industrial waste, 8
Iodine, 266–267
Ion exchange, 73–74

L

Lactic acid bacteria (LAB)
 characteristics, 332
 types and applications, 333–335
Lactococcus lactis, 192
Landfill gas, 44
Landfills, 10–11
Legislation, 132, 134, 150, 350–351
Lignocellulosic biomass, 49–50
Lipases, 95–96
Lipid biofuels
 bio-oils, 47
 definition, 46
 sources of, 47–48
 technologies, process description, 48–49
 vegetable oils, 46–47
Lipid oxidation, 303
Liquefaction, 43
Liquid injection incinerators, 13
Liquid waste, 7
Listeria monocytogenes, 185, 186, 188, 263, 327

M

Maize preservation, 220–221
Maize sieving, 225
Management
 olive oil wastewater (*see* Olive mill wastewater (OMWW))
 waste
  basic principles, 15–16
  description, 9
  public perceptions and attitude, 17
  trends, 16–17
Medicinal plants, 208–209
Membrane technologies, 79–80
Methanogenesis, 84
Methanol, 40–41
Microbial decomposition, 303
Microbial inactivation, food preservation method. *See* Food preservation technologies
Microencapsulation
 consumer trends and market perspectives, 174–177
 energetic aspects of
  emulsification, 170–173
  spray-drying, 173–174
 by extrusion, 167–168
 by gelation and phase separation of proteins and biopolymers, 168–169
 scanning electron micrograph, 160
 by spray-drying
  core material selection, 160–161
  emulsion preparation, 163
  nonencapsulated core material, 166–167
  process conditions physicochemical characteristics, 164–165
  wall material selection, 161–162
 techniques, 159–160
Milling, 43
Modified-atmosphere packaging (MAP)
 argon, 314–315
 carbon dioxide, 312–313
 gas mixture in, effect of, 312–315
 general aspects, 312
 nitrogen, 314
 oxygen, 314
Modified Gompertz model, 241–242
Mold control in foods. *See* Antifungal agents
Monitoring procedures, 25–26
Municipal waste, 8
Mycotoxin contamination reduction
 cleaning procedure, 224–225
 contamination level, 227–228
 economic implication, 228–229
 food safety, 219
 fungal species, toxins, 222
 grain cleaning machine, 226
 maize preservation, 220–221
 reliable and robust analytical methods, 222
 safety issues, 229
 sieve dimensions, 225–226
 variability, 223–224

N

Nanotechnology, 138, 144, 151, 154
Natural antimicrobial preservatives. *See* Antimicrobial preservatives
Natural preservatives. *See* Preservatives, seafood
New product buying, 138–140
Nisin, 192
Nonencapsulated core material, 166–167
Norovirus, 263
Norway lobster *(Nephrops norvegicus)*, 311
Novel food choice
 consumer behavior and, 138–140
 definition, 137
 determinants of, 140–142
 technologies and emerging trends
  consumer acceptance, 144–149
  food policies, 149–152
  overview, 143–144
  recommendations, 154–155
Nutraceuticals. *See* Functional foods and nutraceuticals

Index 419

O

Ohmic treatment, 138
Oleoresins, 209–210, 215
Olive mill wastewater (OMWW)
  biological treatment
    aerobic treatment, 87–89
    agricultural land irrigation, 91–92
    anaerobic digestion, 83–87
    composting, 92
    phytoremediation, 89–90
  characteristics, 68–69
  detoxification of, fungal species for, 88–89
  milling, oil production
    modern method, 67
    traditional method, 66–67
  physicochemical treatment
    adsorption/ion exchange, 73–74
    chemical oxidation, 74–79
    membrane technologies, 79–80
    simple chemical processes, 71–73
    simple physical processes, 70–71
    thermal processes, 80–83
  production from olive oil, 67–70
  valorization
    activated carbon, production of, 100
    biofuels and bioproducts, generation of, 97–99
    enzyme production, 95–96
    organic compounds, in OMWW, 93
    phenolic compounds and pectins, 93–95
  wastewater management technologies, 70
  world olive oil production, 66
Olive oil extraction
  milling
    modern method, 67
    traditional method, 66–67
  waste production
    categories, 67
    characteristics of, 68–70
Ozone (O_3)
  applications, 275
  impacts of, 276–280
  microbial effects, 274–275
  properties, 273–274

P

Palm *(Elaeis)* oil, 47
Pasteurization, 265
Pathogens, food contamination, 263
Pectinase, 96
Pectins, 95
Pediococcus acidilactici, 193
Phenolic compounds, 93–95

Photobacterium phosphoreum, 328
Physicochemical treatment, olive mill wastes
  adsorption/ion exchange, 73–74
  chemical oxidation
    advanced oxidation processes, 74–77
    electrochemical oxidation, 78–79
    wet oxidation, 77–78
  membrane technologies, 79–80
  simple chemical processes, 71–73
  simple physical processes, 70–71
  thermal processes
    combustion, 82–83
    drying, 82
    evaporation/distillation, 80–82
    pyrolysis, 83
Phytoremediation, olive mill wastes, 89–90
Pigment degradation, 347
Plant extracts, antifungals agents
  Alternaria disease control, 213–217
  crude extracts, 207–208
  essential oils and oleoresins, 209–210
  medicinal plants, 208–209
  spices and herbs, 209
  against toxigenic fungi, 210–211
Plasma torch, 14
Polluter pays principle, 16
Polyhydroxyalkanoates (PHAs), 98–99
Polyphenol oxidases (PPO), 338
Polyunsaturated fatty acids, microencapsulation of
  consumer trends and market perspectives, 174–177
  energetic aspects of
    emulsification, 170–173
    spray-drying, 173–174
  by extrusion, 167–168
  by gelation and phase separation of proteins and biopolymers, 168–169
  scanning electron micrograph, 160
  by spray-drying
    core material selection, 160–161
    emulsion preparation, 163
    nonencapsulated core material, 166–167
    process conditions physicochemical characteristics, 164–165
    wall material selection, 161–162
  techniques, 159–160
Preservatives, seafood
  antimicrobial
    biopreservatives, 332–337
    chitosan, 329–331
    organic acids, 326–327
    spices and herbs, 327–329
    uses, 331–332

Preservatives, seafood (cont.)
    antioxidant
        crustaceans preservation, melanosis development, 348–350
        deterioration process, 337
        enzymatic reaction, 338
        evaluation of, 351
        legal issues, 350–351
        lipid oxidation, 337, 339–348
        oxidation, 337
    biopreservatives
        definition, 332
        evaluation of, 336–337
        lactic acid bacteria protection, 333–335
        and probiotics, 335–336
        species and strains, 332
    microbial stress factors, 352
    phenolic compounds, 351
Probiotics, 335–336
Pulsed electric field (PEF)
    with conservation factors, 253–254
    food preservation method, 248
    in novel foods, 137
Purified phage-derived antimicrobials, 188
Pyrolysis, 13–14, 83

Q
Quorum sensing, 195–197

R
Rapeseed oil, 47
Rational management. See Management
Recordkeeping procedures, 27
Recycling, waste, 9–10
Regulation, nutraceuticals and functional foods, 131–134
Rendering, 48
Risk, 22
Rosemary (Rosmarinus officinalis), 342
Rotary kiln incinerators, 13

S
Saccharification, 43
Safety
    of fruits and vegetables, 261, 262
    of functional foods and nutraceuticals
        benefits and risks, 128–129
        biologically active components, 128
        clinical data, 126
        evaluation of, 129–131
        GAO recommendations, 127
        ingredients, 127–128
        requirements, 125–126
        risk–benefit relationships, 127
        specific questions, 126
    HACCP on food safety, 30
    maize milling, 229
Salt treatment, aquatic food products preservation, 305
Sanitizer solution, 268–272
Sardine (Sardina pilchardus), 342
Sedimentation, 71
Settling, 71
Severity, 22
Solid by-products
    fat/oil-containing byproducts, 58, 59
    lignocellulose-containing by-products, 55–56
    starch-and/or sugar-containing by-products, 56–57
Solid waste, 7
Soybean oil, 47
Spices, 209
Spices and herbs, antimicrobial activity, 327–329
Spray-drying
    core material selection, 160–161
    emulsion preparation, 163
    energetic aspects of, 173–174
    nonencapsulated core material, 166–167
    process conditions physicochemical characteristics, 164–165
    wall material selection, 161–162
Staphylococcus aureus, 187
Sterilization, 265
Streptomyces natalensis, 193
Sunflower oil, 47
Synthetic fungicides, 206

T
Thermal desorption, 13
Thermal treatments
    food safety and quality improvement
        blanching, 264–265
        pasteurization, 265
        sanitizer solutions, impacts of, 268–272
        sterilization, 265
    olive mill waste, 80–83
        combustion, 82–83
        drying, 82
        evaporation/distillation, 80–82
        pyrolysis, 83
    waste disposal
        incineration, 12–13
        plasma torch, 14

pyrolysis, 13–14
  thermal desorption, 13
  vitrification, 14
Tomatoes, *Alternaria* disease control
  antifungal activity, 215
  disinfection processes, 215–217
  hurdle technology, 213–214
  tomato products, 214
  treatments assay, 216
Toxigenic fungi, 210–211
T4 phage
  structure, 184
  UV-treated, 189
Transesterification, 48, 49

U
Ultrahigh pressure (UHP), 137
Ultrasound
  with conservation factors, 254–255
  food preservation method, 250
  to fruits and vegetables
    applications, 282–283
    impacts of, 284
    microbial effects, 282
    properties, 275, 281–282
Ultraviolet (UV) light
  of fruits and vegetables
    applications, 285, 288
    impacts of, 286–288
    microbial effects, 283–285
    properties, 283
  physical inactivation of microorganisms, 248–249
US environmental protection agency (EPA), 4

V
Vegetable oils, 46
Verification procedures, 22, 26, 27
Viral predators. *See* Bacteriophages
Vitrification, 14
Voluntary food fortification, 137

W
Wasabi *(Wasabi japonica)*, 329
Waste
  classification, 8–9

concept, 4–5
disposal
  landfills, 10–11
  recycling, 9–10
  thermal technologies, 11–14
generation of, 6, 7
Global Waste Management Market Report 2007, 6
hazardous, 4
management
  basic principles, 15–16
  description, 9
  public perceptions and attitude, 17
  trends, 16–17
real or perceived health risks, 18
recovery, 5
sources of, 8–9
total waste generation, 4
types of
  household, 8
  industrial, 8
  liquid, 7
  municipal, 8
  solid, 7
US environmental protection agency (EPA) reports, 4
Waste recycling, 9–10
Wastewater
  for biofuels
    biogas generation, 50–52
    biogas purification, 52–54
    biogas utilization, 52
    storage, 54–55
  municipal, 7
  Olive mill wastewater (*see* Olive mill wastewater (OMWW))
Weibull distribution, resistances model, 240–241
Wet oxidation, 77–78

Y
Y*ersinia enterocolitica*, 197

Z
Zearalenone, 222